Cost Modelling

Cost models underlie all the techniques used in construction cost and price forecasting, yet until relatively recently industry has been unfamiliar with their characteristics and properties. An understanding of the various types of cost model is vital to enable effective cost control and the development of future forecasting techniques.

This volume brings together more than 25 seminal contributions to building cost modelling and introduces the major landmarks in progress and thinking in this field. The strong techniques bias of this book will appeal to construction professionals involved in estimating, as well as researchers and students of building economics.

Cost Modelling

Martin Skitmore and Vernon Marston

Spon Press
an imprint of Taylor & Francis
LONDON AND NEW YORK

First published 1999 by E & FN Spon
2 Park Square, Milton Park, Abingdon, Oxon OX14 4RN

Simultaneously published in the USA and Canada
by Routledge
52 Vanderbilt Avenue, New York, NY 10017, USA

First issued in paperback 2020

E & FN Spon is an imprint of the Taylor & Francis Group, an informa business

© 1999 this selection and editorial matter E & FN Spon; individual contributions, the original copyright holders

Typeset in Times by Solidus (Bristol) Ltd, Bristol

British Library Cataloguing in Publication Data
A catalogue record for this book is available from the British Library

Library of Congress Cataloging in Publication Data
A catalog record for this book has been requested

ISBN 13: 978-0-367-57915-9 (pbk)
ISBN 13: 978-0-419-19230-5 (hbk)

To our families, colleagues and readers

Contents

Series Preface   xi
Preface   xv
List of sources and acknowledgements   xvii

Introduction to the Readings   1
R. M. SKITMORE AND V. K. MARSTON

PART 1
Strategies and directions   **23**

1.1 Building cost research: need for a paradigm shift?   27
    P. S. BRANDON

1.2 Models in building economics: a conceptual framework for the
    assessment of performance   36
    J. RAFTERY

1.3 A future for cost modelling   45
    D. T. BEESTON

1.4 Developments in contract price forecasting and bidding techniques   53
    R. M. SKITMORE AND B. R. T. PATCHELL

1.5 An agenda for cost modelling   85
    S. NEWTON

1.6 The relative performance of new and traditional cost models in
    strategic advice for clients   101
    C. FORTUNE AND M. LEES

PART 2
Explorations in cost modelling   163

2.1 A new approach to single price rate approximate estimating   166
     W. JAMES

2.2 Predesign cost estimating function for buildings   183
     V. KOUSKOULAS AND E. KOEHN

2.3 Some examples of the use of regression analysis as an estimating
     tool   201
     R. McCAFFER

2.4 Cost optimisation of buildings   215
     A. D. RUSSELL AND K. T. CHOUDHARY

2.5 Predicting the tender price of buildings in the early design stage:
     method and validation   235
     R. McCAFFER, M. J. McCAFFREY AND A. THORPE

PART 3
Cost-product modelling   251

3.1 An investigation into building cost relationships of the following
     design variables: storey height, floor loading, column spacing,
     number of storeys   253
     WILDERNESS GROUP

3.2 The relationship between construction price and height   272
     R. FLANAGAN AND G. NORMAN

3.3 Cost performance modelling   281
     T. MAVER

3.4 The effect of location and other measurable parameters on
     tender levels   293
     I. PEGG

PART 4
Cost-process modelling   307

4.1 Cost consequences of design decisions: cost of contractors'
     operations   310
     D. T. BEESTON

4.2 General purpose cost modelling   324
L. G. HOLES AND R. THOMAS

4.3 Simulation applied to construction projects   331
J. BENNETT AND R. N. ORMEROD

4.4 Cost modelling: a process modelling approach   374
P. A. BOWEN, J. S. WOLVAARDT AND R. G. TAYLOR

4.5 Automatic resource based cost-time forecasts   384
V. K. MARSTON AND R. M. SKITMORE

PART 5
Dealing with uncertainty   **395**

5.1 One statistician's view of estimating   398
D. T. BEESTON

5.2 Accuracy in estimating   412
A. ASHWORTH AND R. M. SKITMORE

5.3 The accuracy of quantity surveyors' cost estimating   438
N. MORRISON

5.4 Experiments in probabilistic cost modelling   460
A. J. WILSON

5.5 The accuracy and monitoring of quantity surveyors' price
forecasting for building work   471
R. FLANAGAN AND G. NORMAN

Selected Bibliography   495
Index   507

Series Preface

The *Foundations of Building Economics* series endeavours to organize the extant knowledge in the field by assembling in one place the seminal papers in building economics written to date. Each volume in the series offers a selection of a score of papers, together with an introductory chapter in which the editors review the area entrusted to them. Most of the papers in the series have already appeared in journals or as chapters in books, but a few papers in each volume may be new.

As a field of economics, building economics is concerned with the best use of scarce resources in the building process. On the one hand, the building process includes all stages in the life of an individual building or constructed facility, from conception to demolition; on the other, the building process is about all the buildings and other constructed facilities that underpin the economic process as a whole. To use the standard terminology, building economics is both about microeconomic and macroeconomic issues. The field is thus potentially vast. But, in the last analysis, building economics is what building economists do. Although it is useful to distinguish between its areas of concentration, it is potentially harmful to cast a sharp boundary around its parts. Many a fine paper included in this series defies unambiguous classification.

As an emerging field, building economics is best served by a series of readers, rather than an attempt at its codification. The series is thus unbounded. It remains open to new ideas and additions to the field of building economics. The hope is that these readers will become essential guides for both students and professionals, both those who are beginning to explore the world of building economics and those who already feel at home in it.

The *Foundations of Building Economics* series is encouraged by the International Council for Research and Innovation in Building and Construction or CIB (from the French acronym standing for the *Conseil International du Bâtiment*). CIB was established in 1953 as an association whose objectives were to stimulate and facilitate international collaboration and information exchange between governmental research institutes in the construction sector, with the emphasis on those institutes engaged in technical fields of research.

CIB has since developed into a world-wide network of over 5000 experts from about 500 organizations active in the research community, in industry, or in education, who collaborate and exchange information in over 50 commissions covering all fields in construction-related research, development, information, and documentation. The members are institutes, companies, and other types of organizations involved in research or in the transfer or application of research results. Member organizations appoint experts to participate in commissions. Individuals also can be members.

The CIB Working Commission W55, 'Building Economics,' has been active in this field since 1970. The Commission has grown to a membership of about 160 experts from some 50 countries. From its inception, the commission has focused on the economic methods central to every phase of building planning, design, construction, maintenance, and management.

The first volume in the series, *Cost Modelling*, edited by Martin Skitmore and Vernon Marston, presents a relatively young field of endeavour. Therefore, the objective of this Reader is not only to introduce building cost models and present their development, but also to generate debate and take first steps toward a common understanding. The editors admit a certain bias in the selection of papers presented towards authors from the United Kingdom. This is in part related to the historical development of this field. The editors hope that a truly international debate will follow.

Martin Skitmore and Vernon Marston are both former lecturers in the Department of Surveying at the University of Salford. Both have extensive experience in quantity surveying practice and academia. They have published many papers on cost modelling and associated topics. They have also jointly conducted a major research project in estimating and cost modelling that has received significant funds from the Science and Engineering Research Council in the UK.

In the end, a few words of thanks. I am grateful to many people who have helped in this project since the CIB Working Commission W55 meeting in Espoo, Finland, in June 1989, where I proposed the series and where I agreed to serve as its Editor-in-Chief. Mrs. Clara Szöke, who was the Coordinator of W55 at the time and Prof. Gyula Sebestyén, the then Secretary General of CIB, come first to mind. Without their enthusiastic support at the Espoo meeting, the series would never have seen the light of day. Prof. Mariza Katavic and Dr. Wim Bakens, who are now serving as the W55 Coordinator and the Secretary General of CIB, respectively, have also been most helpful in this project. Mr. Dan Ove Pedersen, who had served as the Coordinator of W55 in the interim period, deserves praise as well. Of course, I am most grateful to all the editors of the volumes that will comprise the *Foundations of Building Economics* series. Their job has been and will continue to be the most challenging one. Last but far from least, I am indebted to Mr Phillip Read from

E & F N Spon, who has worked with me on the series since 1993, when Spon became involved in the project.

Prof. Ranko Bon   Reading, United Kingdom
Editor-in-Chief   January 1999
University of Reading

Preface

The idea for a book such as this was formed as long ago as 1989. Over the years I had collected publications from a wide range of sources on most things to do with building prices. Latterly I had managed to get the bulk of these assorted papers filed away in alphabetical order of authors' names in a couple of drawers in my office filing cabinet. This little 'library' has become the start of many a student research project. The problem with this system, however is that there is no feel for the importance or timeliness of individual papers, or the historical development of specific topics. That there are no subdivisions in the system is also a problem for users. It seemed that what was needed was to identify a few topic areas and a collection of seminal papers associated with each topic.

At the same time, I noticed that my colleague Vernon Marston always carried two brief cases around with him wherever he went. It turned out that one of the cases contained a smaller version of my library (he had an even larger collection than mine scattered between his home and office) which he carried around on the off chance that he might need to refer to something. As it happened, Vernon was also coming around to thinking that there might be a better way than this!

Having decided to go ahead with our 'collected papers' project, we followed a pretty classical approach. First, we had a couple of hours brainstorming session in which we wrote down every publication we could think of that we liked. During the course of this, it soon became clear that we were going to fill several books unless we specialised the general theme. At first it seemed that this would be 'estimating' but we eventually realised that we were in fact much more interested in 'cost modelling', or what we thought was cost modelling.

Continuing on this basis, we finished up with a list of around eighty titles. Our next stage was to give a score between 0–100 against each of the publications. This we did separately. At the same time, we sent a copy of the list to a few others we knew who were involved in the field. Next we averaged the scores for each of the papers and produced a rank order list.

From this rank order list we eventually identified the five topics or 'parts' that appear in the book. After a few minor changes, we arrived at a reasonable looking balance. Around 85% of the original top rankings were retained.

We then set about writing the introduction to the Readings. In doing this we took the opportunity to use some recent work by Pidd on modelling in the Operations Research field in an attempt to provide a structured summary of our collected papers. Of course this was the most demanding but very interesting part of our contribution. The results of these labours look quite reasonable to us and we think shed a few insights on the cost modelling field, particularly in the relationship between the modelling and implementation processes, elements and functional relationships and validation techniques. Writing the 'Glossary of Terms' was found to be an especially useful exercise in proposing a standardised terminology for cost modelling.

Bearing this work in mind, we then closely read all the papers, making notes in our newly standardised terminology. These notes appear at the start of each of the five parts.

We then extended our original list of publications into a Selected Bibliography. This was done by going through the proceedings of the two major conferences in cost modelling, for both the relevance of the papers themselves and the references contained in the papers of both this book and the proceedings.

Since then, we have been waiting for publishers. During this time, Vernon has 'retired' to his bookshop in Devon. Also several of the original papers included in the book have either been drastically reduced in size or omitted completely. Some of these are quite old and increasingly hard to access. I sincerely hope that these will eventually appear in a subsequent volume.

Of course, we will have made some mistakes and unwittingly overlooked some very important contributions. To the recipients of these oversights we submit our abject apologies and promise to rectify the situation in the next edition.

Finally, I would like to express my sincere thanks to Ranko Bon in particular who has worked hard over a very long period to persuade the publishers to print this book. It has indeed been a long haul but, like most taxing things in life, I think it has been worth it in the end.

Martin Skitmore

List of sources and acknowledgements

Grateful acknowledgement is made to the following for permission to reproduce material in this Reader:

Brandon, P. S. (1982) 'Building cost research – need for a paradigm shift?' in Brandon, P. S. (ed) *Building cost techniques: new directions*, E & F N Spon. Raftery, J. (1984) 'Models in building economics: a conceptual framework for the assessment of performance', in *Proceedings 3rd Int. Symp. on Build. Econ.*, CIB W-55, Ottawa, vol 3. Beeston, D. T. (1987) 'A future for cost modelling', in Brandon P. S. (ed) *Building Cost Modelling and Computers*, E & F N Spon. Skitmore R. M., Patchell, B. (1990) 'Developments in contract price forecasting and bidding techniques', chapter in *Quantity Surveying Techniques: New Directions*, ed P. S. Brandon, BSP Professional Books. Newton, S. (1990) 'An agenda for cost modelling research', *Construction Management and Economics*. Fortune, C., Lees, M. (1996) 'The relative performance of new and traditional cost models in strategic advice for clients', *RICS Research Paper Series* vol 2 no 2. James W. (1954) 'A new approach to single price-rate approximate estimating', *RICS Journal*, vol XXXIII (XI), May. Kouskoulas, V., Koehn, E. (1974) 'Predesign cost estimation function for buildings', *ASCE J. of Const. Div.*, Dec. McCaffer, R. (1975) 'Some examples of the use of regression analysis as an estimating tool', *Quantity Surveyor*, Dec. Russell, A. D., Choudhary, K. T. (1980) 'Cost Optimisation of Buildings'. *American Society of Civil Engineers, Journal of the Structural Division*, January. McCaffer, R., McCaffrey, M. J., Thorpe, A. (1984) 'Predicting the tender price of buildings during early stage design: method and validation', *J. Opl Res. Soc.*, vol 35 no 5. Wilderness Group (1964) 'An investigation into building cost relationships of the following design variables: storey height, floor loading, column spacings, number of storeys', *Report* to the Royal Institution of Chartered Surveyors. Flanagan, R., Norman, G. (1978) 'The relationship between construction price and height', *Chartered Surveyor B and QS Quarterly*, Summer. Maver, T. (1979) 'Cost performance modelling', *Chartered Quantity Surveyor*, vol 2 no 5, Dec. Pegg, I. (1984) 'The effect of location and other measurable parameters on tender levels',

Cost Study F33, BCIS 1983–4–219, Building Cost Information Service, Royal Institution of Chartered Surveyors. Beeston, D. T. (1973) 'Cost consequences of design decisions: cost of contractors' operations (COCO)', *Report*, Directorate of Quantity Surveying Development, Property Services Agency, Department of the Environment. Holes, L. G., Thomas, R. (1982) 'General purpose cost modelling', in Brandon P. S. (ed) *Building Cost Techniques – New Directions*, E & F N Spon. Bennett, J., Ormerod, R. N. (1984) 'Simulation applied to construction management', *Construction Management and Economics*, vol 2. Bowen, P. A., Wolvaardt, J. S., Taylor, R. G. (1987) 'Cost modelling: a process-modelling approach', in Brandon P. S. (ed) *Building Cost Modelling and Computers*, E & F N Spon. Marston, V. K., Skitmore, R. M. (1990) Automatic resource based cost time forecasts, *Transactions* of the 34th Annual Meeting of the American Association of Cost Engineers, *Symposium M – Project Control*, pub. by AACE, Morgantown, USA. Beeston, D. T. (1974) 'One statistician's view of estimating', *Cost Study F3*, BCIS 1974/5–123, Building Cost Information Service, Royal Institution of Chartered Surveyors. Ashworth, A., Skitmore, R. M. (1983) 'Accuracy in estimating', *Occasional Paper No 27*, Chartered Institute of Building. Morrison, N. (1984) 'The accuracy of quantity surveyors' cost estimating', *Construction Management and Economics*, vol 2. Wilson, A. J. (1982) 'Experiments in probabilistic cost modelling', in Brandon P. S. (ed) *Building Cost Techniques – New Directions*, E & F N Spon. Flanagan, R., Norman, G. (1983) 'The accuracy and monitoring of quantity surveyors' price forecasting for building work', *Construction Management and Economics*, vol 1.

Introduction to the Readings

R. M. Skitmore and V. K. Marston

1 INTRODUCTION

Purpose

As a relatively young field of endeavour, building cost modelling is still seeking to establish itself as a worthy scientific enterprise. It does not yet have an accepted theoretical underpinning or established terminology. In different parts of the world, model developers and users are rooted in different disciplines. Architects, design engineers, cost engineers, quantity surveyors and others have their own customs and preoccupations, and generate their own mystique. Communication in the field is poor, with consequent misunderstanding and duplication of effort.

In this Reader therefore, the objective is not only to trace the development of cost models and represent the extent of this diverse field, but also to generate debate and take the first steps towards a common understanding. To achieve these latter objectives, the editorial content seeks to 'grasp the nettle' by making positive assertions, particularly in respect of terminology. There is an admitted bias in the selection of papers presented towards authors from the United Kingdom. Partly this is a result of language convenience and the proliferation of work in the field by UK quantity surveyors, but further it is a result of the editors' desire to make confident statements about work that they know well. It is hoped that debate will follow.

Scope

In order to have a coherent field of study, the Reader has been limited to the consideration of forecasting models used at the project level in the precontract monetary evaluation of building design. Therefore contractors' bidding and accounting models have not been considered. Life cycle costing and other investment appraisal based methods, and macro and meso level economic models have been excluded as these are the subject of other Readers in this series.

Aims

The papers appearing in this Reader have been selected subjectively by the editors on two criteria. First and most importantly, that the papers make a seminal contribution to the field in their own right and in some cases mark a turning point in the progress and thinking in the field. Second, that the papers show as far as possible the full extent of this disparate field and to some extent illustrate the main strands of development. As a result, some older papers have been included to cover aspects that are still valid but that may have fallen temporarily out of favour.

Structure

In general, the papers have been grouped according to their dominant theme. In the absence of an established scientific discipline and sub-disciplines, we have attempted to arrange the papers in way that is most digestible rather than suggestive of a formalism of the subject. There are therefore many overlaps between the content of the sections. In reviewing the papers we have cross-referenced these overlaps wherever possible. The wide variety in terminology contained in the papers has also been addressed by the provision of a glossary of terms which is intended to provide a common vocabulary on the subject to reduce ambiguity. We have also added some notes on modelling generally, taken from some of the OR literature and attempted to address cost modelling from a similar perspective. In doing this difficult task we hope to bring some clarity where confusion might have existed, rather than the reverse!

2 MODELLING: AN OR PERSPECTIVE

What is a model?

A model is defined by the *Fontana Dictionary of Modern Thought* (Bullock and Stalybrass 1977) as 'A representation of something else, designed for a special purpose'. The 'something else' here is what we already know about, or think we know about – called the prototype (after Aris 1978). The 'purpose' can be either to **remind**, **discover** or **explain**. For instance a model aeroplane, a model of Shakespeare's birthplace, or a photograph all represent an original and remind us what the prototype original looks like. A model aeroplane placed in the controlled environment of a wind-tunnel may be used for experiments to discover (forecast) how a real aeroplane prototype built to this design would behave in the sky. The analogous behaviour of the solar system to the atom is used to provide an explanation of the behaviour of the atom prototype.

All models are a result of a mapping of elements of the prototype onto the model. Mapping every element of the prototype results in an **isomorphic model**. Isomorphic modelling however is an expensive business as the costs of building such a model may be disproportionate to benefits derived by the model's use. Invariably, some of the elements are more useful than others and the economic solution is to select a sub-set of elements that gives a satisfactory level of cost–benefit trade-off.

Several issues are raised in this basic description of modelling:

1   what is the 'special purpose' of the model? Will more than one purpose be required?
2   what is the prototype and what 'elements' constitute the prototype. How can they be identified? What form will the elements take when mapped onto the model? How will they be structured?
3   how can elements be assessed for their individual 'usefulness', cost of incorporating into the model and collective benefits in the model's use?

An examination of the special purpose of a model requires the consideration of the needs of the various beneficiaries or **stakeholders** that are the recipients of the model's application. At an individual level, we all need **cognitive models** to provide a meaning to our perceptions of the 'real world'. Human beings never perceive every detail of the world surrounding them. To observe every single book, piece of furniture, thread of carpet, etc in a room would involve an enormous cognitive cost. As children we learn to notice only those aspects of our environment that are relevant to our immediate needs; and these are never more urgent than as children. As they mature, people tend to specialise in their cognitive selectivity according to their interests and, later, their occupational demands. Thus an architect may 'see' a building as comprising a collection of shapes and colours in relation to the surrounding shapes and colours while a window cleaner may only 'see' the same building as comprising a large amount of glass in locations with differing degrees of accessibility. The development of a model for **both** the architect and the window cleaner would necessarily have to satisfy both their respective needs ie shapes, colours and locations of glass.

From this example we can see that these considerations largely address most of the questions posed in modelling. Given that we know who the stakeholders are and their perceptual needs, we can go a long way towards deriving the model's purpose and elements. The final cost–benefit trade-off assessment will then depend on the cost of developing and using the model and the degree to which the model satisfies those needs.

Technical models

Technical models are used to help study the effects of applying different policies, strategies and options in preference to trying them out in practice. They consist of bundles of equations, concepts or computer programs that are abstractions of the most useful elements of the prototype. Fig. 1 shows the three stages involved in the development of technical models.

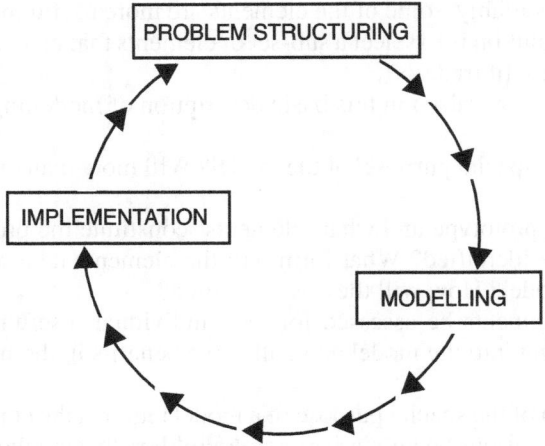

Figure 1 The model development cycle.

Problem structuring

Models can be built on the distinction between messes, problems and puzzles (Ackoff 1979). These three concepts are interlinked and can be viewed as a hierarchy (Fig. 2).

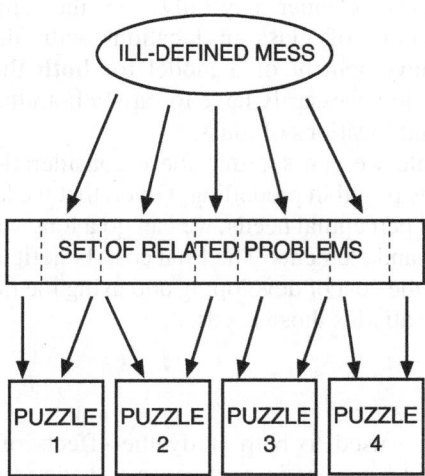

Figure 2 Problem structuring (from Pidd, 1991).

1   A *puzzle* exists when there is a single, well-defined question for which there is a single correct solution.

2   A *problem* exists when there is still only a single well-defined question but when there are several possible solutions which are defensible from different points of view.

3   A *mess* exists when there is no agreement about the questions to be asked and when there is, therefore, no common view about what constitutes an acceptable solution.

By this classification, messes seem to characterise 'real-world' situations more than puzzles. Puzzles, on the other hand, seem to characterise situations amenable to solution by standard textbook, especially mathematical, methods than do messes. To reconcile the two extremes, an obvious approach is to attempt to structure a mess into a series of problems and then structure each problem into a series of puzzles – thus providing a range of solutions. Naturally this can create difficulties for decision-makers faced with the task of making one decision.

An alternative is to develop models able to cope with problems and messes, as well as puzzles (Pidd 1991). The difficulty of multiple solutions is, however, still present. Sometimes the best we can achieve is just to explicate the relevant elements in some intelligible way for the decision-maker to find his own solution – a decision support system. Whilst seeming to avoid the task in hand, this does have the advantage of being an advancement on no action at all and can always be regarded as a temporising measure in advance of further research!

Modelling

As with problem structuring, modelling is best considered as a cycle of activities which can be short-circuited. The modelling cycle is shown in Fig. 3 and consists of four parts.

The type of technical model of most relevance here is a group of models called **symbolic** models. These models represent the elements of the prototype and their relationships in the form of symbols. Symbolic models are the most common type of technical model and the most convenient for analysis. They can be used to test the robustness of solutions by sensitivity analysis through 'what-if' type experiments. They are also useful for extrapolation and experimentation generally. In many way symbolic models are like theories in that they embody relationships which are believed to hold in the prototype. The model should therefore determine what data are needed and should be collected.

The modelling activity itself involves an iterative process of model **identification** and **fitting** (Gilchrist 1984). Identification is the process of finding or choosing an appropriate model containing the set of useful elements

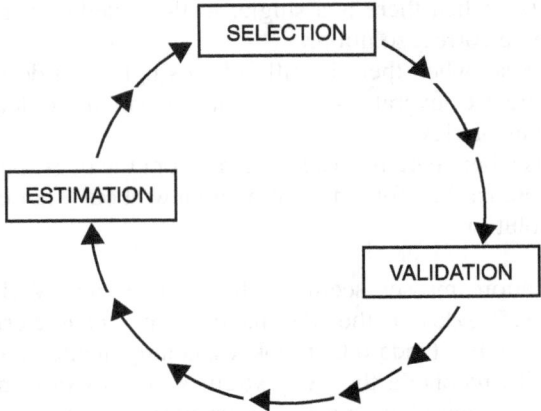

Figure 3 Project cost model.

and their functional relations for the required purpose. The two basic approaches to identification are the **conceptual** approach and the **empirical** approach. At the extreme, conceptual identification seeks a model on the basis of rational argument from some knowledge of the domain, but without reference to any actual data. Empirical identification considers only the data and its properties without reference to the meaning of the data or the situation in which it arose. Model fitting is the process of moving from the general form to the specific numerical form by assigning values to the functional relations in the model.

Once formulated in this way, most technical models are 'programmed' into systems and programs which run on computers. Non-isomorphic models, by definition, are always simplifications of the systems they purport to mimic and so they cannot possibly be valid for all purposes. It is always possible to devise tests which show these models to be deficient in some way or other. This implies that models should be used with care and not pushed beyond the limits of their validity. Two broad approaches to model validation are the **black box** approach and the **white box** approach. Under the black box approach the model's internal structure is assumed to be obscure but its performance can be observed. The idea is thus to compare the output from using the model in response to certain stimuli with the output of the prototype itself under the same stimuli. In contrast, the white box approach involves the detailed comparison of the model and prototype **structures**. Which of the approaches is used depends largely on the purpose of the model – the **performance** approach often being used to test discovery models and the **structural validation** approach for explanatory models.

Implementation

The main purpose of implementation for the modeller is to provide feedback for future refinements to the problem structure and modelling (Fig. 1). The difficulty here is that the feedback may be misleading – short term problems in stakeholder learning being interpreted as basic inadequacies in the model. The danger is that by the time the model has been changed, the stakeholders will have developed a preference for the original model! Genuine improvements can be made however, particularly where forecasts have to be made of the effects of implementation in advance of implementation. At this stage, the forecasts can be replaced by the actual figures and the modelling process iterated as shown in Fig. 1.

3 COST MODELLING

Fig. 4 represents our view of the cost modelling cycle for building projects. In this we have retained many of the general modelling features described, but with certain crucial modifications to allow for the special nature of construction cost modelling. The most important of these results from the need to provide element and functional values during the implementation stage for specific projects. This means that the models provide heuristics for deriving values rather than values themselves (except in regression models, where functional values are provided). One implication of this is that performance validation takes place both in modelling and implementation with estimated accuracy and actual accuracy respectively being used.

The two stages involved are **modelling** and **project estimating**, shown separated by a horizontal broken line. Modelling is done independently of individual projects by a **modeller**, and it is modellers who have contributed most of the papers in this Reader. Building cost modelling is essentially a two-stage process involving structuring and fitting – shown separated by a vertical broken line – of models to enable the estimation of either element values (**q**[uantities] **models**) or functional values (**r**[ates] **models**). Structuring involves a combination of conceptual structuring from theoretical considerations relating to the problem and empirical structuring as a result of the fitting process, with white-box structural validation of process models through comparison with the real-world builders' estimating practices which constitute the prototype* for cost models. Fitting involves the formulation of heuristics or estimation of coefficients to enable the later estimation of project element values (quantities) and project functional values (rates) and, in the case of regression models, estimated black-box performance validation through *ex post* accuracy measures.

---

*Though directly relevant to its theme, lack of space has precluded any papers on builders' pricing behaviour in this volume.

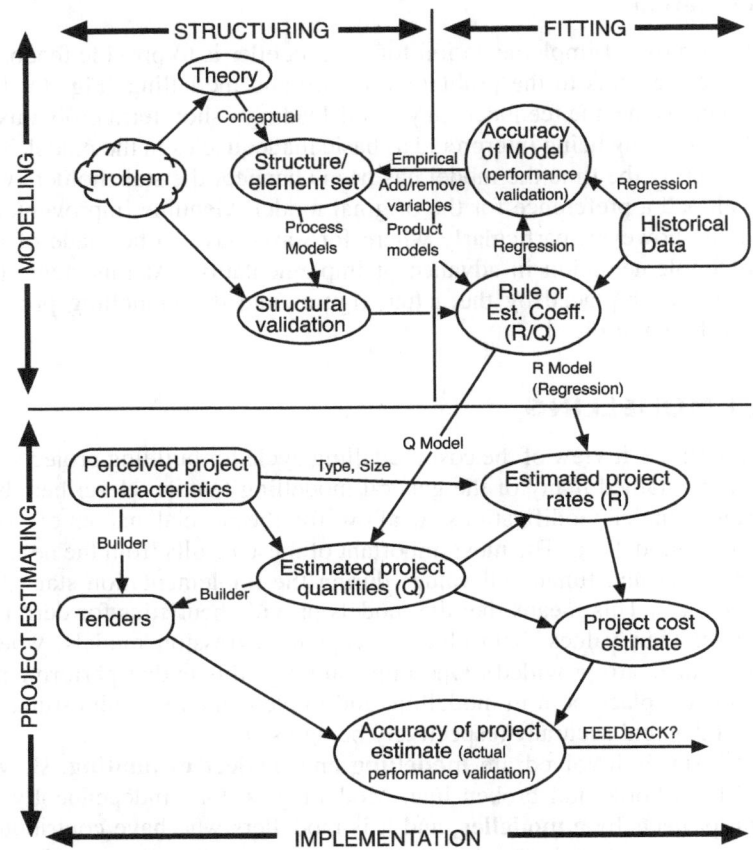

Figure 4 Cost modelling and project estimating.

Project cost estimation is the implementation stage of cost modelling. Implementation is undertaken by a project consultant (implementor) employed for a specific project. The implementor uses the perceived project characteristics to estimate project quantities via a q model, the perceived project characteristics and the estimated project quantities to estimated rates via an r model and these are combined linearly to produce an estimated project cost. This estimated project cost may then be actual black-box performance validated through *ex ante* accuracy in comparison with builders' tenders derived from the perceived project characteristics and estimated project quantities.

There are several dominant aspects that characterise the cost modelling and implementation processes. Firstly, both modelling and implementation are iterative: several cycles are necessary in the development and refinement of models to achieve reasonable degrees of validity, and several cycles of project costings are necessary to examine the likely consequences of different design options in the search for optimal-like solutions. Secondly, both modelling and implementation involve a considerable degree of judgement on the part of the modeller and implementor, due to the inherent uncertainty and complexity of the prototype. This invariably results in a heuristic approach by both modellers and implementors to r and q valuation (here we use the term 'rule' to mean a general regulation or prescriptive convention used by all, and heuristic to denote an arbitrary approach adopted by individual modellers and implementors). As with all estimating activities, estimating bias and consistency are important validation issues and are directly affected by the heuristics employed, and various debiasing techniques are now becoming available (Skitmore and Patchell, ch 1.4; Flanagan and Norman, ch 3.2).

Structuring and fitting

As is clear from Maver's paper (ch 3.3), cost models are technical models used to help in evaluating the monetary consequences of building design decisions. Clients need guidance on the amounts to be included in budgets. Designers need to know if their proposals are within budget and provide good, even optimal, value for money. Society needs to know the cost of maintaining community standards.

The implications are noted in Wilson's paper (ch 5.4). The purpose of cost models generally is to support at least one of the following tasks:

1   forecasting the total price which the client will have to pay for the building, at any stage in the design evolution.
2   comparing a range of actual design alternatives, at any stage in the design evolution.
3   comparing a range of possible design alternatives, at any stage in the design evolution.
4   forecasting the economic effects upon society of changes in design codes and regulations.

The purpose of the models supporting the reactive tasks 1, 2 and 4 is essentially that of discovery whilst for the proactive task 3 to be practically possible, the purpose usually needs to be explanatory, a point made by Bowen *et al.* (ch 4.4).

Two types of cost model structures are identified in several of the papers and both are symbolic. As Wilson (ch 5.4) notes, the first type models the cost, C, as a function of a useful set of *product elements*, p_1, p_2, etc, i.e.

$$C = f_1 (p_1, p_2, \ldots) \tag{1}$$

whilst the second type models the cost as a function of a useful set of production *process elements*, r_1, r_2, etc, required for design realisation, i.e.

$$C = f_2 (r_1, r_2, \ldots) \tag{2}$$

Two basic questions need to be addressed in developing cost models: what elements, functions and values are to be in the models; and how are they to be derived? The nature of cost modelling is such that the element and function values cannot be derived in the absence of a target project and therefore their derivation is necessarily a part of the project cost estimation, or implementation, phase of the project.

There is no general agreement on what constitutes the universally most useful set of elements and functions for each of the model types, nor how they and their values should be derived, although Pegg's (ch 3.4) empirical identification of nine elements is an important contribution in this area. There is also no agreement on the nature of the functions, f_1 and f_2, connecting the cost with the elements although most existing models are of a linear form. Skitmore and Patchell (ch 1.4) survey many of the models in existence and their Table 1.4.1 provides a useful summary in terms of mathematical symbols (all the models examined are in linear form) and other distinguishing features. Newton (ch 1.5) provides a similar survey.

From these studies it is clear that conventions exist for the elements to be used and how they are to be measured (eg RICS, 1988; BCIS, 1969; PSA, 1987), CIOB Estimating Practice Committee, 1983; etc). As a general guide, the terms in normal construction industry usage that are equivalent to the term **Element** include the *product-based* 'design element' (eg walls, floors, windows), 'cost planning unit', 'BQ item', 'functional unit' (eg bed space, theatre seat, car parking space), etc and *process-based* 'operation', 'activity', 'cost centre' etc (these may contain non-design information, eg location, type of client, contract). Fig. 5 shows the relationship between some typical elements and cost models.

The term **Element values** includes the *product-based* item quantities, element unit quantities, storey enclosure units, functional unit, floor area, volume etc and *process-based* resource measures such as the number of man and

Elements	
Product	**Production**
GFA Volume BCIS Elements 'Spon' Approx Q ITEMS SMM7 items	Activities Operations Cost centres
Discover	**Explain**
Models	

Figure 5 Elements and models.

machine hours, quantities of materials etc (Fig. 6). In practice, the element values are never known with complete certainty and estimates of these values have to made either manually in the project cost estimating phase, via the perceived characteristics of the project, or automatically via CAD systems (eg Maver, ch 3.3) or client brief (Holes and Thomas, ch 4.2; Marston and Skitmore, ch 4.5). This is particularly true for the process elements, which are themselves a rather poorly understood function of the building design and whose true values are seldom (if ever) known by cost modellers.

Model	Elements	Values	Functional Rels.	Values
AREA	GFA	m^2	linear	rate/m^2
CUBE	volume	m^3	linear	rate/m^3
ELEMENTAL	BCIS	m^2 etc	linear	rate m^2 etc
APPROX Q	'Spon' items	m etc	linear	rate/item
BQ	SMM7	item quantities	linear	rate/item
CPS	activities	hours etc	linear	rate/hour etc
SIMPLE	activities	hours etc	linear	rate/hour etc

Figure 6 Project model: estimation.

Functional values are invariably expressed in terms of product-based BQ item rates, elemental rates, square metre rates, cubic metre rates etc and process-based hourly rates for men and machines, unit prices of materials etc. They have to be derived either manually in the project cost estimating phase or automatically (Maver, ch 3.3; Marston and Skitmore, ch 4.5; Brandon, 1990; Smith and Skitmore, 1991) from cost information extracted from completed projects. In the case of BQ item type models, there are mandatory functional values imposed in the centrally planned countries and many published guidelines in countries with the market economies (eg Wilderness, ch 3.1; Flanagan and Norman, ch 3.2; Spon, 1985 and 1989; Wessex, 1987a and 1987b; Johnson and Ptnrs, 1987; PSA, 1987). For specific projects, implementors often take the known functional values extracted from a selection of similar completed projects, with suitable subjective adjustments for dissimilarities, such as market conditions, with the perceived target project characteristics. In this case the size and selection of the sample taken is a decision problem in its own right, with the degree of similarity between the sample projects and the target project characteristics (principle of homogeneity) being the key issue.

In yet other cases the functional values are estimated entirely subjectively from the implementor's own experience. Regression (Kouskoulas and Koehn, ch 2.2; McCaffer, ch 2.3; McCaffer et al., ch 2.5) and fuzzy set type models of course generate functional value estimates automatically through the analytical procedures embodied in the approach.

The validity of a cost model effectively depends on three factors – (i) the purpose of the tasks supported by the model and hence the purpose of the model itself, (ii) the usefulness of the elements in the model in the model and (iii) the accuracy with which their values and functional relationships are estimated. For forecasting the total cost to the client (task 1), the validity ofthe model can be assessed indirectly by the black-box approach of measuring performance by its accuracy in use, ie., the difference between the forecasted and the actual cost incurred (Fig. 7). For forecasting the total costs to the client or society of a range of options (tasks 2 and 4,) the validity of the model is less easily determined for, in the absence of any real-world experimentation, only one option is chosen and therefore we will not usually not know what would have been the outcome of the other options. In this case the white box (or at least an opaque box) approach of structural validation is used by comparing the internal workings of the model with the real-world process of contractor pricing. For examining the cost differences between potential building design alternatives (task 3), structural validation is clearly essential for the same reasons as above.

Both Skitmore and Patchell's (ch 1.4) and Newton's (ch 1.5) analysis contain many of these model features, albeit using different terms on occasions. Skitmore and Patchell's 'items' are our 'elements', their 'quantities' are our 'element values', and their 'rates' are our 'functional values'. Newton's 'units of measure' are our 'elements', his 'approach' is our 'purpose', his 'model' is our 'structure', his 'assumptions' are our 'element values' and 'functional values'.

Model	Method	Measure
Product	Black-box	Accuracy
Production	White-box	Structural validation

Figure 7 Project model: validation.

Project cost estimating

Project cost estimating involves choosing an appropriate cost model and applying it to a specific project. The choice of model is dominated by the project chronology (Fig. 8).

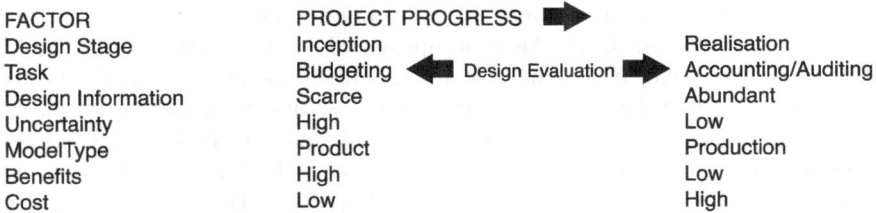

FACTOR	PROJECT PROGRESS	
Design Stage	Inception	Realisation
Task	Budgeting ◄── Design Evaluation ──►	Accounting/Auditing
Design Information	Scarce	Abundant
Uncertainty	High	Low
ModelType	Product	Production
Benefits	High	Low
Cost	Low	High

Figure 8 Project chronology.

The **design stage** largely determines the purpose for which cost models are used. At inception, cost models are used in formulating a budget or target for testing feasibility and financial planning. Cost models are then used in evaluating major alternatives in the early stages of design, through to the evaluation of detailed specification options as decisions are made and the building design progresses.

In the early stages, **design information** is scarce and so assumptions have to be made on likely design decisions that will eventually be made. Later, these assumptions are replaced by information concerning the actual design decisions that are made. Thus the uncertainty surrounding the element values is high in the early stages but virtually disappears by the end of the project. Although there has always been an interest in uncertainty in terms of the accuracy, and hence choice of model, in recent years there has been a con-centrated interest in uncertainty *per se*. The first major contribution in this was Beeston (ch 5.1) whose decomposition of variance, termed 'distribution of variability', attempts to separate the accuracy of the model from other influences on accuracy at late design stage. As a result of this study, Beeston urged modellers to relax the principle of homogeneity much more than is usual in practice and analyse a large sample of previous buildings for estimating the element and function values. Several others have attempted this decom-position analysis with different data (eg Morrison (ch 5.3), with similar results. Similar work by Ashworth and Skitmore (ch 5.2) suggests that accuracy improves surprisingly little through the design period. Special models have also been developed to deal with inherent uncertainty, most notably Wilson's (ch 5.4) probabilistic models, fuzzy models and risk models, as well as uncertainty resulting from implementor bias (Flanagan and Norman, ch 5.5).

As the process elements are an imperfect function of the building design, requiring additional assumptions to convert design information into production information, the compounded uncertainties associated with the design and production are believed to preclude the use of process type models in the early stages in favour of product type models. However, process models are more aligned to the real-world prototype of contractors pricing and therefore considered to be intrinsically more valid than product models. General 'smooth' models have been proposed such as CPS (Bennett and Ormerod, ch 4.3) and CASPAR (Thompson and Willmer, 1985) but is thought by many (eg Drake, 1983)) that it is unlikely that these could be used until the later stages when the design information is less uncertain. At present, more specific models seem more feasible – either for parts of buildings where the production process is more consistent between builders, such as in the structural members (eg Russell and Choudhary, ch 2.4), preliminaries, or in housing modernisation work (Marston and Skitmore, ch 4.5). A more radical view is put forward by Brandon (ch 1.1) who advocates the use of a process model throughout the **whole** of the design phase – a notion explored further by Beeston (ch 1.3) in his 'realistic' modelling approach based on his earlier work with COCO (ch 4.1), and Bowen et al. (4.4) in their concept of 'formal modelling' and its application via nebula theory.

Even amongst product models there is some doubt as to which is the most appropriate. As long ago as 1946 Sir Thomas Bennett (Bennett, 1946) commented that the need to provide cost forecasts at the earliest design stage was not being satisfied by the then popular cube model (James, ch 2.1). By 1954, the floor area method had become popular but was thought by James to be equally as unsatisfactory as cubing, resulting in the 'Storey Enclosure' model described in his paper which tries to reflect differences between buildings by assigning arbitrarily different functional values (weights) for the useful elements (floors, roofs and walls), depending on the storey level and whether or not they are below ground level.

As early stage building design decisions are more cost sensitive than those made later, the benefits to be gained in using good models are that much higher. The shortage of design information and resulting use of the simpler product type models in the early stages means that the costs involved in using the model are relatively low at this time. In later stages, when benefits are low but more expensive models are being used, payoffs (cost–benefits) are certainly likely to be relatively less favourable than in early stages.

Again many of these project estimating features occur in Skitmore and Patchell (ch 1.4) and Newton (ch 1.5) with different terms. The main emphasis is on the treatment of uncertainty, whether by deterministic or probabilistic/stochastic elements and functional values, 'data' (design information), or 'application' (design stage).

4 THE PAPERS

The papers are contained in five parts. Part One **Strategies and directions** contains papers which contribute to the delineation of the field – what cost modelling has been and should be trying to achieve; and how this has been and should be operationalised. By their very nature, these papers are ambitious and highly speculative, motivated by a desire to bring some semblance of order to the chaos perceived by their authors. Collectively, these papers contain what is perhaps the most exhaustive coverage of the factors and issues surrounding cost modelling adumbrated up to their time of publication. Considering that the reference material used by the authors of these papers is likely to have been very similar, it is surprising and instructive how different are their conclusions.

Part Two **Explorations in cost modelling** contains papers in which new models are presented and evaluated. A common factor in all these papers is the high degree of rigour in the analysis and reporting of results – an unusual occurrence in this field – which encourages the view that the best work has been and will continue to be of an empirical nature. This would seem to be a remarkable state of affairs in view of the comparatively poor state of theoretical development in the field.

Part Three **Cost-product modelling** contains only a very small proportion of the extensive work that has been published on this topic world-wide. All these papers examine by various means the cost implications of design alternatives through the use of product models. Outside the market economies, where product element values (prices) are fixed by institutional means, the task is relatively simple and straightforward, the main requirement being a large computer to hold all the data. In a competitive environment however the large variability involved has resulted in a statistical approach to modelling in all except the earliest work. In contrast with the papers concerned with new models in Part Two, the papers in this Part fall well short of the standards of rigour. The main reason for this seems to be that useful product elements are difficult to identify and the relationships sought are extremely weak. The pity is that this is seldom made clear in the papers themselves.

Part Four **Cost-process modelling** in contrast to Part Three contains most of the papers published on process modelling for this domain, reflecting the difficulty of the task which largely depends on finding an aspect of the production process that is common to all contractors. The most recent work relies heavily on the stochastic simulation of the production process by the use of network planning models in combination with knowledge modelling of the planning process. This is perhaps the newest of all fields due to its dependence on the linking of latest developments in information technology for practical applications.

Part Five **Dealing with uncertainty** contains a collection of papers relating the general theme of uncertainty either in terms of validation or non-deterministic cost modelling. The papers dealing with validation use an essentially black box approach to quantify the performance of models through their accuracy, the more recent work concentrating on the human aspects involved. The modelling papers range from the straightforward use of probabilistic and stochastic simulation to the use of dynamic modelling to incorporate feedback.

A glossary of terms follows and finally a **Selected Bibliography** contains a list of further publication relevant to the field. The list is by no means exhaustive but does give a general coverage of the more well known works. The quality of these papers is uneven however and some care is urged, especially where papers have not undergone any review process.

5 THE FUTURE

The purpose of this book is to expose the reader to some milestones in cost modelling. The purpose of this introduction is to coordinate these into some coherent whole – the state-of-the-art. Up to this point we have been concerned with the past, or at most the present. But what of the future?

The barriers to progress are considerable. Cost modelling is a difficult field. Only now are we beginning to fully understand the true purpose of our endeavours. The notion of **pay off** in the modelling process is a relatively neglected but vital aspect and is at the bottom of many criticisms of innovative models, especially from implementors. Known accuracy is not good and not often faced by implementors. The implementation costs involved in using some of the newer and more sophisticated models is likely to be high – a point better appreciated by implementors than modellers.

Cost modelling research is handicapped at both theoretical and empirical levels. Theoretical development in the subject is not far advanced due to the evolutionary and pragmatic development of the subject to date and confusion of terminologies. This has hindered progress in empirical research in guiding the choice of useful elements to study. Empirical research is also plagued by lack of relevant data and communication between researchers in the field as a result of poor reporting of the research method (analysis procedure, validation results, etc) and commercial or governmental constraints. There is a general lack of expertise caused partly by a perceived low status of the subject and partly by insufficient training as a result of undue emphasis in estimating practice. Too few non-qs researchers have been involved. Big gains could be made by the involvement of operations researchers, statisticians, economists, forecasting researchers and information technologists.

Such modelling that has occurred has been piece-meal, with many overlaps and 'reinventions of the wheel'. There has been a disproportional emphasis of research in functional values for BCIS elements and BQ items at the expense of the development and testing of alternative elements and functional relationships. No work at all has been done in non-linear modelling or payoff analysis. Shortage of resources means that we must be more efficient and take advantage of any economies of scale available. This suggests the need for a research programme. No such programme has yet been proposed and there is no general consensus even of its outline.

In developing a research programme for cost modelling, three basic areas need to be addressed – aims, objectives and methodology. To start the ball rolling, we propose the following:

Aims: To find optimal cost models.

Objectives:

1   Identify, define and extend the range of purposes of cost models in use.
2   Define the nature of the messy environment associated with cost modelling
3   Identify, classify and specify potential q and r models in terms of:
    • element lists
    • functional relations
4   Analyse and model the factor relating to project cost estimating and the choice of q and r models, their inter-relationships and changing states over the project chronology, viz:
    • tasks
    • design information
    • uncertainty
    • model estimation and validation
    • benefits
    • costs

Many methodological alternatives exist and each will need careful appraisal for executing the various parts of the programme. Time and space prohibit that exercise here.

In conclusion it should be stated that much of the extant knowledge in the field is of the 'received wisdom' variety. It is a matter of fact, however, that cost modelling is intrinsically interesting to many people and of great potential value to clients and society in general in enabling better decisions to be made on value for money issues in construction.

This could mean building three hospitals for the price of two.

Billions of pounds are at stake.

GLOSSARY OF TERMS

Accuracy   The difference between a cost forecast and the actual cost, usually expressed as a percentage of ratio. Called the error term in regression models

Benefit   See payoff.

Black-box validation   A method of testing how good a model is by comparing the output of the model with its prototype. Also called performance evaluation. In cost modelling, black box validation is represented by the accuracy of the forecast. See also white box validation.

Contractors   The buyers of building contracts. In conventional tendering, contractors' prices are forecasts of the market price of the contract. These prices are based on contractors' estimators' cost models (outside the scope of this book).

Cost-benefit trade-off   See payoff

Cost evaluation   The process of evaluating the design on monetary criteria. This is carried out by the designer along with other evaluations on other criteria such as appearance and function and for different design alternatives. The selection of the best design is a multi-criteria decision-making problem.

Cost forecasting   The process of forecasting the client's cost. Cost forecasting is a part of cost evaluation process.

Cost information   This is supporting material used by cost modellers for estimating functional values. Cost information may take the form of item rates in price books or bills of quantities for completed projects, 'elemental' unit rates of completed projects, etc. Unlike design information, cost information is seldom project specific although some adjustments will usually be made for the special nature of a project before they are used as project functional values.

Cost Consultants   See Implementators.

Cost   The cost of the contract to the client. This is the value of the lowest bid received for the contract, or the contract sum.

Design information   This is information, traditionally in the form of drawings, specifications and schedules, emanating from the designer which communicate his project decisions. This information flows throughout the design process, becoming more and more detailed as the design progresses, with a consequent change of effect from macro to micro design variables. Design information, in contrast, with cost information is project specific.

Design variables   These are aspects of the building design that may be varied under the control of the designer. The values of the variables are estimated by the implementor by inspection of the design information. they include macro variables such as gross and net floor area, number of storeys, and plan shape, and micro variables such as floor and wall finishes. Design variables should not be confused with product elements which are used by cost modellers even though they may sometimes be the same as design variables.

Designers   These are traditionally architects. The modern trend however is to refer to design teams consisting of architects, engineers, surveyors and other consultants.

Element values   These are the physical 'quantities' of the elements for a specific project. In cost model parlance, these include the product-based item quantities, element unit quantities, storey enclosure units, functional unit, floor area, volume etc and process-based resource measures such as the number of man and machine hours, quantities of materials, etc. They are measured from the design information by the implementor. In practice, the element values are never known with complete certainty and estimates of these values have to used instead. This is particularly true for the process elements, which are themselves a rather poorly understood function of the building design and whose true values are seldom (if ever) known by implementors.

Elements   This is a general term used to denote the basic components or building blocks of a model. In cost model parlance, these are equivalent the product-based 'design elements' (eg walls, floors, windows), 'cost planning units', 'BQ items', 'functional units' (eg bed spaces, theatre seats, car parking spaces), etc and process-based 'operations', 'activities', 'cost centres' etc. Cost model elements also include such non-design aspects of projects such as geographical location, type of client, contract, etc. The appropriate elements for the model are determined by the cost modeller.

Estimates   This is the term used for the element values and functional values that are derived by the modeller or implementor.

Forecast   An estimate of some event based on information that is innocent of that event. Therefore an estimate of a future event must by definition be a forecast. This should be compared with a prediction, which is an estimate of an event based on information which contains the event itself. Thus we make the important distinction in regression between ex post simulation predictions (within sample) and ex post forecasts (within sample but not used in model fitting) or ex ante (out of sample) forecasts (*cf.* Pindyck and Rubinfield 1976: 313)

Functional values   These are the unit costs of the elements. In cost models they take the form of product-based BQ item rates, elemental rates, square metre rates, cubic metre rates, etc and process-based hourly rates for men and machines, unit prices of materials, etc. Functional values are obtained by model fitting.

Functional relationships   These are the relationships, at the heart of the model structure, between the element values and the total project cost. In most cost models, these mirror the linear model where the elementsare multiplied by the functional values and the resulting products are summed.

Heuristic   Denotes an arbitrary approach adopted consistently by individual modellers and implementors (ie, 'rule of thumb')

Implementators   People (consultants) who use cost models for making forecasts for specific projects. They usually provide estimates of the element values and maybe functional values, but not the structure or list of useful elements.

Market price   The (unknown) value of the contract to contractors buying on the contract market. Cost models are used to forecast this value. Similarly, contractors' prices are also forecasts of the market price.

Method   A method is a systematic way of doing things and may utilise a model which is structured in a certain way. The terms model and method are often used synonymously in the literature. When in doubt we have used the term model here. The term technique has been found to be redundant and has not been used.

Model fitting   This is the process of estimating the functional values and can take various forms in cost modelling. In the case of BQ item type models, there are mandatory functional values imposed in the centrally planned countries and many published guidelines in countries with the market economies. For specific projects, modellers use cost information. Regression and fuzzy set type models of course generate functional value estimates automatically through the analytical procedures embodied in the approach.

Model structure   The model structure is the overall representation of the prototype in its model form. It contains appropriate elements and functional relationships. The appropriateness of the model structure can be tested by either black box valuation or white box validation. Cost models are either product-based, where the finished building is modelled, or process-based, where the construction production process is modelled.

Modellers   These are people (eg academics) who design models, esp. structure, in contrast with implementors (cost consultants) who use models for forecasting purposes. In cost modelling, the distinction is not always clear as implementors have tended to develop their own models for each project. also, several aspect of cost models have been imposed by institutional prescriptive, and sometimes normative, action, which muddies the waters even more.

Payoff   Some trade-off of costs and benefits. Usually expressed in terms of a benefit/cost ratio or difference. Payoff is a convenient term for use in optimisation models, where it is called an objective function. In this case, the model is said to have reached an optimum when the objective function is maximised.

Product models   See model structure

Process models   See model structure

Prototype   The 'real-world' phenomenon that is being modelled.

Purpose   The purpose of the model is the major influence on its development. The purpose of cost models is to help in cost evaluation through aiding 4 tasks involving cost forecasting. These tasks are either passive, requiring discovery (predictive) models, or active, requiring explanatory models.

Quality   The quality of the model. This has been taken here to be a one-to-one relationship with the level of payoff and has therefore not been used.

Rule   A general regulation or prescriptive convention used consistently by all modellers and implementors (eg SMM rule).

Smooth models   Cost models which can be used without modification at any point in the project chronology. There is some debate whether smooth process models can ever be developed due to the lack of detailed design information in the early stages.

Technical models   A bundle of equations, concepts and computer programs which might be employed to explore the effects of policies and options.

Useful elements   A sub-set of all possible elements associated with the prototype that, when used in a model, will enable an adequate payoff to be achieved.

White-box valuation   A method of testing how good a model is by comparing the structure of the model with its prototype. Also called structural validation. In cost modelling, white box validation is only possible with production process models. See also black box validation.

REFERENCES

Ackoff, R. L., 1979, The future of operations research is past, *Journal of the Operations Research Society*, 30, 93–104.

Ackoff, R. L. and Sasieni, M. W., 1966, *Fundamentally of Operations Research*, John Wiley.

Aris, R., 1978, *Mathematical Modelling Techniques*, Pitman.

Bennett, Sir T., 1946, 'Cost investigation', *Journal of the RICS*, XXV (XI), May, 492–509.

Brandon, P. S. 1990, The development of an expert system for the strategic planning of construction projects, *Construction Management and Economics*, 8(3), 285–300.

Bullock, A. and Stalybrass, O., 1977, *The Fontana Dictionary of Modern Thought*, Fontana/Collins.

B.C.I.S., 1969, *Standard Form of Cost Analysis*, The Building Cost Information Service, The Royal Institution of Chartered Surveyors, Kingston upon Thames, Surrey.

C.I.O.B. Estimating Practice Committee, 1983, *Code of Estimating Practice*, 5th ed., The Chartered Institute of Building.

Drake, B., 1983, *Cost Data*, Paper presented to Research Seminar at the Royal Institution of Chartered Surveyors, London, January.

Gilchrist, W., 1984, *Statistical Modelling*, John Wiley.

Johnson, V. B. and Ptnrs (eds), 1987, *Laxton's National Building Price Book*, 159th ed., Thomas Skinner Directories, Sussex.

Kuhn, T., 1970, *The Structure of Scientific Revolutions*, Chicago Press.

Pidd, M., 1991, Operations research/management science method, in *Operations Research in Management*, 11–31, Prentice Hall.

Pidd, M. and Wooley, R. N., 1980, A pilot study of problem structuring, *Journal of the Operations Research Society*, 31, 1063–9.

Pindyck, R. S. and Rubinfeld, D. L., 1976, *Econometric Models and Economic Forecasts*, McGraw-Hill, New York.

P.S.A., 1987, *Significant Items Estimating*, Department of the Environment.

R.I.C.S, 1988, *SMM7: Standard Method of Measurement of Building Works*, 7th ed., Royal Institution of Chartered Surveyors.

Smith, M. and Skitmore, R.M., 1991, Automatic BQ pricing, *The Professional Builder*, 6(2), 14–21.

Spon, 1989, *Spon's Architects' and Builders' Price Book*, 114th ed., Davis, Langdon & Everest (eds), E & F N Spon.

Spon, 1985, *Spon's Mechanical and Electrical Services Price Book,* 16th ed., Davis, Belfield and Everest (eds), E & F N Spon.

Thompson, P. A. and Willmer, G., 1985, CASPAR – A Program for Engineering Project Appraisal and Management, *Proceedings, 2nd International Conference on Civil and Structural Engineering Computing*, 1, London, Dec. 75–81.

Wessex Database for Building, 1987a, 4th ed., 1:Major Works, Wessex.

Wessex Database for Building, 1987b, 4th ed., 2:Small works, Wessex.

Part 1 Strategies and directions

This Part contains papers which contribute to the delineation of the field – what cost modelling has been and should be trying to achieve and how this has been and should be operationalised. By their very nature, these papers are ambitious and highly speculative, motivated by a desire to bring some semblance of order to the chaos perceived by their authors. Collectively, these papers contain what is perhaps the most exhaustive coverage of the factors and issues surrounding cost modelling adumbrated up to their time of publication. Considering that the reference material used by most of the authors of the papers is likely to have been very similar, it is surprising and instructive how different are their conclusions.

The paradigm referred to in Brandon's **Building cost research: need for a paradigm shift?** (ch 1.1) is one of type adumbrated by Thomas Kuhn (1970) involving a radical change of direction in thinking and application pointing out the current inadequacies in models and theory and arguing against an incremental approach to cost model development – 'what if practice is using the wrong model?'. He maintains that cost modelling is a technology and therefore then the pressure for change is likely to be from other stakeholders, ie clients. He also 'feels' that there should not be several models and advocates the increased use of computer simulation for the development of production type models due to the potential benefits in explaining how costs are incurred in the building process and in occupation. The paper is inspirational and highly perceptive in several respects – the need for small bites of research, building cost is not an easy subject to study, the relevance of Mitroff's papers, the need for parallel research, the relevance of the distinction between scientific and technological (engineering?) research, that cost modelling is a 'late starter'. On the other hand some major points are glossed over – **who** validates the models is examined but **how** the models are validated is completely overlooked, that increased pressure from **client** stakeholders was and is unlikely (ELSIE was not a product of client demand), that production models map site processes rather than contractors' estimators' processes, and why 'simulation' in preference to COCO (Beeston, ch 4.1) or lots of Wilderness (ch 3.1) charts for instance?

Raftery, in **Models in building economics: a conceptual framework for the assessment of performance** (ch 1.2), suggests that cost model development lacks rigour; and predicates a 'consistent and conceptual framework within which the performance of models may be evaluated'. Although tentative and speculative, the paper is a very necessary and timely attempt to focus attention upon the need to improve rigour and consistency of model development and evaluation. A framework is proposed, based on a chain which leads from raw data, through a model and its output, to a decision maker [implementor]. Here the paper makes an important contribution to highlighting the significance of the context in which the model is placed. In developing the idea of a chain, a useful distinction is made between the 'modelling environment' and the 'decision [implementation] environment'; and five points are identified where the chain may be tested: data; data/model interface; modelling technique; interpretation of output; and the implementation decision. The first four of these are commented upon in the paper – disappointingly it is commentary rather than discussion, but nevertheless some feel for the structure of the field does emerge. The fifth point, implementation testing, is wisely left for consideration elsewhere.

Beeston's **A future for cost modelling** (ch 1.3) develops Brandon's (ch 1.1) argument against the conventional use of several product models in prescribing a common basic model for all applications and the rejection of existing product models. Beeston advocates using existing skills more widely and developing models that quantify uncertainty more closely. 'Realistic' production models are recommended for use at an earlier stage than usual because of their enhanced explanatory powers resulting from the introduction of causality into the relationships. To forecast the cost effect of a design change, Beeston argues that all that is needed is to calculate the cost of replacing one component of the design with another with a separate allowance for general plant, labour and time requirements. The new data collection methods necessary to support these models are not considered to be a major problem due to the tendency to a **common** and economical method of planning and execution for a given design. Three types of realistic modelling approaches are proposed (1) simulation of construction processes (eg Marston and Skitmore's intelligent simulation model, ch 4.5), (2) attaching costs to activity networks (eg Bennett and Ormerod's CPS, ch 4.3), and (3) simulation of construction planning methods (eg Beeston's COCO ch 4.1). Beeston's advancement in the field is in his assumption that there is a common construction method for a design, with any residual alternatives being covered by parallel simulation. He overcomes the criticism that contractors do not use this model at the moment by assuming that they will follow the designers' (tendering) model in due course. He also outlines some of the spin-offs that may follow from the use of realistic models. He is essentially proposing an integrated model for a fragmented procurement process – is this practicable or are the disintegrating forces of competition too great? To find answers to these difficult questions may mean taking a closer look at the economists view and considering the relationship between risk and incentive, and the economics of asymmetric information.

Skitmore and Patchell's **Developments in contract price forecasting and bidding techniques** (ch 1.4) section on *Mathematical and topological features* is perhaps the most relevant to the theme of this reader, and the models examined are summarised in their Table 1.4.1 in terms of mathematical symbols (all the models examined are in linear form); relevant contract type; general accuracy; deterministic/probabilistic structure; number and type of items (elements); derivation and deterministic/probabilistic nature of quantities (element values); and derivation, currency, weighting, quantity trended and deterministic/probabilistic nature of rates (functional values). This may be contrasted with Newton's (ch 1.5) 'agenda'. The paper continues to describe the use of regression, CPS and ELSIE (not reprinted here), with some new (to this field) work on the accuracy of the regression forecasts. Also, the bidding model debiaser described in the final section of the paper is new and may be compared with Flanagan and Norman's (ch 5.5) CUSUM approach.

Newton's **An agenda for cost modelling** (ch 1.5) is concerned that there is no adequate way of comparing cost models or research into cost models; and nine 'descriptive primitives' are proposed as classificational features of all cost modelling research, the ones – which are relevant here being – data (design information), units of measurement (elements), approach (purpose), application (design stage), model (structure), technique (regression, network, etc), assumptions (eg element values and functional relationships) and uncertainty. Within each of these descriptive primitives is proposed a set of characteristics (simulation, generalisation, optimisation, etc) to represent the various internal 'levels'. He forecasts a long term trend towards models for optimisation as the field becomes better understood and lodges a plea for explicating the assumptions inherent in the models. As Newton concedes, his work is only intended to provide a supporting framework – to give some order to the way in which we classify, talk and think about cost models – and, as a first attempt, is expected to be replaced by better future versions.

Fortune and Lees' **The relative performance of new and traditional cost models in strategic advice for clients** (ch 1.6) seminal work is of great importance. Theirs has been a labour of love over several years of careful study investigating pre-tender estimating practice. It is a remarkable achievement in several respects. It is the first work to deal comprehensively with the **practice** of pre-tender estimating, it aspires to an utter level of rigour and it is fundamentally and relentlessly concerned with what counts in practice. The objectives of the paper are to establish and explain the incidence of competing forecasting models within the built environment, together with an evaluation of their use in terms of accuracy, reliability and value. An assessment of the degree of judgement involved in the particular forecasting model along with an approach to assess new models forms the remainder of the study's objectives. These objectives are clearly stated and combine to expeditiously orientate the new readers to this field, in terms of the state of its development and future direction. The paper begins by categorising the various currently used models under the headings of traditional models and newer models. The newer models comprise statistical models, knowledge based systems, life cycle costing techniques, resource/process based methods and risk analysis techniques. Many of the models selected, particularly the traditional models, are in fact approximate estimating methods. It is their selection and timing within an overall cost information system which qualify them as 'strategic' from the viewpoint of their user. Having grouped the pre-tender estimating methods into 'traditional' and newer models, they then proceed to find out, by postal questionnaire, how these are getting on in practice. How this is done is a model of good questionnaire design and delivery.

REFERENCES

Kuhn, T., 1970, The Structure of Scientific Revolutions, University of Chicago Press.

1.1 Building cost research: need for a paradigm shift?*

P. S. Brandon

AN EDITORIAL VIEWPOINT ON COST RESEARCH

In 1977 a publication appeared called the Encyclopaedia of Ignorance; it contained fifty essays from prominent writers outlining the lack of knowledge in the natural sciences. Scientists, who are sometimes in the habit of dropping the word 'theory' from their ideas, were confronted yet again with the enormous number of unknowns in a subject area in which measurement is considered a very precise art. As the editors remarked in their foreword (1) 'Compared to the pond of knowledge, our ignorance remains atlantic. Indeed the horizon of the unknown recedes as we approach it'.

If the natural sciences with a formal discipline of research extending over several centuries have problems, then the social sciences with a comparatively short history have even more. The psychological factors that influence the behaviour of human beings have hardly begun to be investigated. (Evidence of the problem can be seen in the variety of results produced by the econometric models which at the moment are predicting the upturn, downturn, stabilisation of the U.K.'s current economic position!) Yet it is this behaviour, whether it be in the production, procedures or procurement of buildings that has a major impact on its cost. With this high degree of 'ignorance' coupled with an observed high degree of variability in measured performance there is obviously a fundamental need to establish a more substantial body of theory, and better models, upon which to base our practice.

PROBLEMS IN INITIATING BUILDING COST RESEARCH

The building professions in general, have shown only a token willingness to invest in research and thereby extend their knowledge and discover an improved methodology. There are a number of reasons for this lack of enthusiasm including:–

*1982, in Brandon, P. S. (ed.) *Building cost techniques: new directions*, E & F N Spon, 5–13.

1   the practices and firms in the industry are widespread, generally small in size, and are competing in a commercial market against each other. The incentive to gain a common pool of knowledge **from which all would benefit** is therefore diminished, and the gathering of funds for substantial research is difficult to achieve.

2   firms investing in research tend to have higher expectations than the researcher can fulfill. As with other complex areas of study it is not possible for one research project to provide a panacea for all the problems faced in say, building cost forecasting. In the natural sciences, very narrow specialisms are the order of the day which when placed together advance the general body of knowledge in the subject discipline. Practitioners should recognise the need for many small bites at the cherry before it is eaten;

3   there is a natural tendency for conservatism and resistance to change, which does not provide a fertile ground for the implementation of research. This subject will be developed later in this paper;

4   the subject matter of specialisms such as building cost falls between a number of well established disciplines e.g. management, design and economics. This has been particularly apparent to those applying for government research grants although credit must be given to the Science and Engineering Research Council for recognising this problem in recent years. The specially promoted programme in encouragement to those involved in management and cost research.

In addition the nature of the production process and its diversity does not make building cost an easy subject to study. This may well discourage many would be researchers from immersing themselves in what is sometimes called a 'messy' problem.

PROGRESS

It appears that the progress that has been achieved has arisen largely from the refinement of current models. This refinement has taken place over a long period of time through the light of experience. Occasionally a research project will speed up the process, although if it proceeds too fast a gulf seems to arise between the new model and current practice which is too great for industry to bridge. There is a wide body of opinion, even found among the research funding bodies, that we should be 'working at the leading edge of current practice'. This is understandable as it simplifies the objectives of a research exercise and more importantly avoids the gulf arising between practice and research.

   The problem is, of course, what if practice is using the wrong model?

   Is there not a danger that we might be pouring our limited resources down a well which can never yield a satisfactory spring of new ideas that will allow us to substantially progress? The methodology of experience has served us well for a number of years but it is an unwieldy inflexible tool which cannot respond quickly to a challenge. Coupled with the inertia towards change it could represent a stumbling block to advancement.

BIAS IN RESEARCH

Experience contains within it an enormous bias towards current practice. Now this is not always undesirable. It has been recognised for some time that as stated by Yerrell (2) 'the commitment and bias of psychologists, and more generally scientists, are necessary for the scientific process'. However this can only apply where there is a considerable body of parallel research which can investigate and test the ideas and bias of a single research group. (The emphasis here must be on research rather than practice i.e. the discovery and generation of new knowledge rather than the implementation of knowledge within well defined rules.)

Where research is sparsely sprinkled across a very complex subject area then there is a great danger of research being isolated and dismissed because it does not conform to conventional wisdom, or, being adopted without being fully tested. Such a problem could exist within the building industry and examples are available which suggest that with hindsight both possibilities have occurred.

A study by Mitroff (3) investigated the question of bias with 42 lunar scientists. He reported the scientists view that 'the objectivity of science is a result ... not of each individual scientists unbiased outlook, but of the scientific communities examination and debate over the merits of respective biases'. In another paper (4) he goes on to say that 'to remove commitment and even bias from scientific inquiry may be to remove one of the strongest sustaining forces for both the discovery of scientific ideas and their subsequent testing'.

Because of the lack of research into building cost, any new model or methodology is tested mainly by practitioners rather than those undertaking parallel research. There is therefore a strong bias towards the status quo and the current stock of models. It is not in the nature of practice to constantly challenge the basis of the techniques which they apply.

SCIENTIFIC AND TECHNOLOGICAL RESEARCH

A qualification should be introduced at this stage. There is of course a subtle difference between the objectives of scientific and technological research. The objective of gaining knowledge in technology is as summarized by Cross *et al* (5) to 'know how, improve performance and establish products of skill and quality. Science on the other hand is directed towards error-free explanation and scientific "truth"'. In this sense technology may more readily be answerable to practice than to the research community. The difficulty lies in persuading practice to apply the rigorous tests for a long enough period to satisfactorily substantiate or reject the research findings. there may be considerable resistance to even the consideration of a new model which threatens to overturn the status quo. As Yerrell (2) points out '. . . within an accepted paradigm the merits of respective biases are readily debated; in contrast to the discussion of biases inherent in the paradigm itself. Between paradigm debate is an uncommon activity for the scientific community'. (The definition of paradigm in this context would no doubt be similar to that of Kuhn (6) i.e. . . . universally recognised scientific achievements that for a long provide model problems and solutions to a community of practitioners'.) If debate of this kind is uncommon for the scientific community it is seldom found between research and practice. Yet this dialogue must exist if rapid progress is to be achieved. As a late starter, building cost research has a long way to go before it has a satisfactory collection of models and theory, upon which to build.

PARADIGM SHIFT

However there are pressures which could well bring about a change in the current techniques used by practitioners and produce what Kuhn called a 'paradigm shift'. In his well known historical study (6) he presented a picture of science (and it may well apply to technology) as a process involving two very distinct phases. The first is 'normal science', a phase in which a community of scientists who subscribe to a shared 'paradigm' engage in routine puzzle-solving by seeking to apply what are agreed-upon theories within the paradigm to problems observed in the physical world. The second phase called 'revolutions' is a very active period in which the paradigm is found to be inadequate in resolving the problems; the discipline is thrown into crisis; and the old paradigm is overthrown and replaced by a new one which then forms the basis of the next phase of 'normal science'. A parallel exist in technology where a change in the tools, machine or technique forces change upon the current paradigm being used. The difference between science and technology in this respect is that science tends to change from within the subject discipline, e.g. a new discovery, idea etc. which alters the scientists view of the world albeit with the assistance of technology; whereas in technology itself the motivation to change is often imposed from outside the immediate discipline. The provision of electricity, the internal combustion engine and the silicone chip are examples of innovations which forced the pace of change in technology over a very wide range of activities. If the study of building cost can be considered as a technology rather than a science it is appropriate to look for factors outside as well as inside the discipline which may provide a catalyst for change.

This brings us back to the central question of 'are we using the right models' and if not, is there any pressure to change.

THE USE OF MODELS

The attractiveness of models lies in their attempt to provide simple and straightforward answers to our very difficult and complex problems. They tend to be rules of thumb which unfortunately sometimes become of more importance than the solution itself. To quote just one example, David Russell (7) writing in the *Chartered Quantity Surveyor* with regard to Engineering services said 'Contractors normally arrive at a lump sum total estimate and then allocate this to the individual items in the Bill of Quantities'. If this is the case then the current model contained within a Bill of Quantities for the pricing of Engineering services is obviously suspect. The situation is probably even more complex than this because the estimators own method of pricing may have little resemblance to the manner in which costs are incurred on site. In addition the person who prepared the cost plan of the services probably used a different model, yet again, based upon a completely different set of criteria. There is a touch of unreality about these models and what the participants in this forecasting process appear to be doing is trying to match one simplified model, with another model that is slightly more complex. Modelling the reality, i.e. the way costs are incurred on site, does not enter into the process until operational costs are considered, usually at the post contract stage.

THE NATURE OF MODELS

The simple nature of models, which in one sense is their attractiveness, is also their weakness. Powell and Russell (8) in a design context state that '. . . to the naive, the persuasive simplicity of categorisations used by most models produces in itself a sort of **model blindness** – an unquestioning and unthinking acceptance of the model because it happens to support the exploitation of a particular idea, concept or system. . . . in their minds, the categorisations used in these models can become almost 'absolute', tight boundaries not to be transgressed. . . . For this group of model users, the model itself does not open up new vistas but encourages particular acts of distinction and boundary definition, which can perpetuate a current stance, or worse prevent thought concerning other meta-level design considerations'.

They define naive as any person unused to thinking in detail about systematic and systematic considerations of design and designing. In building cost research it could be that the slavish referencing of our research data to insensitive deterministic models such as the bills of quantities has resulted in a kind of model blindness which has slowed advancement. It may be that with elemental costs, abbreviated quantities, unit prices and other bills of quantities derivatives, we have reached the highest degree of improvement we can expect from these models. Beeston (9) in his new book Statistical Methods for Building Price Data says that 'there are two distinct ways of representing costs: the realistic and the "black box". Realistic methods are derived from attempts to represent costs in the ways in which they arise. "Black box" methods do not attempt to represent the ways in which costs arise. Their only justification is that they work'. He goes on to discuss their strengths and weaknesses but suggests that with the development of computer technology, simulation i.e. the 'realistic' method, is likely to provide a better basis for development.

BLACK BOX V REALISTIC MODELS

The problem with black box methods is that so many of the assumptions upon which the data is based, are missing. The degree to which the information derived from the model can be analysed is extremely limited. They do not contribute to an understanding of why certain costs arise and because they rely so heavily on descriptive information to explain the numbers, they are subject to the vagaries of language. Descriptive information can seldom be precise and therefore the variability we see in estimates may be largely the result of the inadequate 'word model' found within the black box technique.

On the other hand the 'realistic methods' place much greater reliance on numerical information of site performance. Not only does simulation offer the chance for greater analysis and experimentation because we can now retrace our steps and investigate the point where the cost is incurred but it can in the words of Powell and Russell (7) 'open new vistas'. Human performance is a variable and the interaction of human performance on site has a major impact on cost. With simulation we can investigate the results of this interaction repetitively and provide probability ranges. Unlike the unit cost, element, bill of quantities, operation cost procedure there is not a changing model structure as information is refined. Communication is thereby enhanced because the components of the model have a similar basis from sketch design to construction. The problem of matching models should no longer be a major problem although prior to sketch design the black box method would still need to be used.

What is true for the initial cost of building is also true for the understanding of future costs. established techniques such as cost-in-use, even with sensitivity analysis are essentially deterministic in nature. The interaction of human beings with the building, has major impact on life cycle costs. Simulation may allow a better understanding of this relationship. (Already the technique is being used to help understand the effect of human behaviour in energy usage in buildings.) However the problem of prediction over a long building life, the unreal assumptions of cost-in-use and the lack of post prediction control are still with us! While many scenarios can be postulated it is doubtful whether iterative processes can provide a satisfactory conclusion that can be held to.

SIMULATION, THE WAY FORWARD

It would be foolish to suggest that we are anywhere near providing the simulation models we require for the various degrees of design refinement. However if we can identify a path which will provide a better opportunity for understanding cost then perhaps research can be directed towards this goal. Some of the research papers presented at this conference have already begun to take this particular line.

In terms of traditional quantity surveying practice such a move would certainly be a 'paradigm shift' (although it would be less of a change to a contracts manager). Fortunately any change will occur over decades rather than years and the trauma of a rapid revolution is unlikely to occur. If, as was stated earlier, the motivation for a change in the models in technology arises from outside as well as inside the discipline, then where is the pressure for change to come from? Undoubtedly the major key to change is the computer. Our models that we use today have been tailored to suit the limitations of the human brain and limbs. Both of these limitations have been reduced considerably by machine power, and enhancement of these human faculties is likely to continue for many years to come through the development of computer hardware. The possibility of constructing large data bases; the untiring ability to calculate repetitively various options; the rapid output of results; will tend to force us to look outside the present models, which take their present form in order to cope with mans inadequacies. The difficult problem will, not necessarily be in providing the software to undertake the calculations, but in the collection and inputting of data to feed the new models. It may well result in a highly skilled technical team working parallel to the professionals or a centralised information service which all can access.

CLIENT PRESSURE

Simulation of course will not only be within the province of the cost adviser. Already architects and engineers are using the technique for studying human behaviour in buildings, and relating this to such matters as energy usage, acoustic behaviour and pedestrian movement. As the use becomes more widespread then clients themselves will become familiar with the technique. Some portfolio managers already use simulation in their risk analysis and will begin to demand an equivalent level of investigation from their building cost advisers. It may even be that those responsible for public accountability will also require a more detailed explanation for why certain figures have been presented.

Despite these outside influences, it is likely that cost advisers will themselves wish to improve their service to their client in order to remain competitive. there appears to be a move in Quantity Surveying circles towards project management and this inevitably means a closer involvement with resources. Simulation of those resources will be an important aspect of future development that may well leave the old models behind.

The argument presented above may sound like the presentation of a panacea for all our ills. It is not intended to be so. The problems of handling data in the magnitude required to simulate cost occurrence is enormous. Simulation does however offer opportunities, not available in pre-computer days, to change our model to assist our understanding of how costs are incurred in the building process and in occupation. It must be worth pursuing this line of thought although initial expectations should be low in the early years.

SUMMARY

As all researchers into building cost know there is a considerable lack of knowledge in the subject area. Each research project seems to generate the need for a score or more of new projects. At the present time building cost research is not well funded and this reduces the amount of research activity. This in turn means that the work that is undertaken is sometimes not subject to the same degree of scrutiny by other researchers as found in the natural sciences. Testing is often left to practitioners who have a natural bias towards the status quo. There is a danger that this may result in model-blindness which may prevent potentially more fruitful models being developed. One such technique that has not yet been fully exploited is simulation. By harnessing computer power this method of modelling site and building performance may well hold the key to a better understanding of building cost.

FINALLY

Whatever research is undertaken in the future there is a need for a much greater formal and informal dialogue between researchers. It is hoped that this conference has played its part in this respect and that future conferences will follow. The new refereed journal to be published by Spon is also a most welcome addition to the dissemination and discussion of building cost research at an international level.

REFERENCES

1. Duncan, R. & Weston-Smith, M. (1977) *The Encyclopaedia of Ignorance*, (Pergamon Press), 9.
2. Yerrell, P. (1982) 'Bias in Research – A Methodological Interface', *Portsmouth Polytechnic News*, April, 16–18.
3. Mitroff, I. I. (1974) 'Studying the Lunar Rock Scientist', *Saturday Review World*, 2nd November, 64–65.
4. Mitroff, I. I. (1973) 'The Disinterested Scientist: fact or fiction?', *Social Education*, December, 761–765.

5. Cross, N., Naughton, J. & Walker, D. (1981) 'Design Method and Scientific Method', *Design Studies*, Vol. 2, No. 4, October, 200.

6. Kuhn, T. (1970) *The Structure of Scientific Revolutions*, (2nd edition, University of Chicago Press).

7. Russell, D. (1982) 'Where Computers can help', *Chartered Quantity Surveyor*, May, 304.

8. Powell, J. A., & Russell, B. (1982) *Model Blindness and its Implications for some Aspects of Architectural Design Research and Education*, (Proceedings of the Conference of Problems of Levels and Boundaries, Edited by Professor G. de Zeeuw and A. Pedretti. University of Amsterdam, Netherlands).

9. Beeston, D. (1982) *Statistical Methods for Building Price Data*, (Spon, ch.8 published 1983).

1.2 Models in building economics: a conceptual framework for the assessment of performance*

J. Raftery

ABSTRACT

Insufficient attention has been devoted to the limitations of cost models. It is suggested that decision makers find it difficult to evaluate the performance of models because they lack a consistent framework with which it may be measured. Such a framework is proposed and the criteria discussed. It is hoped that the introduction of a consistent method for comparing models and assessing their limitations, will both expose areas which require further research and assist decision makers in interpreting such models.

INTRODUCTION

This paper is concerned with economic models applied in the building design process. Thus, models for use in the process of building production are not dealt with specifically, although some of the issues discussed may pertain there also. The work described below sets out to be questioning and even sceptical – the reasons for this will become clear. As building economics develops and more research is carried out in universities, polytechnics, state funded institutions and in the private sector, more and more cost models are presented to the public and the industry. The increasing availability of large-capacity, fast computing facilities at lower prices, combined with the high growth rate of the software sales side of the computer industry, has meant that many models are not in fact presented for discussion in academic journals, rather we hear about them from systems purveyors. Even the more traditionally derived and presented models appear to be the independent work of individual workers in a relatively uncoordinated way. Models are reported, their virtues extolled, and their faults given inadequate attention. The work described below has been predicated on the knowledge that models themselves are highly complex entities which need to be carefully designed tosuit their intended purpose. (1)

*1984, *Proceedings 3rd Int. Symp. on Build. Econ.* CIB W-55, Ottawa, 3, 103–11.

In some fields, models do not enjoy a high reputation, there is more than a little justification for this, especially in applied economics. (2,3) Frequently, absent from many presentations of models in building economics is a detailed 'warts and all' assessment of the performance of the model. One reason for this, it is suggested, may be the lack of any consistent conceptual framework within which the performance of models may be evaluated, such a framework is presented below.

TERMINOLOGY

The term 'cost model' as many are aware, is in fact a misnomer. Many models described as such, are in fact models of building price and include mechanisms for considering profit levels and market conditions. This distinction between cost and price is not a pedantic one, it has real implications for the assessment of the model's performance. One of which is that future market conditions are very difficult to predict and any model which attempts to take account of these will necessarily have a higher degree of risk attached to the output than might otherwise be the case. The importance of this distinction derives in a large part from the vagaries and 'touchiness' of western market economies. The term 'economic response model' may be used to describe both types without misleading the reader. (4)

In the context of the economics of building design, two approaches to modelling may be distinguished. Models which use the techniques of statistical inference to deduce relationships between building features or design decision (such as floor area, heating load) and cost are known as deductive models. Inductive models, on the other hand, are causal in nature; the resource implications of design decisions are calculated and aggregated to give the measure of economic performance.

MODELS IN DECISION MAKING

Consider the context of the economic models of various types, used during the building design process. The complexity of the designer's task is well known, the overall design is usually fragmented into several design sub-systems, such as structure, fabric, services, interior decor, etc. The designer is then in the position of juggling many decisions at the same time, and he or she attempts to optimise the overall result of decisions affecting all design subsystems. As the design progresses from inception through to sketches, the designer seeks advice both as to the estimated tender for the design and the difference between alternative schemes or alternative solutions to particular subsystem problems. Examples of the latter would be decisions regarding the relationship between the quality of glazing and the air-conditioning plant, the number and disposition of lifts, staircases, etc. Some models are designed to consider in detail, subsystem problems such as the latter, others seek to predict the tender price expected for the whole building. Clearly, very different characteristics will be required for these tasks, but there are some features of the context which apply regardless of differences of this nature. The model may be viewed as part of a chain which leads from raw data through some kind of model and output and on to a 'decision maker', in this case the designer.

The precise dimensions of this chain give rise to certain metaphysical difficulties, be that as it may, it is presented in Figure 1.2.1. A chain is only as strong as the weakest link, if the chair is dismantled for analytic purposes, we may examine both the links and the interceptions between links. This suggests five points at which the chain might be tested, and these fall into two areas, the modelling environment and the decision environment.

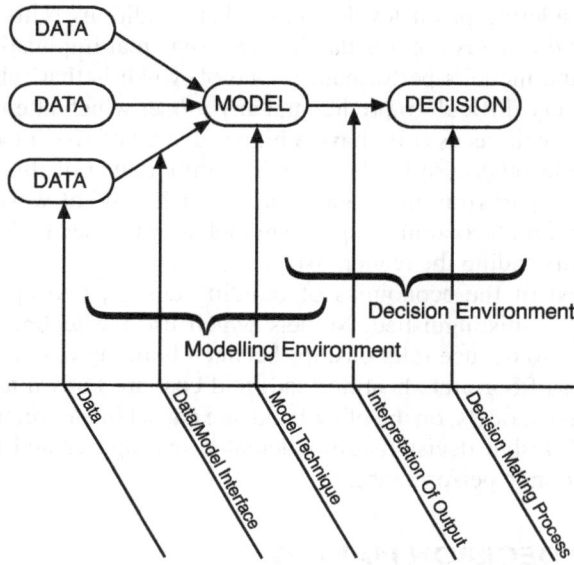

Figure 1.2.1 The decision and modelling environment.

Modelling environment

1   Data: The source and the nature of the data have an impact on the validity of the model.
2   Data/model interface: The matching of the model to its data is dependent upon the degree of development of the design. Data suitable only for the early stages of design should not be used in models intended to be applied later and so on.
3   Model technique: The modelling technique used has an impact on the validity of the model. Each technique has its own strengths and weaknesses.

Decision environment

4   Interpretation of output: How does the designer perceive the output of a model, is objectivity possible? How may the output be interpreted in relation to the age of the model. do the relationships still hold. How has the model been updated?
5   The decision: the internal mechanisms of human decision making are as yet relatively unknown and the subject of much research. The experimental studies are often very situation specific and there are many issues left to be resolved in the area of human decision making and problem solving. Such work is outside the scope of this paper.

CRITERIA OF PERFORMANCE

Four of the criteria identified above will now be examined. Only the briefest of discussions is possible in the space allocated here, for those interested, the issues are presented in detail elsewhere. (5)

(i) Data

Rational decisions are usually made in the light of the information available at the time of the decision. In the construction industry, cost estimation of all types is usually based on 'past experience' projected forward. 'Past experience' is recorded in the various forms of economic data available and also the sum of the personal and professional experience of the decision makers. Whether or not sophisticated economic modelling techniques are used, reliance is based on some form of historic data for input to decision making. The process of cost generation, recording and the eventual derivation of that which is euphemistically called 'data' is muddled with inconsistencies. three of the most important issues will be presented here.

Data transformations

Data, by its very name, purports to be objective. However, there are imponderables and subjective elements involved in data and its collection. This subjectivity, it is suggested, creeps into the data under discussion here in two ways, firstly during the recording of the event, and secondly during subsequent transformations of the data to produce 'information' for various procedural requirements. The most common example of 'misinformation' occurring at the recording stage is that of the records for the time taken to perform various tasks in a project. Tradesmen book in their times on the basis of maximising the value of the work which is bonusable, the total of hours worked is not affected, but tasks with bonus targets tend to be loaded downwards while non-bonus tasks are loaded upwards. The same timesheets may be used for estimating the labour content of units rates.

In the U.K. two major transformations of data occur. The first is when the sum of the resource costs is spread over the unit rates to produce the priced bill of quantities for tendering with the attendant 'loading' and tactical decisions. The second is when the unit rates are subdivided and clustered into element costs. These elemental rates tend to be used by quantity surveyors for cost planning at an early stage in the design when there is little detail. Later on, at sketch design stage, unit rates based on similar bill of quantity items from comparable projects are used in conjunction with approximate quantities to produced more detailed cost plans. Thus, the actual rates used in cost planning using the elemental model of building cost, have been subject to much distortion.

Data incompatible with cost generation

The main inconsistency in the data is that although cost is generated by the use of resources, the economic data generally available and used for cost modelling does not represent cost in this way. It has been shown that it is possible to link economic information directly to resource usage in such a way that there are considerable advantages in designing, planning and controlling construction projects. (6) In the U.K. it is normally only the contractor who has comprehensive files of resource prices and labour rates. These are closely guarded and models used by architects and quantity surveyors usually have to make do with 'unit rates' derived either from price books or old bills of quantities. As well as being incompatible with resource generation, unit rates are subject to a degree of tactical adjustment in the preparation of tenders, which although leaving the tender unaffected, produces a high level of distortion in the actual rates themselves. This can be shown by an examination of the co-efficients of variation for unit rates from priced bills of quantities.

Variability of data

The co-efficients of variation of bills of quantity rates for similarly described items of work are much greater than figures for whole identical buildings. Beeston suggests that the co-efficient of variation for identical buildings in the same location would be about 8 £ %, he calculated that typical figures for the trades within bills of quantities fell in the range 13%–45%.(7)

This variability is exemplified in the published data, identical items of work selected from the various current price books are rarely the same. This would support the idea that rates from bills of quantities are merely a notional breakdown of the overall price and are tempered with so many tactical considerations (forward loading, special discounts with certain suppliers, etc.) as to render them very unreliable for any form of cost modelling.

It is known that prices tend to be distributed about their mean in a skewed way, high prices tending to be further above the mean than low prices are below it. This can be taken into account in a statistical analysis.(8) When variability in prices is due to a combination of several distinct causes, each of which has its own co-efficient of variation, the overall co-efficient may be calculated by adding the squares of the individual co-efficients and taking the square root of the total.

(ii) Data/model interface

In the design of a building, a rapid progression is made from a state of little or no information on the proposed design, to a state of having large amounts of information. This movement towards detail regarding shapes, dimensions, etc., has major implications for the use of deductive and inductive models. Regression based models using independent variables such as floor area, number of rooms, quality factors, etc., are of use in the early stages when little is known about the design. The more detailed data which emerges as sketch designs are produced regarding the approximate quantities of various types of work cannot be used by such models and they begin to become out of tune with the existing data. When this stage is reached, inductive models or those based on quantity surveying techniques become more appropriate and are able to make better use of the existing information.

(iii) Model techniques

Deductive methods

Deductive models are derived from the statistical analysis of sample data. The techniques themselves are well known and are relatively easy to apply in building economics. Three aspects of the theory are relevant, correlation, regression analysis and sampling theory.

The traditional measure of association between two variables has been the correlation co-efficient. This co-efficient is a pure number lying in the range $-1 \ r \ +1$, it takes the value $\square 1$ when all the points lie exactly on a straight line, the sign indicates whether the slope is positive or negative. The drawback of this co-efficient is that it is so well known that its limitations are often forgotten. These are as follows:

1   Linearity, the expression above checks for a linear association between the variables, variables which are associated in a curvilinear way may show a low correlation co-efficient.
2   A value of r close to 1 shows high correlation but does not show causality, both variables may be directly linked to a third unmeasured variable. There are many examples of such spurious correlations.

Inductive methods

Potential weaknesses in inductive methods centre around two issues in general. Firstly, the compatibility of the model with the data, which has been discussed in detail earlier, and secondly, the constraints and assumptions built in to the model. The nature of inductive methods is such that these assumptions are not generally explicit and furthermore, unlike deductive methods, the assumptions are liable to be different in each case, and are built in to the algorithms and routines in computer software. Clearly, the assumptions of this nature are not obvious to the user.

Stochastic models

Knowledge of the fact that models in building design evaluation attempt to predict some future outcome, coupled with the prior assumption that a model is an abstraction of reality and, even a cursory consideration of the quality of the existing economic data, lead inexorably to the conclusion that the phrase 'deterministic model' is a conflict in terms. Whenever we predict the futur we are making an assessment based on human judgement and available data, neither of which is perfect. Notwithstanding that, relatively few models in the literature are in fact, stochastic models. The reasons for this are difficult to find, well developed computational techniques do exist for the measurement of uncertainty and have been used in physics and in applied sciences such as systems analysis and operational research for decades. The strongest brake on the application of these techniques would appear to be the fact that designers and building economists find it difficult to think in terms which imply an acceptance of risk and uncertainty.

The mathematical techniques of risk analysis are quite highly developed. Among the relatively few models which recognise and consider uncertainty in some of the values, the most commonly applied technique in building economics has been that of Monte-Carlo simulation. It is suggested that the technique has often been applied unquestioningly. For example, one its assumptions is that the events simulated are independent events. In building economics this is rarely the case. Secondly, the use of normal and rectangular distributions has frequently been merely assumed as correct.

(iv) Interpretation of output

There are two major areas to be considered, firstly is the problem of subjective interpretation of outputs generally. Secondly is the problem of how output may be interpreted with respect to time, in this case the age of the model. In the field of design optimisation models there has been much debate as to whether formal optimisation techniques should be used. If not, the decision maker should remember that it is known generally that subjective estimates of probability differ from objective estimates by overestimating low probabilities and under-estimating high and compound probabilities. For example, the 'gambler's fallacy', the phenomenon where for many people the subjective probability of getting a head after say, five coin tosses have been heads, is not 0.5. Or, in the words of Tom Stoppard's Guildenstern, 'a weaker man might be moved to re-examine his faith, if in nothing else, at least in the law of probability' (9). The relevance of this to cost models is that the decision maker should be aware that his subjective perception of the relative importance of certain variables may not be entirely rational in that it may not lead to the optimum decision.

Turning now to the issue of the age of models, it is proposed that at least three questions are relevant.

1. Do the relationships embodied in the model still hold? Clearly the answer to this question will be different in each case, depending on the problem modelled. It is suggested also that there may well be generic differences apparent among the various modelling techniques, whether the model is predominantly deductive, inductive or stochastic.
2. Is the data used in the model still relevant? An example of this is where innovations in the manufacture of materials render the previous generation obsolete. Consider the situation of the old lath and plaster technique of constructing ceilings being replaced by the use of large sheets of mass-produced plasterboard. The new material replaces the old because it is a more economic solution. A model containing the old data updated by means of an accurate price index would still be out of step with the current implementation of the 'real world'.
3. How has the model been updated? this question refers to the use of indices. Indices are based on assumptions which gradually become out of date and the accuracy of the index reduces over time.

CONCLUSIONS

It is proposed that, in order to increase the practical applications of the newer and more sophisticated economic response models in building design, we should find some means of consistently and rigorously evaluating models against each other. Insufficient attention has been focused, in the past, on the faults, limitations and assumptions contained within models. These limitations are of prime importance to a decision maker who hopes to incorporate the use of a model. A set of criteria designed to aid the decision maker in assessing the performance and limitations of models has been described. As far as the author is aware, there is little previous work of this nature. Therefore, the conceptual framework proposed is a preliminary one to enable some further work, of a quantitative nature, to be carried out in comparing the performance of models and model types using consistent measures.

The author would welcome discussion on the subject of the paper with others interested in the evaluation of model performance.

ACKNOWLEDGEMENT

The author is pleased to acknowledge the contribution of Dr. Alan Wilson to the work on which this paper is based.

REFERENCES

1. Maver, T. W. 'Appraisal in Design', *Design Studies*, Vol. 1, No. 3 (1980), pp. 160–165.
2. Thurow, L. C. '*Dangerous Currents: the State of Economics*', Oxford University Press (1983).
3. Churchman, C. W. 'Towards a Theory of Application In Systems Science' *Proceedings of the I.E.E.E.*, Vol. 63, No. 3 (1975), pp. 351–354.
4. Wilson, A J. 'Economic Response Models in Computer Aided Building Design', CAPE 83, *Computer Applications in Production and Engineering*, Warman, E. A. (Ed.), North Holland (1983).
5. Raftery, J. J. '*An Investigation of the Suitability of Cost Models for Use in Building Design*', Ph.D. Thesis (CNAA), Liverpool Polytechnic, 1984.
6. Jaggar, D. M., 'Cost Data Bases and their Generation' in '*Building Cost Techniques*, Brandon, P. S. (Ed.), E. & F.N. Spon, London 1982.
7. Beeston, D. T., 'One Statistician's View of Estimating', *The Building Economist*, December 1975, pp. 139–145.
8. Wilson, A J. 'Experiments in Probabilistic Cost Modelling', in Brandon, P. S. (Ed.) *op cit*, 1982.
9. Stoppard, T. '*Rosencrantz and Guildenstern are Dead*', Faber, 1967.

1.3 A future for cost modelling*

D. T. Beeston

ABSTRACT

Estimating methods based on measured in-place quantities seem to have reached the limit of their development with an accuracy insufficient either for estimating or for cost advice at design stage. Regression methods have been widely used by researchers but have not produced satisfactory models. The best hope for improvement in advice to designers and in accuracy of estimating is to calculate costs in the way in which they arise or as closely as possible to this ideal.

Cost modelling has a future but it will require a co-operative effort to forget a tool for measurement of the actual costs of construction. This will benefit the contractor in bidding and cost control and the client in design economy and estimating accuracy.

Basing models on construction costs rather than prices will call for the separate study of market factors to estimate mark-up. This will be a healthy trend.

Keywords:

Realistic models, Simulation, Regression analysis, Estimating accuracy, Cost control, Mark-up.

INTRODUCTION

We are approaching a crucial point in the development of estimating and cost control methods. A comprehensive review of the accuracy of estimating (Ref. 1) led its authors to the conclusion that present methods were inadequate but could probably be improved by exploiting the intuitive abilities of experienced cost forecasters. While this is a valuable line of research which may lead

*1987, Brandon, P. S. (ed.) *Building Cost Modelling and Computers*, E & F N Spon, 15–24.

to simple but useful cost models, this paper proposes that the goal of more accurate representation of cost should also be pursued.

For many years contractors had different ideas of how much to allow for rapid inflation. This made accurate estimating impossible and the existence of high inflation provided a ready excuse for poor estimating. The demand for better techniques was small, but with the recent comparative stability of prices the need for improved methods is easier to perceive and the inaccuracy of methods based on approximate quantities and elements less acceptable. Powerful computers have made it possible to consider more elaborate methods, even if they have been rejected in the past as too complex. In current research new approaches are being used and crucial decisions being made on the best avenues to pursue.

All estimating methods can be described as models. A cost model's task may be to estimate the cost of a whole design or of an element of it or to calculate the cost effect of a design change. It can be classified as in-place quantity based, descriptive or realistic. Although it is possible to use different types of model for different purposes it is a sensible aim to find a common basic method for all applications. Then cost advice at various stages can lead smoothly to the final estimate and cost control by the contractor.

IN-PLACE QUANTITIES

Methods based on in-place quantities seem to have reached the limit of their development with an accuracy insufficient for estimating or for cost advice at design stage. The coefficient of variation of estimating errors is hardly anywhere less than 10% and this is inadequate when modern business methods make increasing demands for accurate financial forecasting and control.

Increasing the detail and complexity of quantity-based methods does not seem to produce greater overall accuracy and this is a sign that the end of that development road has probably been reached. They will have to serve until better methods are available so mean-while they should be used with all possible skill and judgement. It would be wiser to concentrate on spreading existing skills more widely than to expend development resources on them.

Development resources would be better used in a related area which is at last receiving some attention. It is the quantification of uncertainty in the estimate and encouragement of practitioners to give clients numerical guidance on the likely limits of error. (Refs. 2 and 3).

DESCRIPTIVE MODELS

A different approach which had been successful in other fields and which has been used in cost model research for about 15 years is based on the idea of attaching costs to descriptive features of a design instead of to quantities.

Descriptive models are formulae in which the variables describe the design and its environment by measurements of such factors as size, shape, type of construction and location. Most have so far been developed using regression methods so the relationships represented are not necessarily causal. They are related to, but make no claim to cause, variability in the dependent variable. In this respect they are further from reality than methods based on quantities. The coefficients of the factors represent the best combination for estimating the way in which cost depends on the factors.

Although the form of a descriptive model appears to put a number to a variable's effect on cost it is seldom that the effect is so straightforward that it can be used for design stage advice. It would be nice to be able to take a factor's coefficient in the model as measuring the change in cost for unit change in the factor. This is not possible because correlation between the descriptive variables will cause their coefficients to be linked so that a small change in the data can produce large but mutually cancelling effects on the correlated coefficients (Ref. 4).

The effect of a particular choice in a feature of a design can obviously only be measured by a regression model if the factor which describes the feature is present. But even if the appropriate factor is present its effect is likely to have been contaminated by its having to stand proxy for other factors which were not selected during the analysis and which are, even to a small extent, correlated with it. In any case the mechanism producing variability is much more complex than could be represented by the comparatively few variables for which it is practicable to obtain data and which qualify for inclusion in the regression formula.

Although the accuracy of models developed using regression methods is poor it is sometimes better than their critics believe. Regression models suffer in their acceptability from having a built-in measurement of their accuracy. This is sometimes believed to be worse than that for conventional methods even when, as measured, it is not. This is because many practitioners are even now unready to accept that the methods they rely on are performing so badly.

However, the critics are right to be dissatisfied with the performance of descriptive models because their accuracy in practice is less than the published theoretical figures. The calculated accuracy of a descriptive model is usually based on its ability to explain variability in the same set of data from which it was calculated. It ought to relate to its performance when applied to new data in the circumstances in which it is intended to be applied. Judged in this way the performance of descriptive models is poor. They may be suitable for estimating the price of a whole building at early planning stage and are in use for estimating the maintenance costs of estates. They are simple to use and require little data, but their accuracy cannot be expected to reach that of traditional methods.

Descriptive models can be thought of as refinements of the crude £/m2 approach to costing which is applied at some stage, at least informally, to every design. Skitmore (Ref. 5) found that estimates based only on floor area and building type were improved only a little when further information was used. His research gives direct confirmation to indications which have come from attempts to construct useful descriptive models. It supports the impression that descriptive models have intractable limitations on their accuracy.

If there had not already been well-tried quantity-based methods the next logical step in development from unsatisfactory descriptive models would have been to link costs to parts of a building by calculating the cost of each per unit of quantity including the cost of putting the materials in place.

REALISTIC MODELS

The only hope for big improvement in the accuracy of cost models either in the sensitivity of their advice to designers or in their estimating accuracy is to introduce causality into the relationships between factors and cost.

It is attractive to reason that if each component of a design has attached to it the cost of putting it in place the cost effect of design alternatives is easily calculated without the need to recalculate the cost of the whole design. The belief in this property is of course the justification for the use of quantity-based methods for the purpose of cost advice to designers.

If construction cost could be divided into independent parts like this there would be no need to look further for a tool to guide economic design. Even though the total estimate produced by such methods had a poor record for accuracy they would be valued because differences caused by design changes could be accurately measured.

Measured in-place quantity-based costing methods would possess the required property if, whatever the rest of the design was like, each work item had its own fairly constant plant, labour and time requirement which changed proportionately when the quantity of the item changed. Unfortunately this is not so. Construction costs arise from the use of plant and labour on a series of tasks and it is not sensible to try to allocate all their time to them individually. Data produced from such a procedure would not be applicable to the circumstances of another project; yet of course the application of most techniques depends on doing just that.

A proper representation of the effect of a design change can be obtained only by calculating its effect on the whole construction process. This is best done by calculating costs in the way in which they arise, or as closely as possible to this ideal. If this is done the models can be described as realistic.

Because in-place costing methods are highly developed, probably to their limit, the estimating accuracy of realistic models may initially be less, but their potential for improvement is far greater. The limitations are the ability to describe and collect data for construction methods, the ability to handle complexity and the extent to which all contractors can be represented by the same model. These are soft limitations in that there is continual movement towards perfect representation with no definite point at which it must stop. Techniques are rapidly improving to extend the first two limits and although there are still some differences between contractors their methods are already quite similar and further industrialisation, whereby the site process becomes little more than the assembly of large components, should make them more similar still.

The production of realistic models can be tackled in several ways. Three attractive approaches can be described as follows.

(a) the simulation of construction in detail,
(b) attaching costs to activity networks,
or
(c) representing the decision process of a planner when calculating the plant and labour requirements of a design.

SIMULATION OF CONSTRUCTION PROCESSES

Many researchers have concluded that costing methods should take more account of construction processes. Ideas have included operational bills and work packages but all assume that it is impracticable to simulate the construction process itself. Although it may still be beyond the present capabilities of data handling and representation of complexity these capabilities are increasing and it is time to prepare for when they will be sufficient. Also the problems will reduce as the control and monitoring of construction improves. Contractors themselves will want to rehearse the construction of a building using a computer simulation which also calculates costs.

When such models are developed they will calculate the plant, labour and time requirements of a design. The materials involved and the dimensional information about the design will be delivered to them by a program which, by then, will be used by the designer. The interface between the designer's program and the cost model program should be straight-forward but will need to be borne in mind during development.

If industrialisation transforms construction as it has been expected to do for so long, it will be necessary to synthesise the costs of large and complex factory-made components. This should be easier than simulating site processes but will call for decisions on whether to treat a component as an off-the-shelf item with a market price or as part of the construction process which happens to take place off-site. Much will depend on whether the item is commonly used by many contractors or whether only a few contractors buy it ready-made.

ACTIVITY DIAGRAMS

Activity networks or linked bar charts can be made to represent the construction process and can have costs attached to them. Provided that the costs are operationally derived from the costs of plant and labour they can gradually be broken down to approach the ideal of a detailed simulation of the construction process. Research at Reading University is on these lines.

SIMULATION OF CONSTRUCTION PLANNING METHODS

Stepping back still further from what happens on site the simulation of construction planning methods is an expert system approach which is feasible now. A working prototype called 'COCO (Costs of Contractors' Operations)' was devised by the author in 1971 (Ref. 6). It simulated the logical decision process used by several construction planners when selecting plant and allocating labour to a project.

A method based on construction plans rather than on their realisation has the superficial attraction that the bid is based on an estimate which relates to plans and not what finally happens on site. A better objective is for all involved in design and construction to bring plans and realisation closer together. Seen from this point of view a simulation of plans is a step on the way to simulation of the construction process.

MARK-UP

Realistic models aim to estimate a contractor's costs rather than his tender. Estimating costs separately from mark-up is healthy because the two should be unrelated and each can be more accurately measured on its own than when combined in a price estimate. Measuring contractors' costs separately will teach designers how to reduce them and thus make projects more attractive to contractors. This will itself lead to reduced mark-ups. For contractors the study of market variation is vital to formulating a bid. Although clients' estimators also set great store by their feel for the market little has been done to systemise it. Mark-up can be estimated by assessing the market or tendering climate for the construction of the project. The calculation can be divided into two stages as described in Ref. 6 under the headings of 'macro-climate' and 'micro-climate'.

Macro-climate is the general level of bidding keenness and ignores any special features of the particular project. It can be calculated by constructing an index to keep a running measure of the average difference between the lowest tender and an estimate calculated by a consistent method. An adjustment is needed for the number of tenders received. Other adjustments such as for location, type of building and contract value can be made if the index is analysed in the same way as a tender price index.

The estimate of mark-up is completed by a further adjustment for micro-climate. This allows for features of the particular project which make it attractive or otherwise to contractors. A method for adjusting for micro-climate is described in ref. 6 as the D-curve method. It is based on a subjective forecast of the keenness of contractors to win this particular contract.

PROBLEMS IN REALISTIC MODELS

Inevitably the development of realistic methods will introduce new problems. For example there will be a need for new methods of data collection. Attempts to improve the accuracy and coverage of site performance data may be viewed with suspicion by the work-force but once the need for well-organised data collection and analysis is accepted, techniques for doing it efficiently and cheaply will be developed. An expertise will be recognised and new contract documentation will have to be developed to replace the bill of quantities. When documentation is made more relevant to the constructing process contractors' estimators will find more use for site data and will call for an improvement in quality.

There are differences between contractors in the construction methods they use but these differences are less than might be expected. Especially among large contractors there is exchange of staff and ideas which tends to lead to a common, economical method of planning and execution for any given design. The model would seek to represent such a method. Those contractors who would have used a worse one would benefit form being shown it and those using better methods would benefit financially – as they should. If it is desired to represent more than one method there can be alternatives in the program which can be run in parallel simulations. The need for decisions on how to regard factory-made components has already been mentioned.

Facing the problems of developing realistic models will be a salutary experience. Many of them are already present in existing methods but are not being dealt with because they are concealed by the lack of relevance in the costing methods used. They are responsible for unexplained variability and as the reasons for it become known in terms of construction processes and costs of plant, labour and materials so the variability will be reduced.

It may seem that a detailed method of costing, especially one based on site operations, is inappropriate for early design stage advice because the method requires more detail about the design than is available. But any method must make assumptions about the design. For example elemental and approximate quantity methods require the application of data from other designs and implicitly assume that decisions yet to be made in the current design will conform in cost with the average of the contributing designs. In realistic methods such design assumptions are made overtly and specifically and the construction implications spelled out. When the design has progressed the assumptions can be checked and modified if necessary. This process will give valuable insight into how the design differs from others from which the assumptions were drawn. This is easiest to see when they come from a data base of previous designs.

SUMMARY AND CONCLUSIONS

The underlying principle of in-place costing methods renders them unsuitable for design advice. They are more suitable for forecasting the cost of a whole design and for this role they remain the best tools available. All the same, even here, their accuracy is inadequate and cannot be improved.

The inability to forecast costs accurately enough is limiting the success of contractors and clients in controlling expenditure. It introduces a random element into profits which discourages keen tendering and which leads to difficulties with potentially loss-making contracts and disappointed clients who have to be fed from a menu of excuses for estimating errors.

For substantial improvement in estimating, advice to designers and cost control it will be necessary to change fundamentally the basis of calculation so that it corresponds to the ways in which costs arise. There will also be a need for models of the influence of market factors.

A widespread use of communicating computers will make it possible to harmonise contractors' and clients' costing methods by basing both on data from site operations. New contract documentation will be required and the way will be open to new methods of letting contracts in which both clients and contractor share a common method of calculating costs. The valuation of variations will be improved to the benefit of all. The difficulties seem daunting at present but they are becoming less and past experience shows that when tasks such as this are tackled many expected difficulties fall away.

There will still be a need for an easy method of early estimating. This could be provided by simple descriptive models and an expert system encapsulating proved skills. Skitmore's research (Ref. 5) shows how this can be done and further work in this field would provide a valuable thrust of research. It would scarcely overlap that on realistic models but would complement it to make a powerful two-pronged attack: one effort being to produce an expert system using models of the way the best practitioners work and the other to simulate the way costs arise during construction.

REFERENCES

1. A. Ashworth and R. M. Skitmore 'Accuracy in estimating' *CIOB Occasional Paper* No. 27.
2. RICS *QS R&D Forum* 13 March 1986 'Risk Analysis'.
3. D. T. Beeston 'Risk in estimating' *Construction Management and Economics* Vol. 4 No. 1 (Spring 1986)
4. D. T. Beeston 'Cost Models' RICS *QS R&D Forum* December 1977 and *Chartered Surveyor B & QS Quarterly* (1978).
5. R. M. Skitmore *'The influence of professional expertise in construction price forecasts'* Department of Civil Engineering, University of Salford (1985).
6. D. T. Beeston *'Statistical Methods for Building Price Data'* E & F N Spon (1984).

1.4 Developments in contract price forecasting and bidding techniques*

R. M. Skitmore and B. R. T. Patchell

INTRODUCTION AND DESCRIPTION OF TECHNIQUES

The objective of this chapter is to describe and demonstrate the use of four types of design estimating systems that have been developed in recent years to the point of commercial use. These consist of:

- standard regression approach to item identification and pricing;
- CPS simulation for evaluation of item interdependencies;
- ELSIE, the expert system for 'front end' item and quantity generation; and
- a prototype bidding model debiaser developed by the author for 'back end' estimate adjustment for bidder characteristics.

Before considering these systems however, some preliminary observations are provided concerning the type and nature of some currently available estimating techniques.

Designers' and contractors' approaches

The functional separation of design and construction has been reflected in the development of contract price forecasting techniques largely to meet the perceived needs of both these sectors of the construction industry. For designers, the need is to inform on the expenditure implications (to the client) of design decisions to help in achieving financial targets, and value for money.

For contractors, the need is to inform on the income implications (to the contractor) should the contract be acquired. Although the nature of the competitive tendering process is such that both designers and contractors are essentially concerned with the same task – estimating the market price of the contract – the potential for contractors to access production cost information is

*1990, chapter 4 in *Quantity Surveying Techniques: New Directions*, ed. P. S. Brandon, BSP Professional Books, 75–120 (extracts).

a determining factor in the type and reliability of technique used.

Many of the methods used in practice contain a mixture of both market and production related approaches. It is common, for instance for quantity surveyors to build up rates for special items involving new materials or components from 'first principles' by finding out the manufacturer's price and making an allowance for installation and other associated costs involved in a similar manner to the contractors' estimator. It is also common for contractors' estimators to use some 'black book' method which is more intended to generate a competitive price than a genuine estimate of cost. The validity of combining the two approaches, although a popular debating point, is by no means established in academic circles. Clearly, the relationship between market prices and production costs is the determining factor, and research on this aspect has been inconclusive as yet.

It is generally accepted however that the contractors' resource based approach produces very much more reliable estimates than the designers' equivalent. Unfortunately, theoretical research in this area has until relatively recently concentrated entirely on the formulation and solution of a highly simplified mathematical construction of the contractors' bidding decision. Despite the apparent remoteness of bidding research from the real world of quantity surveying, there is an important connection with design estimating as, unlike construction estimating, bidding is directly concerned with market prices.

Strangely, bidding research has been quite opposite in many fundamental ways to estimating research. Firstly, bidding research is firmly founded in economic theory with very little empirical support whilst estimating research has no formal theoretical base at all! As a result, both research fields have run into serious logistical problems. Bidding researchers are now faced with an unmanageable number of theoretical extensions, most of which are not appropriate to real world data, and estimating researchers are now faced with an unmanageable amount of data with no theoretical basis for analysis. Secondly, bidding models are invariably probabilistic in nature whilst estimating has traditionally been treated as a deterministic matter. Perhaps the biggest emphasis in estimating research today is in the reliability of the techniques used, and statistical probability offers the greatest potential for modelling reliability. Clearly then the theories and techniques used in bidding have some relevance in estimating especially in the provision of a mathematical basis for the analysis and developments in the field.

Mathematical and typological features

Traditional estimating models can be generally represented in the form of

$$P = p_1 + p_2 + \ldots + p_N = q_1r_1 + q_2r_2 + \ldots + q_Nr_N \tag{1}$$

or, more succinctly

$$P = \sum_{i=1}^{N} p_i = \sum_{i=1}^{N} q_i r_i \qquad (2)$$

Where P is the total estimated price, p is the individual price, q the quantity of work and r the rate or value multiplier for items 1, 2, . . ., n respectively. So for a bill of quantities based estimate, q_1 q_2 etc. would represent the quantities for items 1, 2 etc. in the bill, r_1 r_2 etc. the rates attached to the respective items, and p_1 p_2 etc. their individual products. For example, an item of say concrete in floors may have a quantity q of 10m^3 and rate r of £60 per m^3, giving an item price p of qr=£600, the sum of all such prices Sp giving the total estimated price P.

The differences that occur between traditional estimating methods are usually in the number and type of items and the derivation and degree of detail involved in estimating their respective q and r values.

From Table 1.4.1 it can be seen that the number of items (N) may range from one single item (unit, functional unit, exponent, interpolation, floor area, cube, and storey enclosure) to very many (BQ pricing, norms, resource).

The unit method is used on all types of contracts. It involves any comparable unit such as tonnes of steelwork or metres of pipeline.

The functional unit method being similar but restricted to use on buildings with units such as number of beds or number of pupils.

The exponent method, used on process plant contracts, involves taking the contract price of an existing contract and multiplying by the ratio f some relevant quantity, such as plant capacity, of a new or existing contract, raised to a power, r, determined by previous analysis.

The interpolation method takes the price per square metre of gross floor area of two similar contracts, one more expensive and one cheaper, and interpolates between them for a new contract.

The floor area method is less specific in that a price per square metre floor area is derived somehow, depending on any information available.

Similarly the cube method involves the calculation of a quantity representing the cubic content of a building for the application of a price per cubic metre.

Perhaps the most sophisticated single item method is the storey enclosure method, involving the aggregation of major building features such as floor, wall, basements and roof areas into a composite index known as a 'storey enclosure unit'.

BQ pricing and norms methods on the other hand involve the detailed quantification of a great number of items, usually prescribed by an external contract controlling institution such as the Joint Contracts Tribunal (in the UK) or Government (in Eastern Block countries).

Several methods fall between these two extremes using either a few items (graphical, parametric, factor, and comparative) or a reduced set of BQ pricing/ norms type items (approximate quantities, elemental).

Table 1.4.1 Resume of estimating techniques

Estimate Technique	Model	Relevant Contract Type	General Accuracy (cv)	Dst/Prob	Quantities (q) Number	Items (i) Type	Quantities (q) Derivation	Quantities (q) Det/Prob	Derivation Data Base	Weighting	Rates (r) Current	Rates (r) Quantity trended	Det/Prob
1 UNIT	$P = qr$	All	25–30%	det	single	any comparable unit, eg tonne steelwork, metre pipeline	Brief	det	averaged price–cost unit	?	direct	none	det
2 GRAPHICAL	$P = f_r(q)$	Process Plant	15–30%	det	few	ditto	Brief	det	trended price–cost	?	interpolated	objective	det
3 FUNCTIONAL UNIT	$P = qr$	Buildings	25–30%	det	single	ditto eg number of beds number of pupils	Brief	det	averaged/rule price–cost unit	?	direct	none	det
4 PARAMETRIC	$P = f_r(q_1; q_2; q_3 \ldots)$	Process	15–30%	det	few	process parameters eg capacity, pressure, temperature, materials, cost index	Brief	det	averaged/rule price–cost parameter	?	direct	none	det
5 EXPONENT	$P_2 = P_1 \dfrac{q_2}{q_2}^{\,r}$	Process Plant	15–30%	det	single	size of plant or equipment eg capacity	Brief	det	averaged/rule price–cost exponent	crude subjective	direct/interpolated	objective	det
6 FACTOR	$P = \sum_{i}^{m} \text{fact}_i \sum_{i} q_i \, r_i$ a) m = 1 (Lang method) b) m > 1, fact_2 etc (Hand method) c) fact_i = U(□_i, B_i) (Chiltern method)	Process Plant	10–15%	det	few	any	Brief/measure	det	averaged/rule price–cost	?	factored	a) none b) none c) subjective	det det det
7 COMPARATIVE	$P_2 = P_1 + \sum (p_{2i} \square p_{1i})$	All	25%–30%	det	few	depends on differences	Brief	det	price–cost items	crude subjective	adjusted	none	det
8 INTERPOLATION	$P = qr$	Buildings	25%–30%	det	single	gross floor area	Brief	det	price/m²	crude subjective	interpolated	–	det

	Price model	Type	Accuracy	det/prob	No.	Measurement	Data source	det	Rate basis				
9 CONFERENCE	$P = f(P_1, P_2 \ldots)$	Process Plant	?	det	any	any	Brief/measure	det	—	—	negotiated	—	det
10 FLOOR AREA	$P = qr$	Buildings	20–30%	det	single	gross floor area	Brief/measure	det	averaged price/m²	crude subjective	direct	subjective	det
11 CUBE	$P = qr$	Buildings	20–45% (based on 86 cases)	det	single	volume	measure	det	averaged price/m²	crude subjective	direct	none	det
12 STOREY ENCLOSURE	$P = qr$	Buildings	15–30% (based on 86 cases)	det	single	floor/wall area/basement/roof	measure	det	averaged price/SE unit	crude subjective	direct	none	det
13 BQ PRICING a) (Conventional)	$P = \sum q r_i$	Construction	10–20% (5–8% for builders)	det	v many	SMM	measure	det	a) averaged BQ's	crude subjective	direct	subjective	det
b) (B Fine)	$P = \sum q r_i$	Buildings	15–20%	prob					b) $r_i - U(r_{min}, r_{max})$		direct	subjective	prob
14 SIG. ITEMS	$P = \sum q r_i$	PSA buildings	10–20%	det	medium	SMM	measure	det	averaged BQ's/rule	crude subjective	direct	objective	det
15 APPROXIMATE QUANTITIES a) (Conventional)	$P = \sum q r_i$	Construction	15–25%	det	medium/	SMM combined few	a) measure	det	a) averaged BQ price	crude subjective	composited	subjective	det
b) (Gleeds)	$P = \sum q r_i$	Buildings	15–25%	det			b) Brief/measure	det	b) averaged BQ/price	crude subjective	composited	subjective	det
c) (Gilmore)	$P = \sum q r_i$	Buildings	15–25%	det			c) Brief/measure	det	c) averaged BQ/price book	crude subjective	composited	subjective	det
d) (Ross 1)	$P = \sum q r_i$	Buildings	25% (based on 17 cases)	det/prob			d) measure	det	d) 50 BQ's averaged	none	direct	none	det
e) (Ross 2)	$P + \sum q r_i$	Buildings	50% (based on 17 cases)	det/prob			e) measure	det	e) 50 BQ's $r_i = a + bq_i + e$ $e - N(c, \sigma^2)$	none	mathematically objective	mathematically objective	prob
f) (Ross 3)	$P + \sum q r_i$ $(P_i = a + bq_i + e,\ e = N(0, \sigma^2))$	Buildings	30% (based on 17 cases)	det/prob			f) measure	det	e) 50 BQ's	none	mathematically objective	mathematically objective	prob
16 ELEMENTAL	$P = \sum q r_i$	Buildings	20–25%	det	medium	BCIS/CI afb entities	Brief/measure	det	averaged BQ's/ BCIS/m²	crude subjective	composited/ direct	subjective	det

17 CPU	$P = \Sigma q_i r_i$	Buildings	20–25%	det	medium	DBE	det	Brief	averaged BQ	crude subjective	composited	subjective	det
18 ELSIE	$p^3 = \Sigma q_i r_i$	Offices	10–20%	det	medium	DBE	det	Brief	averaged BQ/rule	none	direct	none	det
19 NORMS (schedule)	$P^2 = \Sigma q_i r_i$	Buildings	10–20%	det	v many	SMM type eg PSA schedule	det	measure	cost based rules	none	direct	none	det
20 REGRESSION	$P = a + \Sigma q b_i + e$; $e = N(0, \sigma^2)$	All	15–25%	det/prob	few	usually contract characteristics eg floor area, number of storeys	det	Brief	any	crude subjective	mathematically objective		prob
21 LU QIAN	$P = \Sigma q_i r_i$	Buildings	?	det	few	usually contract characteristics eg floor area, number of storeys	det	Brief	any	mathematically	mathematically	none	det
22 RESOURCE (Activity, operational, scheduling)	$P = \Sigma q_i r_i$	All	5–8% (builders)	det	v many	resource eg man hours, materials, plant	det	production plan	average costs	crude subjective	direct/ analytical	subjective/ objective	det
23 PERT-COST	$P = \Sigma p_i$ where $p_i = N(q_i r_i, \sigma^2)$	All	N/A	prob	varies	usually time resources eg man hours	prob (time)	production plan	–	–	–	–	–
24 CPS	$P = \Sigma t_i r_i + \Sigma n_i r_i$; $t_i = F(\mu, \sigma_i^2)$	Buildings	6.5% (based on 4 cases)	prob	usually few	resource eg manhours, materials, plant	prob (time)	production plan	average cost	crude subjective	direct	none	det
25 RISK ESTIMATING	$P = \Sigma q_i r_i$	Con-struction	N/A	prob	usually few	any	det	any	theoretical frequency distribution of cost	crude subjective	random selection $r_i - F(\mu, a^2)$	none	prob
26 HOMOGENISED ESTIMATING (BCIS on line) (BICEP etc)	$P = \Sigma q_i r_i$	Building	N/A	det	any	any	det	any	average BQ	sided subjective	direct	objective	det

Notes

F ( ) some (unspecified) probability function
N ( ) normal probability function
U ( ) uniform probability function

The graphical method, used on process plant contracts, involves plotting the quantities of each of a few items of interest against the contract sum of previous contracts, for a visual analysis of possibly non-linear relationships.

Parametric methods, also used on process plant contracts, adopt a multivariate approach in using a function of several process related items such as capacity, temperature, pressure in combination.

Factor methods, again used for process plant contracts, involve pricing only a portion of the contract which is then multiplied by a factor derived from previous similar contracts. Versions of the factor method include the Lang method, which uses a single factor; the Hand method, which uses different factors for different parts of the contract; and the Chiltern method, which uses factors given in ranges.

The approximate quantities method involves the use of composite groups of BQ pricing items which have similar quantities, e.g. floors and walls, or similar physical functions, e.g. doors and windows, with rates being derived from BQ/price book databases.

The elemental method, perhaps the first of the non-traditional approaches, also uses items representing physical functions of buildings, but with quantities and rates expressed in terms of the building gross floor area.

The reliability of estimates generated by the traditional model is a function of several factors:

1   the reliability of each quantity value; q;
2   the reliability of each rate value, r;
3   the number of items, n; and
4   the collinearity of the q and r values.

This last factor is often overlooked in reliability considerations which tend to assume that q and r value errors are independent.

Traditional thinking holds that more reliable estimates can be obtained by more reliable q values (e.g. by careful measurement), more reliable r values (e.g. by use of bigger data bases) or more items, all else being equal – a proposition questioned by Fine's (1980) radical approach to BQ pricing involving the generation of random values for both q and r values. More detailed theoretical analyses also support the view that traditional thinking may be an oversimplification. Barnes (1971), for example, investigated the implication of the proposition that different r values have different degrees of reliability, specifically that the reliability of $q_i r_i$ is an increasing function of its value. By assuming a constant coefficient of variation for each item, he was able to show that a selective reduction in the number of low valued items would have a trivial effect on the overall estimate reliability. The empirical evidence in favour of Barnes' assumption is quite strong and has culminated in the significant items method now being used by PSA.

Another break with tradition has been to develop entirely new items based on a more conceptual classification of contract characteristics. Elemental estimating is perhaps the first sophisticated example of this, involving as it does the reorganisation of traditional BQ pricing items into composite groups considered to represent mutually exclusive building functions. (A development of this using cost planning units, CPU, provides an alternative.) Most types of approximate estimating methods fall into this category, the single rate methods such as floor area, functional unit, cube, and storey enclosure, or the multiple rate methods such as approximate quantities. A further and most important characteristic of all these methods is that they were all (except for Ross', 1983, alternative approximate quantities methods and the storey enclosure method) developed in the absence of any reliability measures by which to assess their value.

Research over the last 20 years has developed with different emphasis on all of the four factors influencing estimating reliability although systems development has been centred at the item level involving:

1   the search for the best set of predictors of tender price;
2   the homogenisation of database contracts by weighting or proximity measures;
3   the generation of items and quantities from contract characteristics; or
4   the quantification of overall estimate reliability from assumed item reliability.

The first of these is typified by the regression approach, involving the collection of data for any number of potential predictors (floor area, number of storeys, geographical location, etc.) and then by means of standard statistical techniques to isolate a best subset of these predictors which successfully trades off the costs of collection against the level of reliability of estimate. The second is typified by the homogenisation aids provided by the BCIS 'on line' and BICEP systems, and the fuzzy set based automatic procedure contained in the LU QIAN system. The third is typified by the Holes', Calculix and expert systems approach such as ELSIE. This is essentially a 'front end' to a conventional estimating system where items and quantities are derived from basic project information by either a known or assumed correspondence between the two. The fourth approach, typified by the use of probabilistic (statistical) models such as PERT-COST or simulation models such as risk estimating or the construction project simulator (CPS), goes beyond the standard regression approach by introducing more complicated relationships than those assumed by the standard regression method in accommodating some interdependency between and variability within r values (PERT-COST and risk estimating) and q values (CPS).

Bidding models can be represented by

$$B = Cm \tag{3}$$

where B is the value of the bid to be made by a contractor, C represents the estimated production costs that would be incurred should the contract be obtained, and m is a mark-up value to be determined by the bidder. In bidding theory the major interest is in estimating a suitable value of m which will provide the best trade off between the probability of obtaining the contract and the anticipated profit should the contract be obtained. The model for C is similar to the design estimate model for P (equation (2)) in that it consists of the sum of a series of quantified item and rate products, say

$$C = \sum_{j=1}^{k} q_j r_j \qquad (4)$$

where q_j and r_j represent the quantity and rate respectively for the jth item. If the same items are used in providing both design and construction estimates, this simplifies to, in terms of the design estimate model

$$B = \square q_i r_i m = \square q_i r_i = P \qquad (5)$$

as both B and P are essentially estimates of the same value – the market price of the contract.

The idea that bids and profits can be modelled in a probabilistic way, i.e. by treating them as random variables, has a long tradition in bidding theory, but it is generally assumed that it is the actual rather than estimated costs that contain the random component. Like design estimates however, it is becoming increasingly popular to treat the estimated costs also as a random component in the model. Also the similarity of approaches does not end at this point as recent empirical studies strongly suggest the existence of a close relationship between C and P.

STANDARD REGRESSION

Regression, or multiple regression analysis as it is usually called, is a very powerful statistical tool that can be used as both an analytical and predictive technique in examining the contribution of potential new items to the overall estimate reliability. Perhaps the most concentrated research on the use of regression in estimating was in a series of post graduate studies at Loughborough University during the 1970s and early 1980s. The earliest of these studies developed relatively simple models with new items to estimate tender prices of concrete structures, roads, heating and ventilating installations, electrical installations and offices. More recent studies at Loughborough have been aimed at generating rates for standard bill of quantities type items by regressing the rates for similar items against the quantity values for those items.

The regression model usually takes the form

$$Y = a + b_1X_1 + b_2X_2 + \ldots b_nX_n \tag{6}$$

where Y is some observation that we wish to predict and X_1, X_2, ..., X_n are measures on some characteristics that may help in predicting Y. Thus Y could be the value of the lowest tender for a contract, X_1 could be the gross floor area, X_2 the number of storeys, etc. Values of a, b_1, b_2, ..., b_n are unknown but are easily estimated by the regression technique given the availability of some relevant data and the adequacy of some fairly reasonable assumptions about the data.

The regression model in equation (6) is quite similar to the estimate model in equation (1), as Y, X, and b values can be thought of as representing the P, q, and r values. The major difference in approach however is that in applying the regression technique **no direct pre-estimates are needed of the values of item rates**, r, as these are automatically estimated by the technique. This then obviates the need for any data bank of item rates thus freeing the researcher to examine any potential predictor items for which quantities are available. The implications of this are quite far reaching for without the need to have an item rate database the research task is simply an empirical search for the set of quantifiable items which produces the most reliable estimates for Y.

Problems and limitations

Although the task appears to be straightforward, several problems have been encountered. These problems concern the model assumptions and limitations, data limitations, and reliability measures.

Model assumptions

The major assumptions of the basic regression model are that

1   the values of predictor variables are exact;
2   there is no correlation between the predictor variables;
3   the actual observations Y are independent over time; and
4   the error term is independently, randomly and identically normally distributed with a zero mean.

In terms of equation (1) this implies:

•   that the quantities q are exact rather than approximate quantities;
•   that quantities for items do not change in tandem, as does floor area and number of storeys or concrete volume and formwork area for example:
•   that the tender prices for one contract are not affected by the tender prices for the previous contract; and
•   that the differences between the regression predictions and the actual

tenders are purely unaccountable random 'white noise', unrelated in any way to the variables used in the model.

Violation of these assumptions is not necessarily fatal to the technique. The type and degree of effects of violations depends on the type and degree of violation. Unfortunately however this is a rather specialist area in which statistical theory is not yet fully complete. The usual pragmatic approach to this is to try to minimise violations by a combination of careful selection of variables and tests on the degree of resulting violations.

Data limitations

Although there is no theoretical limit to the number of predictor variables that may be entered into the regression model, at least one previous contract is needed for every variable. For reasonably robust results it is often recommended that the number of previous contracts is at least three times the number of variables in the model. Thus for the full traditional bill of quantities model of say 2000 items this means about 6000 bills of quantities would be needed for analysis, even assuming that each bill contains identical items. Fortunately however there is a kind of diminishing return involved in the introduction of each new variable into the model so that there comes a point at which the addition of a further variable does not significantly contribute to the reliability of the model.

This property of regression analysis is often utilised by researchers in a technique called forward regression which involves starting with only one variable in the model. A second variable is then added and checked for its contribution. If it is significant, a third variable is added, checked and so on until a variable is encountered that does not significantly contribute. This variable is then left out of the model and the analysis is completed.

An extension of this method is stepwise regression which leaves out any non-significantly contributing variable already in the model and enters significantly contributing ones until completion. Although stepwise regression works fine if the predictor variables are not correlated as required by model assumption 2 above, violations of this assumption result in different models depending on the order of variables entered and removed from the model. This can be overcome by yet another technique called best subset regression, which examines all possible combinations of predictor variables for their joint contribution, selecting the best set of predictor variables which significantly contribute.

A key issue of course is in specifying a criterion for distinguishing between significant and non-significant contributions to the model. In regression formulations it is usual to concentrate on the behaviour of the error term, i.e. the difference between the actual values of Y and those predicted by the regression model. If for example ten contracts have been used in the analysis, then there will be ten actual lowest tenders (contract values) and ten model predictions. The differences between each of these ten pairs of values is then squared and added together, the resulting total being called the residual sums of squares (RSS). As each new variable is entered into the equation, the RSS decreases a little.

Two possible significance criteria therefore are the minimum total RSS or the minimum decrease in RSS as a new variable is entered. Clearly if this figure is set at zero, then all variables will be entered into the model. If on the other hand the minimum is set at some high level, very few, if any, variables will be entered into the model. Another possibility is to use the proportion of RSS to the total sums of squares that would be obtained if no variables were entered into the equation (TSS). Most standard regression packages use this latter method. For construction price–cost estimates this may not be appropriate. The decision is ultimately an arbitrary one. In most construction price–cost estimating research, the number of variables entered into the model before cut off using various sensible criteria levels is seldom more than ten, which suggests that data limitation is not likely to be a serious problem in practice.

Reliability measures

This has been perhaps the biggest area for problems and misunderstandings in all of regression based research to date. The standard test statistic given by regression packages is the F value which is a measure of the proportion of RSS to TSS mentioned above. This tests the assumption (null hypothesis) that the predictor variables used in the model have no real predictive ability, such apparent ability revealed by the regression technique being more attributable to chance than some underlying correlation. Thus if we use the F value to test a one variable model regressing contract value against gross floor area it is incomprehensible that the null hypothesis should hold, and there is no practical advantage in testing against the incomprehensible.

Another very common failing has been to confuse measures of a model's fit with measures of the model's predictive ability. Measures of the extent to which the model fits the data are readily available in most standard regression packages, the most popular measures being the F test mentioned above which offers evidence on whether the model's fit is due to chance, and the multiple correlation coefficient which indicates the degree to which the model fits the data. However, as the model has been derived from the same data by which it is being tested for fit, it is not at all surprising that the fit should often be a good one.

The real test of reliability of the regression model is to see how it performs in predicting some new data. The obvious way of doing this is to obtain the regression a and b coefficients from the analysis of one data set, collect some more X data for some more contracts, apply the old a and b values to obtain estimates of Y, and then use these estimates against the actual values of Y as a means of measuring the model's predictive ability. A more subtle approach, called jackknife validation, is to omit one contract from the data, calculate the a and b coefficients, estimate the Y value of the omitted contract and compare this with the actual Y value for that contract. This procedure is then repeated for all the contracts and then the residual analysis is conducted as before.

Applications

The regression method involves six operations:

1   data preparation and entry;
2   selection of model;
3   selection of predictor variables;
4   estimation of parameters;
5   application of parameter estimates to specific task;
6   reliability analysis.

In practice, operations 2 and 3 are executed concurrently.

Data preparation and entry

Data are prepared in the form of a matrix in which the rows correspond to contracts and columns correspond to the lowest tender (contract value) and values of potential predictor variables. Most regression packages can handle a few missing data which need to be flagged by the use of a special number such as 0 or 999. Contract values are normally updated by one of the tender price indices although this is not strictly necessary (the tender price index applicable to each contract could be entered as a predictor variable for instance). Another possibility is to include some approximate or even detailed estimate of the contract value as a potential predictor variable also. If the latter is used then the regression is essentially a **debiasing** rather than estimating technique.

It is advisable to carefully check that the data is correctly prepared and entered before analysis as errors may not be immediately apparent. If errors do exist in the data, it is quite likely that a great deal of time and effort could be wasted in abortive analysis prior to their correction. Most regression packages offer a facility to reproduce the data in hard copy form for checking purposes.

Selection of model and predictor variables

These operations are very well documented in many intermediate level statistics texts and regression package manuals. It is usual to carry out both operations concurrently, seeking the best model and subset of predictor variables together. Two fundamentally different approaches exist that have great significance for academic work. One, called the deductive approach, involves the proposition of specific pre-analysis (*a priori*) hypotheses based on some theoretical position derived from an examination of extant ideas on the subject. The consequent models and predictor subsets are then tested against each other for primacy, any potential new models and subsets, however obvious from the data, being strictly excluded.

The other approach, called inductive, is strictly empirical in that the intention is to find patterns in the data that can be used to generate some future hypotheses and, hopefully, theoretical foundations. For all practical purposes the separation of these approaches is hardly relevant except as a stratagem for dealing with the logistical problems that are invariably encountered in regression analysis.

Construction price–cost estimation lacking any formal theoretical base tends to preclude the deductive approach and we are usually left to look for patterns in the data. This means trying out many possible models using not only the simple additive models described here but those involving transformations of variables (e.g. log, powers, reciprocals) together with combinations of several variables (e.g. products, powers) in either raw or transformed states. Even with only two predictor variables the number of combinations of transformations and combinations are quite substantial, with a large number of variables, some simplification is needed. In the absence of theory such simplification is bound to be an arbitrary process. As a result, research in regression is far from comprehensive and, because of the arbitrary means employed, has not been reported well enough to allow any incremental progress to be made. Also, with the confusion over reliability measures, it has not been possible to properly evaluate the progress that has been made.

Experience suggests that the best and easiest starting point is the simple regression model with raw (untransformed) values and no interaction terms (combinations of variables). Then, by using stepwise or best subset regression, this model can be trimmed down to significant predictors. The next stage is to try to find any other variable set or transformation that will significantly improve the model.

Parameter estimation

Once a satisfactory model is fitted to the data, the standard regression package will automatically calculate values of the a and b coefficients for use in estimating. This an extremely simple process by which the value of the predictor variables for the new contract are just multiplied by the b coefficients and added together with the a coefficients. The resulting total is the regression estimate for the contract.

Reliability analysis

The reliability of the regression model can be estimated in terms of both bias, i.e. the average error of the forecast, and variability, i.e. the spread of the forecast errors. The regression technique is designed to give unbiased predictions and therefore the average error of prediction is assumed to be zero. Two relevant measures of the likely variability of the forecast are the 95% confidence limits and the coefficient of variation (Appendix A).

Table 1.4.2 Regression data

Contract number	Standardised contract sum*	Gross floor area (m²)	Number of storeys	Air-conditioning (0 = no) (1 = yes)	Contract period (months)	Number of tenders
CASENO	CONSUM	GFA	STOREY	ACOND	PERIOD	BIDDERS
1	1085.95	452	2	0	8	6
2	5042.91	1601	7	0	11	8
3	2516.59	931	3	1	11	7
4	18290.60	6701	7	1	17	6
5	3195.81	219	3	0	12	1
6	8894.68	3600	6	0	15	9
7	932.06	490	2	0	7	6
8	979.93	415	1	0	8	8
9	1684.94	504	3	0	9	6
10	1896.39	320	2	0	7	7
11	8789.05	372	2	0	6	7
12	2445.12	837	2	0	9	4
13	1501.91	491	3	0	6	6
14	1114.31	496	1	0	6	6
15	943.48	430	2	0	99	99
16	3670.98	1368	4	0	12	4
17	1094.75	469	2	0	6	4
18	4584.87	1260	2	0	8	5
19	10942.28	2994	8	1	15	1
20	760.29	312	2	0	6	6
21	3002.67	1225	2	0	9	7
22	2720.44	1230	2	0	10	8
23	58365.39	23089	7	1	20	7
24	11323.40	4273	4	1	20	7
25	37357.91	11300	5	1	18	6
26	46309.12	14430	3	1	30	6
27	1704.17	437	3	1	10	9
28	6792.04	2761	5	1	12	6

*Contract sum □ tender price index □ location factor e.g. contract number 2 contract sum = £1514385 in London (location factor 1.30) on Sept 1987 (TPI 231). Standardised contract sum = 1514385 □ 231 □ 1.30 = 5042.91

Example

Table 1.4.2 contains a set of data extracted from the RICS Building Cost Information Service's Brief Analysis files. A total of 28 contracts are included for office blocks from a variety of geographical locations in the UK for the period 1982 to 1988. The dependent variable contract sum (CONSUM) was standardised by dividing by the all-in tender price index and location factor to remove inflationary and location effects. The independent (predictor) variables chosen were the gross floor area (GFA) in square metres, the number of storeys (STOREY), air-conditioning (ACOND) valued at 1 if present and 0 if not present, contract period (PERIOD) in months, and the number of tenders (BIDDERS) received for the contract. The contract period and number of tenders received was not known for contract number 15 and these were given a special 'missing' value of 99.

The problem now is to provide an estimate for a new contract, not included in the data set, which has a gross floor area of 6000 square metres, 5 storeys, no air-conditioning, an 18 month contract period and 6 tenders.

The first task is to fit a suitable model to the data set. From experience and a few trials a reasonable model was found by (1) using the price per square

Table 1.4.3 Stepwise regression results

Variables entered	Regression coefficient	Constant	Forecast	95% confidence limits for new forecast		cv
Step 1						
BIDDERS	−0.4355	1.8298	1.0495	□0.6844		38.31
Step 2						
BIDDERS	−0.4251					
GFA	−0.0395	2.0929	0.9873	□0.7090		39.10
Step 3						
BIDDERS	−0.2987					
GFA	−0.3071					
PERIOD	0.8814	1.7040	1.0609	□0.6011		32.05
Step 4						
BIDDERS	−0.2665					
GFA	−0.3293					
PERIOD	0.8480					
STOREY	0.1032	1.7887	1.0634	□0.6075		33.99
Step 5						
BIDDERS	−0.2646					
GFA	−0.3253					
PERIOD	0.8848					
STOREY	0.1108					
ACOND	−0.0667	1.6847	1.1166	□0.6836		35.06

21 Mar 89   example of regression analysts estimating for bc is offices
09:55:05   University of Salford   Prime 9955   rev 20.1.0

＊＊＊＊M U L T I P L E   R E G R E S S I O N＊＊＊＊

Listwise Deletion of Missing Data
Selecting only Cases for which   CASENO LE   28:00

	Mean	Std Dev	Label
PRICE	1.088	.395	
GFA ($\bar{x}_2$)	7.103	1.279	
STOREY	1.082	.566	
ACOND	.333	.480	
PERIOD($\bar{x}_2$)	2.336	.436	
BIDDERS($\bar{x}_1$)	1.703	.536	

N of Cases = 27

Figure 1.4.1 Regression output at step 3.

metre value (CONSUM/GFA ratios), and (2) taking the natural logs of all the variables except ACOND.

The stepwise procedure was used to enter the independent variables into the regression model one at a time. The resulting sequence of independent variables entering the model was firstly BIDDERS, then GFA, PERIOD, STOREY, and finally ACOND. Table 1.4.3 gives the results obtained as each variable was entered into the model.

Up to step 5 the results are very much as expected with the number of bidders and gross floor area producing a negative regression coefficient indicating a drop in price per square metre as these variables increase due to intensity of competition and economies of scale respectively. Similarly the contract period and number of storeys produce a positive regression coefficient indicating a rise in price per square metre as these variables increase due to the effects of complexity of design and construction. The negative value for air-conditioning at step 5 however is not expected, as it indicates that air-conditioned buildings are cheaper. This, together with the increase in variability as measured by the 95% confidence limits and coefficient of variation suggests that the ACOND effect is spurious. The best looking model here seems to be that obtained at step 3, which has a better standard error of forecast and better coefficient of variation than the other models.

The model at step 3 is

$$\log(CONSUM/GFA) = a + b_1\log(BIDDERS + b_2\log(GFA) + b_3\log(PERIOD) \quad (7)$$

where a b_1 b_2 b_3 are the constant and regression coefficients respectively. Thus the forecast of log (CONSUM/GFA) for the new contract is, from Figure 1.4.2:

$$
\begin{aligned}
\log \text{(CONSUM/BIDDERS)} &= 1.7040 - 0.2897 \,\square\, \log(6) - 0.3071 \,\square\, \log(6000) + \\
&\quad 0.8814 \,\square\, \log(18) \\
&= 1.7040 - 0.2897(17918) - 0.3071(8.6995) + \\
&\quad 0.8814(2.8904) \\
&= 1.0609
\end{aligned}
$$

The antilog of 1.0609 is 2.8890 which is the standardised pounds per square metre estimate, and the standardised total estimate is therefore 2.8890 x 6000 = 17334. This can now be converted into a current estimate in a given location by multiplying by the current tender price index and location factor. If the new

21 Mar 89   example of regression analysts estimating for bc is offices
09:55:07   University of Salford   Prime 9955   rev 20.1.0

* * * * MULTIPLE REGRESSION * * * *

Equation Number 3   Dependent Variables. .   PRICE

Descriptive Statistics are printed on Page   3

Beginning Block Number   1.   Method: Enter   BIDDERS   GFA   PERIOD

Variable(s)   Entered on Step Number   1. .   BIDDERS
                                         2..   PERIOD
                                         3..   GFA

Analysis of Variance

	DF	Sum of Squares	Mean Square
Regression	3	2.28993	.76331
Residual	23	1.75910	(S^2) .07648

F =   9.98020   Sign if F =   0.0002

Var-Cover Matrix of Regression Coefficients (B)
Below Diagonal: Covariance   Above: Correlation

		BIDDERS		PERIOD		GFA
BIDDERS	(S^2c_{11})	.01211		.37795		−.38235
PERIOD	(S^2c_{13})	.01126	(S^2c_{17})	.07333		−.88752
GFA	(S^2c_{12})	.00390	(S^2c_{27})	.02226	(S^2c_{22})	−.30858

- - - - - - - - - - - Variables in the Equation - - - - - - - - - - - -

Variable		B	SE B	Beta	T	Sig T
BIDDERS	(b_1)	−0.289683	.110059	−.393112	−2.632	.0149
PERIOD	(b_3)	.881413	.270786	.974902	3.255	.0035
GFA	(b_2)	−.307112	.092623	−.995027	−3.316	.0030
(Constant)	(a)	1.704044	.357162		4.771	.0001

End Block Number   1   All requested variables entered.

Figure 1.4.2 Regression output at step 3.

21 Mar 89   example of regression analysis estimating for bc is offices
09:55:07   University of Salford   Prime 9955   rev 20.1.0

* * * * M U L T I P L E   R E G R E S S I O N * * * *

Equation Number 3   Dependent Variable. .   PRICE

Casewise Plot of Standardized Residual

*: Selected   M: Missing   X: Unselected

Case #	PRICE	*SDRESID	*DRESID	*ADJPRED	EXP ADJPRED/PRICE =	EXP -DRESID
1	.88	-.9860	-.2823	1.1588		1.33
2	1.15	.7307	.2103	.9370		0.81
3	.99	-5.942	-.1737	1.1681		1.19
4	1.00	.1031	.0308	.9733		0.97
5	2.68	3.0287	1.1730	1.5075		0.31
6	.90	-.1312	-.0392	.9437		1.04
7	.64	-1.3530	-.3909	1.0239		1.46
8	.86	-.8505	-.2501	1.1093		1.28
9	1.21	-.0137	-4.0407E-03	1.2109		1.00
10	1.78	3.0960	.7677	1.0117		0.46
11	.75	-.5638	-.1671	.9190		1.18
12	1.07	-.3665	-.1064	1.1785		1.11
13	1.12	.9900	.2922	.8259		0.75
14	.81	-.1842	-.0555	.8650		1.06
15	.79					—
16	.99	-1.0747	-.3048	1.2919		1.36
17	.85	-.5630	-.1722	1.0198		1.19
18	1.29	1.6465	.4668	.6248		0.63
19	1.30	-1.7940	-.6667	1.9628		1.95
20	.89	-.4103	-.1216	1.0123		1.13
21	.90	.0121	3.523HE-03	.8930		1.00
22	.79	-.5588	-.1615	.9553		1.18
23	.93	.9884	.3222	.6052		.72
24	.97	-.9481	-.2892	1.2637		1.34
25	1.20	1.3244	.3945	.8013		0.67
26	1.17	-.3167	-.1056	1.2716		1.11
27	1.36	.5284	.1692	1.1927		0.85
28	.90	-.1523	-.0444	.9446		1.05
29						—
	-13.30	-49.4444	-14.3656	1.0609		
Case #	PRICE	*SDRESID	*DRESID	*ADJPRED		

$\bar{x}$ 1.0419
s_{n-1} 0.3339
cv.   32.05%

Figure 1.4.3 Regression output at step 3.

***** MULTIPLE REGRESSION *****

Equation Number 3   Dependent Variable..   PRICE

Residuals Statistics:

Selected Cases:	CASENO LE	28.00			
	Min	Max	Mean	Stc Dev	N
*PRED	.6952	2.2392	1.0880	.2965	27
*ZPRED	-1.3234	3.8792	.0000	1.0000	27
*SEPRED	.0650	.2184	.0994	.0388	27
*ADJPRED	.6052	1.9628	1.0693	.2608	27
*RESID	-.3548	.6954	.0000	.2691	27
*ZRESID	-1.2829	2.5145	.0252	.9405	27
*SRESID	-1.7132	2.6420	.0187	1.0928	27
*DRESID	-.6667	1.1730	.0586	.3751	27
*SDRESID	-1.7940	3.0960		1.1846	27
*MAHAL	.4749	15.2555	2.8889	3.5218	27
*COOK D	.0000	2.8055	.1564	.5474	27
*LEVER	.0183	.5868	.1111	.1355	27

Unselected Cases:	CASENO GT	28.00			
	Min	Max	Mean	Std Dev	N
*PRED	1.0609	1.0609	1.0609	.0000	1
*ZPRED	-.0913	-.0913	-.0913	.0000	1
*SEPRED	.0891	.0891	.0891	.0000	1
*ADJPRED	1.0609	1.0609	1.0609	.0000	1
*RESID	-14.3656	-14.3656	-14.3656	.0000	1
*ZREDSID	-51.9448	-51.9448	-51.9449	.0000	1
*SRESID	-49.4444	-49.4444	-49.4444	.0000	1
*DRESID	-14.3656	-14.3656	-14.3656	.0000	1
*SDRESID	-49.4444	-49.4444	-49.4444	.0000	1
*MAHAL	1.5725	1.5725	1.5725	.0000	1
*COOK D	.6164	.6164	.6164	.0000	1
*LEVER	.0582	.0582	.0582	.0000	1

Total Cases = 28

Durbin-Watson Test = 1.00664 (Sig. pos. corr.)

Figure 1.4.4 Regression output at step 3.

contract is for January 18989 (TPI = 300) in Salford (Location Factor = 0.90), the estimate will be 17334 □ 300 □ 0.90 = £4680180.

The 95% confidence limits of £2 565 918 and £8 536 266 (Appendix A) are very large – a reflection to some extent on the limited amount of data used in this example.

BIDDING MODEL DEBIASER

Although not strictly estimating techniques, estimate debiasers constitute a collection of very new 'back end' techniques, still in the research phase, aimed at improving or fine tuning estimates generated by other techniques. Three types of debiasers are under current development:

1   Regression debiasers, which are identical to the usual regression estimating techniques except that the estimate is specifically included in the predictor variable list.
2   Control chart debiasers, using dynamic time series detrending techniques to detect real time biasing of recent estimates.
3   Bidding model debiasers, which utilise bidding theory to assess bias and reliability in estimates.

This section describes one of the bidding model debiasers being developed by the authors.

The purpose of bidding models is to enable a bidder to assess the best mark up value to use in an auction given some information concerning the likely bids to be entered by his competitors. This information can range from a knowledge of the identity and past bids of all the competitors in the auction through to virtually no information at all. Many formulations have been proposed to model this situation. The model adopted here is the multivariate model (Appendix B). As the probability of entering the lowest bid with a given bid is exactly the same as the probability that the lowest bid will be less than an estimate of the lowest bid, the model can be applied without modification.

Problems and limitations

The problems and limitations are similar to those of the regression estimating approach in concerning model assumptions and data limitations.

Model assumptions

Three major assumptions concerning the validity of equation (B.1) need to be addressed.

1   the log normal assumption;

2   the independence assumption; and
3   the consistency assumption.

The first of these assumptions is dealt with in Appendix B.

Violations of the independence assumption may have much more severe consequences. Independence is however very difficult to establish with data of this kind, and possible violations tend to be ignored for this reason. It is probably now a truism that possible lack of independence is the reason for many models of this kind failing to achieve commercial status.

The consistency assumption, i.e. that bidders behave in much the same way irrespective of the type, size and other characteristics of contracts, is a reflection on the simplicity of the model (of which the independence assumption is a special case). Some research is currently proceeding on this aspect.

Data limitations

Although most bidding models have very heavy demands on data, particularly on the frequency with which certain specified bidders compete against each other, the multivariate model is relatively undemanding in this respect.

The data consist of any previous designer's estimates together with contractors' bids. It is not necessary that designer's estimates are available for all the contracts in the data, nor that the designer's estimates and the specified contractors' bids are recorded for the same contract. All that is necessary is an indirect link between the designer and the specified bidders. For example, if contractors A, B, and C are bidding for a new contract, it is not important that the designer has produced estimates for previous contracts on which A, B, and C have entered bids, nor that any of the people involved have bid against each other before. All that is required is that all the people involved have bid at least once against another bidder who has bid at least once against the current competitors, or have bid at least once against another bidder who has bid at least once against another bidder who has bid at least once against the current competitors, etc.

There is a price to pay however for the relative lack of data restriction, reflected in the independence and consistency assumptions mentioned above.

Applications

The bidding model debiaser involves three stages

* data preparation and entry;
* estimation of parameters; and
* calculation of probability of lowest bid values.

Data preparation and entry

The data consists of all available designers' estimates, bids and associated bidders' names for a set of historical construction contract auctions. These are entered into an auction database in the form of a contract number, designer's estimate value/designer code, and bid value/bidder code.

Estimation of parameters

The computer program automatically calculates the required model parameters from equations (B.2) and (B.3) and stores the results in a computer file. This operation is only necessary when new historical auction data is entered into the auction data base, and takes a few seconds of computer time.

Calculation of the probability of the lowest bid values

The computer program automatically calculates the unbiased estimated probability values for each of a sequence of m values and plots the resulting curve in terms of P□ Further probability estimates are then obtained via estimates of □ and □^{2} obtained by stochastic simulation, each iteration generating a different probability value. These additional values provide an indication of the variability of the probability estimates and are plotted as points on the graph. The resulting graph therefore enables the user to gain an impression not only of the reliability of the debiased estimate but also of the reliability of the reliability!

Example

This example contains data collected from a London building contractor (coded number 304) for an incomplete series of construction contracts auctioned during a 12 month period in the early 1980s (Table 1.4.4). For the purposes of this example, the bids entered by bidder 304 are treated as designer's estimates.

The program requests the value of the estimate for the new contract together with the identity of the bidders. In this example the estimate of the new contract is £3m and the bidders are code 55, 73, 134, 150 and 154. The program then automatically proceeds as follows:

1   transforms the data to the log values $y_{ik} = \log(x_{ik} - \Box x_{(1)k})$ in this case $\Box = 0.6$;
2   calculates the required model parameter estimates, $\Box_i$, $\Box_k$, and s^2_i;
3   calculates the probability of code 'underbidding' the other bidders with m = −0.70, −0.69, etc. by substituting $\Box_1$ and s_i for $\Box_i$ and $\Box_1$ in equation (B.6);

Table 1.4.4 Data for incomplete series of construction contracts

PROJ	BID	BDR	BID	BDR	BID	BDR	BID	BDR	BID	BDR	BID	BDR	BID	BDR	BID	BDR	BID	BDR	BID	BDR
1	1454515.	150	1514865.	55	1475398.	304	1468775.	134	1447867.	154	1457977.	73								
2	535608.	304	502042.	291	529744.	154	516376.	157												
3	1333142.	75	1331156.	217	1366863.	304	1266892.	281	1276787.	115	1277652.	280	1865545.	360						
4	696743.	304	696972.	292	701062.	237	637815.	79	697826.	361	637815.	157								
5	404110.	55	422297.	304	413224.	97	389196.	117	389848.	157										
6	2116877.	134	2169966.	99	2187991.	293	2161120.	304	2198655.	221	2296108.	137	2165611.	8	2153344.	117	2133608.	294		
7	3065742.	304	3119689.	150	3141641.	170	3153800.	134	3249927.	191	3269768.	55	3335993.	187						
8	7925257.	221	7351929.	304	7374650.	247	6900000.	20												
9	871520.	118	899935.	137	902378.	304	914393.	291	950737.	83	996483.	221								
10	1063337.	304	1154023.	251	1102272.	173	1079657.	201												
11	1759614.	154	1792123.	281	1838532.	157	1918066.	170	1947733.	304	1784215.	308								
12	1126816.	304	1146398.	201	1169795.	154	1227296.	24	1312527.	280	1399472.	221								
13	698005.	304	625501.	268	630288.	308	666545.	55												
14	588810.	364	584833.	365	639229.	79	646341.	145	682802.	304	691474.	154								
15	1429218.	303	1493849.	291	1511033.	304	1521628.	12	1526377.	366	1717715.	55								
16	842319.	6	870894.	304	883617.	185														
17	284947.	367	292692.	356	294694.	368	303700.	152	307282.	85	313203.	134	315727.	369	333597.	118	334353.	370	348969.	304
18	461444.	150	483862.	304	482241.	308	447021.	55	493417.	154	455480.	311								
19	2858191.	280	2947007.	371	2950723.	134	2999999.	304	3093587.	60	3099528.	6	3278229.	266	3325198.	170	3333793.	55		
20	7831865.	276	7837276.	304	7859122.	256	7904172.	55	8047230.	152	8145323.	293	8279564.	117	8657685.	134				
21	3971051.	55	3854074.	304	4724785.	372	3955009.	154	3944772.	373	3731543.	79	4001188.	237						
22	573485.	292	596737.	291	597730.	134	613528.	201	615015.	304	621223.	170								
23	1610942.	304	1623447.	163	1646286.	173	1663742.	268	1700000.	152										
24	1196036.	64	1199328.	374	1206837.	187	1226589.	304	1262082.	291	1271000.	152	1295954.	170	1302161.	254				
25	2636397.	137	2654728.	150	2673906.	187	2685127.	55	2762123.	304	2845567.	152								
26	469663.	24	476784.	268	485870.	286	486485.	55	504026.	122	529468.	263	540814.	304						
27	1526553.	201	1533719.	152	1698797.	148	1876612.	304												
28	2106139.	201	2175928.	304	2210065.	308	2223710.	280	2255246.	221	2296623.	117	2331830.	266						
29	499888.	102	559596.	55	592026.	217	602042.	170	608957.	304	619065.	134								
30	2639525.	304	2842407.	308	2874130.	280	2861665.	55	2736300.	152	2770720.	256								

31	732572.	304	599429.	365	623906.	145	691759.	79	744332.	154	607065.	364						
32	546641.	134	539565.	268	608242.	55	538382.	24	599934.	170	559351.	304						
33	792966.	221	811788.	99	819971.	308	847621.	55	847892.	137	853793.	304						
34	2085151.	152	2130217.	107	2150583.	280	2203956.	115	2219653.	137	2241687.	154	2325900.	304				
35	821617.	268	844579.	115	848459.	303	871927.	304	872215.	106	935765.	375						
36	792474.	304	747374.	24	778559.	217	743788.	252	808345.	268	835465.	170						
37	7279854.	304	7650271.	60	7029448.	308	6631664.	150	7089879.	193	7230120.	170	6986341.	247	7143710.	191	6794551.	266
38	592096.	304	573997.	150	518613.	217	508985.	121	544480.	376								
39	538600.	348	567031.	377	621365.	378	699839.	268	825451.	72	991468.	190	1001254.	304				
40	2087946.	154	2104017.	276	2183122.	186	2205359.	304	2212382.	280	2267987.	112	2332476.	221	2400000.	294		
41	1503739.	247	1536654.	24	1576905.	304	1583595.	154	1616432.	294	1704995.	157						
42	3624453.	191	3694803.	221	37372133.	304	3751115.	170	3773967.	193	3866339.	55	3922937.	134	4122448.	281		
43	629164.	157	695284.	173	723315.	311	729305.	266	743578.	304	768189.	379						
44	2252833.	304	2264310.	24	2274380.	112	2323385.	191	2384494.	55								
45	1202916.	163	1268733.	55	1291365.	221	1294986.	304										
46	2968891.	217	2772626.	280	2822857.	186	2972189.	134	2821600.	276	2857275.	304	2793000.	221				
47	1398400.	286	1401500.	152	1427140.	237	1436804.	304	1453070.	301	1511643.	55	1591986.	371	1665760.	83		
48	698161.	294	709676.	24	758565.	291	789355.	304	797926.	134	842684.	55	751677.	252				
49	248733.	31	2510077.	291	251415.	252	261286.	380	264933.	304								
50	358840.	293	362370.	217	386983.	304	421797.	381	456272.	154								
51	527692.	317	870874.	311	588854.	75	609221.	173	636451.	308	694297.	304						

4   generates a value for $\square_{304}$ $\square^2_{304}$, $\square_{55}$, $\square^2_{55}$, $\square_{73}$, $\square^2_{73}$, $\square_{134}$, $\square^2_{134}$, $\square_{150}$, $\square^2_{150}$, $\square_{154}$, $\square^2_{154}$, by stochastic simulation;
5   calculates the probability of code 304 'underbidding' the other bidders with m values obtained by stochastic simulation;
6   repeats 4 and 5 600 times;
7   detransforms the m values to P$\square$by equation (B.10) and plots the curve resulting from step 3 and points resulting from steps 4 to 6.

The resulting graph is shown in Figure 1.4.5. The graph is interpreted by drawing a horizontal line at the 50, 2.5 and 97.5 percentage probability points across to the curve and thence down to the estimate axis as shown to obtain the unbiased estimate (£2 827 300) and 95% confidence limits (£2 566 700 to £3 138 000, i.e. £2 827 300 + 10.99% − 9.22%) due to the variability of the designer's estimates and contractors' bids *as predicted by the model*.

The surrounding points indicate the effect of the size of the database on the reliability of the parameter estimates in the model − the true curve will be contained somewhere within these points. With the small amount of data used in this example, the points are quite widespread. The existence of a larger database should have the effect of decreasing this spread.

CONCLUSION − THE FUTURE?

Contract price forecasting techniques clearly comprise a large topic area worthy of a book in its own right. As Table 1.4.1 indicates, the field is rapidly developing out of the older deterministic approaches into methods which specifically accommodate the inherent variability and uncertainties involved in forecasting the price of construction work. One result of this is that the traditional distinction between 'early stage' or 'conceptual' estimating and 'later stage' or 'detailed' estimating is being replaced by the more fundamental distinctions concerning the reliability of forecasts and their components − items, quantities and rates. This has focussed attention on the means of modelling and predicting reliability − statistically for simplicity and stochastically for complexity. The construction project simulator, for instance, contains stochastic elements for item quantities, whilst risk estimating utilises both statistical and stochastical techniques. Little has been done to treat the items themselves in this way, although some relatively new quantity generation systems, such as ELSIE, are clearly capable of extension. The logical conclusion of these approaches will be a technique which combines all three elements of the forecasting equation into the same non-deterministic, item, quantity, rate, (NDIQR) system.

Although a somewhat daunting prospect for practitioners, the development of NDIQR systems will mark a new and exciting phase in the evolution of construction price forecasting systems generally. Firstly, current deterministic requirements will still be accommodated as deterministic forecasts are simply

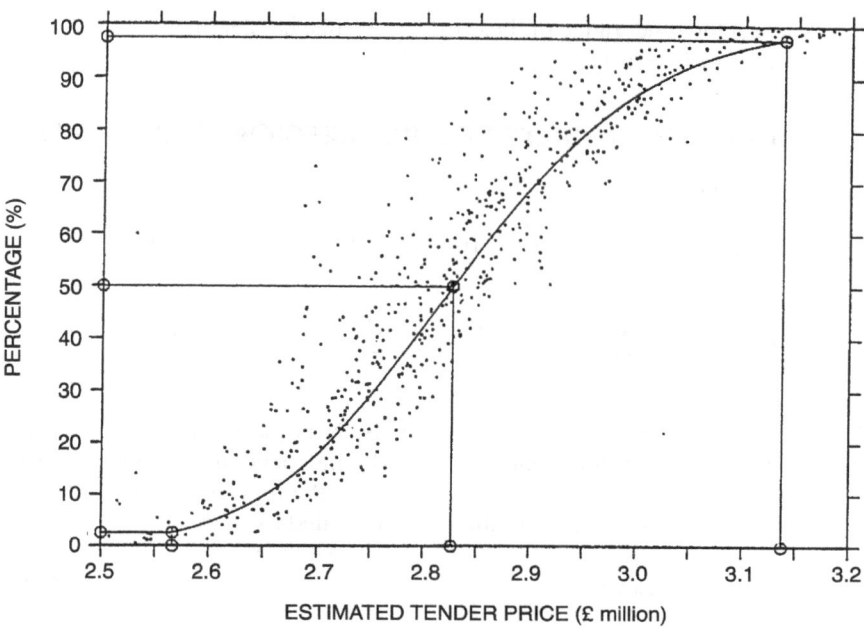

Figure 1.4.5 Bidding model estimate debiaser results.

a special case for a non-deterministic system. A rate with mean say £5 and standard deviation £0 is effectively a deterministic rate. Also the 'best guess' of a non-deterministic system is a deterministic answer. Thus the range of forecasts provided by a non-deterministic system can be regarded as secondary information to the deterministic forecast, to be divulged or not as the user wishes. Secondly, NDIQR systems will allow forecasts to be made at any stage of the design process. Treating the items themselves as random variables, for example, means that the standard deviation simply reduces as we become more certain that the item will be appropriate. Thus a BQ PRICING NDIQR system will commence with a notional bill of quantities that will gradually firm up as the design progresses. Thirdly and perhaps most importantly, the reliability measures provided by NDIQR systems will enable comparisons to be made between alternative systems. For example, if a practice uses several systems to provide price forecasts for the same contract and obtains the following results:

System	Forecast range
A	£3.2m to £3.9m
B	£3.5m to £4.0m
C	£3.7m to £4.5m

we may select the inner range of these three systems, i.e. £3.7m to £3.9m, to be the best range of forecasts.

APPENDIX A: RELIABILITY OF REGRESSION FORECASTS

95% confidence limits

This is obtained from the standard error of the forecast SE (Y) where

$$SE(Y)^2 = S^2(1 + 1/N) + \sum_{i=1}^{k}(x_i - \bar{x}_i)^2 S^2 c^{ii} + \sum_{i=1}^{k-1}\sum_{j>i}^{k} 2(x_i - \bar{x}_i)(x_j - \bar{x}_j) S^2 c^{ij} \quad (A.1)$$

where S^2 is the mean square of the residuals, N is the number of previous cases, $\bar{x}_1, \bar{x}_2, \bar{x}_3$, etc. are the mean values of the independent variables, x_1, x_2, x_3, etc., and $S_3 c^{ij}$ are the variance–covariances of the regression coefficients.

The 95% confidence limits are then approximated by

$$\pm\, t_{(0.025)\,(N-n-1)}\, SE(Y) \quad (A.2)$$

where $t_{(0.025)\,(N-n-1)}$ is obtained from the students t distribution tabulated in most elementary statistical texts.

Coefficient of variation

The coefficient of variation, cv, can be obtained from the distribution of jack-knife deleted residuals as follows

$$cv = 100S_d/(x_p - \bar{x}_d) \quad (A.3)$$

where x_d and S_d represent the mean and standard deviation of the deleted residuals and x_p represents the mean prediction. This statistic, though not conventionally used in regression applications, has the advantage of being directly comparable with the variability measures associated with other techniques and studies.

Reliability of forecast for Salford Offices example

The standard error of forecast is obtained from equation (A.1) as

$$SE(Y)^2 = S^2(1 + 1/N) + (x_1 - \bar{x}_1)^2 S^2 c^{11} + (x_2 - \bar{x}_2)^2 S^2 c^{22} + (x_3 - \bar{x}_3)^2 S^2 c^{33} +$$
$$2(x_1 - \bar{x}_1)(x_2 - \bar{x}_2)S^2 c^{12} + 2(x_1 - \bar{x}_1)(x_3 - \bar{x}_3)S^2 c^{13} +$$
$$2(x_2 - \bar{x}_2)(\bar{x}_3 - \bar{x}_3)S^2 c^{23} \quad (A.4)$$

where x_1, x_2, x_3, represent the log values of bidders GFA and period respectively. The regression output (Figures 1.4.1 and 1.4.2) at step 3 shows that

$S^2 = 0.07684$, $N = 27$, $\bar{x}_1 = 1.703$, $\bar{x}_2 = 7.103$, $\bar{x}_3 = 2.336$, $S^2c^{11} = 0.01211$, $S^2c^{22} = 0.00858$, $S^2c^{33} = 0.07333$, $S^2c^{12} = 0.0390$, $S^2c^{13} = 0.01126$, $S^2c^{23} = 0.02226$.

Substituting into the above equation gives:

$$
\begin{aligned}
SE(Y)^2 = {}& 0.07648\,(1 + 1/27) + (\log 6 - 1.703)^2 0.1211 + (\log 6000 - 7.103)^2 \\
& 0.00859 + (\log 18 - 2.336)^2 0.07333 - 2(\log 6 - 1.703) \\
& (\log 6000 - 7.103)\,0.00390 + 2\,(\log 6 - 1.703)\,(\log 18 - 2.336) \\
& 0.01126 - 2(\log 6000 - 7.103)\,(\log 18 - 2.336)\,0.2226 \\
= {}& 0.0793 + 0.0001 + 0.0219 + 0.0225 - 0.0011 + 0.0011 - 0.0197 \\
= {}& 0.0844
\end{aligned}
$$

so $SE(Y) = 0.2905$.

The 95% confidence limits for the standardised log forecast is then, from equation (A.2)

$$
\begin{aligned}
Y \pm t_{(0.025)\,(27 - 3 - 1)}\,0.2905 &= 1.0609 \pm 2.069 \pm 0.2905 \\
&= 1.0609 \pm 0.6010 \text{ i.e. } 0.4599 \text{ to } 1.6619
\end{aligned}
$$

which is 1.5839 to 5.2693 for the standardised forecast, and £2 565 918 to £8 536 266 for the Salford January 1989 contract.

The coefficient of variation is obtained by dividing the standard deviation of the forecast error ratios by their mean. As a log model has been used, the antilog of the deleted residual will give the required ratio. At step 3 the standard deviation is 0.3339 and the mean 1.0419, giving a coefficient of variation of 32.05%.

Some statistical packages make the calculations easier by providing a direct forecast for the new contract. In this example we have entered the new con-
^tract into the data base with a dummy value of £0.01 but excluding it from the model by a special select instruction. As Figure 1.4.3 shows, the correct forecast of 1.0609 is obtained. This package also gives the standardised deleted residual SDRESID, which is the deleted residual DRESID divided by its standard error. The standard error of the forecast can therefore be obtained very quickly as

$$
\begin{aligned}
SE(Y) &= DRESID/SDRESID \\
&= -14.3656/-48.4444 = 0.2905 \quad\quad\quad\quad (A.5)
\end{aligned}
$$

The residual statistics (Figure 1.4.4) also enable a quick approximation of the coefficient of variation to be made by

$$cv = 100 \times \text{std dev DRESID}/(\text{mean PRED} - \text{mean DRESPID})$$
$$= 100 \times 0.3751/1(1.0880 - 0.0187) = 35.08 \tag{A.6}$$

Although this figure is the coefficient of variation of the log values, it is proportional to the raw coefficient of variation and therefore indicative of the relative values. Also, as the model becomes more reliable, then the coefficient of variation of the logs values becomes closer to the coefficient of variation of the raw values

APPENDIX B: THE MULTIVARIATE BIDDING MODEL DEBIASER

The multivariate bidding model is

$$\log(x_{ik}) = y_{ik} \sim N(\mu_i + \delta_i, \sigma^2_i) \tag{B.1}$$

where x_{ik} is bidder i's bid (i = 1, 2, . . ., r) entered for auction k (k = 1, 2, . . ., c), the log of which is normally distributed with mean μ_i, a bidder location parameter, plus δ_k, an auction size datum parameter, and with a unique variance parameter σ^2_i for each bidder. These parameters can be estimated quite easily by an iterative procedure solving

$$\delta_k = \hat{\delta}_k = \sum_{i=1}^{r} \delta_{ik} (y_{ik} - \mu_i)/n_i \tag{B.2}$$

$$\mu_i = \hat{\mu}_i = \sum_{k=1}^{c} \delta_{ik} (y_{ik} - \delta_k)/n_i \tag{B.3}$$

and thence

$$S^2_i = \hat{\sigma}^2_i = \sum_{k=1}^{c} \delta_{ik} (y_{ik} - \mu_i - \delta_k)^2 / \left\{ (n_i - 1) \left(1 - \frac{c-1}{N-r} \right) \right\} \quad \text{for } n_i > 1 \tag{B.4}$$

$$= \sum_{k=1}^{c} \delta_{ik} (y_{ik} - \mu_i - \delta_k)^2 / (N - c - r + 1) \quad \text{for } n_i = 1 \tag{B.5}$$

where Kronecker's δ_{ik} = 1 if bidder i bids for auction k
$\phantom{where Kronecker's \delta_{ik}\ } = 0$ if bidder i does not bid for auction k

$$n_i = \sum_{k=1}^{c} \delta_{ik}, \text{ the number of bids made by bidder } i$$

$$N = \sum_{k=1}^{c} n_i, \text{ the total number of bids by all bidders}$$

The probability of a bidder, say $i = 1$, underbidding a set of specified competitors on a contract is then given by

$$\Pr(y_1 + m < y_i, i \square 1) =$$

$$\int_{\square\square}^{\square} \left\{ (2\square)^{1/2} \right\}^{\square 1} \exp\left(\square \frac{1}{2} y_1^2\right) . \left\{ \prod_{i=2}^{n} \int_{\square}^{\square} \left\{ (2\square)^{1/2} \right\}^{\square 1} \exp\left(\square \frac{1}{2} y_i^2\right) d y_i \atop y_i = (\square_1 y_1 + \square_i + \square - \square_i)\square_i^{-1} \right\} d y_1 \quad (B.6)$$

where m is a 'decision' constant used by bidder 1 to bring about a desired probability state (the usual approach is to use bidder 1's cost estimates in the analysis in preference to his bids on the assumption that the m value will reasonably approximate his likely profit should be acquire the contract). To observe the effects of the limited accuracy of the parameter estimates due to the sample size, values for $\square$ and $\square$ are obtained from $\square$ and s^2 from their sampling distributions

$$\square_i \sim N(\square_i, s^2_i/n_i) \tag{B.7}$$

$$\square^2_i(n_i - 1)s^2_i \sim \square^2(n_i - 1) \qquad \text{for } n_i > 1 \tag{B.8}$$

$$\square^2_i(N - c - r + 1)/s^2_i \sim \square^2(N - c - r + 1) \qquad \text{for } n_i = 1 \tag{B.9}$$

Now if we substitute the designer's estimator for bidder 1 in the above formulation, the value of m which results in a probability of 0.5 represents the bias in his (log) estimate and thus the amount that needs to be added to his estimates to give an unbiased estimate of the lowest bid. Also the values of m which result in probabilities of 0.025 and 0.975, will give the 95% confidence limits.

The log normal assumption has been tested with three sets of UK construction contract bidding data indicating that a three parameter log normal model may be more appropriate than the general two parameter model proposed here. The modifications necessary to convert the formulation are quite straightforward however, involving a prior transformation of the data before applying the iterative procedure. The resulting m values have however to be detransformed before plotting the final probability graph. Thus, for data transformed by $y_{ik} = \log (x_{ik} - \square x_{(1)k})$, the unbiased estimate $P\square$ is given by

$$P^{\square} = \frac{\square Pe^{\square}(1 \square e^{m})}{1 \square\square + \square e^{\square}} + Pe^{m} \qquad (\square = m/Pr = 0.5) \qquad (B.10)$$

REFERENCES

Barnes, N. M. L. (1971) *The design and use of experimental bills of quantities for civil engineering contracts*, PhD thesis, University of Manchester Institute of Science and Technology.

Bennett, J. and Ormerod, R. N. (1984) Simulation applied to construction projects. *Construction Management and Economics*, 2, 225–263.

Brandon, P. S., Basden, A., Hamilton, I. W., Stockley, J. E. (1988) *Application of Expert Systems to Quantity Surveying: the Strategic Planning of Construction Projects*. London: Surveyors Publications, RICS.

Fine, B. (1980) *Construction Management Laboratory*, Fine, Curtis and Gross.

Gilmore, J. and Skitmore, M. (1989) A new approach to early stage estimating. *Chartered Quantity Surveyor*, May.

Lu Qian (1988) Cost estimation based on the theory of fuzzy sets and predictive techniques. *Construction Contracting in China*. Hong Kong Polytechnic.

Ross, E. (1983) *A Database and Computer System for Tender Price Prediction by Approximate Quantities*. MSc project report, Loughborough University of Technology.

Southgate, A. (1988) Cost planning – a new approach. *Chartered Quantity Surveyor*, November, pp. 35–36.

1.5 An agenda for cost modelling*

S. Newton

INTRODUCTION

The human is a neat and tidy world-builder. There is a need in everything we do for some order and purpose (what Katz (1986) refers to as a 'pattern'), otherwise our mind will begin to reject it. Where is the pattern in current cost modelling research? On what common structure are those consigned with understanding cost modelling, both as a science (theory) and a technology (practice), supposed to orientate their creative thinking?

This paper is an attempt to identify an appropriate pattern from the mist of current research activities. It is intended to proffer a framework of pegs which both set out the overall topology of cost modelling as a subject, and provide the points of reference on which individual research contributions can be located. It is in this sense an ambitious undertaking. The pattern being proposed will undoubtedly need to change over time. The point of an agenda however, is not to act as a straight-jacket, but as a way of promoting some semblance of order.

PREVIOUS AGENDA

Previous authors have recognised the need for some kind of structure and direction for cost modelling research. Some have produced position papers in this regard. (It should be noted however, that the particular aspect of 'structure' may not have been the singular regard of each paper). Because the proposed agenda draws from those papers, and because their different perspectives might tend to blur some of the issues, some key position papers will be reviewed first.

Brandon (1982)

Building cost modelling became demonstrably of age with the Building Cost

*1990, *Construction Management and Economics*, E & F N Spon.

Research Conference held at Portsmouth Polytechnic in September 1982. In the editorial to the conference proceedings Brandon (1982) addressed the need for some research direction, calling for what he termed (after Kuhn (1970) a paradigm shift.

Brandon argued for 'a more substantial body of theory, and better models, upon which to base our practice'. The case was made that progress had 'arisen largely from the refinement of current models', built on an unsatisfactory collection of theory. A shift away from current models was called for, representing a fundamental change in the nature of cost models. The impetus for the shift would come from advances in the various computer technologies, the goal being a more widespread application of 'realistic' simulation techniques.

Without establishing a specific agenda for future research, the paper does act as an emergency flare; to draw attention to an impending problem and flood light a potentially fruitful direction of response. It falls short both of describing the limitations of the existing paradigm (c.f. Ferry and Brandon (1980), and of considering in any real depth a possible replacement.

Bowen and Edwards (1985)

The two short-comings in Brandon's paper stimulated a response paper by Bowen and Edwards (1985). Their much fuller consideration of the existing paradigm concluded that its strengths lay in being structured to parallel the design process, and being relatively easy to understand and follow. It 'may begin with a relatively crude model such as the rate per m^2; progress through the ascending order of the elemental cost planning system; and conclude with the quantity surveyor's pricing of the bills of quantities'. What the authors considered a predominantly 'historical–deterministic' approach.

The so-called 'new' paradigm was seen as evolving through such mathematically-based techniques as regression (termed 'inferential–relational' methods), towards a concept of 'stochastic simulation'. The goal being 'to incorporate more explicit considerations of the uncertainty and variability' in cost estimating. Expert systems were considered as the key to stochastic simulation.

Raftery (1984)

In an alternative approach, an attempt was made by Raftery (1984) to produce a conceptual framework for the assessment of model performance. From an essentially 'systems' point of view, models were defined in terms of their data, data/model interface, model technique, interpretation of output, and decision making process. Later the framework was used to describe a selection of cost modelling approaches (Raftery 1986). Unfortunately, while the framework establishes useful criteria (the basis on which a particular model might be judged), it fails to give any specific guidance on what possibilities exists within each dimension. It is left to the individual to determine how data, data/

model interfaces, etc. might be classified. The descriptive power of this approach is limited when placed in the context of an overall paradigm.

Skitmore (1988)

Just how disadvantaged cost modelling research is by having no formal means of describing one cost model relative to another is highlighted in a paper by Skitmore (1988). In this paper Skitmore examines the current state of research in bidding. (Arguably bidding is merely a subset of cost modelling, but in reality it tends to be kept separate and distinct.) The range of descriptive primitives available in that domain is well established. (A 'descriptive primitive' is some generally accepted label or term.) Using such a base, the existing paradigm can be better described, and targets for future research are more easily determined.

On the other hand, what we have summarised in the first two papers (and these are fairly representative) are agendas which focus too closely on individual techniques and systems. (The unfortunate trend towards a 'technique-bound' problem domain has been discussed elsewhere – Newton 1987). Regardless of how appropriate an individual technique or approach might actually be, its supporting argument is considerably diluted when the technique is presented as a paradigm within itself. Individual techniques, a paradigm do not make.

In fairness, authors have had little choice in the matter. The only real descriptive primitives available to them, have been the labels given to particular techniques and systems. This has had a stultifying effect on cost modelling research. Without some means of describing cost modelling in an abstract, or commonly recognised way, the only recourse possible is to adopt the language of a specific technique, or invent some set of temporary labels. Naturally, over the course of time certain, more abstract descriptors have evolved – deterministic, inferential, etc. However, there has been no previously published attempt to combine or relate these primitives as a general classification for cost modelling. Such is the intention of this paper.

DESCRIPTIVE PRIMITIVES

The descriptive primitives being proposed in this paper are the 'pegs' referred to in the introduction. They are intended to give some formal basis to the way in which alternative approaches to cost modelling might be classified.

For the purposes of this paper, the descriptive primitives are categorised under the following nine headings:

- Data – whether it relates to a specific design proposal.
- Units – the units of measurement.
- Usage – the intended purpose.
- Approach – the level at which modelling is applied.

- Application – when during the design process.
- Model – general classification of technique.
- Technique – specific classification.
- Assumptions – can they be accessed.
- Uncertainty – how is it treated.

These nine criteria are used to classify alternative approaches to cost modelling, and are described in more detail below. In addition, importantly, a basic taxonomy of classification for each criteria is also presented in the headings, following the form 'criteria (classifications)'.

Data (Specific, Non-Specific)

This describes whether the cost data being used is intended to relate specifically to an individual project, or to be more descriptive of some 'standard' proposal.

All cost data is based on historical information of some sort, and is arguably therefore to some degree non-specific. However, there is an important distinction between cost data generated for an actual design proposal, and cost data generated for some characteristic design proposal. In the former case, design would have progressed past a description in generic terms, such as office, high quality, cladded external walls, stud partitions, and so on.

Thus the data criteria speaks largely of the transition from adopting general design descriptions, to the quantification of a particular and uniquely distinguishable proposal. It indicates the general level of design detail at which the model operates.

Units (Abstract, Finished Work, As-Built)

Describes whether the units of measurement are abstract descriptions, units of finished work, or units which describe the actual construction process.

An abstract descriptor would use something like cost per student place, number of hospital wards or cost per m² of restaurant. It does not deal with a tangible building product. The constituents of student place, hospital ward or restaurant are rarely completely defined.

Units of finished work are those which cost in terms of the items left behind when construction is complete. It is the original bill of quantities approach. Cost is equated directly to the final product: area of external wall, number of windows, and so on. Temporary items, such as scaffolding, large items of plant, etc., are deemed to be included.

As-built refers to operations as they occur on-site. Thus the costing structure would follow such activities as might appear on a construction network; such as raising brickwork to dpc, dismantling and removal from site of the tower crane, etc.

Usage (Cost, Price)

Describes whether the estimate produced is intended as a basic cost, or of basic cost plus mark-up.

The bidding process is usually held to comprise two stages:

- the basic cost estimate – which is intended to establish the cost to the builder.
- the mark-up – which is an amount added to the basic cost to cover items such as overheads, profit and variations in market conditions.

The basic cost and mark-up together comprise the bid price. Because of the added complexities of estimating potential mark-up, certain models target solely on the basic cost.

Approach (Micro, Macro)

Not all models are intended to address an estimate in its totality, at the macro level. Certain models target at the micro level of an individual cost element, such as external walls, tower crane or lift installation. There are considerable differences in approach, and philosophy, between the two alternative views.

Application (feasibility, sketch, detailed, tender, throughout, non-construction)

To a certain degree, this reflects the criteria on data. As the requirements of the model move from non-specific to specific project data, so the application of the model must generally come later in the design process.

The classification within this criteria comes largely from the RIBA Plan of Work (RIBA 1967). In addition however, some models are intended for use throughout the design process, and others are non-construction related. The non-construction related cost models come principally from the chemical and industrial engineering fields, and generally have less relevance to design problems in the building arena.

Model (simulation, generation, optimisation)

This particular classification of models is non-standard – for a more detailed explanation, see Newton (1988).

The simulation model provides only a formal representation of how a particular problem might be structured. (In this sense, structure refers to how the problem is conceptualised in terms of problem boundary, the variables considered, and the inter-relationships between variables.) The better a problem is understood, the 'more' it can be simulated. The user provides a set of inputs, and has then to evaluate the outcome based on a range of other considerations.

As more of the problem is understood however, and the model contains more and more of the linkages between variables, so also does the use of the model become largely mechanical. For a range of starting values, the model is capable of generating an entire collection of potential solutions. In effect, the model both simulates the structure of the problem, and provides a range of alternative candidate solutions. Note, that in this sense, the technique of Monte Carlo Simulation would be classified as a 'generation' model because it 'proposes' many, many candidate solutions.

It is usual that when a candidate solution to some problem is evaluated, it is compared with other candidate solutions to establish if it is some way 'better'. An extension of this would suggest that when there exists a large number of candidate solutions, some will indeed be better than others. Further, there is likely to be one, or at least a much smaller subset, which can be described as the best, or optimum solution. Optimisation models are those which seek to identify, from a set of candidate solutions, that which best fits some given criteria (Wilson 1987).

Naturally, the intention of any model could be to promote the best solution. This, however, is informal optimisation. In the formal modelling sense, simulation represents the structure of a problem, generation also produces a set of candidate solutions, optimisation also evaluates the set of solutions.

Technique (Various)

Within the general classification of model, there are a variety of techniques applied. Those considered here include:

- Dynamic Programming (Woodward 1975)
- Expert Systems (Lansdown 1982)
- Functional Dependency (Newton and Logan 1988)
- Linear Programming (Cohen 1985)
- Manual – where the model is applied manually
- Monte Carlo Simulation (Spooner 1974)
- Networks (Harris and McCaffer 1977)
- Regression Analysis (Bathurst and Butler 1977)

Assumptions (explicit, implicit)

A model carries with it considerable baggage. Baggage in the sense of a set of assumptions about problem boundaries, about what is or is not significant, about how the user might best conceptualise the problem. The more a model is expected to analyse, evaluate and appraise, the more critical it becomes that users have access to, and appreciate (in order to question and understand) those assumptions.

Where the assumptions are built-into a model, as computer coding or unstated assumptions, that critical access is denied. Certain technologies, such as Spreadsheets or Expert Systems, enable assumptions to be made explicit. The alternative is extensive documentation, such as in the case of the BLCC model (Petersen 1984).

Uncertainty (deterministic, stochastic)

The nature of cost is known to be uncertain. The only question is whether that uncertainty is formally assessed in the model, or dealt with intuitively by the user.

The classification here distinguishes between those models without a formal measure of uncertainty (deterministic), and those with (stochastic). Formal measures of uncertainty may be such metrics as the associated co-efficient of variation (as in regression) or the cumulative frequency distribution (as in Monte Carlo Simulation).

A REVIEW OF COST MODELLING

Using the descriptive primitives proposed above, it is possible to review the literature on cost modelling research to date. The review is arranged in chronological order, and classified on this authors own interpretation of the individual models.

The review is not intended to be exhaustive of very cost model ever proposed. Especially in the engineering fields, a variety of cost models have been developed, but are considered of limited relevance to building construction. Further exclusions include models where capital cost is but one consideration, such as life cycle costing, value engineering and time management. Similarly, most bidding models have been excluded to maintain some degree of generality (alternatively Skitmore (1988)). Finally, purely conceptual models of cost have also been excluded – the papers reviewed all describe some tangible product. Making these exclusions here, does not preclude their inclusion in some later system of classification.

The review uses the following, highlighted abbreviations of the descriptive primitives:

- Data   – *Specific*
             – *Non*-Specific
- Units   – *Abstract*
             – *Fi*nished Work
             – As-*Built*
- Usage   – *Cost*
             – *Pri*ce
- Approach   – *Mi*cro
             – *Ma*cro

- Application   – *Feasibility*
                – *Sketch* Design
                – *Detail* Design
                – *Tender*
                – *Throughout*
                – *Non-Construction*
- Model   – *Simulation*
                – *Generation*
                – *Optimisation*
- Technique   – *Dynamic* Programming
                – *Expert Systems*
                – *Functional* Dependency
                – *Linear* Programming
                – *Manual*
                – *Monte* Carlo Simulation
                – *Networks*
                – *Parametric* Modelling
                – *Probability* Analysis
                – *Regression* Analysis
- Assumptions – *Explicit*
                – *Implicit*
- Uncertainty – *Deterministic*
                – *Stochastic*

Dunican (1960)	S	A	P	Mi	S	S	Man	I	D
RICS (1964)	S	A	P	Ma	S	S	Man	I	D
Thomsen (1965)	S	A	P	Ma	F	G	Func	I	D
Nadel (1967)	N	A	P	Ma	S	S	Para	I	D
Meyrat 91969)	N	A	P	Ma	S	S	Para	I	D
Barrett (1970)	N	F	P	Ma	S	S	Man	I	D
Gould (1970)	N	A	P	Mi	S	S	Reg	I	S
DOE (1971)	N	A	C	Mi	S	S	Func	I	D
Buchanan (1972)	N	A	P	Mi	S	S	Reg	I	S
Regdon (1972)	N	F	P	Ma	F	S	Reg	I	S
Tregenza (1972)	N	A	P	Ma	F	S	Para	I	D
Kouskoulas and Koehn (1974)	N	A	P	Ma	F	S	Reg	I	S
Braby (1975)	N	A	P	Ma	F	S	Reg	I	S
McCaffer (1975)	N	A	P	Ma	F	S	Reg	I	S
Wilson and Templeman (1976)	N	A	P	Mi	S	O	DP	I	D
Bathurst and Butler (1977)	N	A	P	Ma	F	S	Func	I	D
Bathurst and Butler (1977)	N	F	P	Mi	F	S	Reg	I	S
Brandon (1978)	N	F	P	Ma	S	S	Para	I	D
Flanagan and Norman (1978)	N	A	P	Ma	F	S	Func	I	D
Townsend (1978)	N	A	P	Ma	S	S	Func	I	D

Moore and Brandon (1979)	N	A	P	Mi	S	G	Func	I	D
Russell and Choudhary (1980)	N	F	P	Ma	S	O	LP	I	D
Powell and Chisnall (1981)	N	A	P	Ma	S	G	Func	I	D
Gray (1982)	S	F	P	Mi	D	S	Man	I	D
Holes and Thomas (1982)	S	F	P	Ma	Th	S	Func	E	D
Mathur (1982)	N	F	P	Ma	S	G	Mont	I	S
Pitt (1982)	S	B	P	Ma	S	G	Mont	I	S
Scholfield et al (1982)	N	F	P	Mi	S	S	Func	I	D
Sierra (1982)	N	A	P	Ma	S	S	Reg	I	S
Skitmore (1982)	N	A	P	Ma	Th	S	Prob	I	S
Wilson (1982)	S	B	C	Mi	S	G	Mont	I	S
Zahry (1982)	N	A	P	Ma	S	S	Prob	I	S
Langston (1983)	N	A	P	Ma	S	G	Func	I	D
Newton (1983)	N	A	P	Ma	S	G	Func	I	D
Bennett and Ormerod (1984)	S	B	P	Ma	S	S	Mont	I	S
Sidwell and Wottoon (1984)	S	B	C	Ma	Th	S	Func	I	D
Cusack (1985)	N	B	P	Ma	S	O	LP	I	D
Gehring and Narula (1986)	N	A	P	Ma	NC	G	Mont	I	S
Atkin (1987)	N	A	P	Ma	S	O	DP	E	D
Berny and Howes (1987)	N	A	P	Ma	Th	S	Func	I	D
Bowen et al (1987)	N	B	P	Ma	Th	G	Net	I	D
Brown (1987)	N	A	P	Ma	S	S	Prob	I	S
Cusack (1987)	N	B	P	Ma	S	O	LP	I	D
Holes (1987)	S	F	P	Ma	Th	S	Func	E	D
Kiiras (1987)	N	F	P	Ma	Th	S	Man	I	D
Meijer (1987)	N	A	P	Ma	F	S	Func	I	D
Pegg (1987)	S	F	P	Ma	S	S	Prob	I	S
Weight (1987)	S	A	P	Ma	S	S	Func	E	D
Woodhead et al (1987)	S	F	C	Ma	Th	S	Func	I	D
Brandon (1988)	S	F	P	Ma	S	S	ES	E	S
Dreger (1988)	S	F	P	Ma	Th	S	Man	I	D
Khosrowshahi (1988)	N	A	P	Ma	Th	S	Reg	I	D
Park (1988)	N	A	P	Ma	NC	S	Para	I	D
Selinger (1988)	N	F	P	Ma	F	S	Para	I	D
Walker (1988)	N	A	P	Ma	T	S	Mont	I	S
Yokoyama and Tomiya (1988)	N	A	P	Ma	S	S	Reg	I	D

The review of cost models is summarised in Figures 1.5.1–1.5.9. These graphs indicate clearly where the emphasis has lain in cost modelling research to date. No doubt there has been good reason for this particular pattern. It remains to consider if certain aspects might provide more fruitful grounds for research in the future.

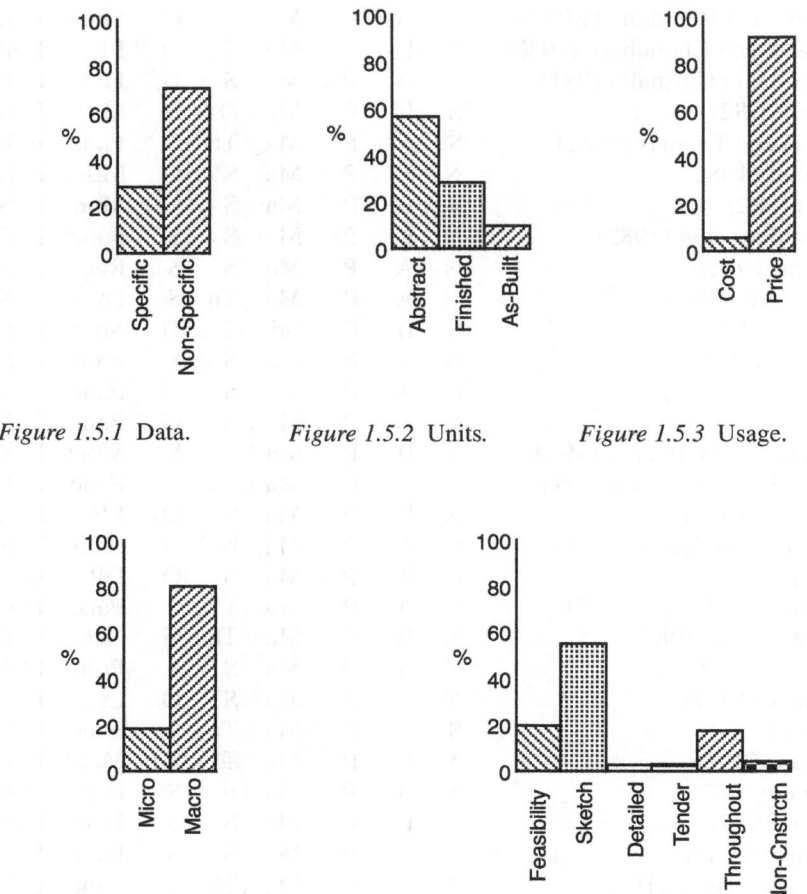

Figure 1.5.1 Data.   *Figure 1.5.2* Units.   *Figure 1.5.3* Usage.

Figure 1.5.4 Approach.   *Figure 1.5.5* Application.

CURRENT TRENDS – A SUGGESTED AGENDA

The relative weightings illustrated in Figures 1.5.1–1.5.9 indicate that the most likely cost model is currently one which applies non-specific data; using abstract units of measurement; prices the project; at the macro level; applied at sketch design; using a simulation model; based on functional dependencies; with implicit assumptions; and a deterministic outcome (N A P Ma S S Func I D). It transpires that actually only one model matches this classification completely.

There is nothing to suggest that the most likely classification is necessarily

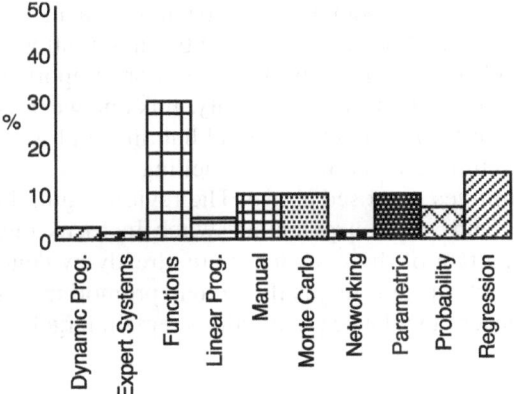

Figure 1.5.7 Technique.

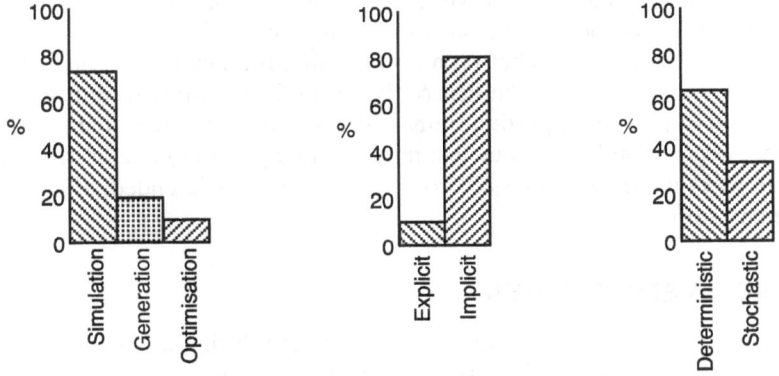

Figure 1.5.6 Model.   *Figure 1.5.8* Assumptions.   *Figure 1.5.9* Uncertainty.

also the preferred approach. Indeed on current trends, the preferred models of the future will use as-built units of measurement, and make considerably more use of stochastic techniques.

It is the authors opinion that the existing focus on non-specific, price models, at the macro level, is well founded. As the cost problem is better understood there will be a natural drift from simulation towards optimisation models, and this will reflect in the choice of formal techniques.

The criteria on which there appears far too little concern, is in making the assumptions explicit. At this stage in the development of cost modelling research it is critical that model builders begin to focus on the various assumptions being made. Focus on the tangible assumptions, such as the function used to link rate to quantity. Perhaps more importantly, focus on the inherent assumptions about model validity, links between cost and other measures of design performance, professional liability, and so on.

The agenda for future research in cost modelling is thus structured around the descriptive primitives proposed above. The review highlights where current emphasis lies, and where a potential switch in focus might be made. Moves to achieve certain of those switches are already evident. Conversely, for others, there would appear little justification in promoting change. There is reason for some concern that basic assumptions remain largely implicit.

CONCLUDING REMARKS

Each of the cost models reviewed in this paper have proven difficult in some way to classify under one primitive rather than another – the primitives undoubtedly fall short of being mutually exclusive. In other ways, a further expansion in the available primitives might be argued, better to describe each model uniquely. For the present, such issues are incidental.

There is often a thin line between a supporting framework and one which overly restricts expression. The descriptive primitives proposed in this paper are intended to act as a supporting framework; to give some order to the way in which we classify and talk about cost models. It may be only one of the many possible ways in which to classify cost models; but it is intended, at least, to set the ball rolling.

ACKNOLWEDGEMENTS

The formative work for this paper was undertaken during a Visiting Senior Research Felowship at the University of Technology, Sydney.

REFERENCES

Atkin, B. L. (1987) A Time/Cost Planning Technique for Early Design Evaluation, in *Building Cost Modelling and Computers*, by P. S. Brandon (ed.), London: E & FN Spon, pp. 145–153.

Barrett, A. C. (1970) Preparing a Cost Plan on the Basis of Outline Proposals, in *Chartered Surveyor*, May 1970, pp. 517–520.

Bathurst, E. P. and Butler, D. A. (1977) *Building Cost Control Techniques and Economics*, Heinemann.

Bennett, J. and Omerod, R. N. (1984) Simulation Applied to Construction Projects, in *Construction Management and Economics*, Vol. 2, pp. 225–263.

Berny, J. and Howes, R. (1987) Project Management Control Using Growth Curve Models Applied to Budgeting, Monitoring and Forecasting Within the Construction Industry, in *Managing Construction Worldwide*, Volume One, Systems for Managing Construction, by P. R. Lansley and P. A. Harlow (eds), London: E & F N Spon, pp. 304–313.

Bowen, P. A. and Edwards, P. J. (1985) Cost Modelling and Price Forecasting: Practice and Theory in Perspective, in *Construction Management and Economics*, No. 3, pp. 199–215.

Bowen, P. A., Wolvaardt, J. S. and Taylor, R. G. (1987) Cost Modelling: A Process-Modelling Approach, in *Building Cost Modelling and Computers*, by P. S. Brandon (Ed), London: E & FN Spon, pp. 387–395.

Braby, R. H. (1975) Costs of High-Rise Buildings, in *The Building Economist*, Vol. 14, No. 2, pp. 84–86.

Brandon, P. S. (1978) A Framework for Cost Exploration and Strategic Cost Planning in Design, in *Building and Quantity Surveying Quarterly*, Vol. 5, No. 4, pp. 60–63.

Brandon, P. S. (ed.) (1982) *Building Cost Techniques: New Directions*, by P.S. Brandon (ed.), London: E & FN Spon.

Brandon, P. S. (1988) Expert Systems for Financial Planning, in *Transactions of the American Association of Cost Engineers*, Morgantown: AACE, pp. D.4.1–D.4.9.

Brown, H. W. (1987) Predicting the Elemental Allocation of Building Costs by Simulation with Special Reference to the Costs of Building Services Elements, in *Building Cost Modelling and Computers*, by P. S. Brandon (ed.) London: E & FN Spon, pp. 397–406.

Buchanan, J. S. (1972) *Cost Models for Estimating: Outline of the Development of a Cost Model for the Reinforced Concrete Frame of a Building*, London: RICS.

Cohen, S. S. (1985) Operational Research, London: Edward Arnold.

Cusack, M. M. (1985) Optimization of Time and Cost, in *Project Management*, Vol. 3, No. 1, pp. 50–4.

Cusack, M. M. (1987) An Integrated Model for the Control of Costs, Duration and Resources on Complex Projects, in *Proceedings of the Fourth International Symposium on Building Economics, Session C: Resource Utilisation*, by D. O. Pedersen and J. Lemessany (eds). Copenhagen: SBI, pp. 56–65.

Department of the Environment (1971) *Cost Consequences of Design Decisions: Cost of Contractors Operations*, London: HMSO.

Dreger, G. T. (1988) Cost Management Models for Design Application, in *Transactions of the American Association of Cost Engineers*, Morgantown: AACE, pp. J.4.1–J.4.6.

Dunican, P. (1960) Structural Steelwork and Reinforced Concrete for Framed Buildings, in *The Chartered Surveyor*, August 1960, pp. 74–77.

Ferry, D. J. and Brandon, P. S. (1980) *Cost Planning of Building*, 5th Edition, London: Granada.

Flanagan, R. and Norman, G. (1978) The Relationship Between Construction Price and Height, in *Building and Quantity Surveying Quarterly*, Vol. 5, No. 4, pp. 69–71.

Gehring, H. and Narula, S. (1986) Project Cost Planning with Qualitative Information, in *Project Management*, Vol. 4, No. 2, pp. 61–65.

Gould, P. R. (1970) *The Development of a Cost Model for H & V and A. C.*

Installations in Buildings, MSc Thesis, Department of Civil Engineering, Loughborough University of Technology.

Gray, C. (1982) *Analysis of the Preliminary Element of Building Production Cost in Building Cost Techniques: New Directions*, by P. S. Brandon (ed.). London: E & FN Spon, pp. 290–306.

Harris, F. and McCaffer, R. (1977) *Modern Construction Management*, London: Granada.

Holes, L. G. and Thomas, R. (1982) General Purpose Cost Modelling, in *Building Cost Techniques: New Directions*, by P. S. Brandon (ed.). London: E & FN Spon, pp. 221–227.

Holes, L. G. (1987) Holistic Resource and Cost Modelling, in *Proceedings of the Fourth International Symposium on Building Economics, Session B: Design Optimisation*, by D. O. Pedersen and J. R. Kelly (eds). Copenhagen: SBI, pp. 194–204.

Katz, M. J. (1986) *Templets and the Explanation of Complex Patterns*, London: Cambridge University Press.

Khosrowshahi, F. (1988) Construction Project Budgeting and forecasting, in *Transactions of the American Association of Cost Engineers*, Morgantown: AACE, pp. C.3.1–C.3.5.

Kiiras, J. (1987) NCCS, Normal Cost control System for Finnish Building Projects, in *Proceedings of the Fourth International Symposium on Building Economics, Session B: Design Optimisation*, by D. O. Pedersen and J. R. Kelly (Eds) Copenhagen: SBI, pp. 79–94.

Kouskoulas, V. and Koehn, E. (1974) Predesign Cost Estimation Function for Buildings, in *American Society of Civil Engineering Proceedings, Journal of the Construction Division*, Vol. 100, No. CO4, pp. 589–604.

Kuhn, T. (1970) *The Structure of Scientific Revolutions*, 2nd Edition, University of Chicago Press.

Lansdown, J. (1982) *Expert Systems: Their Impact on the Construction Industry*, London: RIBA.

Langston, C. A. (1983) Computerised Cost Planning Techniques, in *The Building Economist*, Vol. 21, No. 4, pp. 171–173.

McCaffer, R. (1975) Some Examples of the Use if Regression Analysis as an Estimating Tool, in *The Quantity Surveyor*, December 1975, pp. 81–86.

Mathur, K. (1982) *A Probabilistic Planning Model, in Building Cost Techniques: New Directions*, by P. S. Brandon (ed.). London: E & FN Spon, pp. 181–191.

Meijer, W. J. M. (1987) Cost Modelling of Archetypes, in *Building Cost Modelling and Computers*, by P. S. Brandon (ed.). London: E & FN Spon, pp. 223–231.

Meyrat, R. F. (1969) Algebraic Calculation of Cost Price, in *BUILD International*, November 1969, pp. 27–36.

Moore, G. and Brandon, P. S. (1979) A Cost Model for Reinforced Concrete Frame Design, in *Chartered Quantity Surveyor*, October 1979, pp. 40–44.

Nadel, E. (1967) Parameter Cost Estimates, in *Engineering News-Record* March 16, 1967, pp. 112–123.

Newton, S. (1983) *Analysis of Construction Economics*, PhD Thesis, School of Architecture and Building Science, University of Strethclyde.

Newton, S. (1987) Computers and Cost Modelling: What is the Problem?, in *Building Cost Modelling and Computers*, by P. S. Brandon (ed.). London: E & FN Spon, pp. 41–48.

Newton, S. (1988) Cost Modelling Techniques in Perspective, in *Transactions of the American Association of Cost Engineers*, Morgantown: AACE, pp. B.7.1–B.7.7.

Newton, S. and Logan, B. S. (1988) Causation and its effect: the blackguard in CAD's clothing, in *Design Studies*, Vol. 9, No. 4, pp. 196–201.

Park, R. E. (1988) Parametric Software Cost Estimation with an Adaptable Model, in *Transactions of the American Association of Cost Engineers*, Morgantown: AACE, pp. G.11.1–G.11.7.

Pegg, I. A. (1987) Computerised Approximate Estimating from the BCIS On-Line Data Base, in *Building Cost Modelling and Computers*, by P. S. Brandon (ed.). London: E & FN Spon, pp. 243–249.

Petersen, S. R. (1984) *A Users' Guide to the NBS BLCC Computer Program*, Washington: National Bureau of Standards.

Pitt, T. (1982) The Identification and Use of Spend Units in the Financial Monitoring and Control of Construction Projects, in *Building Cost Techniques: New Directions*, by P. S. Brandon (ed.). London: E & FN Spon, pp. 255–262.

Powell, J. and Chisnall, J. (1981) Getting Early Estimates Right, in *Chartered Quantity Surveyor*, March 1981, pp. 279–281.

Raftery, J.J. (1984) Models in Building Economics: A Conceptual Framework for the Assessment of Performance, in *Proceedings of Third International Symposium on Building Economics, Ottawa: CIB Working Commission W55*, Vol. 3, pp. 103–111.

Raftery, J. J. (1987) The State of Cost/Modelling in the UK Construction Industry: A Multi Criteria Approach, in *Building Cost Modelling and Computers*, by P.S. Brandon (Ed), London: E & FN Spon, pp. 49–71.

Regdon, G. (1972) Pre-Determination of Housing cost, in *BUILD International*, March/April 1972, pp. 94–99.

RIBA (1967) *Handbook of Architectural Practice and Management*, London: RIBA Publications.

Royal Institution of Chartered Surveyors (1964) *The Wilderness Cost of Building Study Group: An Investigation into Building Cost Relationships of the Following Design Variables – Storey Heights, Floor Loadings, Column Spacings, Number of Storeys*, London: RICS.

Russell, A. D. and Choudhary, K. T. (1980) Cost Optimization of Buildings, in *American Society of Civil Engineering, Journal of the Structural Division*, Vol. 106, No. ST1, pp. 238–300.

Schofield, D., Raftery, J. and Wilson, A. (1982) An Economic Model of Means of Escape Provision in Commercial Buildings, in *Building Cost Techniques: New Directions*, by P. S. Brandon (ed.). London: E & FN Spon, pp. 210–220.

Selinger, S. (1988) Computerized Parametric Estimating, in *British-Israeli Seminar on Building Economics*, Haifa: The Building Research Station, pp. 160–167.

Sidwell, A. C. and Wootton, A. H. (1984) Operational Estimating, in *Organizing and Managing Construction, Volume 3: Developing Countries Research*, by V. K. Handa (ed.). Ontario: University of Waterloo, pp. 1015–1020.

Sierra, J. E. E. (1982) A Statistical Analysis of Low Rise Office Accommodation Investment Packages, in *The Building Economist*, Vol. 20, No. 4, pp. 175–178.

Skitmore, M. (1982) A Bidding Model, in *Building Cost Techniques: New Directions*, by P. S. Brandon (ed.).London: E & FN Spon, pp. 278–289.

Skitmore, M. (1988) fundamental Research in Bidding and Estimating, in *British Israeli Seminar on Building Economics*, Haifa: The Building Research Station, pp. 136–160.

Spooner, J. E. (1974) Probabilistic estimating, *American Society of Civil Engineers, Journal of Construction Division*, March 1974, p. 65–77.

Thomsen, C. (1965) How High to Rise, in AIA Journal, April 1965, pp. 66–68.

Townsend, P. R. F. (1978) The Effect of Design Decisions on the Cost of Office Developments, in *Building and Quantity Surveying Quarterly*, Vol. 5, No. 4, pp. 53–56.

Tregenza, P. (1972) Association Between Building Height and Cost, in *The Architects Journal*, Vol. 156, No. 44, pp. 1031–1032.

Walker, D. H. T. (1988) Using Spreadsheets for Simulation Studies, in *The Building Economist*, Vol. 27, No. 3, pp. 14–15.

Weight, D. H. (1987) Patterns Cost Modelling, in *Building Cost Modelling and Computers*, by P. S. Brandon (ed.). London: E & FN Spon, pp. 257–266.

Wilson, A. J. and Templeman, A. B. (1976) An approach to the optimum thermal design of office buildings, in *Building and Environment B*, Vol. 11, No. 1, pp. 39–50.

Wilson, A. J. (1982) Experiments in Probabilistic Cost Modelling, in *Building Cost Techniques: New Directions*, by P. S. Brandon (ed.). London: E & FN Spon, pp. 169–180.

Wilson, A. J. (1987) Building Design Optimisation, in *Proceedings Fourth International Symposium on Building Economics*, Keynotes Volume, Copenhagen: CIB Working Commission W55, pp. 58–71.

Woodhead, W. D., Rahilly, M., Salomonsson, G. D. and Tait, R. (1987) A Integrated Cost Estimating System for House Builders, in *Proceedings of the Fourth International Symposium on Building Economics, Session B: Design Optimisation*, by D. O. Pedersen and J. R. Kelly (eds). Copenhagen: SBI, pp. 142–151.

Woodward, J. F. (1975) *Quantitative Methods in Construction Management and Design*, Macmillan.

Yokoyama, K. and Tomiya, T. (1988) The Integrated Cost Estimating Systems Technique for Building Costs, in *Transactions of the American Association of Cost Engineers*, Morgantown: AACE, pp. B.6.1–B.6.6.

Zahry, M. (1982) *Capital Cost Prediction by Multi-variate Analysis*, MSc Thesis, School of Architecture, University of Strathclyde.

1.6 The relative performance of new and traditional cost models in strategic advice for clients*

C. Fortune and M. Lees

ABSTRACT

Clients of the construction industry need reliable strategic cost advice to enable them to assess a project's viability. The quality of advice depends, in part, on the actual forecasting model used. This study established current practices in the formulation of strategic advice, determined factors that influenced current practices and obtained a practitioner assessment of the relative performance of the differing forecasting models currently in use.

For the results to be of general application it was essential that a large sample was used. Data was collected by postal questionnaire with a sample of 675 organisations based in the north of England and north Wales. Particular attention was given to the development and design of the measuring instrument to ensure a high response rate. The overall response rate achieved was 61%.

The research has allowed conclusions to be drawn on a number of issues which include: practitioners make little use of newer computer based cost models, factors that influenced the incidence in use of the models listed were educational, organisational and operational in nature, a practitioner evaluation of forecasting techniques in terms of accuracy, reliability and value, the role of judgement in the use of the forecasting models and a preliminary assessment model that may be used by researchers seeking to assess the potential value of new cost models in practice.

Key words:

Cost models, incidence in use, accuracy, reliability, judgement, preliminary assessment model.

*1996, *RICS Research Paper Series* Volume Two, Number Two, March, © Royal Institution of Chartered Surveyors.

INTRODUCTION

Aims and objectives

Clients of the construction industry need reliable early stage cost advice to enable them to assess a project's viability as soon as possible. Studies undertaken by Ellis and Turner (1986) and Proctor *et al.* (1993) have shown that clients have been generally dissatisfied with the quality of strategic construction cost advice forecasts provided by their professional advisors. Skitmore (1991) identified the factors affecting the 'quality' of such forecasts as being *'the nature of the target, the information used, the forecasting technique used, the feedback mechanism used and the person providing the forecasts.'* Previous work in this area by Morrison (1984), Skitmore *et al.* (1990), Mak and Raftery (1992) and Proctor *et al.* (1993), has concentrated on issues such as, the nature of the forecast target, assessment of forecast accuracy levels, human expertise and bias. Skitmore *et al.* (1990) acknowledged that the earlier studies have been *'informational'* in nature and both he and Raftery (1991) have called for further empirically based studies to be undertaken on the quality of forecasting models in use. This work seeks to start that process by obtaining a practitioner assessment of the relative performance of the forecasting models used in practice.

Consequently the main objectives of this study were to:

- Establish the incidence in use of the forecasting models currently used by practitioners.
- Evaluate the forecasting models in use in terms of the practitioners' assessment of their accuracy, reliability and value.
- Determine the factors which influence the incidence in use of particular strategic forecasting models.
- Assess the amount of judgement used by practitioners in connection with the forecasting models in use.
- Advance a preliminary assessment model which may be used by researchers in the development and assessment of new cost forecasting models.

The capture of such data would also establish whether the 'paradigm shift' away from single point deterministic forecasting techniques, called for by Brandon in 1982, has now been achieved in practice. The first stage in the work involved a review of literature in order to establish the strategic cost forecasting models potentially available to practitioners as well as the use made of 'judgement' by construction professionals providing such advice to clients.

Literature Review

Strategic cost forecasting models

Brandon (1982) and Bowen and Edwards (1985) signalled some dis-satisfaction amongst academics with the existing 'traditional' methods of pro-viding strategic cost advice to clients and they called for newer cost modelling theories to be developed. Indeed Brandon called for a new paradigm that encouraged a 'shift away from current traditional models' and towards 'a more widespread application of realistic simulation techniques'. Bowen and Edwards indicated that they thought the pressure to consider new models emanated 'almost entirely from the academic pursuit of knowledge'. Bowen and Edwards termed the traditional models used to provide strategic cost advice as being of a '*historical – deterministic nature*'.

Many of the traditional techniques thought to be in common use were reported in standard textbooks on the subject by Seeley (1985) and others. Previous studies by Ashworth and Skitmore (1986), Skitmore and Patchell (1990), Skitmore (1991), Newton (1991) and Raftery (1991) collected all such forecasting models together and set out a taxonomy of traditional as well as newer cost forecasting models. They have listed the following models as being of a 'historical – deterministic nature' (ie. traditional), namely: judge-ment without quantification, the unit method, the functional method, interpolation, the floor area, the principal items method, the cubic method, the storey enclosure method, BQ pricing, approximate quantities, the elemental method and the significant items method.

A survey of traditional cost forecasting models in use carried out by Fortune and Lees (1989) revealed that less than 2% of its respondents made use of the cubic and storey enclosure methods to provide strategic cost advice to clients. These results confirmed Ashworth's (1988) suspicions that the cubic and storey enclosure methods were not in general use. So it was decided to omit the cubic and the storey enclosure methods from this survey of traditional models in use and to include the following with their definitions:

Traditional models

Judgement: the use of experience and intuition without quantification
Functional unit method: a single rate applied to the amount of provision (eg. £/pupil/school)
Cost per M2 method: a single rate applied to the gross floor area
Principal item method: a single rate applied to the principal item of work
Interpolation method: the interpolation of the total cost from previous projects
Elemental analysis method: the use of element unit rates to build up an estimate
Significant items method: the measurement of significant items of work which are individually priced

Approximate quantities method: the measurement of a small number of grouped items priced in groups
Detailed quantities method: the use of priced detailed quantities (eg BQ)

Over fifty potential cost models were identified by Newton (1991) in his encyclopedic work that sought to set the agenda for future work on cost modelling. This work was given a construction focus by Skitmore (1991) who listed the following as potential models that were in use: regression analysis, BCIS on line and ELSIE expert systems. A model advanced by Bennet and Ormerod (1984) called 'construction project simulator' was also quoted as being potentially in use. Raftery (1991) corroborated the existence of all the previous newer models and added monte carlo simulation, causal methods, time series models as well as resource and process based models. In addition Raftery indicated that there were models that took account of the total costs in use of a building over its useful or economic life. Flanagan *et al.* (1989) had considered these methods and identified the pay back method, the discounted cash flow method and the investment appraisal technique (NPV) as being the methods thought to be currently in use. It was decided to use the classification framework suggested by Raftery (1991) as the structure within which the following newer models, and their definitions, could be classified and included in this survey:

Statistical Models

Regression analysis: cost models derived from statistical analysis of variables
Time series models: based on statistical analysis of trends
Causal cost models: models based on algebraic expression of physical dimensions

Knowledge Based Systems

ELSIE: as marketed by Imaginor Systems
Life Cycle Costing Techniques
Net present value: discounting future costs to present values
Payback method: calculating the period over which an investment pays for itself
Discounted cash flow: calculating the discounted rate of return

Resource/Process Based Methods

Resource based: the use of schedules of materials plant and labour
Process based: the use of bar charts and networks
Construction cost simulator: A proprietary system

Risk Analysis Techniques

Monte carlo simulation: the aggregation of probabilities for parts of an estimate to give a probability for the estimate as a whole

Judgement and strategic cost modelling

The need for construction professionals to exercise 'judgement' when providing cost forecasts to clients has been identified by Mak and Raftery (1992), Al-Tabtabai and Diekmann (1992), and Skitmore et al. (1990). Al-Tabtabai and Diekmann suggested that 'a good price forecasting technique needed to include both historical trend based data and competent "judgements" based on construction experience and knowledge'. Skitmore et al. (1990), in their work on accuracy of construction price forecasts, indicated that the exercise of judgement was the responsibility of the expert construction professional involved in making the strategic cost forecast rather than the model itself. This view was supported by Raftery (1992) when he asserted that a reliable strategic cost forecast required the input of human judgement, but he also warned of 'errors and bias' being introduced into the forecast due to the 'cognitive processes of human beings'. Tversky and Kahnemann (1974) in their work on judgement under uncertainty pointed out that such errors or biases could be due to differences in the attitude to project risks adopted by the construction professionals involved in formulating the strategic cost forecast. The reasons for such errors and biases were set out by Woods (1966) and Mak and Raftery (1992) when they suggested that in most organisations a 'knowledge gap' existed between the forecast compiler and the forecast transmitter. This 'knowledge gap' caused errors and biases to be introduced into forecasted prices which could lead to inaccuracies and client dissatisfaction. Therefore, if this study could identify the amount of judgement that each of the previously identified cost forecasting models required then it could be possible to indicate which of them had the greatest potential for such a 'knowledge gap' to exist.

Having identified the potential models that were suspected as being in current use and the existence of error or bias due to the judgemental decisions taken by construction professionals utilising such models it was now necessary to determine a research methodology that would collect data that would after analysis, allow the aims of the study to be addressed.

RESEARCH METHODOLOGY

Data collection

Dixon et al. (1987) highlighted the need to determine an appropriate method of collecting information. The methods they identified were, 'from documentary sources, by direct observation, by mailed questionnaires and by interviewing'. It was decided that the most appropriate method of collecting data for this study would be to use the 'mailed questionnaire survey'. This was because the main aims of the work were to collect quantitative data on cost forecasting models in actual use as well as the collection of practitioner's opinions on the value of the models that they currently used. Another consideration was that if the results of the survey were to be taken as indicative of the profession as a whole then the sample size needed to be large. This proposed method of data collection had been used by Fortune and Lees (1989) in a previous study and had facilitated the efficient and economic collection of information from 384 quantity surveying organisations in the north of England. The earlier survey had achieved a return rate of 60% and it was hoped that this survey would achieve a similarly high return rate by giving attention to the factors identified as favouring a high response rate, namely, questionnaire design, extensive piloting, effective issue and thorough follow up procedures (Scott 1961). If a high response rate could be achieved then it would allow the results of the survey to be generalised and also mitigate the disadvantages involved with this method of collecting data, namely, uncertain respondent quality, the potential for misinterpretation and the inability to supplement respondents answers with observational or additional data.

Questionnaire design

The design of the questionnaire form itself was given particular attention. Scott (1961) stated that, *'a high response rate to mailed surveys was achieved by keeping the questionnaire brief and the amount of work needed to be done by the respondent to a minimum'*. Therefore it was agreed that the length of the questionnaire would have to be limited to permit completion within a fifteen to twenty minute time period. A selection of the literature on questionnaire design was reviewed to achieve this aim.

Scott (1961) identified the following factors as being instrumental in designing an effective mailed questionnaire, namely:

> *The design of the questionnaire form must raise the interest of the respondent, the type and style of the questions asked within the questionnaire, the use of effective follow up procedures so as to convince the recipient that his/her response is really needed, a cover letter which indicates official sponsorship which is attached to the back of the questionnaire form itself, the inclusion of a stamped addressed envelope for the return of the completed questionnaire.*
> *Scott (1961)*

Moser and Kalton (1971) indicated that the key factors in the successful use of self administered postal questionnaires were the actual length and style of the questionnaire form itself. The number of potential strategic cost forecasting models were limited to those identified in the literature search above and rationalised by reference to the previous survey work on models currently in use. Attention was also given to the forma, phrasing and arrangement of the actual questions within the questionnaire form in order to minimise any potential misunderstandings.

The aims of the survey indicated that both quantitative data on incidence of model use and qualitative data on reliability, judgement and value had to be captured by the questionnaire. Kerlinger (1969) in his work on questionnaire design indicated that the types of questions included in the survey to capture such data could be described as being either fixed alternative, open ended or scaled item in form. It was decided that the scaled item type of question would be the format which would facilitate an optimum response rate to questions which called upon subjects to evaluate between differing strategic cost advice models in terms of reliability, judgement and value. In addition the fixed alternative type of question as also included in the questionnaire to allow subjects to make responses to factual questions such as, type of organisation, computing facility, incidence of technique use, understanding and capacility, which could then be statistically analysed. Kerlinger went on to indicate that there were three types of response scales which could be used within the questionnaire and these were the 'agreement-disagreement, rank order and forced choice' types of response. It was suspected that the 'likert like' type of response scale stretching between 1–5 would assist the subjects responding to the survey. It was resolved to address both of these matters within the piloting exercises associated with the study.

The style and layout of the questionnaire form itself were given particular attention and a number of preliminary designs were investigated within the study's pre-test and piloting exercises. Such exercises were highlighted by Panten (1950) who emphasised the importance of 'piloting' as a process that could provide guidance on sampling frame, suitability of the method of collecting data, adequacy of the questionnaire form itself, the length of time taken for the form to be completed and the non-return rate to be expected. The necessity of adopting rigorous 'piloting' procedures was emphasised by Allen and skinner (1991) who stated that such pilot studies should 'test the applicability of the proposed measuring instrument', before data was collected form the main sample.

Pre-test and piloting

The pre test was carried out with four work colleagues who were each chartered quantity surveyors with some experience of providing strategic cost forecasts for clients. As a result of this pre-test it was discovered that the constraint of using a single side of A4 paper, which was the format used in a previous survey on this topic, was not now satisfactory as its brevity caused confusion and misinterpretation amongst the pre-test subjects. As a consequence the pre-test subjects took longer than was anticipated to complete the draft questionnaire form as they had to ask several explanatory questions. Therefore, the draft questionnaire form was expanded to cover four sides of A4 paper and revised to include precise instructions, definitions and explanatory notes to aid respondents using the form.

The revised draft questionnaire form was then used in an initial pilot study with a group of thirty part-time degree students at the final level of their degree programme in quantity surveying. This initial pilot study resulted in modifications being made to the questionnaire's style and it suggested the introduction of an early exit point for those respondents who did not offer an early cost advice service to clients. The length of time taken to respond to the form was between fifteen and twenty-five minutes and this appeared to be a reasonable timescale given the type of subject and the fact that the questionnaire form itself was in its developmental stage. It was felt that a further pilot study should be undertaken with a revised questionnaire form and a subject group drawn from the likely sample for the survey.

A revised pilot questionnaire form together with an explanatory letter was sent by mail to a sample of ten quantity surveying organisations. The subjects for this pilot study were selected to reasonably reflect the mix and proportions of organisational types anticipated as being within the sample frame for the main study. This second pilot study achieved a 70% return rate within two weeks of the questionnaire forms being issued. Each respondent returned a feedback questionnaire on the revised pilot form itself and this indicated that the form could be completed within a fifteen to twenty minute period. The only amendment made to the main study questionnaire form as a result of this second piloting exercise was the addition of a separate group of models called 'value related techniques'. This caused the authors to reconsider the models listed and it was decided to include as a standard 'any other technique' within each of the newly emerging knowledge-based and risk analysis groups of forecasting models. These amendments brought the total number of models included within the main survey to twenty four. The speed of response and high return rates of the questionnaires used in this second pilot study con firmed that this self-administered postal questionnaire was capable of generating data that could be analysed. Therefore, it was resolved to amend the revised questionnaire form as indicated above and proceed with the main study. A copy of the final questionnaire form used in the survey can be fond in Appendix A.

Issue and follow up procedures

A sample was established that comprised all the separate organisations listed by the RCIS (Quantity surveying division) as operating in the north of England (i.e. Northumberland, Cumbria, Yorkshire, Humberside, Lancashire, Greater Manchester, Merseyside, Cheshire, Staffordshire, Derbyshire, Nottinghamshire) and North Wales. The number of contracting organisations included within the sample frame was expanded by including all employing organisations who were sponsoring part-time quantity surveying degree students at the authors respective higher education institutions. A final total of 675 separate organisations were included as subjects within the sample frame. For the purposes of this study an organisation was defined as 'the part of a company/practice (eg. branch, office, department) for which the respondent had responsibility'. This definition was included within the 'notes for guidance' section of the mailed questionnaire form.

The follow up procedures advocated by Scott (1961) and Moser and Kalton (1971) were adopted in order to maximise the potential return rates. In particular it was resolved to allocate a unique reference number on the stamped and addressed return envelope provided for each organisation included within the sample frame. In so doing it was possible to identify the non-responding organisations so that they could be contacted again should any necessary follow up procedures prove necessary. Although 60% of the responses to this initial postal survey were returned within two weeks of the initial mailing (confirming suggestions on speed of survey response rates made by Kerlinger and Scott), the overall response rate to this initial mailing was only 36%. Moser and Kalton indicated that the results of a postal survey could be considered as biased and of little value if the response rate was lower than 30% – 40%. Therefore it was resolved to activate a follow up procedure to increase the rate of return.

Each non-responding organisation was contacted again four weeks after the initial mailing with a reminder note and a further copy of the survey form. The second mailing increased the final total response rate to 61%. This overall response rate can be considered to be high compared to other recently reported mailed questionnaire survey response rates that ranged between 24% and 28% (O'Brien and Al-Soufi 1994).

RESULTS – SUMMARY OF RESPONSES TO THE QUESTIONNAIRE

Generally

The mailed questionnaire (Appendix A) used in the main survey had a total of five separate questions and in the 'notes for guidance' section subjects were asked to respond to each question by circling the most appropriate option listed. In addition, definitions of 'organisation' (see above) and 'early cost advice' were also included on the form to avoid misunderstanding. Strategic cost advice, for the purposes of this study, was defined as 'any cost advice given to the client prior to a formal offer to contract being made'. A total of 374 (61%) organisations responded to the survey. The results for each of the questions listed on the form is given below.

Table 1.6.1 Distribution of types of organisation in the sample

Type of Organisation	No	%
Quantity Surveying practice	234	62.6
Multi-disciplinary practice	27	7.2
Architectural practice	2	0.5
Consulting Engineering practice	0	0.0
Local or Public Authority	38	10.2
General Contracting	2	0.5
Management Contracting	2	0.5
Design and Build Contracting	6	1.6
Specialist Contracting	4	1.1
Project Management	20	5.3
Civil Engineering	6	1.6
None of these	7	1.9
Total	374	100.0

Type of organisation (question 1)

Each respondent was asked to classify their appropriate category. The responses to this organisation under one of the twelve categories listed. This question ensured that the sample of respondents captured in the survey was not biased as each respondent determined their own most appropriate category. The responses to this question would also enable 'organisational' trends and biases to be identified in the subsequent statistical analysis of the data generated by the survey. The distribution of the responses to this question can be seen in Table 1.6.1. The most popular types of organisation included in the survey were: quantity surveying practices (62.5%), local or public authorities (10.2%), and general contractors (7.5%). The results indicated that the categories of organisation type suggested on the questionnaire form were appropriate for the sample frame surveyed. It can be seen that only seven of the respondents (1.9%) were unable to identify themselves under any of the fixed alternatives and only one category (consulting engineers) failed to attract any respondents.

Computing facility (question 2)

This question asked respondents to identify the computer facilities that were available to staff involved in the preparation of strategic cost advice for clients. A simple question asking respondents to indicate whether computers were available to them in the workplace was discounted as this would have collected only quantitative data whereas it was thought that the quality of provision was more likely to be a factor determining the level of individual model use.

Table 1.6.2 Computing facility of organisations in the sample

Computer Facility	No in sample
1 staff to 1 workstation	76
2 staff to 1 workstation	86
3 staff to 1 workstation	69
4 staff to 1 workstation	93
No facility	48
Did not respond to question	2
Total	374

Furthermore, the responses to this question would also enable trends and biases in terms of 'computing facility' to be identified in the subsequent statistical analysis of the data. Consequently, the question that was asked required the respondents to assess the ratio of work stations to staff within their organisation. Responses to this question established the degree of access to computer facilities an individual staff member would have. The results are set out in Table 1.6.2 and as can be seen some 49% of respondents had a 2:1 or better ratio of workstations to staff and only 12% of respondents indicated that they had no computer facilities at all. However a further 24% of respondents had a ratio of staff to workstation of more than 3:1 which made more marginal the actual availability of computing facilities to individual staff members concerned.

Strategic advice provision (question 3)

This question asked respondents to state whether their organisation was in fact involved in preparing strategic cost advice for clients. If an organisation was unable to respond positively to this question then it was advised to answer no more of the questions listed and return the part completed questionnaire to the sender. This question was included following the suggestions made by the subjects involved in pilot study two who stressed that it would be of benefit to respondents to have an early opportunity of declining to complete the survey. The question was placed after questions on organisation type and computing facility as it was anticipated that the responses to these questions by those organisations that did not actually provide strategic cost advice to clients would still be of value to the survey. The results of this question revealed that 91.7% of the respondents to the survey did actually provide strategic cost advice to clients. This result indicated the appropriateness of the sample frame and reinforced the relevance of the responses received.

Size and work distribution (question 4)

This question was divided into two parts. Part A asked respondents to enumerate the number of staff actively involved in the early advice function. Part B asked respondents to assess the proportion of the strategic advice function that was undertaken by staff at trainee, surveyor, senior surveyor or partner/director level. The information gathered in part A would allow the relationship between size of operation and use of forecasting models to be analysed. The question was designed in this way as some organisations, particularly contractors, may have large numbers of quantity surveyors working for them but may only have a small number of those surveyors involved in the giving of advice. The information gathered in part B reveals the level or grade of staff involved in the production of strategic sot advice for clients to be established. The results were summarised in Table 1.6.3. As can be seen some 74% of respondents indicated that they had between one and five members of staff actively involved with the strategic advice function whereas only 5% indicated that they had more than fifteen members of staff involved. The results of the question in part B showed that the strategic advice function was more likely to be undertaken by staff employed at the partner/director level (59%) rather than staff employed at the trainee/surveyor grades (9%), Table 1.6.4.

Table 1.6.3 Size of organisation in the sample

Size of Organisation No. of staff involved with strategic advice	No in sample
0–5	225
6–10	52
11–15	18
15+	11
Did not respond to question	68
Total	374

Table 1.6.4 Work distribution

Level of staff	Mean % of total function	Minimum value	Maximum value	Standard deviation
Partner/Director	59.32	0.00	100.00	34.69
Senior Surveyor	31.37	0.00	100.00	29.14
Surveyor	7.69	0.00	100.00	15.76
Trainee	0.92	0.00	50.00	4.01

Strategic advice models (question 5)

Generally

This question asked whether the respondents had ever used each of the models listed on the form (question 5a). If respondents answered negatively to any of the models listed then they were asked to indicate, firstly, whether they understood how the model worked (question 5f) and secondly, whether their organisation had the capability to perform the model (question 5g). These questions were included on the form in an attempt to identify the factors that were preventing the individual model from being used. If the respondents did actually use any of the individual models listed in the questionnaire then they were asked to score each of them between 1 (low) and four (high) for their reliability and accuracy (question 5b), use of judgement (question 5c), value as a tool (question 5d) and current use of the model (question 5e). The responses to these questions would generate data that would allow the study to determine the use and practitioner assessment of individual strategic cost advice models. Tables 1.6.5 and 1.6.6 and Figs 1.6.1 and 1.6.6 report and illustrate the overall results to questions 5a to 5f.

Incidence of Use (Question 5a)

Traditional Models. The traditional group of forecasting models had the highest incidence of use. The most popular of the traditional models were the cost per m^2 (97%) and approximate quantities (96%). The least well used models were the principal item (41%) and detailed quantities (69%). The use of the principal item method is thought to be particularly relevant to engineering projects that centre around a single item of work. The same frame illustrated above did not include many engineering organisations and this could explain the low incidence of use.

Statistical Methods. The results indicated that statistical methods have a low incidence of use amongst practitioners. The lowest incidence of use was recorded by 'causal techniques' which 93% of respondents had not ever used.

Knowledge Based Systems. A total of 53 organisations (15%) indicated that they used ELSIE when preparing strategic cost advice for clients. This knowledge based system seems to be the most well known and used but the survey did reveal that 41 organizations (13%) are making use of one or more other knowledge based systems. This result indicated that the questionnaire form was adequately designed in that it identified the most significant system in use.

Life Cycle Costing. The survey indicated that the differing models used to provide clients with early life cycle cost information were used evenly by the

Table 1.6.5 Summary of responses to questions 5a to 5c

Technique	5a Has your organisation ever used the technique		5b Mean value rating for reliability and accuracy	5c Mean value rating for use of judgement	5d Mean value rating for valuable tool	5e Currently use the technique	
	Yes						
	No	% of Respondents				No	% of users
Traditional							
Judgement	291	85.8	2.30	3.67	2.61	259	90.6
Functional unit	253	75.5	2.29	2.83	2.51	216	86.7
Cost per m²	319	97.3	2.73	3.10	3.10	319	98.2
Principal item	156	47.3	2.28	2.97	2.27	123	80.4
Interpolation	294	87.0	2.74	3.11	2.80	274	94.2
Elemental analysis	299	88.7	3.15	2.79	3.15	265	89.5
Significant items	243	73.0	2.98	2.85	2.99	206	86.2
Approximate quantities	325	96.2	3.45	2.70	3.48	311	97.8
Detailed quantities	228	68.7	3.71	2.28	3.40	210	93.8
Statistical							
Regression analysis	36	10.6	2.53	3.14	2.69	23	65.7
Time Series Models	49	14.5	2.49	2.94	2.50	37	77.1
Causal Cost Models	25	7.5	2.40	3.04	2.46	16	66.7
Knowledge Based							
ELSIE	53	15.7	2.82	3.04	2.92	45	88.2
Other knowledge based	41	12.7	3.03	2.77	2.98	37	90.2
Life Cycle Costing							
Net Present Value	136	40.5	2.52	2.92	2.51	92	69.2
Payback method	125	37.2	2.67	3.04	2.64	96	78.0
Discounted Cash-flow	111	32.9	2.65	2.81	2.70	85	76.6
Resource/Process							
Resource based	171	50.4	2.89	2.78	2.66	115	70.1
Process based	107	31.8	2.61	3.06	2.63	83	79.0
Construction Cost Sim	18	5.4	2.81	3.13	13.06	14	82.4
Risk Analysis							
Monte Carlo Simulation	19	5.6	2.61	3.33	2.50	15	83.3
Other risk analysis	57	17.1	2.71	3.29	2.60	43	87.8
Value Related							
Value Management	78	23.1	2.79	3.25	2.88	61	81.3
Other value related	28	8.6	2.92	3.25	3.13	23	85.2

Table 1.6.6 Summery of responses to questions 5f and 5g

Technique	5f Do you understand how the technique works Yes		5g Does your organisation have the capability to perform the technique? Yes	
	No	% of non users	No	% of non users
Traditional				
Judgement	42	93.3	42	95.5
Functional unit	73	96.1	63	81.8
Cost per m²	8	100.0	8	100.0
Principal item	107	67.3	113	73.9
Interpolation	37	90.2	34	85.0
Elemental analysis	35	97.2	32	88.9
Significant items	70	92.1	66	89.2
Approximate quantities	11	100.0	11	100.0
Detailed quantities	88	97.8	84	96.6
Statistical				
Regression analysis	90	31.9	68	26.3
Time Series Models	81	30.8	62	24.9
Causal Cost Models	54	19.0	51	19.5
Knowledge Based				
ELSIE	130	49.4	49	19.9
Other knowledge based	57	25.0	38	17.5
Life Cycle Costing				
Net Present Value	131	70.4	100	55.2
Payback method	136	70.5	108	58.7
Discounted Cash-flow	128	61.5	102	51.3
Resource/Process				
Resource based	93	58.5	66	44.6
Process based	106	49.1	72	35.8
Construction Cost Sim	66	22.8	33	12.2
Risk Analysis				
Monte Carlo Simulation	45	15.3	41	14.9
Other risk analysis	39	15.7	29	12.6
Value Related				
Value Management	86	36.1	61	27.5
Other value related	46	18.0	42	17.5

practitioners responding to the survey. As a group of models this group had the second highest incidence of use.

Resource/Process Based. The relatively high incidence of use of resource based models indicated that practitioners do use basic prices of labour, materials and plant rather than always basing their strategic advice to clients solely on tender price data. The lower incidence of use of process based models may be explained by the lack of production programming information at the early stage of a project. The construction cost simulator technique recorded the lowest incidence of use (5%) of all the models listed in the survey.

Risk Analysis.   The results of the survey indicated that risk analysis had a low incidence of use amongst practitioners. The lowest incidence of use was recorded by Monte Carlo simulation which 94% of respondents did **not** use. The results of the survey revealed that 17% of respondents are making use of one or more other risk analysis techniques. This result indicated that the questionnaire form was flawed in that the identified risk model was not the most significant in use.

Value Related Models.   The use of value management models (23%) by the respondents indicated that this was the most used value related model. The results of the survey indicated that there was one or more other value related models (9%) currently in use.

Reliability and accuracy (question 5b)

Traditional Models   The individual models listed as traditional were scored in a range between 2.28 and 3.71. This range included models that were scored as the highest (detailed quantities – 3.71, approximate quantities – 3.45, elemental analysis – 3.15) and lowest (judgement – 2.30, functional unit – 2.29, principal item 2.28) of all the models listed in the survey for reliability and accuracy.

Statistical Models   The individual models included as statistical achieved the lowest mean value rating of all the groups of models within the survey. Causal cost models (2.40) was the lowest and regression analysis (2.53) was the highest of the individual models included within this group.

Knowledge Based   The models listed within this group achieved the highest collective mean value rating scores of all the groups of models listed in the survey. The highest mean value score was achieved by 'other knowledge based techniques' (3.03).

Life Cycle Costing   The models listed within this group were scored below the collective mean scores for value ratings of models for reliability and accuracy. Net Present Value (2.52) was the lowest scored and Discounted Cash Flow (2.65) was the highest scored of the models listed within this group.

Resource/Process Based This group of models achieved the mean of the collective value rating scores of the respondents to the survey. Resource based (2.89) and Process based (2.61) were the individual models that were scored as the highest and lowest in this group.

Risk Analysis This group of models included Monte Carlo Simulation and 'Others'. The highest scored model was 'Others' (2.73).

Value Related This group of models were collectively scored highly by the respondents for reliability and accuracy. The models listed as 'Other' were scored the highest (2.92).

Use of judgement (question 5c)

Traditional Models The individual models listed as traditional were scored in a range between 2.28 and 3.67. This range included techniques that were scored as the highest (judgement – 3.67) and lowest (detailed quantities – 2.28, approximate quantities – 2.70 and elemental analysis – 2.79) of all the models scored in the survey for the use of judgement.

Statistical Models The individual models included as statistical achieved the mean collective value rating of all the groups of models within the survey. Time Series Models (2.94) was the lowest and Regression Analysis (3.14) was the highest scored of the individual techniques included within this group for use of judgement.

Knowledge Based The models listed within this group achieved the lowest collective mean value rating scores of all the groups of models listed in the survey. The lowest mean value score was achieved by 'other knowledge based techniques' (2.77).

Life Cycle Costing The models listed within this group were scored below the collective mean scores for value ratings of techniques for the use of judgement. Discounted Cash Flow (2.81) was the lowest scored and the Payback Method (3.04) was the highest scored of the models listed within this group.

Resource/Process Based This group of models were collectively scored above the mean of the collective value rating scores of the respondents to the survey. Resource based (2.78) and Construction Cost Simulator (3.13) were the individual models that were scored as the lowest and highest in this group.

Risk Analysis This group of models included Monte Carlo Simulation and

'Others' and they were collectively scored as the group of models that called for the highest amounts of judgement in use. The highest scored model in this group was Monte Carlo Simulation (3.33).

Value Related   This group of models were collectively scored highly by the respondents for the use of judgement. Both Value Management and Other Value Related techniques were scored equally at 3.25 by respondents to the survey.

Value as a tool (question 5d)

Traditional Models   The individual models listed as traditional were scored in a range between 2.27 and 3.48. This range included models that were scored as the highest (Approximate Quantities – 3.48 and Detailed Quantities – 3.40) and lowest (Principal Item – 2.27) of all the models scored in the survey for value as a tool.

Statistical Models   The individual models included as statistical achieved the joint lowest mean value rating of all the groups of models within the survey. Causal Cost Models (2.46) was the lowest and Regression Analysis (2.69) was the highest scored of the individual models included within this group for value as a tool.

Knowledge Based   The models listed within this group achieved a high collective mean value rating score when compared to other groups of models listed in the survey. The highest mean value score was achieved by 'other knowledge based techniques' (2.98).

Life Cycle Costing   The models listed within this group were scored below the collective mean scores for models of value as tools. Discounted Cash Flow (2.70) was the highest scored and Net Present Value (2.51) was the lowest scored models listed within this group.

Resource/Process Based   This group of models were collectively scored as the mean of the collective value rating scores of the respondents to the survey. Process based (2.63) and Construction Cost Simulator (3.06) were the individual models that were scored as the lowest and highest in this group.

Risk Analysis   The individual models included in this group achieved the joint lowest collective rating of all the groups of models within the survey for value as a tool. The highest scored model was 'Others' (2.60).

Value Related   This group of models were collectively scored as the highest by the respondents for value as tools. The models listed as 'Other' were scored highest (3.13).

Retention in use (question 5e)

Traditional Models   The individual models included as traditional achieved the highest collective retention percentage of all the groups of models listed within the survey. The individual traditional models were scored in a range between 80.4% and 98.2%. This range included models that were scored as the highest (Cost per m^2 – 98.2%, Approximate Quantities – 97.8% Interpolation – 94.2% and Detailed Quantities – 93.8%) of all the models scored in the survey as being retained in use by the respondents. The traditional model that achieved the lowest retention percentage was the principal item method (80.4%).

Statistical Models   The individual models included as statistical achieved the lowest mean retention percentage of all the groups of models listed within the survey. Time Series Models (77.1%) was the highest scored model within this group for retention in use. Regression Analysis (65.7%) was the lowest scored of the individual model included within the survey for retention in use.

Knowledge Based   The models listed within this group achieved a high collective mean percentage retention when compared to other groups of models listed in the survey. The highest retention percentage was achieved by 'other knowledge based techniques' (90.2%).

Life Cycle Costing   The models listed within this group were scored below the collective mean retention percentages for models listed in the survey. Payback Method (78.0%) was the highest scored and Net Present Value (969.2%) was the lowest scored of the models listed within this group.

Resource/Process Based   This group of models were collectively scored below the mean of the collective retention percentages of respondents to the survey. Resource based (70.1%) and Construction Cost Simulator (82.4%) were the individual models that were scored as having the lowest and highest retention percentage in this group.

Risk Analysis   The individual models included in this group achieved a collective retention percentage above the mean of all the groups of models within the survey. The model with the highest retention percentage was 'Others' (87.8%).

Value Related   The individual models included in this group achieved a collective retention percentage above the mean of all the groups of models within the survey. The models listed as 'Other' (85.2%) achieved the highest retention percentage.

Questions 5f and 5g asked for responses from those subjects that had not ever used the strategic cost advice models listed in question 5a.

Understanding of models (question 5f)

Traditional Models The individual models included as traditional were generally understood by those respondents that had not ever made any use of them. The individual traditional models were scored for understanding in a range between 67.3% and 100.0%. This range included models that were scored as the highest (Cost per m^2 – 100.0%, Approximate Quantities – 100.0% and Detailed Quantities – 97.8%) of all the models scored in the survey as not used yet still understood. The traditional model that recorded the lowest understanding percentage was the principal item method (67.3%).

Statistical Models The individual models included as statistical were generally not understood by those respondents that had not ever made any use of them. Causal Cost models (19.0%) was the least well understood and Regression Analysis (31.9%) was the most well understood of the statistical models not used by the respondents to the survey.

Knowledge Based 49.4% of respondents who had not used ELSIE indicated that they understood how the technique worked. However, only 25.0% of respondents had an understanding of any other Knowledge Based System.

Life Cycle Costing The individual models included as life cycle costing methods were generally understood by those respondents that had not ever made any use of them. The individual life cycle costs models were scored for understanding in a range between 61.5% and 70.5%. The most well understood and least well understood models were the Payback Method (70.5%) and the Discounted Cash Flow Method (61.5).

Resource/Process Based 58.5% and 49.1% of respondents who did not use resource and process based models understood how the techniques worked. However only 22.8% of respondents that had not used the construction cost simulator understood how it worked.

Risk Analysis The individual models included as risk analysis methods were generally not understood by those respondents that had not made any use of them. Monte Carlo (15.3%) and other risk analysis (15.7%) recorded the lowest levels of understanding of all the models listed as not being used by respondents to the survey.

Value Related   The individual models included as value related were generally not understood by those respondents that had not made any use of them. Other value related (18.0%) was the least well understood and value management (36.1%) was the most well understood of the value related models not used by the respondents to the survey.

Capability of performing models (question 5g)

Traditional Models   The individual models included as traditional were found generally to be capable of being performed by those respondents that had not made any use of them. The individual traditional models were scored for capability in a range between 73.9% and 100.0%. This range included models that were scored as the highest (Cost per m^2 – 100.0%, Approximate Quantities – 100.0%, Detailed Quantities – 96.6% and Judgement 95.5%) of all the models scored in the survey as not used yet still being capable of being performed. The traditional model that recorded the lowest capability percentage was the principal item method (73.9%).

Statistical Models   The individual models included as statistical were generally found to be not capable of being performed by those respondents that had not made any use of them. Causal Cost models (19.5%) was the lowest and Regression Analysis (26.3%) was the highest scored statistical models for capability of being performed.

Knowledge Based   The individual models included as knowledge based were generally found to be not capable of being performed by those respondents that had not made any use of them. ELSIE (19.9%) and other knowledge based models (17.5%) were the highest and lowest of the models rated by the respondents for capability.

Life Cycle Costing   The individual models included as life cycle costing methods were generally capable of being performed by those respondents that had not made any use of them. The individual life cycle costs models were scored for capability in a range between 51.3% and 58.7%.

Resource/Process Based   44.6% and 35.8% of respondents who did not use resource and process based models still had the capability to perform them. However only 12.2% of respondents that had not used the construction cost simulator had the capability of performing that model.

Risk Analysis   The individual models included as risk analysis methods were generally not capable of being performed by those respondents that had not made any use of them. Monte Carlo (14.9%) and other risk analysis

(12.6%) recorded the lowest collective levels of capability of all the models listed as not being used by respondents to the survey.

Value Related   The individuals models included as value related were generally not capable of being performed by those respondents that had not made any use of them. Other value related (17.5%) was the least capable and value management (27.5%) was the most capable of being performed by the respondents to the survey.

ANALYSIS AND DISCUSSION

Generally

The results set out in the previous chapter showed that the incidence in use of several of the listed models was below 50%. Some of the newer models have been used by less than 20% of organisations included in the survey. In general, the newer models were not in use. The results generated by this survey suggest that the 'paradigm shift' called for by Brandon (1982) has not been reflected in practice.

The results provided information which allowed the analysis of the relationships between a number of factors and the incidence in use of a particular technique. Other relationships were explored through analysis and these have also been presented with the analysis of factors influencing the incidence in use. This analysis and associated discussion is organised under the headings:

- educational issues
- organisational issues
- operational issues

These three areas, which cover issues such as knowledge and understanding of models, capability to operate, and use in practice, are important to the successful introduction and use of a new technique. This is because a new technique must firstly be understood before the question of its use can be considered. Only with knowledge can an organisation take a view whether it has the capability to use the technique in practice. Finally, when the technique has been used the operational issues of value, reliability and judgement will influence its continued use. This arrangement of the factors which may influence the use of models – moving from knowledge through resources to practical use – is a hierarchy of factors. All the factors identified are important but the later ones cannot influence use until the earlier factors have been encountered and resolved.

Educational issues

The incidence in use of the models has been summarised in Figure 1.6.1. The 50% line has been shown on the diagram as an indicator of the point at which incidence in use moves from the 'majority' to the 'minority' position. Of the nine traditional models only one (principal item) has an incidence in use less than 50%. There are fifteen new models. Only one (resource based) reaches the 50% mark, all the others fall below it.

In the hierarchy of factors knowledge comes first. All models or practices which have been developed through research have to be popularised and the results disseminated to those who would benefit form their implementation. The question is – has there been a failure in this area with regard to some of the newer models contained in the survey? The questionnaire included a question which was designed to gather information on this issue. The question asked those who were in a position of managerial responsibility for the strategic cost advice function to indicate whether they understood how a technique worked. The respondents were only required to answer the question if they had identified themselves as a non-user of a technique. Analysis of the results of incidence in use show the extent to which lack of knowledge was a factor. The results can be seen in Figure 1.6.2.

The results showed that lack of understanding, the 'Iggy Factor', was a major issue for many of the newer models. The traditional models, as expected, did not have any significant problem with understanding. As knowledge comes first then where the Iggy Factor was greater than 50% this means that lack of understanding is the single most significant factor influencing those who have not used a technique. Figure 1.6.2 showed that ignorance or lack of understanding was the significant reason for non use for ten out of the fifteen newer models which were identified as generally not being used by the survey's respondents.

For the statistical models group all three models produced Iggy Factors in excess of 50%. The actual figures ranging from 68% to 81%. Lack of understanding was a significant factor influencing the use of these models with up to four out of five organisations who have not used the models not understanding how the technique works.

The result in the knowledge based category is less conclusive but lack of understanding was still significant. ELSIE, the system marketed by Imaginor Systems Ltd, had an Iggy Factor of 49%. Therefore, education would not necessarily be the most significant factor although it is probable that it was. For other knowledge based systems the figure was 75%, with three out four non users indicating a lack of understanding.

The trend of significant Iggy Factors is broken with the results for the life cycle costing group. Here, none of the three models considered had an Iggy Factor greater than 38%. Both net present value and payback method had 30%. It seems unlikely that ignorance of these models was the most significant

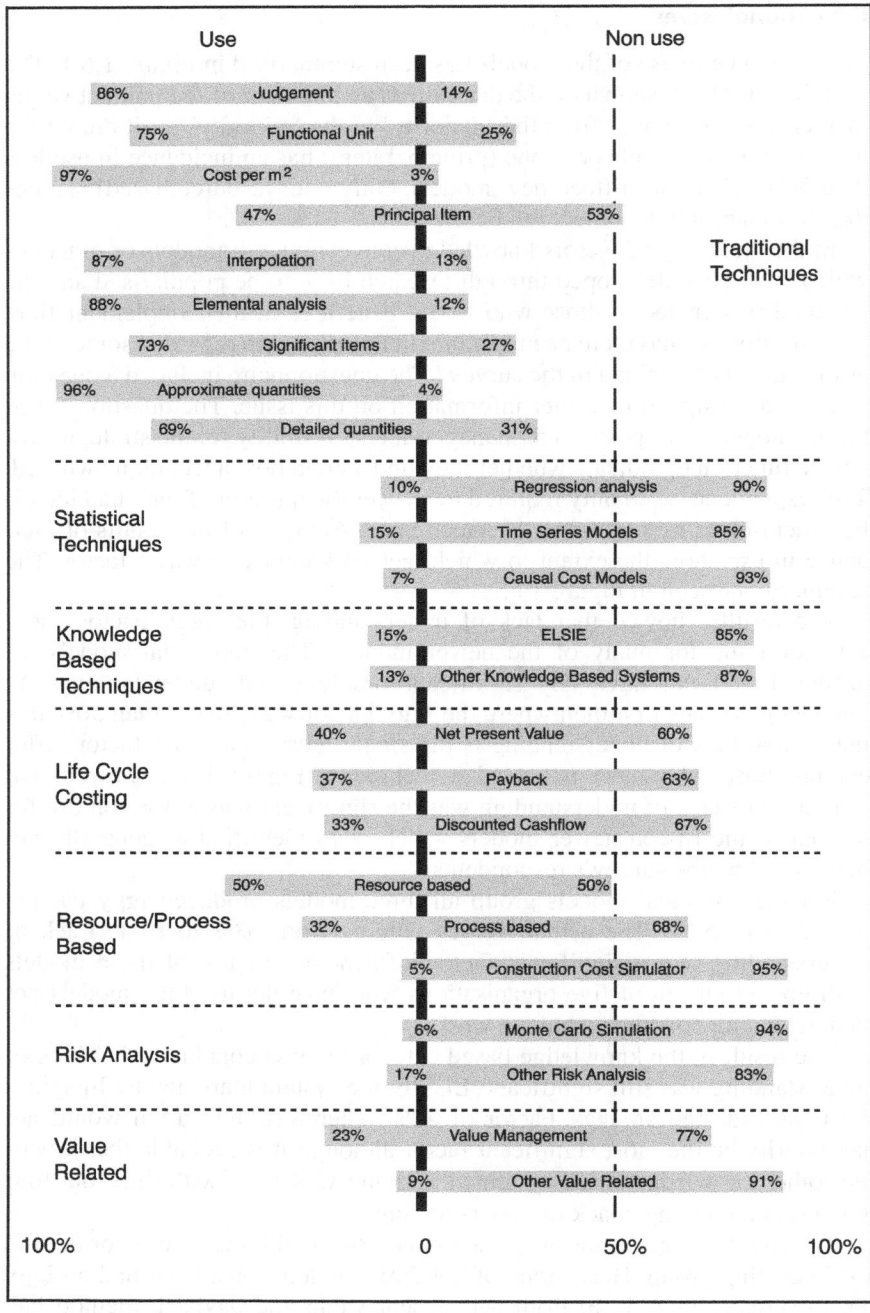

Figure 1.6.1 Incidence of use of techniques (% of respondents who use or do not use the techniques).

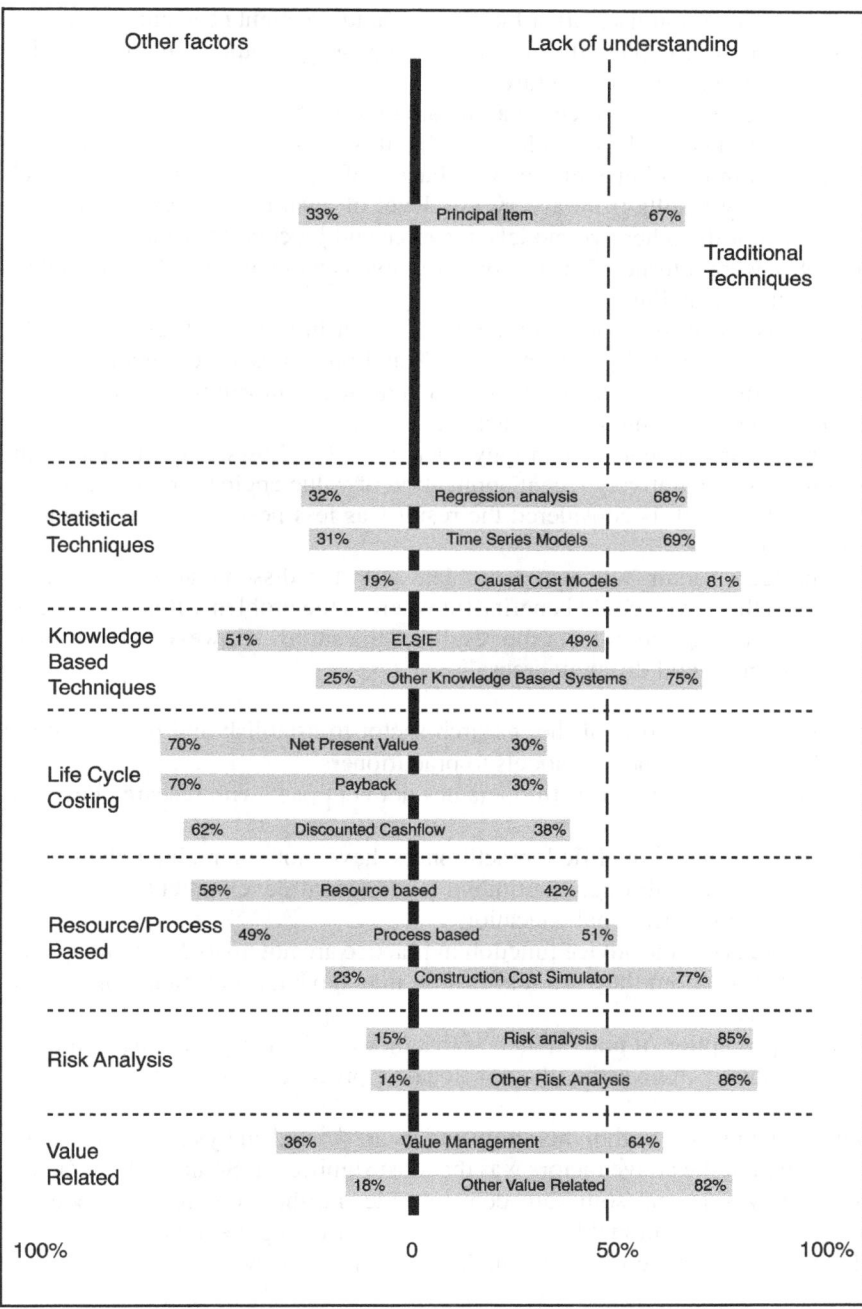

Figure 1.6.2 Extent of lack of understanding as a factor influencing incidence of use (% of respondents who do not use the technique and identified a lack of understanding).

reason for their non-use. Other factors surrounding client need and/or value of information may explain the non-use. Further study would be required in this area to establish a clearer picture.

In the resource and process-based category the construction cost simulator was not well known. It is not clear whether this is due to its uniqueness or to its lack of promotion but with an Iggy Factor of 95% it suggests that its lack of popularity results from a very low level of awareness of its existence in practice. For the other two models, resource and process, Iggy Factors of 50% and 68% were returned. The use of both models was significantly affected by lack of understanding.

The risk analysis group had the worst result in terms of Iggy Factor. The Monte Carlo simulation returned a 94% and other risk-based models a 83% Iggy Factor. Educational issues were a significant impediment to the proper evaluation of these models in practice.

Value management with an Iggy Factor of 77% shows a significant result. But if the comparatively recent application of value engineering principles to this field of work is considered the result was less pessimistic than some of those above.

The Iggy Factor indicated a problem with the dissemination of research results to those who could benefit. But where is the problem? It is not possible to answer this question from the results of the survey. However, the problem could be in several different areas:

- failure on the part of the research sector to establish and popularise the benefits of the newer models to practitioners
- basic education of practitioners has not kept pace with research developments
- professionally qualified practitioners have not maintained their level of awareness through continuing professional development courses and continuing vocational education
- managers of the advice function in practice are not up to date although the staff for whom they are responsible may have an understanding of new models
- the perception of practitioners is that they are doing a good job and there is no interest in investing the new ideas or practices

The data collected within this study and its associated analysis does not establish which of the above factors was the most significant. Some of these factors are supply side and some are demand side. Further research is needed to prioritise the educational factors suspected as affecting the incidence in use of the newer strategic cost advice models listed in the survey.

Organisational issues

Under this heading there were three particular issues which were considered as potentially affecting the incidence in use of the newer models listed, namely, type of organisation, size of organisation and computing facility. The SPSS/PC + statistical package (Nie *et al.* 1975) was used to analyse the survey's results by using crosstabulations and statistical significance tests. The crosstabulations which were undertaken set out the incidence in use of the various models with type, size and computing facility. For each crosstabulation a non-parametric test (chi square) was carried out. The results of this test would indicate the strength of the null hypothesis that there was no relationship between the independent variable (type etc) and the dependent variable (use of the technique). If the chi square test returned a value of less than 0.001 then the null hypothesis was almost certainly false and some relationship must have existed between the two variables under test. In these circumstances lambda values were computed. The lambda value is a 'measure of association' which gives an indication of the strength of the relationship. It works by calculating the reduction in the error of predicting the dependent variable result from the independent variable value. For example (in organisational type analysis) a lambda value of 0.010 would mean that a 10% better guess can be made about whether an organisation uses a technique if its organisational type is known. Details of the outcome of the statistical tests are given in Appendix B.

A graphical assessment of the relationship can also be drawn from the charts which were created by comparing the expected count in the crosstabulation (ie if the null hypothesis were true) and the actual count, this difference is then expressed as a percentage of the number of organisations of that type. This standardised the data and allowed comparisons to be drawn from type to type. There is a criticism of this approach in that when a group has a small population then a deviation from the expected by a small number may appear exaggerated when expressed as a percentage of the group. The authors took the view that provided the data was read with this problem in mind the advantages of the comparison outweigh the disadvantages of overemphasis for small groups. In certain instances the statistics did not indicate a significant result and it should be borne in mind that these indicators focus on the null hypothesis. The lack of statistical significance does not mean that there was no relationship between the variables. The graphs are more useful in these cases and provide the opportunity to observe patterns of interest.

Type of organisation

Type of Organisation – computing facilities.   As some of the types of organisation had very low frequencies it was necessary to manipulate the data so as to avoid problems in the statistical analysis. Consequently the categories of organisation were regrouped according to their generic type. This meant that all types of contracting organisation were grouped together as were architectural and multi disciplinary practices. The none of these group was eliminated since it was not representative of any specific type of organisation. The revised population distribution is shown in Table 1.6.7.

Figure 1.6.3 shows the relationship between type of organisation and computing facility of any type measured on a quantitative basis. It is perhaps surprising to find that the quantity surveying practices have the highest

Table 1.6.7 Distribution of types of organisation in the sample

Type of Organisation	No	%
Quantity Surveying practice	234	63.8
Multi-disciplinary practice	29	7.9
Local or Public Authority	38	10.4
Contracting	46	12.5
Project Management	20	5.4
Total	367	100.0

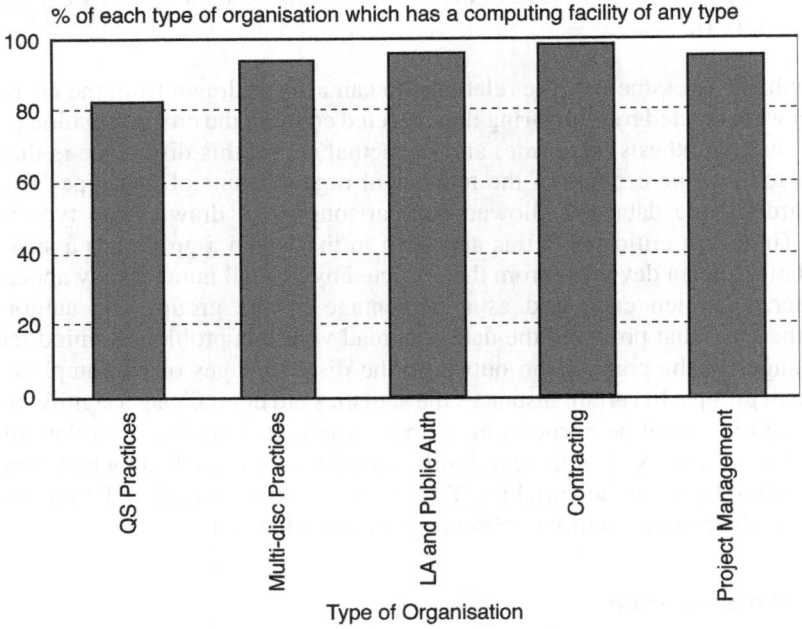

Figure 1.6.3 Cross-tabulation of type of organisation and computer facilty (of any type).

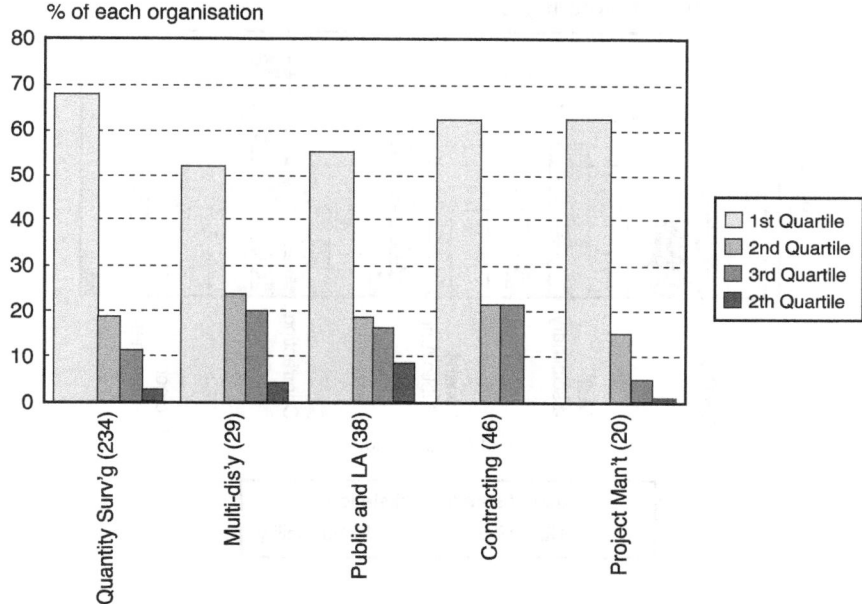

% of each organisation

Figure 1.6.4 Cross-tabulation of type of organisation and size of organisation (Quartiles calculated on total staff number of staff in the sample).

proportion of organisations without a computing facility and that contractors have the lowest. This outcome can be explained by the fact that the quantity surveying practices were the largest group int he sample and therefore covered both large and small organisations. Whereas for the other types of organisation they tend to be larger organisations, this can be seen from Figure 1.6.4. There is a possible criticism of the sample selection technique in this result but it must be remembered that the sample was chosen on the basis that the organisations were, or were likely to be, involved with strategic advice for clients.

The qualitative position can be seen in Figure 1.6.5. This showed that the quantity surveying practices have the highest quality of provision with over 50% of organisations having a 2:1 ratio or better, and the contractors the worst with less than 20% in a similar position.

Type of Organisation – incidence in use. The traditional models were generally well used by all types of organisation. There were two points worthy of note. Firstly, that contractors were significantly less likely to use functional unit and elemental analysis with deviations from the expected count of 17.4% and 16.7%. Secondly, that this negative incidence was mirrored by the project management group which had an increase usage of both models of 19.5% and 11.5% respectively.

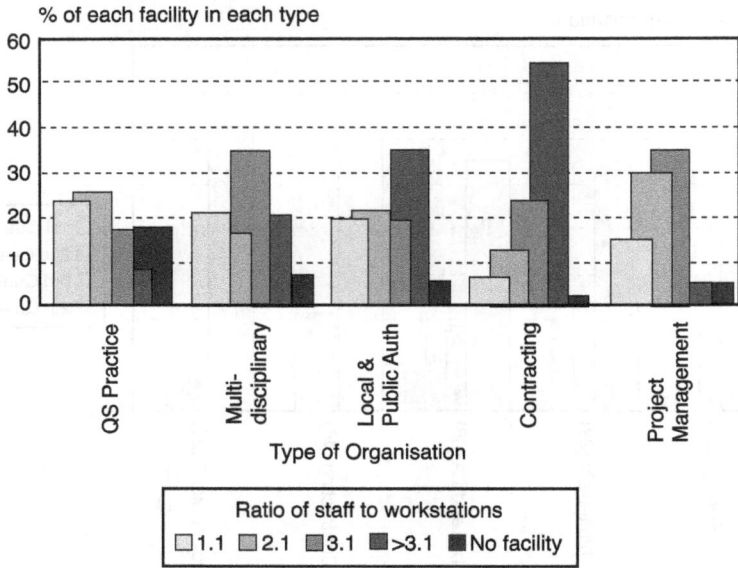

% of each facility in each type

Type of Organisation

Ratio of staff to workstations
☐ 1.1 ☐ 2.1 ☐ 3.1 ■ >3.1 ■ No facility

Figure 1.6.5 Cross-tabulation of type of organisation and type of facility qualitative.

For the statistical models there was no overall pattern which was found to be statistically significant. However it could be seen that contractors used these models less than would be expected, with both public and local authorities and project management types using the models more than expected. It may be that the more focused workload and strict financial targets of public and local authorities allowed them to make more use of statistical models.

There was a greater incidence in use of knowledge based systems amongst quantity surveying practices with public and local authorities and multi-disciplinary practices using them less than expected.

The life cycle costing models showed a result which was statistically sig nificant. The lambda value was 0.081 if use of the models was the dependent variable. This analysis showed that project management and multi disciplinary practices were high users with contracting and quantity surveying practices being low users. A possible cause of multi-disciplinary practices making use of life cycle costing models may be as a result of the close working relationship between designer and financial controller which may promote a demand for this type of advice. These results can be seen in Figure 1.6.6.

The resource/process based group of models also showed statistically sig-nificant results within a 95% confidence limit. Project management and contracting being high users and quantity surveying practices, public and local

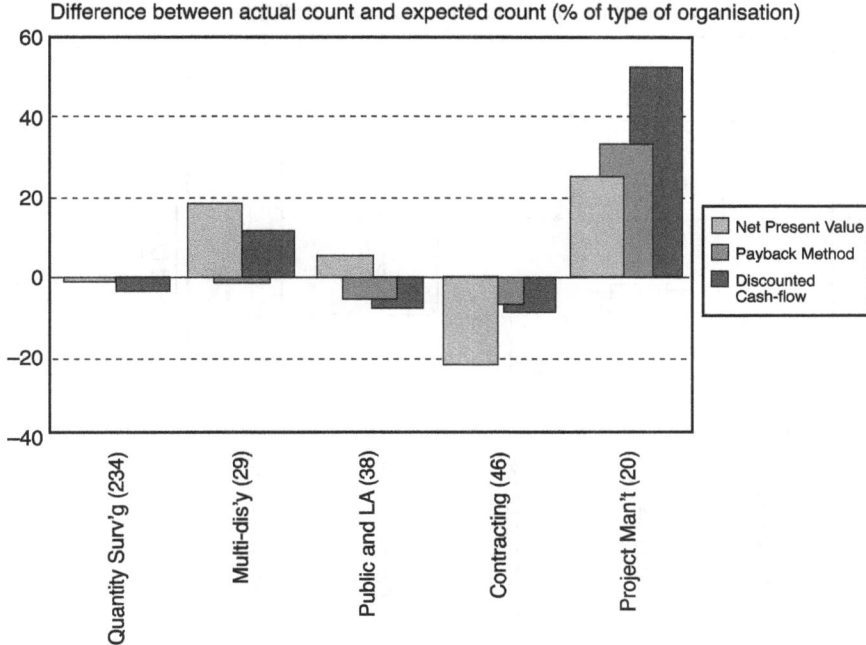

Difference between actual count and expected count (% of type of organisation)

Figure 1.6.6 Cross-tabulation of type of organisation and life cycle costing techniques.

authorities being low users. The result can be seen in Figure 1.6.7. An explanation could be that access to data on resources and processes could influence the result. However, this would be inconsistent with the result for local and public authorities, as they should have access through their direct works organisations but use these models less than expected. Another explanation is found if the data that is generally available to the organisation is considered. The traditional source of data for the advisor is information on prices taken from bills of quantities or other price sources. These sources are normally readily available to quantity surveying practices and their public sector equivalent and are less available to the other types of organisations. This could explain why the contractors and project managers have a higher than expected incidence in use in resource and process based models where the type of information is more accessible.

In the risk analysis group the project management and multi-disciplinary practices use the models more than expected. All other types of organisation used them less. The direct relationship of the project manager and designer with the client and in particular the clients needs could explain this result.

For value related models it was seen that project management type

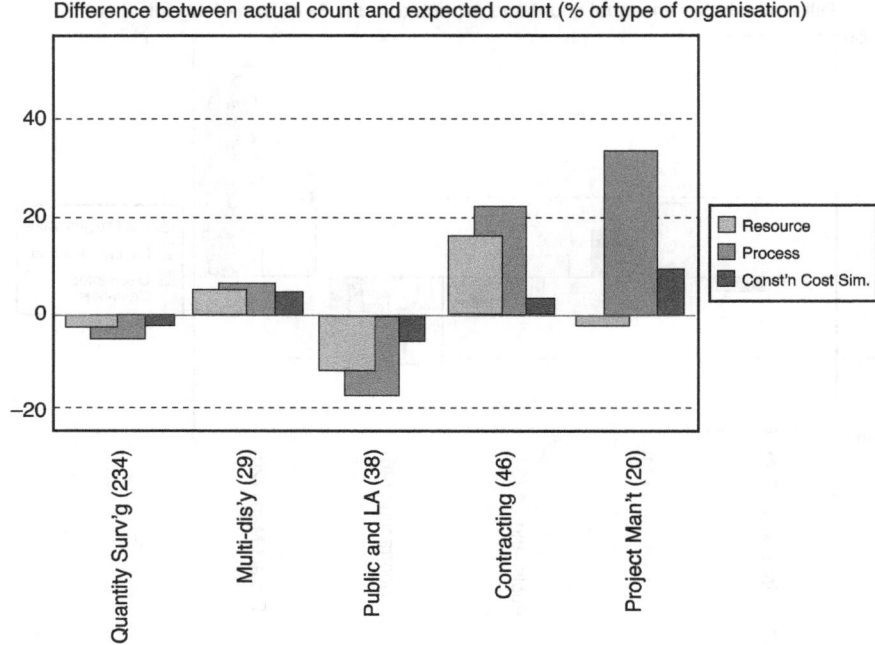

Figure 1.6.7 Cross-tabulation of type of organisation and process resource based techniques.

organisations were a high user whilst all other types of organisations produced lower than expected incidence of usage.

The above analysis and discussion indicates that project management type organisations have a consistently higher incidence of using newer cost advice models. This may be due to this type of organisation being more prepared to be a leading organisation in terms of adopting new models in practice due to perceived client expectations. Other potential explanations may relate to the size and scale of projects which employ a project management function, which may present problems that can be more easily solved by using the newer cost advice models available.

Size of organization

Size of Organization – computing facilities.   The relationship between size and computing facility can be seen in Figures 1.6.8 and 1.6.9. Figure 1.6.8 shows the qualitative and Figure 1.6.9 the quantitative relationship. The results were as expected with the largest organisations having the least proportion of practices without a facility and the smallest ones having the highest proportion with no facility.

% of each quartile in each category of facility

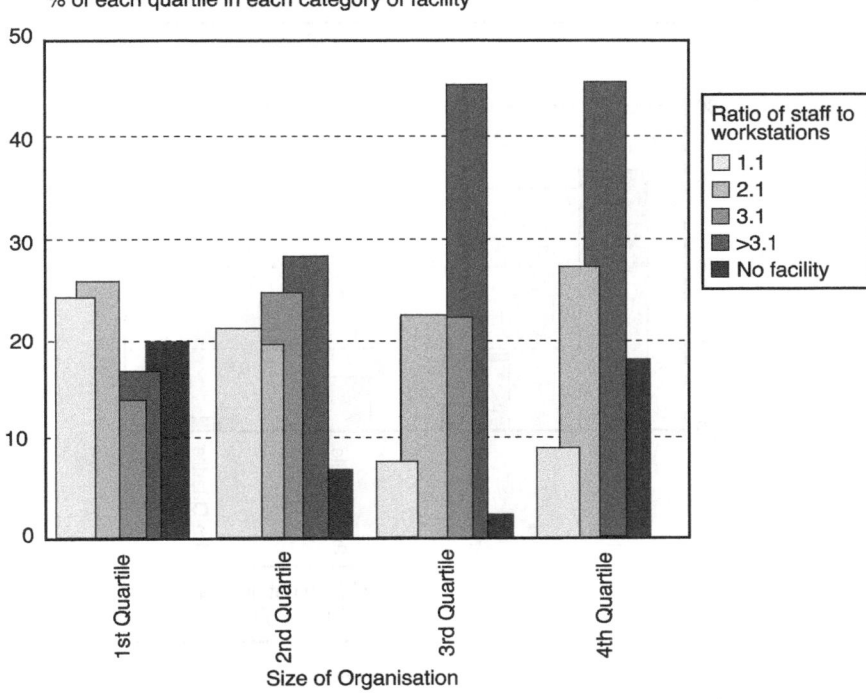

Figure 1.6.8 Cross-tabulation of size of organisation and computer facility.

On a qualitative level it can be seen that the best facilities, measured in terms of ratio of staff to work-station, were in the small organisations. Both the 3rd and 4th quartiles showed a particularly high proportion of practices with greater than 3:1 ratios. This could significantly affect their capability to adopt newer models (see later discussion).

Size of Organisation – incidence of models. The analysis of the relationship between size of organisation and incidence in use of models was performed using crosstabulations and chi-square tests of statistical significance. The results in Table 1.6.3 showed that the sample contained a large number of small organisations and a small number of large organizations. This is typical of the popular view of the structure of the profession. The question is does the size of the organisation affect the models it uses? The data collected about the number of staff involved with the early advice function was categorised into quartiles. Each organisation was placed according to its number of staff, with the small organisations in the first quartile and the largest in the fourth quartile. The quartiles were calculated on a total population of staff

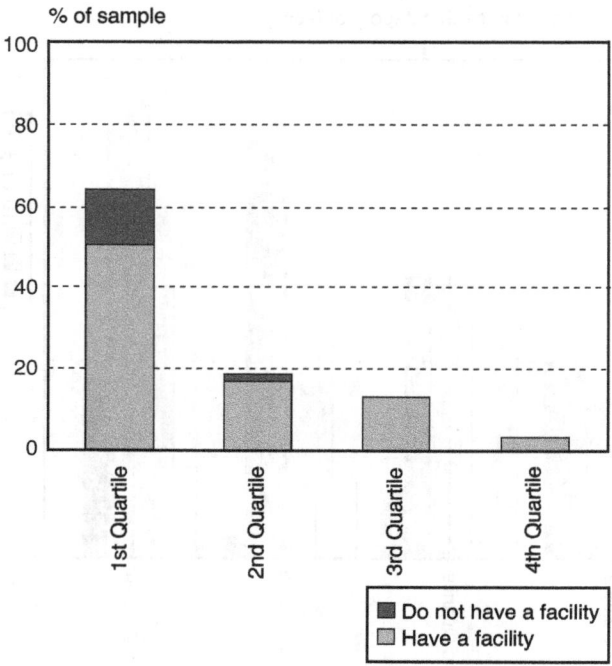

Figure 1.6.9 Cross-tabulation of size or organistion and computing facility (of any type).

involved in the early advice function rather than a population of organisations basis. This was done in response to criticism made by practitioners at conferences who were concerned that the latter basis would under-emphasise the position of large organisations who were using new models and who would only count as one incidence in the analysis. The way the analysis was actually performed the large practices count for each member involved with the early advice function. The concerns of the practitioners are not supported by the outcome of the analysis in that the largest organisations were generally not heavy users of new models. In fact, the authors performed the analysis both ways and the results are not significantly different from each other.

As with type of organisation the results of the analysis for traditional models were not significant with all sizes of organisation using the models. There was one exception and that was that detailed quantities was used significantly less than expected by the 1st quartile organisations. This may be due to that fact that this technique can only be used when a bill of quantities has been prepared and the practices in the smallest quartile are least likely to be involved in projects that require the production of such documents.

For the statistical group the results can be seen in Figure 1.6.10. The results were not statistically significant but the pattern is clear from the chart. The smallest organisations used the models less than expected and the larger ones more than expected. This was as expected but the fact that the 4th quartile organisations do not have a higher incidence than the 3rd quartile is surprising.

In the knowledge based category the result was similar to that above (see Figure 1.6.11). This time the result was considered to be statistically significant, with chi-square values of 0.000 and 0.004. The chart again displayed the pattern shown in Figure 1.6.10 where the 4th quartile under performs against the 3rd quartile.

For the life cycle costing group all the results were significant and lambda values were returned between 0.060 and 0.113 with the use of the technique as the dependent variable. Here again the small practices used the models less and the larger ones more than expected. Unlike the two groups above the 4th

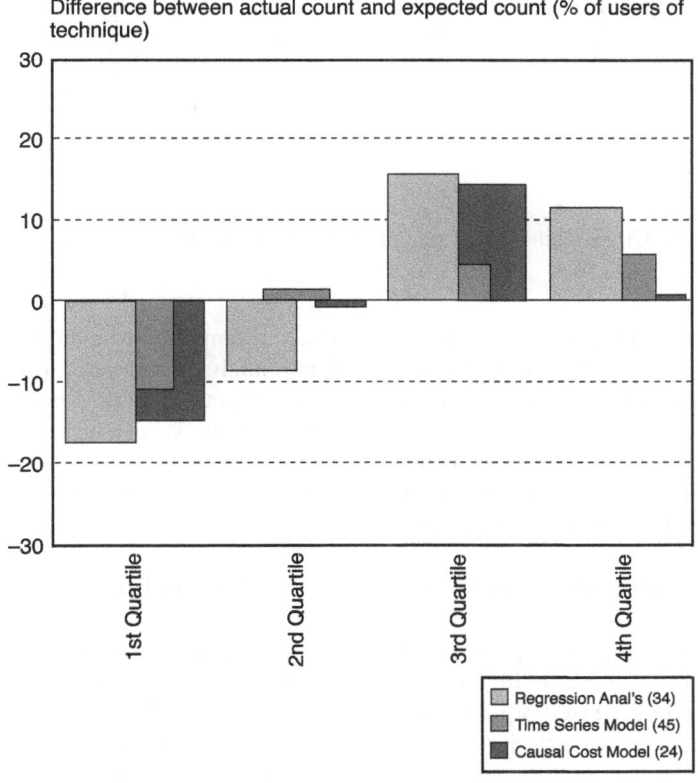

Figure 1.6.10 Cross-tabulation of size of organisation and statistical techniques.

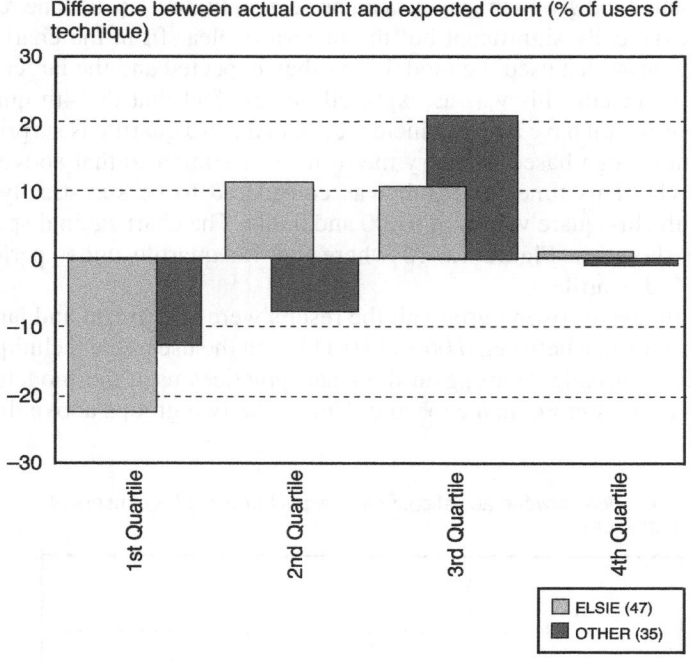

Figure 1.6.11 Cross-tabulation of size of organisation and knowledge based techniques.

quartile performed as well as, but still not better than, the 3rd quartile.

The results for resource/process based, risk analysis and value related are similar to those for the life cycle group and are set out in Table 1.6.7.

The pattern of results for size of organisation was consistent across virtually all the newer costing models listed and showed that the larger the organisation the more likely that organisation was to use the newer models. Explanations for this could include:

• greater diversity in quality of staff ie higher level of topmost qualified staff
• greater variety of experience amongst staff in larger sized organisations
• potential for structured continuing professional development activity within larger sized organisations
• wider range of activity, recognising that some of the models are more suited to certain areas of operation, within larger sized organisations

The data collected within this study did not establish which of the above factors was the most significant. The pattern identified above where the 4th quartile practices demonstrate use of models inconsistent with the general pattern across the quartiles is more difficult to explain. All the factors explaining the greater use of new models in larger organisations should apply to the very largest practices. But the results show that something else must affect these organisations. The respondent to the questionnaire could be some way divorced from the strategic function and less able to respond accurately than in other organisations. The strategic function could be fragmented into a number of branches so that the practice appears large but is in fact a series of small ones. The questionnaire was designed to minimise the effect of misinterpretations of this type so unless there was a misunderstanding by the entire population of the 4th quartile these explanations seem unlikely. An alternative explanation could be that the diseconomies of scale affect the operation at this level. Perhaps it is more difficult to manage the strategic advice function at this level or that large organisations develop their own inertia and are less willing or able to change.

Computing facility

Computing Facility – incidence of models. Information was gathered in the survey about the quality of the organisation's computing facility. The respondents were asked to indicate the ratio of staff to work-stations or, if appropriate, to indicate no facility. The results of the responses to this question were crosstabulated with the incidence in use and subjected to the same statistical testing as before. Graphs were produced showing the difference between the expected and actual count expressed as a percentage of users of a particular technique.

For the traditional group there were no results with statistical significance and no discernable patterns. This would suggest that there is no relationship between the incidence of traditional models and the availability of computers. This is a expected result since most of the traditional models were developed before the advent of computers.

The statistical category shows a pattern which is not statistically significant but is nonetheless clear. Those organisations with low ratios of staff to work-stations had a higher than expected incidence in use of the models, typically in the range 2–9% higher. Whilst those practices without a computer had much lower incidence in use (11–14% lower).

This pattern is repeated in the knowledge based group as shown in Figure 1.6.12. The result for the ELSIE technique was statistically significant. The chart also displayed another interesting pattern. This secondary pattern was that the 2:1 category out performs the 1:1 category of staff to work-station.

Figure 1.6.13 shows the result for the life cycle category. The pattern was

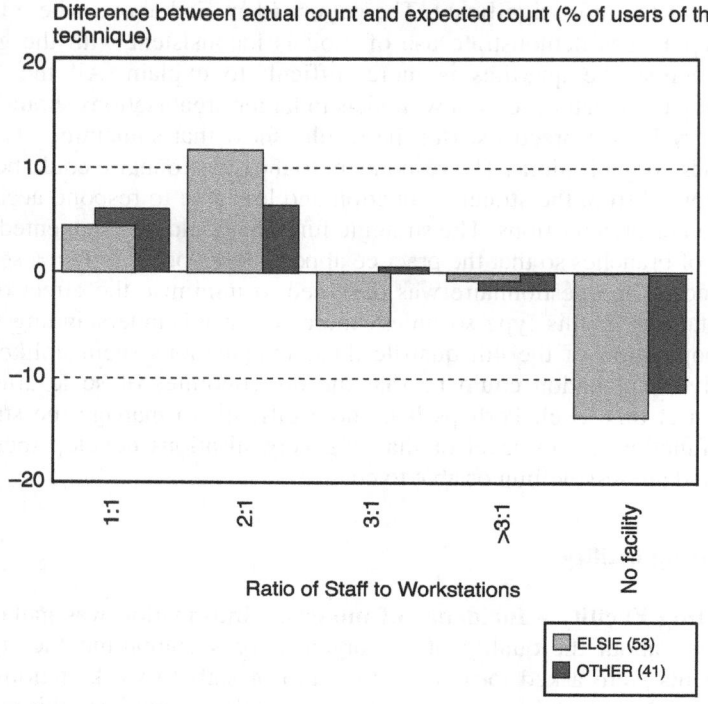

Figure 1.6.12 Cross-tabulation of computer facility and knowledge based techniques.

much the same as for the knowledge based group. Both primary and secondary patterns are displayed. All the results in this category were statistically significant within 95% confidence limits.

The resource and process based category does not show any pattern and was more consistent with the traditional models group.

The results for the risk analysis group can be seen in Figure 1.6.14. This group demonstrated the strongest pattern in both the primary and the secondary modes. The value management group followed the same pattern as the risk analysis category but not as strong.

This analysis of the crosstabulation of computing facility with use of models showed two patterns – the primary pattern where low ratios of staff to work-stations had higher incidence in use of newer models and the secondary pattern where the 2:1 ratio category has a higher incidence in use than the 1:1 ratio.

In general it can be said that the better the facility the more likely that a newer cost forecasting technique will be used. This is as most people would

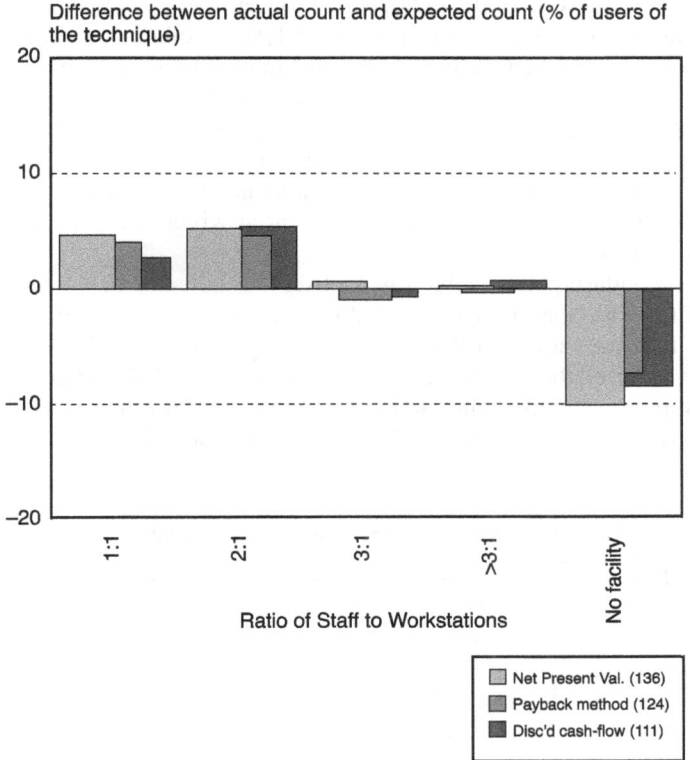

Figure 1.6.13 Cross-tabulation of computer facility and life cycle costing techniques.

have predicted. The question about cause and effect remains to be answered, ie is this relationship because having the facility encourages the use of newer cost forecasting models or is the facility a response to demand for the technique? The computer facility could be aiding the use a particular technique or conversely it could be that a computer was necessary to perform a technique for which there was an existing demand. Further research in this area would be required, however, the authors have the view that the development of computing facilities probably passes through at least two stages.

In the first stage a demand for a newer cost forecasting technique is perceived and the benefit of a computing facility identified. Once the facility is provided and the initial demand for the new models satisfied the existence of the facility becomes a positive stimulus to its continued use. The degree of continuing stimulus would depend upon the extent and quality of the facility. The results showed some support for this view. Figure 1.6.13 showed the results of the analysis of the incidence in use of life cycle costing models and computing facility. The pattern for the analysis of other models was similar. The figure also showed the basic relationship with better than expected use in the 1:1 category and less than expected use in the no facility category. The figure also showed that there is a distinct change in the actual use once the ratio of staff to work-stations exceeds 2:1, indicating that the absence of a quality facility may be a factor in the incidence in use of newer models. It further suggests that there is a critical point at which a facility is genuinely useful ie 2:1 or better.

The secondary pattern which was observed indicates that where the ratio of staff to work-station is 1:1 then this inhibits the use of new models. The authors take the view that this is an unlikely outcome and that other reasons may have caused the effect. Figure 1.6.8 showed the relationship between size and computing facility. The largest group of 1:1 ratio facility is in the first quartile of size of organisation. The first quartile organisations were the least

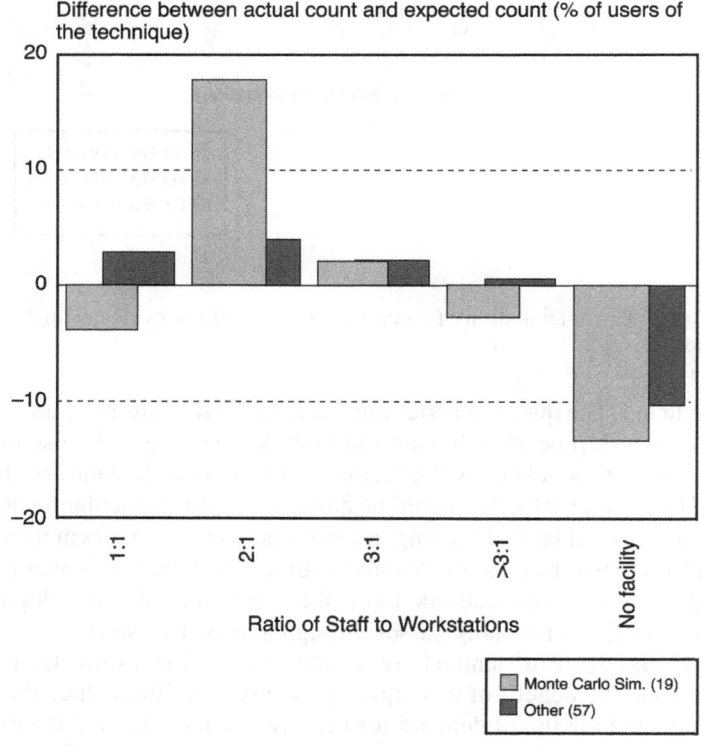

Figure 1.6.14   Cross-tabulation of computer facility and risk analysis techniques.

users of new models. This further indicates that there is a significant group of practices with quite small numbers of staff each of which has a computer. In fact, it is likely that these organisations are sole principal practices who use a computer for administrative purposes only. There is no evidence from the survey to support this view but it fits with the authors' experience and is a satisfactory explanation for the secondary pattern.

Operational issues

Generally

The final area of concern in the implementation of new models is the area of operational issues. In this area the survey looked at indicators which would give an insight into the relationship between the use of a technique and operational factors. These factors were:

- reliability and accuracy
- value as a tool to the advisor
- continued use of technique (retention)
- use of judgement

The main results for each of these issues have been presented in the Results section. However, each will be discussed here to allow for expansion on the findings of the survey.

Reliability and accuracy

The results for this issue can be seen in Figure 1.6.15. The purpose of this question was to allow respondents to score the models on the basis of whether the information produced was reliable and accurate. The result for the traditional models was interesting in that there was a marked difference between the scores for the single rate methods (judgement, functional unit, cost per m^2 and interpolation) and the quantitative methods (significant items, principal item, elemental analysis, approximate quantities and detailed quantities). The latter group were scored much higher than the former. This is as expected and would suggest that the respondents fully understood the question and therefore how to respond.

The result for the newer models is quite significant. Most of these models returned rating scores at least as good as the single rate traditional models. This would indicate that the practitioner is unlikely to reject the models out of hand. It can be argued that for a new technique to be adopted it must perform to a significantly better standard then the traditional models. At this stage the results of this survey show that the new models do not perform significantly worse then most of the traditional ones. It could be a disappointment that the

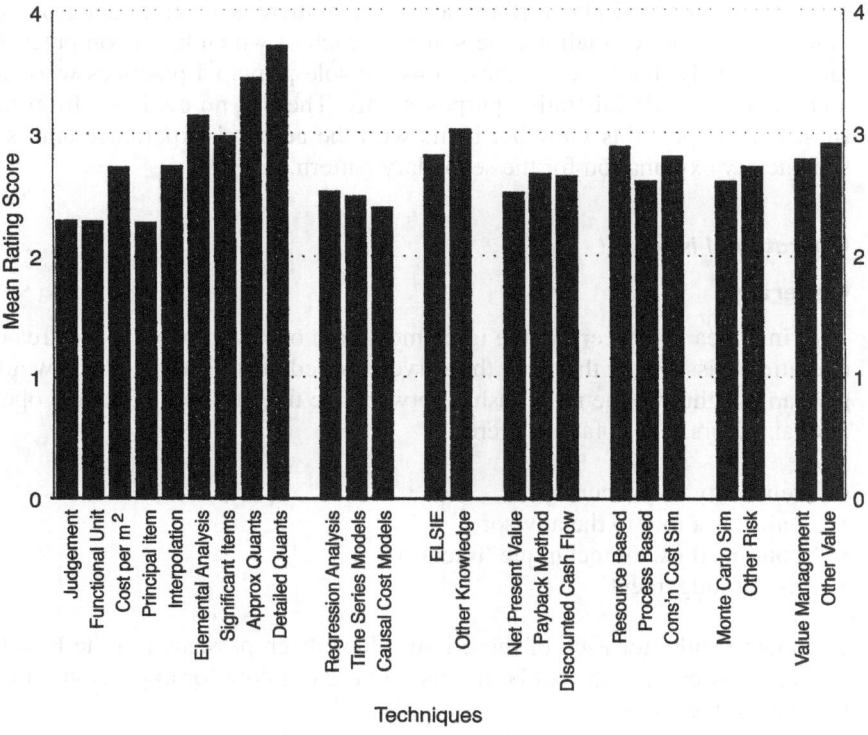

Figure 1.6.15 Mean rating score for reliability and accuracy.

new models do not perform better but there may be reasons other than the models are poor. These could include:

- a lack of refinement of the technique
- difficulty in adapting to the technique in use
- inadequate sources of data available for use in the technique

Value as a tool for the advisor

The question on how the respondents rated a technique as a valuable tool was intended to assess the operational value of a technique. This issue was considered important and distinct from the reliability and accuracy issue because being reliable and accurate in itself is not enough, a new technique must also be useful. The reliability and accuracy must be capable of being attained with reasonable effort. A technique which is reliable but requires too much effort will not be adopted by practitioners. The results for value as a tool were given in Figure 1.6.16.

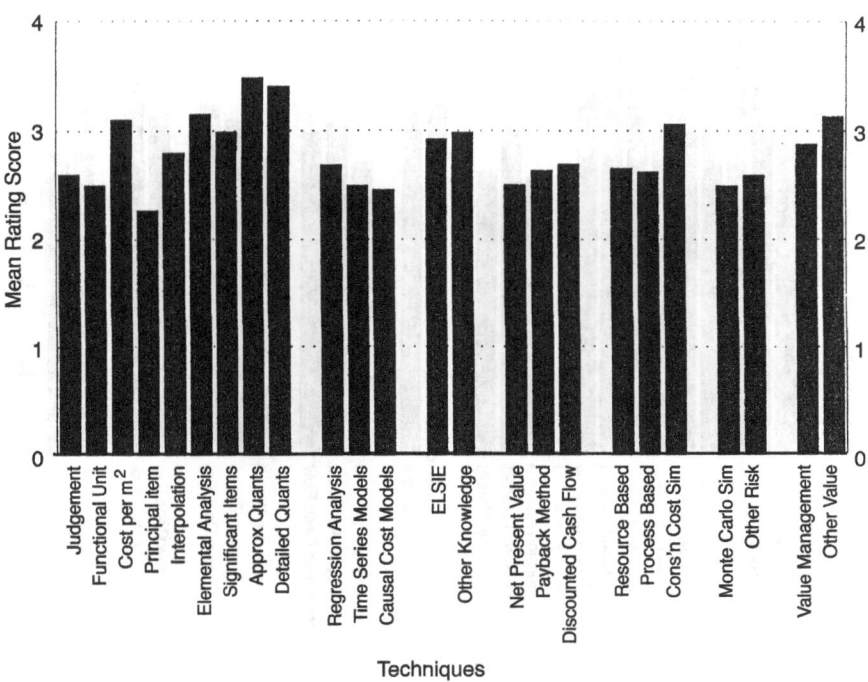

Figure 1.6.16 Mean rating score for valuable tool.

The results for the traditional models again show that the respondents understood the question. The single rate method are considered less valuable than the quantitative methods, but this time the margin is considerably less than in the previous section. This is the expected result.

For the new models the position is more encouraging than indicated by the analysis for reliability and accuracy. The new models have consistently better scores than the single rate traditional methods. This would seem to indicate that there is potential for the further development of the models which could result in their wider acceptance.

Continued use of models (retention)

The survey included a question for those who had used a technique on whether they continued to use it. The purpose of this question was to attempt to gather objective data rather than opinion data on the respondents view of the models. It was considered the best proof of a technique's worth if the organisation continued to use it. The results can be seen in Figure 1.6.17.

The traditional models had generally high retention rates whereas the

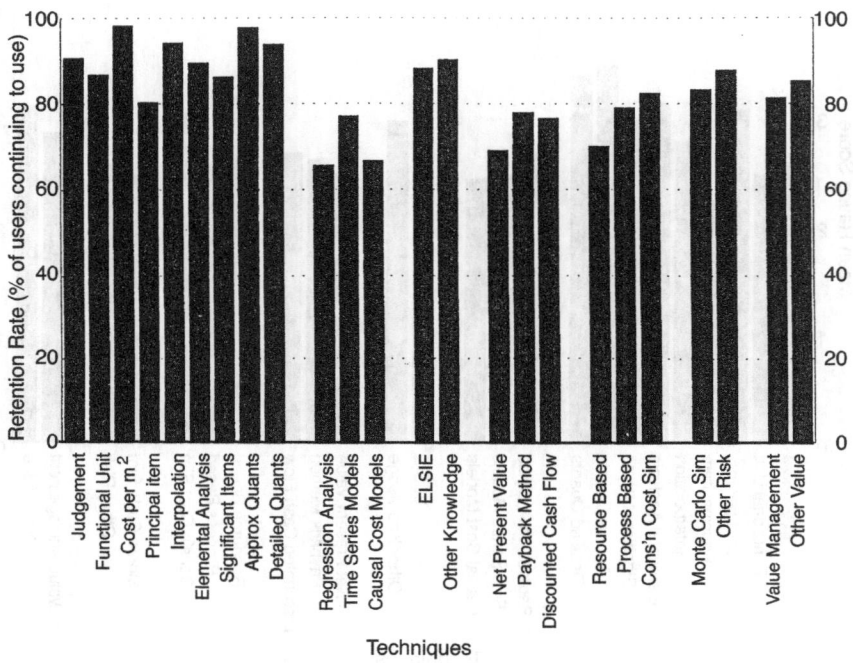

Figure 1.6.17 Retention rate.

statistical, life cycle and process/resource based were marginally lower. The differences are real but are not considered significant since they could be caused by a number of factors:

- changes in the level of demand during recession
- lack of refinement in a technique's application
- difficulties encountered in adapting to a new technique

Use of judgement

Role of judgement. In the Literature Review section the importance of taking into account the extent to which judgment was required in operating a technique was established. The survey gathered opinion from respondents on this issue. The results can be seen in Figure 1.6.18. The survey results reinforce the view that none of the models can be used without the application of judgement by the user. This relationship can be summarised as follows:

[technique] + [judgement] = [advice]

Judgement is an important part of any decision making process. A proper understanding of the role judgement plays in the development of strategic advice is essential if new models are to enhance rather than detract form the process. This survey has produced information which may advance the understanding of the role of judgement.

There are two opposing views about the role of judgement in the process of developing strategic advice for clients.

The first view is that if a technique relies heavily on the use of judgement this is a negative indicator. The reliance suggests that the technique is inaccurate, or unreliable, or requires considerable effort on the part of senior surveyors in interpreting the results and formulating advice. The argument follows that for a new technique to be better than a traditional one it must allow a more accurate result to be arrived at in a more efficient manner and, if possible, by less well qualified staff. If respondents indicated that a new technique relies heavily on the use of judgement then this means that none of the requirements had been achieved.

The second view is that a high use of judgement score indicates that the technique is receptive to the exercising of judgement, keeping the user in

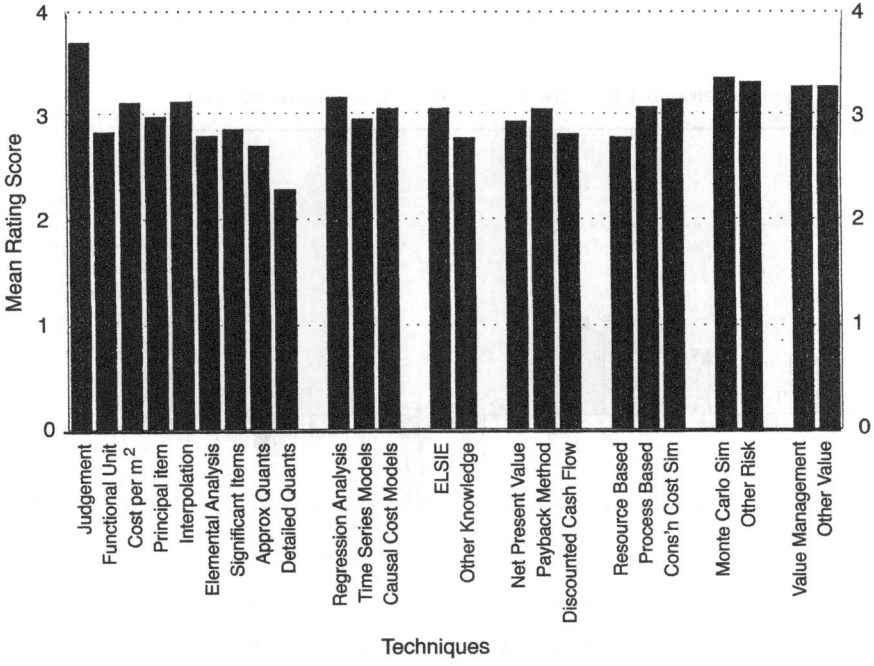

Figure 1.6.18 Mean rating score for use of judgement.

touch with the data and its manipulation, and therefore allowing expert inter-pretation and the formulation of good advice. This second view is a positive indicator for judgement and is based on the concept that if expert advice is required then the expert must have an appropriate opportunity to demonstrate expertise, with models that attract high judgement scores providing this opportunity.

The authors considered whether the data from the survey would help in determining which of these contrasting views is more likely to be true. The retention rate is an objective measure of the users view of the models (see earlier discussion). The hypothesis was presented as

there is a relationship between the objective measure of performance (retention rate) and the extent to which a models relies on judgement

The retention rate was cross-tabulated with the rating score for the question on the use of judgement. The results can be seen in Figure 1.6.19. There was a significant difference in the retention rates of those who gave the use of judgement question a low score (ie 1 or 2) and those who gave high scores (3 and 4). The group giving the high scores were more likely to retain the technique. This indicates that this group of organisations considered that there was a link between the quality of a technique and the ability to apply

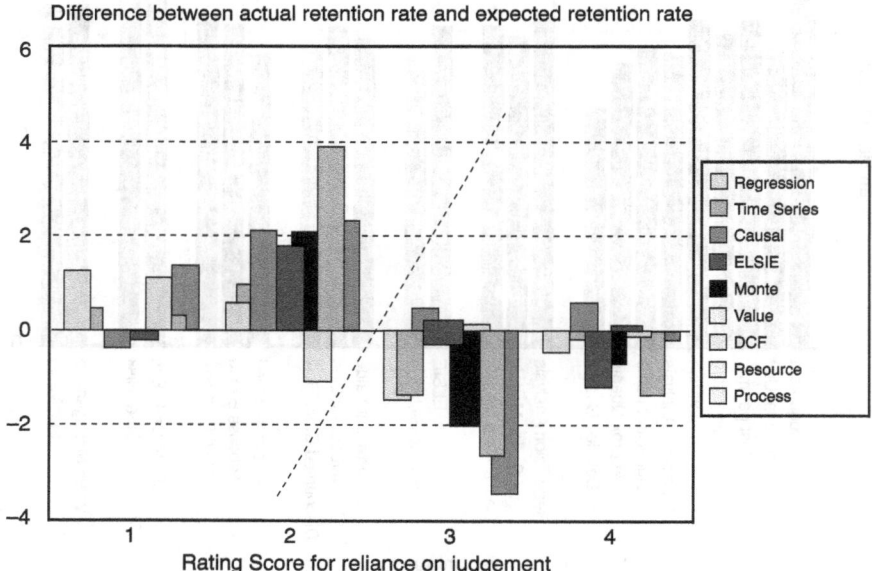

Figure 1.6.19 Cross-tabulation of reliance on judgement with rentention rate.

judgement to it. This finding supports the second view above ie that judgement is a positive indicator.

This finding is interesting in that it would indicate that the future development of models should pay careful attention to the interface with the user. Past developments have often set an objective of de-skilling ie models which are capable of being used by less qualified staff, but the result of this analysis suggests that the market place is not receptive to this approach. What is required is the opportunity to perform existing tasks more efficiently and apply new models to produce advice that was not previously available.

Preliminary Assessment Model.   There is a need for researchers to be able to test the product of their study, and therefore offer it to the profession in the knowledge that it will be of real benefit to users.

This study has established that for many of the newer costing models lack of understanding is the largest factor influencing the technique's incidence in use. However, the question remains whether, when this educational problem is resolved, the new models will benefit the practitioner? It can be argued that this is a matter for the practitioners themselves to determine having had experience of using the models, but equally it can be argued that researchers should be able to establish and demonstrate the benefits of their research before it is launched in the 'real world'. Pilot testing new models with practitioners working in real situations is useful, but the assessment of the results of such trials can present a problem. The usual method is to elicit opinion from the users about the technique and its value to them as practitioners. However, knowing what questions to ask and whether the opinions expressed can be relied upon are the difficulties.

The survey collected data on the opinion of users of new models about reliability and accuracy, use of judgment, and value as a tool. The data collected is based on the practitioners perception of the models. The survey also posed the question to those who had used a particular technique whether they continued to do so. Expressing the response to this question as a percentage of users allows the calculation of a 'retention rate' for each technique. The retention rate is an objective measure of the importance and value of the models, with the better models having higher retention rates than the less valuable models. The retention rate provides an objective measure of the performance in practice for each technique.

The authors were interested in whether there was a relationship between practitioner perception – ie the response to reliability and accuracy, judgement, and valuable tool questions – and the retention rate for a particular technique. If there is a relationship and it could be modelled, it will be possible to predict the retention rate for a new technique from perception data collected from practitioners' involved in trials. Researchers would also be able to have greater confidence in the trial results. This would allow researchers to assess the likely benefit to practice and, if necessary, refine the models until a satisfactory result had been achieved. It would also give potential users of a new technique a clearer indication of its value.

The basic thesis is

> *that there is a relationship between the practitioners' verdict based on perception ('verdict value') and the objective measure of performance in practice ie the retention rate*

The survey results contain data about three key factors – reliability and accuracy, judgement, and valuable tool. The main problem in developing the thesis is how do these key factors combine to give a single verdict value.

It is a reasonable assumption that if a technique produces reliable and accurate information this will be considered a positive indicator. Similarly, if a technique receives a high valuable tool score this will also be considered a positive indicator. The difficulty is with the judgement factor. As discussed above the problem lies with how to asses the judgement factor ie is it positive or negative? It was determined to advance both views and to examine the data to see if there was any evidence as to which view was correct. The calculation of verdict value was undertaken by two possible methods:

method 1   verdict value = [reliability & accuracy]
                  + [valuable tool]
                  – [judgement]
method 2   verdict value = average [reliability &
                  accuracy . . . valuable
                  tool . . . judgement]

The thesis was tested using linear regression analysis. However, before the analysis could take place it was considered appropriate that the data was manipulated. Some of the categories of models did not represent a single known technique, but were 'catch-alls' for the respondents benefit. These categories were eliminated. The construction cost simulator was also eliminated since it had a particularly low incidence in use. Two of the traditional models were also eliminated from the analysis – functional unit and principal item. Both these models had low incidence in use for traditional models and were considered therefore to be unrepresentative of the sample as a whole.

The data collected for the components of the verdict value was in the form of a range score, the range being from 1 to 4. This score was converted to a scale of 0 to 100 by making 1 = 0 and 4 = 100.

The analysis plotted the retention rate and verdict value (calculated by both method 1 and 2) on appropriate axes. A regression line was determined and the r^2 value and Pearson correlation coefficient calculated.

Based on method 1 for the calculation of verdict value the r^2 value is 0.31

and the Pearson correlation coefficient is 0.56. This indicates that there is a relationship between retention rate and verdict value but that the relationship is not strong.

Method 2 produced an r^2 value of 0.66 and correlation coefficient of 0.81 which indicates a much stronger relationship than with method 1. The results of the analysis are summarised in Figure 1.6.20.

The conclusion is that, based on method 2, there is a strong relationship between verdict value and retention rate and that this conclusion is statistically significant. This indicates that practitioner perception of the use of judgement in a technique is a positive indicator. It also means that in the trial of new models if data on practitioner perception of the value is gathered, then some prediction of the future retention rate for the technique can be made.

This model for assessing the future benefit to practice of new models is preliminary. However, a more refined model may produce a more precise prediction of future benefit but that greater precision may not necessarily be

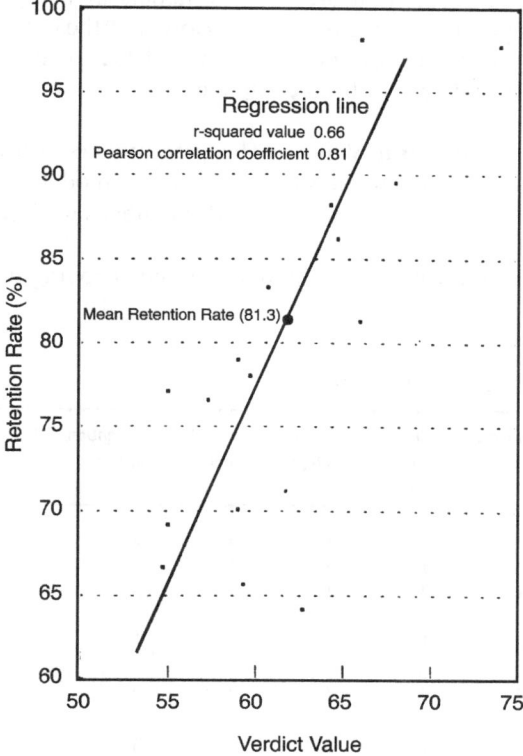

Figure 1.6.20 Regression analysis on verdict value with retention rate.

useful. The preliminary model at least allows researchers to assess whether a new technique is likely to have an above or below average retention rate, or whether it will be marginal. This information should be extremely beneficial and may allow more thorough testing of new models, or software packages based on them, before they are launched in the industry.

CONCLUSIONS AND RECOMMENDATIONS

Conclusions

The results of the survey and the analysis of the data are summarised in this section in relation to the objectives established in the Introduction. For detailed discussion of the issues see the previous section.

Incidence in use of Models.   The incidence of use of models has been established for the sample from whom data was collected. The general conclusions can be seen in Table 1.6.8.

The conclusions would indicate that new models are being tried and used to formulate strategic cost advice for clients. However, the extent of use is small and the paradigm shift called for by Brandon (1982), if it is taking place, is occurring slowly and has not been completed.

Reliability, Accuracy and Value.   The practitioners' verdict on the reliability and accuracy of the models is that the newer models generally perform as well as the single rate traditional methods but not as well as the quantitative traditional methods.

The verdict on the value of the models is more encouraging for researchers

Table 1.6.8 Summary of conclusions

	Incidence in use	Reliability and accuracy	Use of judgement	Valuable tool	Retention in use	Understood by non users	Capability to perform
Traditional	3	2	1	3	3	3	3
Statistical	1	1	2	1	1	1	1
Knowledge Based	1	3	1	3	3	1	1
Life Cycle Costing	2	1	1	1	1	3	2
Resource Process	2	3	1	2	2	2	1
Risk Analysis	1	2	3	1	3	1	1
Value Related	1	3	3	3	2	1	1
Key	3 – High		2 – Medium		1 – Low		

as the new models were scored as more valuable than the single rate traditional methods but still less valuable than the quantitative methods. This would indicate that there is potential for these new models at the very early stage of project planning when detailed information is not normally available.

Factors affecting Incidence in Use.   The research concludes that the factors set out below influence the incidence of use of the new models within a hierarchy of factors. Each group is important but they are listed in the order in which the factors occur in the development and implementation of a new technique:

* Educational issues. The practitioners knowledge and understanding of the technique.
* Organisational issues. The type, size and computing facilities available all affect whether a particular organisation will use a new technique.
* Operational issues. The way a technique interfaces with the user and the role judgement plays in the formulating of advice will influence the use a technique.

Use and Role of Judgement.   The results for the extent to which the models rely on judgement indicate that all the models – traditional and new – rely on the use of judgement. This supports the contention that advice is formulated by using a technique to produce an answer and then applying the practitioner's judgement to that answer before giving any advice.

The conclusion of this study in respect of the role of judgement is that it should be viewed as a positive part of the advice formulation process. Practitioners continue to use models which allow them to exercise judgement more than models which do not.

Preliminary Assessment Model – A preliminary assessment model based on data which can be collected from the practice trials of new developments is advanced in the previous section. The model allows the research to predict the likely retention rate of a technique form opinion data gathered from the trial users about its reliability, accuracy, value and use of judgement.

Criticisms of the study

Any study based on a mailed questionnaire suffers from weaknesses connected with uncertain respondent quality, potential for misinterpretation and inability to supplement respondents answers with additional or observational data. The analysed responses to the questionnaire showed that widespread respondent misinterpretation was not a problem in this study. However, the quality of respondent could not be guaranteed and the data captured could not be supplemented by additional or observational information due to the research design adopted. These latter criticisms can only be deflected by ensuring that any follow up study adopts a more controlled experimental research design rather than the quantitative data collection methods adopted in this work.

The study sought to obtain results that could be generalised to apply to the profession at large. This called for the development of a reliable measuring instrument as well as the establishment of a representative sample drawn from the population of surveyors involved in providing strategic cost advice to clients. The design of the questionnaire form itself was given particular attention and several versions were piloted. The results of the survey showed that the final form did not fully identify all the risk analysis, knowledge based and value related models currently used by practitioners. This issue should be addressed in any subsequent work in this area.

The actual sample that was established could be flawed due to potential bias in its geographic location. The study was centred on the north of England and North Wales, and although large in size, did not obtain data from across the U.K. The conclusions are valid for the geographic locations studied. As the majority of the large practices and contracting organisations had at least one branch within the geographical area surveyed, the authors take the view that the findings are indicative of the position across the U.K.

The small number of contracting organisations included in the sample was a further cause of potential bias in the results obtained. Given the changing pattern of procurement in the towards design and build it is apparent that further studies in this area must seek to establish a more balanced sample of respondents in terms of organisational types.

The incidence in use of certain models was small. Consequently when analysis data drawn from small populations of respondents bias can be introduced due to exaggerated percentage differences being observed. The final source of potential bias in the work was due to the nature of the subjective assessments that respondents were asked to make. This potential for human bias was unavoidable given the research design chosen. Any further work into specific should seek to establish results on am ore objective basis that takes into account the assessments made by the users as well as the providers of strategic cost advice.

Recommendations for further study

Any future work in this area should address the following issues,

- The establishment and evaluation of the additional knowledge based, risk analysis and value related strategic cost advice models currently used by practitioners but not fully identified in this survey.
- The establishment of the reasons for the non use of life cycle costing models by organisational types other than project management and multi disciplinary practices.

- The investigation and prioritisation of the educational factors suspected as affecting the incidence in use of the newer models listed in the survey, especially the identification of the location of the suspected 'Iggy Factor' within organisations.
- The investigation of the organisational factor of size. In particular, the issue of the largest organisations (in the fourth quartile) having a lower incidence in use of the newer models than the medium sized organisations.
- The investigation of the organisational factor of computing facility. In particular, the confirmation that the existence of a computer facility is a positive stimulus to the continued use of a newer model.
- The objective calibration of the cost advice models in use in terms of their accuracy and reliability in use to both users and providers of strategic cost advice.
- The objective investigation of the role and extent of judgement in the use of the modelling techniques and the confirmation that its role is a positive factor in the value of the models.
- The refinement of the preliminary assessment model developed in the study in order to ensure that it is of use to researchers and others involved in the development of new strategic cost models.

APPENDIX A: QUESTIONNAIRE FORM

Early Advice to Clients

A study into current practices

Notes for Guidance

1 **Early advice** is defined as any cost advice given to the client prior to a formal offer to contract being made.
2 **Organisation** is defined as the part of the company/practice (eg branch, office, department) for which you have responsibility.
3 Please indicate your response by circling the appropriate response in the shaded areas.

Question 1 – Type of organisation

Which of the following best describes your **organisation**'s field of operation?

1	Quantity Surveying practice	7	Management Contracting
2	Multi-disciplinary practice	8	Design and Build contracting
3	Architectural practice	9	Specialist Contracting
4	Consulting Engineering practice	10	Project Management
5	Local or Public Authority	11	Civil Engineering
6	General Contracting	12	None of these

Question 2 – Computing Facility

Within your **organisation** what is the ratio of staff to computer workstations?

1 1 staff to 1 workstation	**2** 2 staff to 1 workstation	**3** 3 staff to 1 workstation	**4** more than 3 staff per workstation	**5** no computing facility

Question 3 – Early advice

Does your **organisation** give **early advice** to clients?   Yes/No

If the answer is **No** you do not need to respond to the remaining questions, please return the questionnaire and thank you for cooperation.
If the answer is **Yes** please continue with the questionnaire.

Question 4 – Work distribution

a How many members of your staff are actively involved with
 early advice function?

                                                                     staff

b What proportion of the **early advice** function is undertaken at
 the following levels?

Partner/Director	%
Senior Surveyor	%
Surveyor	%
Trainee	%
	100%

Question 5 – Early Advice Techniques

The following questions relate to techniques for developing **early advice** to clients. If your response to 5a is **Yes** then continue with 5b to 5e, if **No** then continue with 5f and 5g.

Traditional Techniques	5a Has your organisation ever used the technique?	Yes to 5a (1 – least extent, 4 – most extent)				No to 5a	
		5b To what extent does the technique produce reliable and accurate information?	5c To what extent does the technique rely on the use of judgement?	5d To what extent is technique a valuable tool?	5e Does your the organisatin currently use the technique?	5f Do you understand how the technique works?	5g Does your organisation have the capability to perform the technique?
Judgement – the use of experience and intuition without quantification	Yes/No	1 2 3 4	1 2 3 4	1 2 3 4	Yes/No	Yes/No	Yes/No
Functional unit – single rate applied to the amount of provision (eg £/pupil schools)	Yes/No	1 2 3 4	1 2 3 4	1 2 3 4	Yes/No	Yes/No	Yes/No
Cost per m2 – single rate applied to the gross floor area	Yes/No	1 2 3 4	1 2 3 4	1 2 3 4	Yes/No	Yes/No	Yes/No
Principal item – single rate applied to the principal item of work	Yes/No	1 2 3 4	1 2 3 4	1 2 3 4	Yes/No	Yes/No	Yes/No
Interpolation – interpolating the total cost from previous projects	Yes/No	1 2 3 4	1 2 3 4	1 2 3 4	Yes/No	Yes/No	Yes/No
Elemental analysis – the use of Element Unit Rates to build up an estimate	Yes/No	1 2 3 4	1 2 3 4	1 2 3 4	Yes/No	Yes/No	Yes/No
Significant items – measurement of significant items of work and priced individually	Yes/No	1 2 3 4	1 2 3 4	1 2 3 4	Yes/No	Yes/No	Yes/No
Approximate quantities – measurement of a small number of grouped items priced in groups	Yes/No	1 2 3 4	1 2 3 4	1 2 3 4	Yes/No	Yes/No	Yes/No
Detailed quantities – priced detailed quantities (eg BQ)	Yes/No	1 2 3 4	1 2 3 4	1 2 3 4	Yes/No	Yes/No	Yes/No

Question 5 contd – Early Advice Techniques

The following questions relate to techniques for developing **early advice** to clients. If your response to 5a is **Yes** then continue with 5b to 5e, if **No** then continue with 5f and 5g.

	5a Has your organisation ever used the technique?	Yes to 5a (1 – least extent, 4 – most extent)				No to 5a	
		5b To what extent does the technique produce reliable and accurate information?	5c To what extent does the technique rely on the use of judgement?	5d To what extent is technique a valuable tool?	5e Does your the organisatin currently use the technique?	5f Do you understand how the technique works?	5g Does your organisation have the capability to perform the technique?
Statistical Techniques							
Regression analysis cost models derived from statistical analysis of variables	Yes/No	1 2 3 4	1 2 3 4	1 2 3 4	Yes/No	Yes/No	Yes/No
Time Series Models – based on statistical analysis of trends	Yes/No	1 2 3 4	1 2 3 4	1 2 3 4	Yes/No	Yes/No	Yes/No
Causal Cost Models – Models based on algebriac expression of physical dimensions	Yes/No	1 2 3 4	1 2 3 4	1 2 3 4	Yes/No	Yes/No	Yes/No
Knowledge Based Systems							
ELSIE – as marketed by Imaginor Systems	Yes/No	1 2 3 4	1 2 3 4	1 2 3 4	Yes/No	Yes/No	Yes/No
Other knowledge based systems – any other systems	Yes/No	1 2 3 4	1 2 3 4	1 2 3 4	Yes/No	Yes/No	Yes/No
Life Cycle Costing Techniques							
Net Present Value – discounting future costs to present values	Yes/No	1 2 3 4	1 2 3 4	1 2 3 4	Yes/No	Yes/No	Yes/No
Payback method – calculating the period over which an investment pays for itself	Yes/No	1 2 3 4	1 2 3 4	1 2 3 4	Yes/No	Yes/No	Yes/No
Discounted Cashflow – calculating the discounted rate of return	Yes/No	1 2 3 4	1 2 3 4	1 2 3 4	Yes/No	Yes/No	Yes/No

Question 5 contd – Early Advice Techniques

The following questions relate to techniques for developing **early advice** to clients. If your response to 5a is **Yes** then continue with 5b to 5e, if **No** then continue with 5f and 5g.

	5a Has your organisation ever used the technique?	Yes to 5a (1 – least extent, 4 – most extent)				No to 5a	
		5b To what extent does the technique produce reliable and accurate information?	5c To what extent does the technique rely on the use of judgement?	5d To what extent is technique a valuable tool?	5e Does your the organisation currently use the technique?	5f Do you understand how the technique works?	5g Does your organisation have the capability to perform the technique?
Resource/Process Based Techniques							
Resource Based the use of schedules of materials' plant and labour etc	Yes/No	1 2 3 4	1 2 3 4	1 2 3 4	Yes/No	Yes/No	Yes/No
Process based – The use of bar charts and networks	Yes/No	1 2 3 4	1 2 3 4	1 2 3 4	Yes/No	Yes/No	Yes/No
Construction Cost Simulator – a proprietary system	Yes/No	1 2 3 4	1 2 3 4	1 2 3 4	Yes/No	Yes/No	Yes/No
Risk Analysis Techniques							
Monte Carlo Simulation – based on estimated price and chance deviation	Yes/No	1 2 3 4	1 2 3 4	1 2 3 4	Yes/No	Yes/No	Yes/No
Other risk analysis – any other form of risk analysis	Yes/No	1 2 3 4	1 2 3 4	1 2 3 4	Yes/No	Yes/No	Yes/No
Value Related Techniques							
Value Management – the use of function appraisal in budget formulation	Yes/No	1 2 3 4	1 2 3 4	1 2 3 4	Yes/No	Yes/No	Yes/No
Other Techniques – any other form of value related techniques	Yes/No	1 2 3 4	1 2 3 4	1 2 3 4	Yes/No	Yes/No	Yes/No

APPENDIX B: ADDITIONAL STATISTICAL INFORMATION

Table 1.6.9 Cross-tabulation of type of organisation and use of techniques (showing statistical indicators)

Technique % difference(actual) to expected count	QS Pract's	Organisational type				Statistical indicators		
		Multi- Disc	LA & Public	Cont- racting	Project Man't	Chi- Square Pearson	Lambda with dependent type	tech'que
Traditional								
Judgement	1.6	−6.9	−1.8	0.9	6.0	0.674	0.000	0.000
Functional unit	3.0	0.0	−7.9	−17.4	19.5	0.001	0.000	0.012
Cost per m^2	1.7	−4.1	0.9	−7.0	2.5	0.001	0.017	0.000
Principal item	−1.1	4.8	−3.4	2.0	8.0	0.871	0.000	0.019
Interpolation	0.4	−4.5	2.1	−4.6	8.0	0.527	0.000	0.000
Elemental analysis	2.6	1.0	−2.1	−16.7	11.5	0.000	0.000	0.000
Significant items	1.5	1.0	−19.2	4.6	7.0	0.059	0.000	0.000
Approx quantities	0.8	−3.4	−1.8	−2.0	3.5	0.501	0.000	0.000
Detailed quantities	0.8	−0.7	−1.8	1.7	−8.5	0.921	0.000	0.000
Statistical								
Regression analysis	1.1	−3.8	−2.6	−5.0	9.5	0.329	0.000	0.000
Time Series Models	−0.4	−4.1	6.8	−5.2	10.5	0.271	0.000	0.000
Causal Cost Models	−0.2	−0.3	3.4	−0.7	−2.0	0.931	0.000	0.000
Knowledge								
ELSIE	1.4	−5.5	−7.6	−1.3	9.0	0.405	0.000	0.000
Other knowledge	1.8	−9.0	−6.8	3.0	−2.5	0.252	0.000	0.000
Life Cycle								
Net Present Value	−0.8	17.9	5.3	−22.2	24.5	0.000	0.000	0.081
Payback method	−0.5	−1.4	−4.7	−7.0	32.5	0.026	0.000	0.064
DCF	−3.3	11.7	−5.8	−8.5	52.0	0.000	0.000	0.126
Resource/Process								
Resource based	−1.8	4.8	−12.1	16.1	−0.5	0.041	0.000	0.005
Process based	−5.4	7.2	−17.6	22.8	33.5	0.000	0.000	0.151
Const Cost Sim	−1.2	4.8	−5.3	3.0	9.5	0.057	0.000	0.000
Risk Analysis								
Monte Carlo	0.3	−2.1	−5.5	0.7	9.5	0.211	0.000	0.000
Other risk based	−1.1	7.5	−8.9	−2.2	23.5	0.017	0.000	0.000
Value Related								
Value Management	−1.1	−2.1	−11.3	−2.0	42.0	0.000	0.000	0.078
Other value related	−0.4	−4.5	0.3	−0.7	12.0	0.299	0.000	0.000

Table 1.6.10 Cross-tabulation of computing facility and use of techniques (showing statistical indicators)

Technique	Computing facility (ratio of staff to work-stations)					Statistical indicators		
% difference (actual to expected count)	1:1	2:1	3:1	>3:1	None	Chi-Square Pearson	Lambda with dependant facility	tech'que
Traditional								
Judgement	−1.2	0.1	−0.1	0.8	0.4	0.678	0.012	0.000
Functional unit	1.3	0.3	−0.1	−1.3	−0.4	0.770	0.012	0.000
Cost per m^2	0.2	0.1	0.2	−0.6	0.1	0.648	0.008	0.000
Principal item	−4.0	1.0	1.7	0.9	0.4	0.537	0.000	0.013
Interpolation	−0.8	−1.8	0.1	1.9	0.7	0.092	0.039	0.000
Elemental analysis	1.1	0.0	0.3	−0.1	−1.3	0.232	0.000	0.000
Significant items	−1.0	−1.9	0.3	1.2	0.9	0.538	0.012	0.000
Approx quantities	−0.1	0.0	0.4	−0.3	−0.1	0.864	0.004	0.000
Detailed quantities	−2.1	0.1	1.2	−0.1	0.9	0.634	0.000	0.000
Statistical								
Regression analysis	2.2	0.8	6.1	1.7	−10.8	0.352	0.000	0.000
Time Series Models	3.1	0.4	−2.9	6.9	−7.6	0.389	0.011	0.000
Causal cost models	8.8	11.6	−11.2	4.8	−14.0	0.079	0.000	0.000
Knowledge								
ELSIE	4.9	11.7	−1.9	−0.9	−13.8	0.011	0.020	0.000
Other knowledge	6.1	6.6	0.2	−1.2	−11.5	0.181	0.000	0.000
Life Cycle								
Net Present Value	4.5	5.1	0.6	−0.1	−10.1	0.000	0.024	0.000
Payback method	4.0	4.8	−1.1	−0.2	−7.3	0.024	0.020	0.000
DCF	2.8	5.6	−0.8	0.7	−8.3	0.024	0.016	0.000
Resource/Process								
Resource based	3.2	1.1	−0.6	−1.2	−2.5	0.423	0.012	0.073
Process based	−0.1	0.1	0.7	6.3	−6.9	0.082	0.024	0.000
Const Cost Sim	7.8	3.3	−2.2	−1.1	−8.3	0.789	0.000	0.000
Risk Analysis								
Monte Carlo	−3.7	17.9	2.1	−2.6	−13.7	0.222	0.008	0.000
Other risk based	3.2	4.2	2.3	0.7	−10.4	0.170	0.008	0.000
Value Related								
Value Management	1.8	14.2	−4.7	−1.5	−9.9	0.002	0.039	0.000
Other value related	5.4	3.9	−0.4	1.4	−10.0	0.575	0.000	0.000

Table 1.6.11 Cross-tabulation of size of organisation and use of techniques (showing statistical indicators)

Technique % difference (actual to expected count)	Size of organisation				Statistical indicators		
	1st qtr	2nd qtr	3rd qtr	4th qtr	Chi-Square Pearson	Lambda with dependence facility	tech'que
Traditional							
Judgement	−0.9	0.4	−0.1	0.6	0.532	0.000	0.000
Functional unit	−4.3	1.6	2.0	0.6	0.028	0.000	0.000
Cost per m²	−0.5	0.2	0.3	0.1	0.635	0.000	0.000
Principal item	−5.8	3.4	0.5	1.9	0.125	0.000	0.084
Interpolation	−1.5	0.7	0.6	0.2	0.587	0.000	0.000
Elemental analysis	−3.2	1.8	1.0	0.4	0.010	0.000	0.000
Significant items	−3.7	2.9	0.9	0.0	0.147	0.000	0.000
Approx quantities	−0.5	0.7	−0.3	0.1	0.298	0.000	0.000
Detailed quantities	−7.1	2.8	3.1	1.2	0.001	0.000	0.000
Statistical							
Regression analysis	−17.9	−8.8	15.6	11.5	0.000	0.000	0.000
Time Series Models	−11.6	1.3	4.7	5.6	0.068	0.000	0.000
Causal cost models	−14.6	−0.4	14.2	0.8	0.138	0.000	0.000
Knowledge							
ELSIE	−22.6	11.1	10.4	0.9	0.004	0.000	0.000
Other knowledge	−13.1	−7.1	20.9	−0.6	0.000	0.000	0.000
Life Cycle							
Net Present Value	−13.3	4.9	5.2	3.1	0.000	0.000	0.113
Payback method	−11.0	2.6	6.4	1.9	0.004	0.000	0.078
DCF	−10.5	1.4	4.2	4.7	0.002	0.000	0.060
Resource/Process							
Resource based	−2.8	2.2	0.3	0.3	0.748	0.000	0.053
Process based	−6.5	0.6	0.9	4.9	0.018	0.000	0.054
Const Cost Sim	−31.3	0.7	29.3	3.3	0.001	0.009	0.000
Risk Analysis							
Monte Carlo	−20.6	−8.1	14.4	15.9	0.001	0.000	0.000
Other risk based	−17.8	1.4	11.4	4.9	0.003	0.000	0.000
Value Related							
Value Management	−21.0	0.3	14.5	6.2	0.000	0.000	0.085
Other value related	−22.7	5.8	8.8	8.1	0.016	0.000	0.000

REFERENCES

Al-Tabtabai, H. and Diekmann, J. E. (1992) Judgemental Forecasting in Construction Projects, *Construction Management and Economics*, 10, 19–30.

Allen, G. and Skinner, C. (1991) *Handbook for Research Students in the Social Sciences*, The Falmer Press.

Ashworth, A. and Skitmore, M. (1991) *Accuracy in Estimating*, Occasional Paper, Nr. 27, The Chartered Institute of Building.

Bennet, J. and Ormerod, R. N. (1984) Simulation Applied to Construction Projects, *Construction Management and Economics*, Vol. 2, 225–263.

Bowen, P. A. and Edwards, P. J. (1985) Cost Modelling and Price Forecasting: Practice and Theory in Perspective, *Construction Management and Economics*, Vol. 3, 199–215.

Brandon, P. S. (1982) Building Cost Research – Need for a Paradigm Shift, in *Building Cost Techniques – New Directions*, E & FN Spon, London, pp. 5, 15.

Brandon, P. S. (1990) The Challenge and the Response, *Q. S. Techniques – New Directions*, BSP Professional Books.

Dixon, B. R., Bourma, G. D. and Atkinson, G. B. J. (1987) *Handbook of Social Science Research*, Oxford University Press.

Ellis, C. and Turner, A. (1986) Procurement Problems, *Chartered Quantity Surveyor*, April, p. 11.

Flanagan, R., Norman, G., Meadows, J. and Robinson, G. (1989) *Life Cycle Costing – Theory and Practice*, BSP Prof. Books.

Fortune, C. J. and Lees, M. A. (1989) *An Investigation into Methods of Early Cost Advice for Clients*, Unpublished Research Report Salford College of Technology.

Kerlinger, F. N. (1969) *Foundations of Behavioral Research*, Holt, Rinehart and Winston.

Mak, S. and Raftery, J. (1992) Risk Attitude and Systematic Bias in Estimating and Forecasting, *Construction Management and Economics*, 10, 303–320.

Morrison, N. (1984) The Accuracy of Quantity Surveyors' Cost Estimating, *Construction Management and Economics*, 2, 1, 57–75.

Moser, C. and Kalton, G. (1971) *Survey Methods in Social Investigation*, Heinemann.

Newton, S. (1991) An Agenda for Cost Modelling Research, *Construction Management and Economics*, Vol. 9, April, 97–112.

Nie, N., Hadlai Hull, C., Jenkins, J. G., Bent, D. H. and Steinbrenner, K. (1975) *Statistical Package for the Social Sciences*, 2nd. Ed., McGraw Hill, U.S.A.

O'Brien, M. and Al-Soufi, A. (1994) A Survey of Data Communications in the UK Construction Industry, *Construction Management and Economics*, 12, 457–465.

Panten, M. B. (1950) *Surveys, Polls and Samples: Practical Procedures*, Harper, New York.

Proctor, C. J., Bowen, P. A., Le Roux, G. K. and Fielding, M. J. (1993) Client and Architect Satisfaction with Building Price Advice: An Empirical Study, *CIB W55/ W95 International Symposium on Economic Evaluation and the Built Environment*, Lisbon, 213–226.

Raftery, J. (1991) Models for Construction Cost and Price Forecasting, *Proceedings of the First National Research Conference*, E & F N Spon.

Scott, C. (1961) Research on Mail Surveys, *Journal of the Royal Statistical Society*, XXIV, Part A, 143–195.

Seeley, I. H. (1985) *Building Economics*, 2nd. Ed., McMillan.

Skitmore, M. and Patchell, B. (1990) Developments in Contract Price Forecasting and Bidding Techniques, in *Q.S. Techniques – New Directions*, ed. P. S. Brandon, BSP Professional Books, 75–120.

Skitmore, M., Stradling, S., Tuohy, A. and Mkwezalamba, H. (1990) *Accuracy of Construction Price Forecasts*, The University of Salford.

Skitmore, M. (1991) *Early Stage Construction Price Forecasting – A Review of Performance*, Occasional Paper, Royal Institution of Chartered Surveyors.

Tversky, A. and Kahnemann, D. (1974) Judgement under Uncertainty: Heuristics and Biases, *Science* 185, 1124–1131.

Woods, D. H. (1966) Improving Estimates that Involve Uncertainty, *Harvard Business Review*, Vol. 44, 91–98.

Part 2 Explorations in cost modelling

This part contains papers in which new models are presented and evaluated. A common factor in all these papers is the high degree of rigour in the analysis and reporting of results – an unusual occurrence in this field – which encourages the view that the best work in the field has been, and will continue to be, of an empirical nature. This would seem to be a remarkable state of affairs in view of the comparatively poor state of theoretical development in the field.

James' **A new approach to single price rate approximate estimating** (ch 2.1) is an early paper reiterating Sir Thomas Bennett's (1946) comments on the need to provide cost forecasts at the earliest design stage, that there is considerable amount of uncertainty surrounding functional values due to the great variation in the costs of similar buildings, that there is an increasing need for more accurate early stage budget/target forecasts of cost commitment, and that using popular cube model did not necessarily produce the most accurate results for all types of buildings and circumstances. By 1954, the floor area method had become popular for use with certain types of building but was thought by James to be equally as unsatisfactory as cubing 'for universal application' as neither model seemed able to reflect the effects of different design variables (building shapes, vertical positioning of the floor areas, storey heights, and basements). The 'Storey Enclosure' model described in this paper tries to reflect these differences by assigning arbitrarily different functional values (weights) for the useful elements (floors, roofs and walls), depending on the storey level and whether or not they are below ground level. Some 90 buildings were analysed, by reference to the original drawings and priced bills of quantities, and the results compared with those obtained by using the floor area and cube models. Judged by the rather crude method of analysis used, the new model certainly appears to fare very well – its use producing more accurate results than did the use of the two competing models [the results turn out to be statistically significant (chi square 5.99, 2df)]. The major interest in this work is the bold and well conceived approach to the work, with refreshingly little reliance on conventional wisdom (although extensively quoted in the text). The comparison of accuracy levels achieved through the use of this and other models is a new feature for this type of work (a feature often lacking even today!) although an Analysis of Variance might have been beneficial. Also, the functional values (weighting factors) could have been derived by least squares (the drawbacks of the arbitrary approach are pointed out in the Wilderness paper (ch 3.1)). It is clear, to us at least, that the model has considerable potential for further development by statistical means.

Kouskoulas and Koehn's **Predesign cost estimating function for buildings** (ch 2.2) is an early paper offering an easy introduction to regression modelling for **engineers**. Described in basic terms (residuals etc) it gives a worked example with useful elements (independent variables) – building locality price index, type, height, quality and technology on a set of 3 contracts giving a cv of 3.0%. All variables are indexed (measured at interval level) and tested on 2 *ex post* projects but with no real means of measuring *ex ante* accuracy. It is particularly interesting in its use of **raw** dependent variable with inflation index as independent variable (with fourth highest beta value).

Significantly, McCaffer's later paper (ch 2.3) improves on this by mentioning the deterioration in *ex post* accuracy of models applied *ex ante*.

McCaffer's **Some examples of the use of regression analysis as an estimating tool** (ch 2.3) refers to eight postgraduate studies covering reinforced concrete, services, schools, houses, old peoples homes, passenger lifts and electrical services in buildings which use regression models, resulting in regression models with accuracy levels with between 5 and 40% within model coefficient of variation. It is claimed that in several cases the use of regression models provided significantly more accurate predictions than implementor' early forecasts using conventional manual models and therefore, in appropriate cases, the use of regression models can provide a substantial improvement in accuracy on existing methods of forecasting. The paper provides a nice and easy introduction to the use and potential of regression models.

Russell and Choudhary's **Cost optimisation of buildings** (ch 2.4) considers the optimisation of building **structure** costs (cf Wilderness study, ch 3.1) through the use of process models on the grounds that the cost models adopted 'must reflect a costing procedure similar to that employed by contractors' (cf Brandon ch 1.1, Beeston ch 1.3). It is shown in this study that different design configurations are optimal, 'depending on the form of model used'. Linear regression type models are used for activity times obtained 'after consultation with a contractor'. The drawback to this approach, however, was found to be the requirement that implementors have access to contractors' cost data and have knowledge of construction operations. They then go on to use a fancy method of optimising this now suitably tricky looking problem – their cost contour map showing lots of local optima problems. The main benefit of this paper is to show that even a relatively simple task such as the optimisation of the cost of structural members is pretty difficult to do automatically.

McCaffer *et al.*'s **Predicting the tender price of buildings in the early design stage: method and validation** (ch 2.5) contains an analysis of data using BCIS elements (39 maximum useful elements). There is one quantity/rate [element/function value] pair for each element. Data comprised 107 buildings used to forecast the cost of a further 35 buildings. The 107 building database and 35 targets were subdivided into four sub-groups of (1) Middle and secondary schools (8 targets), (2) Primary, middle and secondary schools (13 targets), (3) Health centres (8 targets), and (4) Offices (6 targets). The numbers in the database sub-divisions are not given. 32 models were tried for each useful element through the database and the best predictive model selected for each element. This was then applied to the target contracts to obtain measures of **forecast** accuracy (average 15% cv). All 32 models provide estimates of the expected value of the element unit rate (function value). The 32 models comprised 16 models using all the data and the same 16 models using elemental data arbitrarily 'segmented' into quantity ranges. The 16 models comprised: 4 univariate models, providing estimates of the expected element unit rate (function value) assuming model structures with (1) normal distributions, (2) log normal distributions, (3) triangular distributions, and (4) log triangular distributions; 4 bivariate models, providing estimates of the expected element unit rate (functional value) from the element unit quantity (element value) by (1) least squares regression, (2) absolute error regression, (3) L1.5 regression, and (4) weighted (chi-square) regression. Each bivariate model was used with (a) raw data, (b) log transformed data, and (c) double log transformed data, making 12 models in all. They used all 32 models to forecast the value of each target useful element in each target contract and recorded the model associated with the most accurate result each time. This failed to show the consistent superiority of any one model. The final part of the study involved developing accuracy selection criteria for the models from the results of fitting to the database – resulting in a minimum median absolute residual criterion. The mix of models thus derived was then applied to the target data. The results are then compared with a few known accuracy rates occurring with conventional manual models (cf Ashworth and Skitmore ch 5.2 for a more comprehensive account of that literature).

2.1 A new approach to single price rate approximate estimating*

W. James

The following paper was read at a General Meeting of Quantity Surveyors, held on 14th April, 1954. A summary of the discussion will be published later. The Annexes referred to in the paper are available from the Institution on request.

In speaking to you to-night I am, of course, breaking one of the golden rules which should be observed by all honorary secretaries, who are always enjoined to comply with the first part of that north-country maxim which begins 'see all, hear all, say nowt'; but in extenuation, I can plead that I am performing a task quite proper to a Secretary: namely that of expressing, however imperfectly, the theories of a number of colleagues. For I must emphasise that what I have to say is the result of teamwork, and therefore no single individual, least of all myself, can claim all the credit, or suffer all the criticism, which is evoked by the ideas I shall endeavour to explain. I am, however, responsible for any views which I specifically state to be personal to myself.

The Institution's Junior Organisation – which is, I believe, the oldest officially sponsored body of professional juniors in the world – is a wonderful breeding ground for dining clubs, formed mainly of past junior committee men; and the object of this Paper is to give you a report on the work of a study group of about a dozen members which is sponsored by one of the youngest of these clubs. It is known as 'The Wilderness,' and has the honour to be presided over by the doyen of the quantity surveying profession, our well-loved Mr. John Theobald.

INTRODUCTION

By way of introduction to our subject, I would like to remind members of a paper entitled 'Cost Investigation,' which was read to a General Meeting of quantity surveyors in April 1946, by that foresighted thinker, Sir Thomas Bennett.

*1954, *RICS Journal*, XXXIII (XI), May, 810–24.

In the first part of his paper – the only part with which we are concerned to-night – Sir Thomas made, among many others, the following points

1   It was only possible to prepare estimates of cost in principle after the architects had made a certain number of drawings, and in so far as the cost picture might cause either the abandonment of the scheme or very substantial modifications, that cost picture must be prepared at the earliest possible moment.
2   The cube was the accepted basis at that time.
3   An examination of the actual costs of a large number of similar buildings brought to light a remarkable variation in the cost per unit.
4   He himself did not for one moment believe that the present cube rules for estimating costs of pitched and flat roofs was right.
5   The public was becoming more and more conscious of the necessity of spending money well and this meant that methods of ascertaining costs which were sufficiently accurate for past requirements no longer met the needs of the moment and would not meet the needs of the future.

These statements are, in my view, just as applicable to-day as when they were made. I must confess, however, that until I began to gather the material for this paper, I did not appreciate how well these points reflected the views of the study group when it set to work.

The figures reported to the client at the preliminary stage of approximate estimating, in which cubing now has the prime place, provide him with his first real indication of the likely capital outlay involved in the building project. As a consequence, any sizable errors in the estimates then made can have serious repercussions later; for, like murder, they will out – accompanied, it may be, by both the architect and the quantity surveyor. It was for these reasons that our group decided to study this aspect of approximate estimating first.

DEFECTS OF THE CUBING AND FLOOR-AREA METHODS

I think that perhaps the simplest way to indicate how the group began its task is to list the principal failings of the two single price-rate systems at present in use.

These failings seem to me to be as follows:

1   Neither the cube nor the floor-area system is satisfactory for universal application.
       This is indicated by the fact that cubing has been largely superseded by the floor-area method for buildings of certain types.
2   With the sole exception of the differentiation made under the R.I.B.A. cubing rules between the measurement of pitched and flat roofs – a differentiation upon which I think most surveyors share Sir Thomas's doubts

– neither cubes nor floor supers reflect, in a realistic way, the shape of the building: and shape has an important bearing on cost.

In this respect, the cube method in effect requires only the void enclosed within the building to be measured and not the structural enclosure which contains this void: and, although the floor area method is more realistic in its approach to this aspect of estimating, in so far as it requires direct measurements of the basic horizontal parts of the building to be taken, this method completely fails to account in these measurements for the variations which occur in the ratio which the external wall perimeters of differently-shaped buildings bear to their floor plan areas.

This point may perhaps be appreciated more clearly by saying that we can all visualise two buildings of the same cube, floor area and storey heights, each of broadly similar specification – the plans of one being of a square shape whilst those of the other are of a long and thin rectangular shape. The costs of these will differ appreciably owing to the greater external wall area of the rectangular-shaped building.

3   Under neither of the established methods do the measurements effectively reflect the number of storeys among which its total floor area is distributed, or any variations in the heights of these storeys.

In the case of cubes, this is best illustrated by recalling that two identical type building may have the same cubic content when one is of six and the other of seven storeys: whilst the floor-area system gives an identical square footage for every building which has the same total floor plan area, quite irrespective of storey heights and/or of whether the area is contained on one or a dozen floors.

4   The only practicable way of estimating, by single price-rate methods, the cost of a building with a large basement is to cube or square the floor area of the basement separately from the remainder of the building.

This solution is really foreign to the whole idea of these methods for estimating building costs. Moreover, its application becomes very difficult when one is faced with semi-basements, or with buildings partially sunk into rising ground so that the ground floor on one side is the basement on the other.

5   The price rate to be attached to the total cube, or floor area, thus has to take into account not only the type of structure and standard of finish required – points which I think every surveyor will acknowledge are, in themselves, sufficiently intricate to tax his skill – but also all the various imponderable factors of shape, number of storeys, storey heights and extra cost of any storeys below ground. In addition, under the cubing method the integral unit (the foot cube) is so comparatively small, and the collective total so comparatively large, in relation to the cost of building, that a penny difference in estimating the rate may sometimes be a serious matter.

For these reasons in our lighter moments, my office sometimes refers to the assessment of cube rates as the 'crystal and black velvet' stage of the estimate.

THE REQUIREMENTS ANY NEW METHOD SHOULD ENDEAVOUR TO SATISFY

In the light of the above considerations, the study group decided to see whether it was practicable to evolve a new, universally applicable, single price-rate system, which would accomplish what the cubing method purports to do, without the obvious drawbacks of the latter – in other words a system which, whilst leaving the type of structure and the standard of finish of the building to be assessed in the price rate, would take into account *in the measurements*, the following factors:

1 the shape of the building;
2 the vertical positioning in the building of the floor areas it contains;
3 the storey height, or heights, of the building; and
4 the extra cost of sinking usable floor area below the ground level.

Within the practicable limits of the problem, the group believes that it has substantially accomplished this task, although it fully appreciates that a great deal more research and exploration by a much larger body of practitioners is necessary before this statement can be generally accepted at its face value. Since, however, the investigation made to date appear to indicate an appreciable advance over the results provided by the cube and floor-area methods, it seemed only right that the theories explored and developed to date by the group should now be published for general examination. But I must emphasize strongly that these theories are not an Aladdin's lamp which will throw light upon and spirit away all your approximate estimating difficulties. They will not, for instance, indicate how you value a proposed building's construction and finish, factors which affect price rates to a far greater extent than any of those I have dwelt upon earlier; not will they serve to assist you much in the evaluation of projects of an abnormal character or those concerned with works of alteration, maintenance or redecoration, for all of these are outside the scope of any single price-rate system.

In my view, single price-rate estimating has two main uses. The first – sometimes known in quantity surveyors' slang as 'guestimating' – consists of applying to the sum total of the measurement products an inclusive, overall price rate which aims to give a rough guide to the likely total cost of all the contract works. Inclusive rates of this kind are of course the ones which architects so frequently expect the quantity surveyor to be able to give them over the telephone, before any scale drawings have been made and before any of the costly pitfalls in the project have come to light. However misleading surveyors may consider such rates to be, they have a definite place in design technique and therefore any system which aims to supersede cubing must take some account of them. The second, which if the quantity surveyor is lucky comes a little later in point of time, is the more carefully considered estimate to which Sir Thomas referred; the one, you will remember, which 'cannot be made until the architect has prepared a number of drawings'; drawings which presumably the surveyor can take away and study, at least for the period between 5 p.m. on one day and 10 a.m. on the next.

Clearly, when devising a new method, the approach should be by way of examining its application to estimates of the second type, since only in this way can it be properly tested and proved: once this high hurdle is cleared, price rates of the overall variety will automatically follow. At any rate this is the assumption upon which the study group proceeded, and therefore, with the exception of one further passing reference to inclusive rates appropriate to estimates of the first kind, the rest of this paper is devoted entirely to considered estimates of the second type.

At an early stage, the group found it essential to decide what part of the cost of a new building project could fairly be expected to be assessed by means of any single price-rate method – a method designed, as it must of necessity be, for the application of empirically deduced rates to mere outline particulars of proposed new schemes. The group reached the conclusion that only the value of the structure and its finishings could be assessed in this way, and that, in consequence, the following categories of work would require to be separately estimated:

1   *Site works, including roads, paths, etc., drains, service mains and external ducts.*

Since these works vary so extensively from one scheme to another, the group considers that the only realistic way of estimating them is to proceed by way of priced approximate quantities.

2   *The extra cost of foundations that are more expensive than those which would normally be provided for the type of building concerned.*

This item has the same extensive cost variability between one site and another as applies to external works, and such extra foundation costs are therefore best estimated in a similar way.

3   *The category of works which is sometimes loosely described as engineering services, comprises:*
(*a*) sanitary plumbing and cold water;
(*b*) hot water;
(*c*) central heating and ventilation;
(*d*) electrical installations;
(*e*) gas, compressed air and similar items; and
(*f*) lifts.

The cost of these services is seldom constant between one project and another and is therefore a matter which requires separate attention when preparing a considered estimate.

Many surveyors value the first four of these engineering services from experience, by way of a carefully judged percentage addition to the estimated net building cost as assessed by the cube or floor-area methods, and this principle commends itself to the group. Where past experience of similar types of installation is lacking, valuation must be accomplished by way of priced approximate quantities or with the aid of a helpful specialist consultant – methods which also apply to other services such as gas, compressed air, wireless, lifts and similar work.

Sir Thomas drew attention to the fact that building owners were often in a position to provide one or other of these service installations for themselves, and this is another cogent reason for separately assessing the value of such items – at any rate in factory projects.

4   *Features not general in the structure as a whole*

These include items such as boiler flues, dormers, canopies and the like, which it is only practicable, in the group's opinion, to value by way of priced enumeration.

5   *Circular work*

This can vary so greatly according to the radius, the particular material and the form of construction concerned, that the group has not, at any rate as yet, been able to evolve a method of reflecting the extra cost of circular work of all kinds, by way of appropriate measurement allowances. Any constructive help on this matter will be much welcomed.

BASIS OF THE PROPOSED NEW SYSTEM

By now I hope you will be curious to know the basis of the group's proposed new system, so I will state it without further ado. Fundamentally, this consists of measuring the areas of the external walls, the floor and the ceiling which enclose each storey of the building. It is for this reason that the group has called its proposal 'The Storey-Enclosure System' of approximate estimating. However, as you would expect, this simple basic concept does not, by itself, provide the whole answer to the problem and it is therefore necessary to complicate it slightly by incorporating certain modifications.

The modifications comprised in the latest set of rules prepared by the group are as follows:

1   *To allow for the cost of normal foundations*, an additional timesing factor of 1 is applied to the area of the lowest floor; this, when added to this single timesing under the basic rule, requires the lowest floor area to be twiced.

2   *To provide for the extra cost of upper floors*, an additional timesing factor is applied to the area of each floor above the lowest. This factor consists of an arithmetical progression of 15 per cent., a term which sounds rather frightening but which only means that the additional timesing for the first

suspended floor is .15, for the second .30, for the third .45, and so on.

3   *To cover the extra cost of storeys wholly or partly below the finished ground level*, as additional timesing factor of 1 is applied to the basic wall and floor areas of any such storey to the extent that its external walls and lowest floor adjoin the earth face.

A little thought about the application of this system will make it evident that the required superficial figure is ascertained by totalling the sum of the following measurements of the building:

First – twice the area of the lowest floor to the extent that this is above the finished ground level, and three times the area of this floor to the extent that it is below this level;

Secondly – once the area of the roof, which the groups considers should be measured on the flat plan area in all cases;

Thirdly – once the area of the upper floors multiplied by a composite timesing factor made up of a constant of 2 (namely, once for the storey below and once for the storey above the floor concerned); plus the appropriate positional factor to which I referred earlier; and

Fourthly – once the area of the external walls of storeys or parts of storeys above the ground, plus, in addition, twice the area of external walls below the finished ground level to the extent that these latter walls enclose any storey or storeys wholly or partly below this level.

The current set of detailed rules, embodying these principles, which the group is now using is set out in Annex 1. (*Note:* The Annexes to the paper are available from the Institution, on request.) As with everything of an experimental nature, these rules should not be regarded as the final word on the subject, and the group has an open mind for any amendments which further experience in their application may show to be desirable. In any case, I am sure that you will wish me to employ the limited time available to-night in explaining the basic theory of the system rather than the individual clauses of the rules themselves.

Since the aim of the Storey-Enclosure System is to provide a total superficial figure to which a single price-rate can be attached, the effect of the principles embodied in these rules is to apply a 'weighted' cost factor to each of the main components of the building. The factors so applied are as follows:

To lowest floors (except those to parts of the building which are below ground) and to roofs: a timesing factor of $1\frac{1}{2}$ to each. Since these two areas are nearly always identical, the rules are, for convenience, framed to indicate a factor of 2 for the floor and 1 for the roof.

To upper floors: a basic timesing factor of 2 plus the requisite multiple of 15 per cent, which is appropriate to their vertical situation in the building.

To external walls (except those of storeys below ground): a timesing factor of 1.

In the case of multi-storey buildings where the cost of the lower parts of external walls and stanchions is greater, the group considers it inappropriate to increase the timesing factor to walls, because this extra cost applies equally to internal structural walls, stanchions, etc., which are not measured under the rules. This extra cost is therefore incorporated in the progressionally increasing timesing factor applied to upper floors.

To enclosures below ground: a times factor of 2 to walls and $2\frac{1}{2}$ to floors.

As a general average – and I must emphasize that no single price-rate system can reflect anything but a general average from which almost all individual projects will differ in some respects – I think these factors provide a reasonably logical cost 'weighting' on the following counts.

1   The lowest floor 'weighting' factor of 2 allows for the cost of:
internal partitioning, finishings, fitments, doors, etc., on the floor;
a non-suspended floor;
finishings on one side of it; and
normal foundations to all vertical structural members in a single story building including those of its external walls.

2   The roof 'weighting' factor of $1\frac{1}{2}$ provides for the cost of:
a suspended roof and its (lighter-than-floor) load;
finishings on both sides of it (one weatherproof);
horizontal structural supports to it (such as beams and trusses); and
vertical structural supports to it (such as walls and columns).

3   The upper floor 'weighting' factors of 2 *plus* cover the cost of:
internal partitioning, finishings, fitments, doors, etc., on the floor;
a suspended load-carrying floor;
finishings on both sides of it;
horizontal structural supports to it;
vertical structural supports to it; and
the further cost which arises, in the case of vertical structural floor supports to the lower floors of multi-storey buildings, from the need to support the additional transmitted load of all superimposed floors and the roof above them.

4   The external wall 'weighting' factor of 1 allows for the cost of:
a wall with weatherproof qualities;
finishings on both sides of it;
windows and external doors, etc.; and
normal architectural features.

5   The below ground 'weighting' factors of 2 for walls and $2\frac{1}{2}$ for floors provides for the cost of:
displacement and disposal of earth;
waterproof tanking and the loading skins to keep it in position;

members of heavier construction than those required in equivalent positions above ground;
finishings on one side of these members;
internal partitioning, finishes, fitments, doors, etc.; and
normal (in the basement sense) foundations to all vertical structural members in a single basement-storey building.

For the reasons I have just outlined, the study group maintains that, as nearly as can be expected from any system based on general averages, a total superficial figure ascertained in accordance with the measurement rules of the Storey-Enclosure System of approximate estimating satisfies the four requirements I listed earlier, that is to say:

- *Shape* is reflected because all the surrounding surfaces of the building are measured;
- *Vertical positioning of floor areas* is taken into account by reason, firs,t of the greater measurement product given to suspended floors as compared with non-suspended, and secondly, of the increasingly greater emphasis applied to their areas according to the numbers of floors below them;
- *Storey Heights* are embodied because the lower the height of the storey, the higher will be the proportion of the total figure taken up by floor and roof measurements, as compared with those of external walls – moreover, measurements taken in this way allow each storey height in the building to be given its relative importance in making up the total figure, an attribute which is particularly useful, for instance, when dealing with those awkward things known as mezzanine floors which are so difficult to value under the cubing method; and
- *Cost of sinking storeys below ground* is allowed for by means of the proportionately larger area allocated in making up the total, to those enclosing walls and floors which are underground.

These are bold assertions, and when the basic theory of the new system was first mooted, the majority of the group was just as sceptical as the major part of my audience no doubt is now: indeed this unbelief was clearly justified until the original proposal of measuring the structural enclosure of the whole building (plus only once the area of its suspended floors) was modified to that of applying this principle separately to each of its storeys.

EXAMPLES OF STOREY-ENCLOSURE PRICE RATES CALCULATED FROM ACTUAL CONTRACT DOCUMENTS

Before reaching its present stage of confidence in the principles of the Storey-Enclosure System, the individual members of the study group analysed some 90 tenders for new buildings of various kinds, of a total value of about £5 million, by reference to the original drawings and priced bills of quantities, and its members began to test the system, in double harness with cubes on floor areas, in their own offices on a number of proposed new projects. To enable the price rates ascertained by these analyses to be compared and studied, two preliminaries were essential.

- *First:* since little could be gained by a comparison between the rate of a dutch barn with that of a Buckingham Palace, it was necessary to separate the various buildings into a number of general types and to study only those types of which a reasonable number were likely to be available. The types of building selected by the group were flats, schools and industrial buildings. More recently some attention has been given to the application of the system to houses built, under contract, for letting.
- *Secondly:* in this era of fast-changing prices, it was also necessary to devise a table of conversion factors which, by taking account of the general price levels ruling at the dates of any of the tenders under examination, could be applied to all the individual rates to reduce them to a common price-time datum. The table finally adopted by the group is set out in Annex 2 to this paper. The factors in this table are merely the group's empirical opinion to the nearest $2^{1}/_{2}$ per cent.; they are, of course, anybody's guess since, no one can demonstrably prove what is essentially a matter of opinion. I have, however, checked that these factors correspond reasonably well with the index of building costs used by the Board of Inland Revenue for assessing allowances for the cost of construction of industrial buildings under the post-war income tax acts, to the extent that this index has been published in the *Journal.*
- The selected datum of June, 1951, has no special significance, for it is merely the time at which the group began to compile the conversion factor table.

Before drawing your attention to the results of the analyses made, I would like to clear the one further point of why the group used tender figures and not final account figures. This lies primarily in the fact that the aim of an approximate estimate is to forecast the amount of the lowest tender; for it is against this sum, and at that stage against this sum alone, that the calibre of the estimate will be judged by the client. Divergencies between the contract sum and the total of the final account are generally due to reasons entirely unsuspected at the time the forecast is made and cannot, therefore, properly be taken into account when devising any system of approximate estimating.

With this preamble, I refer to you Annexes 3A to 6A and 3B to 6B inclusive, which set out in tabulated form the main results of the group's investigations. The No. 3 sheets relate to flats, the No. 4 to schools, the No. 5 to industrial buildings and the No. 6 to houses.

- The 'A' papers give the comparative cube, floor area, and storey-enclosure price rates of the buildings concerned. The left-hand rates are those relating to the carcass and finish only of the buildings exclusive of all the engineering services, site works and other abnormal extra costs to which I have referred earlier; and the right-hand rates are the inclusive overall ones for the same buildings obtained from the related total tender figures; these latter rates include all the services, site works, and abnormal costs except in those cases where any of these items were excluded from the building contract. All the price rates on these 'A' sheets have been converted to the 1951 price-time datum by the application of the conversion factors given in Annex 2, so as to enable a proper comparison to be made between all the rates concerned.
- The 'B' papers give explanatory particulars connected with application of the Storey-Enclosure System to these same projects. The conversion factors have not been applied to any of the price rates on these 'B' sheets, but the date of the tender is given in each case. When using the system in practice, it would not, of course, be necessary to examine and record a large part of the information shown on the 'B' sheets, which has been included in this paper to assist you in studying the new system and to give initial data to those who propose to try out the new system in their own practice.

In the light of the very short acquaintance you now have with the principles of the new system, it is obviously impracticable for me to take you through the details of these Annexes. So I must confine my remarks entirely to those points on the sheets which seem to me to be of particular interest and leave you to study their remaining contents at your leisure.

POINTS OF INTEREST ARISING FROM THE FIGURES IN ANNEXES 3, 4, 5 AND 6

First, it seems important, before proceeding, to see whether the figures support the group's contention that separate assessment of the value of engineering services makes the estimating of the main price rate a little easier.

I think that a glance at the 'B' sheets will show that this particular point is a sound one owing to the extreme variability of the relationship between the cost of the building work on the one hand and that of the services on the other; for examples, see Annex 5B, references 29 and 31 where the total cost percentages of the engineering services are $38^1/_8$ per cent. and 14 per cent. respectively.

I am, of course, assuming that the need for estimating separately the cost of site works, lifts and abnormal foundations is undisputed when one is preparing a carefully considered cost forecast. For where all these matters are thrown in, to be assessed in a flash in one price rate (of the kind indicated in the right-hand rates of the 'A' and 'B' Sheets), the possibility of real accuracy is about as great as those of wining a treble at the races. It is curious to note that overall inclusive rates usually have to be given, at once, in casual conversation; while a sight of the drawings and a day or two to consider the problem are usually allowed for estimates based upon net rates. Thus is the time element reversed in the old maxim that surveyors can do the impossible at short notice, but require a little longer for miracles!'

Secondly, it is vital to examine the net price rates on the 'A' sheets to see whether the new method effects a reduction in the *range of variation* of the price rates, by comparing those obtained by the cube, floor-area, and Storey-Enclosure systems respectively. For this purpose, I have applied two tests, the first being to ascertain how many times the highest rate is of the lowest, and the second, how many of the rates fall within a similar range group above and below the common average. For flats, schools and houses a 10 per cent. plus and minus grouping has been adopted; for industrial buildings, which are not subject to such closely applied ministerial planning and financial restrictions, a wider grouping of 20 per cent. plus and minus is necessary to provide a parallel kind of test.

I am sure it is not necessary for me to stress the advantages which would flow from the adoption of a single method of approximate estimating for all types of new buildings – provided of course that the method was an improvement on cubing and that it at least holds its own in the particular field where the floor area method has already tended to oust the cube.

From the results I have just given, it is clear that the storey-enclosure price rates are in all cases nearer to the common average in respect of the projects which the group has examined; and in all cases the range of price variation is in fact reduced. The gains are not perhaps as great as might have been hoped, but they nevertheless represent an appreciable advance on present methods. More-over, it must be remembered that type of structure and finish of the building are bound to have by far the largest influence on price rates; this can be verified by looking first at the price rates of the buildings numbered 2 and 24 on Annex 5A – No. 2 being a heavy steel-framed explosion-resisting building, tiled throughout with acid resisting tiles, and No. 24 being a plain customs shed; and secondly at the price rate of

Table 2.1.1 The results of these tests

Type of Building	Total Number of Cases Examined	Timesing of Lowest Rate to equal Highest Rate			Number of Rates within Percentage Grouping		
		Cube	Floor Area	Storey-Enclosure Area	Cube	Floor Area	Storey-Enclosure Area
Flats	16	2.96	2.63	2.09	9	10	12
Schools	14	1.41	1.36	1.34	9	8	12
Industrial Buildings	39	4.42	2.61	2.48	16	24	26
Houses	17	1.45	1.44	1.42	8	9	10

building No. 10 on Annex 3A, which is a block of West End flats, compared with those of the remainder of the flats which are for local authority housing.

Thirdly, you will no doubt wish me to point to something in the annexes which illustrates that the additional timesing factor of 1 is an appropriate allowance for foundations. This I regretfully found that I could not do without giving you a great many more charts and figures – and in this connection I had in mind the saying that a certain amount of livestock is good for a dog since it keeps it energetic, but too many may kill it. I can, however, give you the main reasons for selecting this factor, which were:

1   the application of such a factor to the lowest floor area, to allow for the cost of all normal wall and stanchion foundations, is an improvement on the way in which the cube method tackles this problem, since it enables all wall measurements to be stopped **at the lowest floor level** – a point at which most sketch plans tend to become very vague; and
2   without such a timesing addition, in the case of houses, the system showed less advantage over the cube and floor-area methods; the relationship between the price rates of bungalows and houses was out of alignment; and the improvement was not so appreciable as it now is in respect of flats and schools.

Fourthly, you will naturally be concerned with the propriety of the additional timesing allowance for taking into account the vertical positioning of the upper floors. Here, I ask you to turn to Annex 3A and look at the rates in column 5 of the flats, building numbered 3,4,5 and 6. All these buildings are part of a single contract designed by the same architect and priced by the same contractor; building No. 3 is of eight storeys, whilst the others are of only three storeys. As you will observe, the respective storey-enclosure rates are:

- 9s 7¼d. for the eight-storey building which consists of three bedroom flats; and
- 9s. 9d., 10s 1d. and 9s. 5½d. for the three-storey buildings which consist of old peoples', two bedroom and four bedroom flats respectively.

In this respect the 10s. 1d. rate is slightly out of step with the others – a fact which appears to be due to some peculiarity of balcony arrangements which the group is still examining – but otherwise the cumulative timesing factor provides a proper relationship between these buildings of widely different numbers of storeys – a relationship which is absent, not only in the case I have just mentioned, but in the general run of all the figures in the various annexes, if the factor is increased, reduced or omitted altogether.

Fifthly, I will comment on the timesing factor for enclosing floor space below the ground.

This was originally devised from the starting point that in the case of three buildings, for which we had separately costed bills of the basement work, the cube rates for the basements worked out at a little less than twice the super-structure cube rates.

As far as the application of the storey-enclosure rules is concerned, I suggest that you look at Annex No. 3A, buildings Nos. 14 and 15, which have fairly large basements. You will see that the rates in column 5 appear to be in reasonably good alignment with the rates of other buildings in this annex, most of which have no basements. The same observation applies equally to building No. 34 on Annex No. 5A, which is a 500-foot long, 60-foot wide, industrial building with a sub-storey under half its plan area, this sub-storey being let into an embankment so that half its volume only is below the finished ground level. All these three examples relate to projects with which I am personally familiar, and I can only say that the rates obtained by the application of the present rules seem to me to be about right in these cases.

Sixthly, I refer to the storey-enclosure method of measuring roofs and whether the present rules are realistic in making no measurement distinction between flat, pitched or any other shape of roof. I must confess that the group is not yet entirely satisfied even whether it is proper to treat flat and pitched roofs similarly, for the reason that sufficient projects of a kind suitable for testing this question have not yet been available. For you will appreciate that, for a proper test, it would be necessary to have analysed storey-enclosure rates for several buildings which were similar in general construction, finish, floor loading, shape, storey heights, etc., some of which had pitched roofs and some flat roofs.

The difficulties are well illustrated by an investigation carried out by the group under which a pitched roof and a flat roof to a building 117 feet 5 inches long and 27 feet $10^{1}/_{2}$ inches wide were each measured in detail and billed, priced at analogous rates and totalled for comparison. The types of roof selected were:

1   **the pitched roof**, consisting of 40□pitch roman tiling, battens and felt on 4□by 2□rafters with normal collars, struts, etc.; two intermediate party walls complete with two chimney stacks; 4□by 2□ceiling joists and skim-coated plaster-boarded ceiling, overlaid with strips of asbestos wool between the joists; the eaves being constructed with sprocket pieces , fascia, soflite board and gutters; and

2   **the flat roof**, formed of a 6□hollow tile structural slab covered on top with foamed slag screed and felt roofing and with two coat plain face plastering beneath; there were two chimney stacks and two small tank roms with half brick walls and 4□reinforced concrete roofs, above the general roof level; and a 12□projection eaves slab $4^{1}/_{2}$□thick was taken all round the roof.

The comparative price rates for the roof works alone in these two cases were found to be:

- *Cube rates:* 1s. $2^1/_2d$. for the pitched and 3s. $2^1/_2d$. for the flat roof.
- *Storey-Enclosure Rates:* 6s. 8d. for the pitched and 8s. 2d. for the flat roof (these rates do not, of course, cover the cost of vertical supports beneath the roof and so are less than normal rates).

In the group's opinion, this test points to two main conclusions:

1   that the R.I.B.A. cubing rules for the measurements of pitched and flat roofs are unrealistic; and
2   that because it can logically be maintained that the lighter pitched roof construction should be somewhat cheaper than the heavier fire-resisting flat roof construction with which it is in this case being compared, the storey-enclosure rules as now put forward by the group appear to be on reasonable lines when applied to pitched and flat roofs of similar construction.

The group, however, is well aware that these deductions rest upon only a single detailed comparison of two isolated types of pitched and flat roofs, and that what we are really after is that unattainable goal, a comparison between the **average** cost of **all** types of flat roof and that of **all** types of pitched roof. In this connection it is worth looking at the rates for blocks of flats with pitched roofs on Annex 3A as compared with those of similar buildings with flat roofs; for if the cube basis was right all the 'pitched roofer' rates in column 5 would tend to be higher than the 'flat roofer' rates. The 'pitched roofers' are buildings Nos. 2, 8, 10 (the West End flats), 12 and 16, so you can see that this is not the case. As regards vaulted and barrel type roofs, the group has not yet made any attempt to study and devise a method of allowing for their extra cost in the measurements and this factor, under the present rules, must still be assessed in fixing the price rate.

Lastly, the group is well aware that the new system calls for measurements which are less simple than those required under the cube and floor-area methods: but in this case, where a choice lies between simplifying measurement (which is factual) and simplifying pricing (which is an art), the group has no hesitation in recommending the latter. Nevertheless, its members are continuing their researches with the aim of devising improvements in the system to make the measurements easier, without sacrificing the pricing advantages obtained under the present rules, and in this connection the group will welcome the wider research and constructive criticism which only the profession as a whole can give.

APPLICATION OF THE STOREY-ENCLOSURE SYSTEM

As I mentioned earlier, various individual members of the study group, including myself, have for some time been using the new system for the preparation of approximate estimates in our own offices and we have found it practical and helpful, particularly after some experience in its use. But, when one has been used to roller skates for a long time, it is foolish to expect that the first time one dons ice skates perfect figure skating will be possible. There is a difference in technique between assessing cube and storey-enclosure prices and before any of you put the Storey-Enclosure System into practice, I suggest you will find it helpful to apply the rules to several past contracts and study the resulting price rates so as to get the general 'feel' of the system. Thereafter, the new and old methods can be applied side by side – the resulting totals, ascertained respectively by cube or floor-area methods and by the Storey-Enclosure System form a useful crosscheck on each other – until you find you have gained confidence enough in the new system to use it alone.

Before leaving this aspect of our subject, I would like to address a word of warning to any members and visitors here to-night who are not quantity surveyors. The prices set out in the annexes to this paper have been published solely to illustrate the theories I have endeavoured to explain *and these prices are not intended to form the nucleus of a cost information bureau for general use.* In my view, and I think most of those with practical experience of approximate estimating will agree with me here, the compilation and publication of any such cost information by a central body would be fraught with risks and impossibly expensive. For, to be of any practical value, the figures would have to be accompanied by the most exhaustive description of every aspect of the project and a set of the drawings concerned: these descriptions would have to be framed by someone who had a wide experience of approximate estimating; the copyright of the drawings would have to be purchased; and probably an 'all risks' insurance policy would have to be taken out to cover the claims of the inevitably large number of people inexperienced in estimating who would say that they had been misled by the information which had been published. The jet-propelled fighter is a wonderfully designed aeroplane which, as we all heard last autumn, can travel at an enormous pace in the hands of skilled test pilots; I think it likely, however, that if the designers had been at the helm of Duke's and Lithgow's machines, their design ability would have been of little help towards the task in hand and, in consequence, something other than the speed record would probably have been broken. By the same token, approximate estimating, even with the best of aids, is a process which, in my view, is only safe to entrust to a quantity surveyor – and then only to the most experienced men in the quantity surveying firm concerned; so do not hesitate to consult a quantity surveyor on cost problems at this early stage of the project. He will be glad to help you on these vital questions, for the quantity surveying profession considers this service to be one of the most important it can render to the building owner. I do not, of course, guarantee that the quantity surveyor's estimate will always be right, but I maintain against all comers that it is far more likely to be right than anyone else's.

Apart from the direct application of the storey-enclosure rules to estimating, the group ventures to suggest that architects should find them helpful in solving at least one type of design problem, namely which of a number of possible planning solutions to a particular project is likely to be the most economic. In such cases, provided the general construction, finish and site works of the alternatives are similar and provided a reasonable adjustment allowance is made for any differences there are likely to be in the cost of engineering services (in particular for lifts), a straight comparison of the total super-footages of each, measured in accordance with the rules, should generally give a good guide; the one with the lowest total being likely to be the cheapest.

THE PERSONNEL OF THE STUDY GROUP

In conclusion, I think you should know the names of those who have formed the backbone of the study group. They are: A. Prichard (Chairman), E. N. Harris (Honorary Secretary), L. H. Assiter, E. A. Baker, J. B. Cannell, J. W. Cloux, J. F. Green, J. G. Osborne and M. O. Sheffield.

Whether or not you agree with their conclusions, I think your thanks are due to them for their efforts. These efforts are now continuing in pursuit of a much more complex aspect of approximate estimating that of an analysis of the comparative costs of various types of structural floors, framings and walls, in an endeavour to see if any basic principles exist governing the cost relationship between live floor loads on the one hand and various methods of constructing, spanning and supporting the floors to carry them on the other hand in which they have been joined by two most helpful structural engineers, F. J. Brand and G. W. Granger.

This particular investigation appears to have the main characteristic of a commuted death penalty namely, it is likely to be almost a life sentence because of the vast amount of detailed work involved, all of which has to be fitted into the so-called spare time of busy principals. Whether there will be any productive conclusions on the subject to report to you in the years to come is at present an open question; for the simile of a blindfold man with a butterfly net chasing a gnat in the dark in the Albert Hall is peculiarly apposite to basic cost research.

2.2 Predesign cost estimating function for buildings*

V. Kouskoulas and E. Koehn

INTRODUCTION

The final cost estimate of construction works must be very close to actual cost since it is generally used as a basis for bidding. Detailed quantity takeoffs, material, and labor costs are both time consuming and expensive. Indeed, the price or cost for such detailed estimates may vary between 0.6% and 3% of total project costs for an accuracy between +3% and -12% of the actual costs (6).

Before arriving at the final estimation process for a project, preliminary or predesign estimates are usually required. Decision-making, initial appropriations, and economic feasibility studies are based upon such estimates. In the absence of design plans and specifications, preliminary estimates lacking a rational basis may prove detrimental to the decision maker and the owners financing a project. On the other hand, any such estimates must be inexpensive, quick, and reasonably accurate. The owner and those financing a project cannot afford preliminary estimate costs anywhere near the aforementioned order of magnitude; much less the designer or contractor can afford such service costs for a potential client. A method for preliminary or predesign estimation is necessary for all types of projects and should be available whenever feasible to those concerned with the conception, design, and construction of projects.

The need for predesign cost estimation techniques has certainly been realized and such methods have been developed utilizing past information on costs as functions of variables characteristic of classes of projects. For example, the cost of sewage treatment plants as a function of waste-flow, in million gallons per day, the cost of steam-electric generating plants as a function of plant-capacity, in megawatts, and the costs of public housing projects and schools as functions of the number of rooms and the square feet per student have been studied in Ref. 13 utilizing information contained in Refs. 4, 9, 12, and 15. Similar techniques may be developed for several classes of buildings employing the cost history for each class whenever available. It

*1974, *ASCE J. of Const. Div.* Dec., 589–604.

will be necessary, however, that the costs be related to more variables for more reliable preliminary estimates. In view of the diversity of buildings with reference to function, locality, quality, etc., it is expected that cost depends on all these factors.

This work attempts to solve the predesign cost estimation of buildings by deriving a single cost-estimation function that applies to several classes of buildings and defines cost in terms of several other measurable variables. The end result is a predesign multilinear cost-estimation function. This function is tested by means of several completed projects demonstrating that the cost of proposed new buildings projects can be estimated with an accuracy that is satisfactory and reasonable for many purposes. The resulting predesign estimation technique is simple, fast, and applicable to a wide variety of building construction projects. But more important, the methodology may be generalized in a global sense.

It is not the intention of this work to replace human experience and judgment by mathematical formulas. On the contrary, it is its primary objective to provide a framework for the very best utilization of experience and in the absence of sufficient experience with a certain project to introduce objectivity by inference to other similar or related projects. This is immediately evident from the structure and content of the predesign cost-estimation function that is applicable with a variety of building projects.

PREDESIGN COST-ESTIMATION FUNCTION

The cost of a building is a function of many variables. The first and basic problem is the selection of a set of independent variables that may describe a project and define its cost. Such variables must be measurable for each new building project. The second problem is the determination of the cost function in terms of the selected variables in closed mathematical form. Whatever technique one may choose for the construction of the cost function, its immediate use dictates that a criterion for the selection of the variables is the availability of data on such variables from building projects completed in the past. these data are not only required for the construction of the cost functions, but also for solving the third problem, i.e., testing the reliability of the derived function.

Guided with the preceding principles, and after extensive search regarding the availability of data, one may define the cost function by the relation

$$C = f(V_1, \ldots, V_6) \tag{1}$$

in which C = some cost measure characteristic of buildings; and V_k = six independent variables, which are in sequence specifying building locality, price index, building type, building height, building quality, and building technology. Each one of these variables will be examined individually. However, consider first the cost dependence on the variation of these variables and assess their importance.

The type variable classifies the buildings as schools, apartments, stores, offices, etc. The costs per square foot for two buildings of the same floor area but of different types are different. The building quality refers to the quality of workmanship and materials used in the construction process. The better the workmanship and materials for a building are, the higher are the costs of construction. The building height introduces the variation of building costs with height primarily due to cost increases in labor and equipment. The variable of locality accounts for changes in building costs with building location. Any cost differences due to location must be the result of labor wage differentials and proximity of the location to the sources of building labor and materials. The building technology variable is a measure of cost variations with respect to the techniques and methods of construction. A new piece of equipment for handling materials horizontally and vertically or a new procedure of scheduling and cost control not previously employed will certainly affect costs and must be reflected in the structure of the cost function. Likewise, extra building features imply higher costs. The price index, finally, reflects a multitude of pressures on the prices of construction labor, materials, and equipment. Any time-measure of prices including inflationary and wage pressures may be used as a price index. This is an important determinant of the money value of construction costs.

A reflection on the nature of the variables reveals the generality of the cost-estimation function and suggests ideas regarding the measurement of the variables. The function is applicable in as many types of building projects as may be assigned to the domain of the type variables, at any moment in time and for any place. At this point, it may be defined as a global predesign cost-estimation function. Any limitations resulting from its construction rather than definition will be examined later.

Each variable can be measured with reference to some basis to which the value of one is assigned. After this, one is dealing with nondimensional indices and may choose C to measure cost per some convenient building unit. The bases for the measurement of the variables and the selection of units for C depend on the availability of data. The generality of the constructed function, f, will also depend on these data.

Of the six variables, two are very subjective while the other four are quite objective. The quality and technology variables are not only subjective but seem to be highly correlated with the cost. The type variable that classifies buildings on the basis of the service functions they perform must and does influence the value of cost to a great extent. On the other hand, the locality variable, as long as its domain is confined within the boundaries of the United States, must not and does not greatly affect the value of cost. Differences between localities must balance in view of the existing freedoms of mobility and information.

DESCRIPTION AND MEASUREMENT OF COST VARIABLES

In preliminary cost estimating one may identify a building by its location, time of realization, function or type, height, quality and technology. To these correspond the variables previously identified as locality index, price index, type index, height index, quality index, and technology index. Consider now these variables in the order specified with the emphasis on their measurement.

The locality variable, $V_1 = L$, identifies the differences in construction costs as a consequence of differences in the style and cost of living between different cities as well as of wage differentials resulting from differences in labor structures.

Table 2.2.1 introduces typical locality cost indexes for 10 cities. It was taken from the city cost index tabulation contained in Ref. 7. One hundred major cities are considered as the basis for the compilation. The construction cost index figures are averaged from several major appraisal and construction indexes commonly used.

The price index variable, $V_2 = P$, is time independent. The problem, therefore, is to project the price index in the future when construction will take place or to construct the function

$$P = g(t) \tag{2}$$

utilizing past historical data. Table 2.2.2 provides a sample time history of the average construction costs, in dollars per cubic foot, for various apartment buildings. It was extracted from data released by the Department of Buildings and Safety Engineering of the City of Detroit, which have been tabulated by that office for various building types since 1915. Such data represent a rather accurate picture of the movement of building costs in the Detroit area and may well serve as the price index (1).

Table 2.2.1 Typical locality cost indexes for 10 cities

City (1)	Index, L (2)
Boston, Mass.	1.03
Buffalo, N. Y.	1.10
Dallas, Tex.	0.87
Dayton, Ohio	1.05
Detroit, Mich.	1.13
Erie, Pa.	1.02
Houston, Tex.	0.90
Louisville, Ky.	0.95
New York, N. Y.	1.16
Omaha, Meb.	0.92

Table 2.2.2 Sample time history of average construction costs, in dollars per cubic foot, for various apartment buildings

Year (1)	t (2)	Index, P (3)
1963	0	1.66
1964	1	1.71
1965	2	1.76
1966	3	1.80
1967	4	1.93
1968	5	2.09
1969	6	2.30
1970	7	2.49
1971	8	2.76
1972	9	2.95

Table 2.2.3 Range of classes of buildings with corresponding relative cost values (1)

Type (1)	Index, F (2)
Apartment	2.97
Hospitals	3.08
Schools	2.59
Hotels	3.08
Office building (fireproof)	2.95
Office building (not fireproof)	1.83
Stores	2.43
Garages	1.99
Factories	1.20
Foundries	1.49

The type variable, $V_3 = F$, specifies the types of buildings. A good measure for this variable is provided by the average cost per cubic foot for different types of buildings. Upon dividing the different values of the class variable by the smallest value corresponding to a certain type of building, the discrete range of V_3 provides also the relative cost among various types of buildings. Notice that each value in the range corresponds to a particular type of buildings. Table 2.2.3 provides a range of classes of buildings with their corresponding relative cost values as provided by Ref. 1.

The height index, $V_4 = H$, measured by the number of stories in a building needs no explanation. This along with the class and quality variables describe the building. If to these one associates the total number of units, in square feet, one gets a good description of a building at least for preliminary study purposes.

The quality variables, $V_5 = Q$, stands for what it specifies. It is a measure of:

(1) The quality of workmanship and materials used in the construction process; (2) the building use; (3) the design effort; and (4) the material type and quality used in various building components. To define a measure for such a variable let C_i $i = 1, =13=$, k denote the average cost portions of the total building cost distributed among the k building components. Assign an integer index, I_i, from 1 to 4 to each component (corresponding to fair, average, good, and excellent) and compute the values of the variable, V_5, from

$$V_5 = \frac{1}{k} \sum_{i=1}^{k} I_i C_i \tag{3}$$

A more subjective way would be to identify the components of the building, assign to each the values of 1 to 4 and, on this basis, pass a judgment on the quality of the total building by assigning to it one of the integer values 1 to 4. It must be realized that the range of integers 1 to 4 is arbitrarily selected and it is only good as far as it can provide a reliable estimation function. In other

Table 2.2.4 Quality index, Q

Component (1)	Fair (2)	Average (3)	Good (4)	Very good (5)
Use	Multitenancy	Mixed, single tenant, and Multi-tenancy	Single tenant	Single tenant with custom requirements
Design	Minimum design loads	Average design loads	Above average design loads	Many extra design loads
Exterior wall	Masonry	Glass or masonry	Glass, curtain wall, precast concrete panels	Monumental (marble)
Plumbing	Below average quality	Average quality	Above average quality	Above average quality
Flooring	Resilient, ceramics	Resilient, ceramics and terrazzo	Vinyl, ceramic terrazzo	Rug. terrazzoRw marble
Electrical	Fluorescent light, poor quality ceiling	Fluorescent light, average quality, suspended ceiling	Fluorescent light, above average quality ceiling	Fluorescent light, excellent quality ceiling
Heating, ventilating, and air conditioning	Below average quality	Average quality	Above average quality	Above average quality
Elevator	Minimum required	Above required minimum	High speed	High speed deluxe

words, any other range of values may be used as long as it leads to a more reliable function. Table 2.2.4 identifies general building components and on the basis of their qualitative description rates them accordingly from fair to excellent. For any building subject to preliminary cost estimation once the components are selected and rated, the total building quality index may be either selected by judgment or computed from Eq. 3. The latter, however, presupposes a knowledge of the relative costs that go to different components from previous experiences.

The technology index variable, $V_6 = T$, takes into account the extra cost that must be expended for special types of buildings or the labor and material savings resulting from the use of new techniques in the process of construction. For the usual construction situation, this index should be given the value, $V_6 = 1$, with $V_6 > 1$ for extra costs as a result of special features and $0 < V_6 < 1$ for savings as a result of new technology. This variable clearly provides the engineer or estimator the flexibility to utilize the finally constructed cost function for the most unusual cases and furthermore to consider in his preliminary cost estimation a wide selection of technology alternatives with minimum expended time and effort. With extended use the estimator should be able to develop an understanding and knowledge of the magnitude of V_6 to be utilized for specific circumstances. Some data regarding this variable are given in Table 2.2.5.

DERIVATION OF PREDESIGN ESTIMATION FUNCTION

In view of the definitions of the variables and the exposed ideas, one may say on a purely theoretical basis, that each variable V_k assumes a nondimensional value in the region, a_k, b_k, in which a_k, b_k for $k = 1, \ldots, 6 =$ the domain limits whose magnitudes are immaterial. Furthermore, assume that for each variable V_k one can obtain a set of values, V_{ki}, in a_k, b_k for $i = 1, \ldots, N_k$. In addition,

Table 2.2.5 Data regarding magnitude of V_6

Technology (1)	Index, T (2)
Bank-monumental work	1.75
Renovation building	0.50
Special school building	1.10
Chemistry laboratory building	1.45
Telephone building-blast resistant	1.60
County jails	1.20
Dental school	1.15
Hospital addition	1.05
Correctional center	1.20
Home for aged	1.10

let C denote cost per square foot of building floor area. This choice is supported by experience, judgment on a good cost measure of buildings, and data availability. The problem now is to construct $f(V_1, \ldots, V_6)$ utilizing the data available and any intuitive behavioral relationship between C and the variables, V_k.

Of all possible forms of $f(V_1, \ldots, V_6)$ select the linear relationship

$$\hat{C} = \sum_{k=1}^{6} A_k V_k + A_0 \tag{4}$$

in which A_k = constants to be determined from the available data, V_{ki}. Once A_k are known, the cost estimate, $\hat{C}$, can be computed for any new project identified by a set of values V_k. It follows that the next problem is the computation of the constants, A_k. The problem is not simply to arrive at Eq. 4 but to understand also its meaning and limitations. Intermediate relations that lead to C are not a means to an end but they are themselves part of the goal.

In Eq. 4, $\hat{C}$ = an estimate of the actual cost, C. Now determine the constants, A_k, $k = 0, 1, \ldots, 6$ so that

$$\left. \begin{aligned} e &= \sum_{l=1}^{n} e_i = \sum_{l=1}^{n} (C_i - \hat{C}_i)^2 \\ &= \sum_{i=1}^{n} \left(C_i \ \Box \ A_0 \ \Box \ \sum_{k=1}^{6} A_k V_{ki} \right)^2 = \min \end{aligned} \right\} \tag{5}$$

i.e., determine the constants, A_k, so that the sum of the squares of the differences of the estimated costs, $\hat{C}_i$, from the actual costs, C_i, for the n data sets $(C_i, V_{1i}, \ldots, V_{6i})$ is a minimum. This is clearly the method of least squares (8).

For Eq. 5 to be satisfied it is necessary that the first derivatives of e with respect to A_k are zero. This leads to the seven linear algebraic equations

$$\sum_{i=0}^{6} a_{kl} A_l = B_k; \quad k = 0, \ldots, 6 \tag{6}$$

in which a_{kl} are defined by

$$\left. \begin{aligned} a_{kl} &= \sum_{i=1}^{n} V_{ki} V_{li}; \quad k, l = 0, \ldots, 6 \\ B_k &= \sum_{i=1}^{n} C_i V_{ki}; \quad k = 0, \ldots, 6 \end{aligned} \right\} \tag{7}$$

for $V_{oi} = 1$. This certainly specifies $a_{oo} = n$ as it follows from the first equation of Eq. 7. A simultaneous solution of Eq. 6 gives the desired coefficients, A_k, and completely determines the estimate in Eq. 4.

To compute the constants, A_k, utilize the historical data shown in Table 2.2.6. The variables were assigned the values as described in the previous section. Inserting these data in Eq. 7 one gets a_{ki} and B_k. Using these values in Eq. 6, solving for A_i, and inserting their values in Eq. 4, one arrives at the desired function

$$
\left.
\begin{aligned}
\hat{C} &= -81.49 + 23.93\ V_1 + 10.97\ V_2 + 6.23\ V_3 + 0.167\ V_4 \\
&\quad + 5.26\ V_5 + 30.9\ V_6 \\
\hat{C} &= -81.49 + 23.93\ L + 10.97\ P + 6.23\ F + 0.167\ H \\
&\quad + 5.26\ Q + 30.9\ T
\end{aligned}
\right\}
\tag{8}
$$

in which L = location index; F = type index; P = price index; H = height, in stories; Q = quality index; and T = technology index.

For any building project subject to estimation, its cost, in dollars per square foot, if obtained from Eq. 8. It is only necessary to identify the values of the indexes for the building in question. The price index, however, is defined by Eq. 2, which is not yet known. The variation of data, P, versus time suggests the relation

$$
\left.
\begin{aligned}
\hat{P} &= \exp\ (A + Bt) \\
\ln \hat{P} &= A + Bt
\end{aligned}
\right\}
\tag{9}
$$

For this relation, Eqs. 6 and 7 along with the 10-yr historical data of Table 2.2.2 define the values, A, B, and produce

$$
\ln P = 0.192 + 0.029t
\tag{10}
$$

in which $t = 0$, in 1963. With this, the value of the price index at any time int he future is determined.

This completes the solution of the second problem and leads to testing the reliability or rather selecting criteria for testing the reliability of the developed functions.

ANALYSIS OF PREDESIGN ESTIMATION FUNCTION

Knowing the regression or estimation relation one would like to know how accurately $\hat{C}$ estimates the actual cost, C, for any building project in question. For the evaluation of accuracy one may choose some measure of the deviation of the actual values of C from the corresponding estimated values of $\hat{C}$

Table 2.2.6 Historical data

Description (1)	C, in dollars per square foot (2)	V_1 (3)	V_2 (4)	V_3 (5)	V_4 (6)	V_5 (7)	V_6 (8)
Office building	36.00	0.90	2.76	2.95	40	2	1.0
Office building	25.00	0.87	2.49	2.95	18	1	1.0
Bank and office	68.50	1.02	2.76	2.95	6	4	1.75
Housing apartment	31.90	1.03	2.49	2.97	8	2	1.0
College	36.50	1.10	2.49	2.59	11	3	1.0
Renovated office building	23.30	1.13	2.95	2.95	5	2	0.5
Health science building	40.00	0.95	2.30	3.08	14	4	1.0
Telephone center	56.00	1.13	1.93	2.95	3	4	1.60
Hospital addition	40.00	1.05	2.09	3.08	5	4	1.00
Small garage	21.70	1.13	2.09	1.99	1	2	1.00
Office building	42.00	1.00	2.76	2.95	4	4	1.00
College building	45.81	1.16	2.49	2.59	1	4	1.10
Chemistry laboratory	62.00	1.16	2.95	2.59	7	4	1.45
	62.00	1.16	2.95	2.59	7	4	1.45
Hospital	85.00	1.00	2.95	3.08	6	4	2.25
Dental school	47.50	1.00	2.49	3.08	7	4	1.15
Home for aged	34.30	1.13	1.93	3.08	3	3	1.10
Office building	37.00	1.13	2.76	2.95	24	1	1.00
Office building	31.90	1.13	2.30	2.95	10	2	1.00
Office building	40.00	1.13	2.30	2.95	22	3	1.00
Office building	49.50	1.13	2.95	2.95	27	3	1.00
Medical school	36.20	1.13	2.09	3.08	10	3	1.00
Union hall	24.00	1.13	2.76	1.83	1	1	1.00
Hospital addition	38.80	1.13	2.09	3.08	1	4	1.05
Office addition	20.00	1.08	1.93	2.95	4	1	1.00
College building	18.80	1.13	1.93	2.59	2	1	1.00
Office building	34.70	1.13	2.09	2.95	5	3	1.00
Office building	15.10	1.13	1.93	1.83	2	1	1.00
School, high	18.10	1.13	1.22	2.59	3	2	1.00
County correctional center	39.00	1.13	2.30	3.08	4	2	1.20
County jail	36.00	1.13	2.09	3.08	2	2	1.20
College dormitory	21.10	1.07	1.66	2.59	6	2	1.00
College dormitory	24.30	1.07	1.93	2.50	6	2	1.00
College building	30.00	1.13	1.93	2.59	6	3	1.00
Hospital addition	27.50	1.13	2.30	3.08	2	1	1.00
Foundry	11.30	1.00	2.09	1.49	1	1	1.00
Factory	14.50	1.02	2.09	1.20	1	2	1.00
Factory	10.00	1.05	2.09	1.20	1	1	1.00
Factory	14.75	0.92	2.30	1.20	1	2	1.00

or of the magnitude of error, e, defined by Eq. 5. Indeed, let one choose

$$\sigma_c^2 = \frac{1}{n-7} \sum_{i=1}^{n} (C_i - \hat{C}_i)^2 \tag{11}$$

which is clearly the average of e as defined by Eq. 5 and which measures the standard error of estimate. (The reason for choosing $n - 7$ in place of n rests on the fact that this makes Eq. 11 an unbiased, consistent, and sufficient estimate of the variance of C, which may be computed when the probability distribution of C is known. This, however, is of no concern in the context of the present work.)

The value of σ_c^2 may now be computed using Eq. 8 and the data in Table 2.2.6. Its magnitude gives an idea of the goodness of the estimate but, as such, it is not very informative. Therefore, look at it from a different angle. Consider the identity

$$\sum_{i=1}^{n} (C_i - \bar{C})^2 = \sum_{i=1}^{n} (\hat{C}_i - C) + \sum_{i=1}^{n} (\hat{C}_i - \bar{C})^2 \tag{12}$$

in which $\bar{C}$ = the arithmetic mean of C; and introduce the sample coefficient of determination

$$r_c^2 = \frac{\displaystyle\sum_{i=1}^{n} (\hat{C}_i - \bar{C})^2}{\displaystyle\sum_{i=1}^{n} (C_i - \bar{C})^2} \tag{13}$$

From this it follows that

$$0 \le r_c \le 1 \tag{14}$$

since max $(r^2) = 1$ when $C_i = \hat{C}_i$; and min$(r^2) = 0$ when $\hat{C}_i = \bar{C}$.

When $r_c^2 = 1$, the regression plane (see Eq. 8) is the plane of the actual cost and the error is zero ($e = C_i - \hat{C}_i = 0$); and when $r_c^2 = 0$, the fitting of the regression plane has not reduced the error at all ($e = C_i - \hat{C}_i = $ max). Then r_c^2 is a measure of the improvement in terms of reducing the total error as a result of fitting the regression plane. Otherwise, it may be viewed as a measure of the closeness of fit. There is a prefect fit when $r^2 = 1$, and there is not fit at all when $r_c^2 = 0$, since then the C_i are so scattered that the regression plane becomes horizontal ($\hat{C}_i = \bar{C}$). Thus, the closer the fit of the regression plane, $\hat{C}$, to the points, the closer r_c^2 is to 1. Otherwise, r_c^2 may be viewed as a measure of the assumed linearity of $\hat{C}$ since the closer r_c^2 is to 1 the closer the points are the assumed plane.

In conclusion, r_c^2, which is also known as the coefficient of multiple correlation, measures the amount of improvement in terms of: (1) Reducing the total error by fitting the regression plane; (2) measuring the closeness of fit of the regression or estimation plane to the points; and (3) the degree of linearity of the scatter of the points. For the regression equations, Eqs. 8 and 9, the corresponding correlation coefficients were found to be $r_c^2 = 0.009$ and $r_p^2 = 0.980$, respectively. They are, of course, computed from Eq. 13 utilizing Eq. 8 along with Table 2.2.6, and Eq. 10 along with Table 2.2.2, respectively.

Aside from the accuracy of the cost estimation, $\hat{C}$, one would like to know the degree of correlation of $\hat{C}$ with its variables V_k, i.e., to what degree the cost data support the relation of $\hat{C}$ with all and each of its variables V_k. By introducing the coefficient, r_c^2, one has in a sense answered part of this question. Indeed, r_c^2 may also be considered as a measure of the correlation of variability between $\hat{C}$ and the group of variables, V_k. For the case in question, $r_c^2 = 0.998$ indicates almost perfect correlation. However, the correlations of variability of $\hat{C}$ with each variable V_k are equally meaningful. They provide a measure of the relative importance of the independent variables as indicators of cost movements.

The correlations, r_{cv}^2, of $\hat{C}$ with each of its variables V_k are defined by Eq. 13. It should be understood, however, that $\hat{C}_i$ in this case are defined by

$$\hat{C} = A_{ok} + A_k V_k; k = 1, \ldots, 6 \tag{15}$$

in which each time A_{ok} and A_k are computed, as previously, employing the values, C_i and V_{ik}, given in Table 2.2.6. But in this case, it is preferable to use the relation

$$r_{cvk} = \frac{\Box(C \Box \bar{C})(V_k \Box \bar{V}_k)}{[\Box(C \Box \bar{C})^2 \Box(V_k \Box \bar{V}_k)^2]^{1/2}} \tag{16}$$

which follows from Eqs. 15 and 13, rather than to define Eq. 15 each time and then compute r_{cv}^2 using Eq. 13.

From Eq. 16 and from the data in Table 2.2.6 the following correlation coefficients between cost $\hat{C}$ and its variables V_k were obtained: $r_{cL} = 0.02$; $r_{cP} = 0.53$; $r_{cF} = 0.58$; $r_{cH} = 0.21$; $r_{cQ} = 0.68$; and $r_{cT} = 0.78$. These coefficients immediately specify the relative value of each variable and indicate that a simpler expression with fewer variables but with still high correlation results by eliminating the variables, L and H. However, this will give a poor model in comparison with the original one. This is in view of the fact that a change in the sample to involve a greater diversity of locality and a greater diversity of tall buildings towards the construction of a truly global predesign cost-estimation function may give more value to the variables, L and H.

Further information regarding the goodness of the predesign cost-estimation function is provided by analysis of the errors, $e_i = C_i - \hat{C}_i$ known as

Table 2.2.7 Computed values

Description	C, in dollars per square foot	$\hat{C}$ in dollars per square foot	Residual $e = C - \hat{C}$	Residual $100\ e/C$, as a percentage
(1)	(2)	(3)	(4)	(5)
Office building	36.000	36.804	−0.80362	−2.2323
Office building	25.000	24.178	0.82239	3.2896
Bank and office	68.500	67.687	0.81301	1.1869
Housing apartment	31.900	31.721	0.17922	0.56181
College	36.500	36.796	−0.29579	−0.81038
Renovated office building	23.300	23.081	0.21924	0.94093
Health science building	40.000	39.942	0.05788	0.14470
Telephone center	56.000	56.080	−0.08019	−0.14321
Hospital addition	40.000	38.525	1.478	3.6869
Small garage	21.700	22.452	−0.75186	−3.4648
Office building	42.000	43.6999	−1.6988	−4.0448
College building	45.810	44.912	0.39814	1.9606
Chemistry laboratory	62.000	61.775	0.22464	0.36231
Hospital	85.000	85.551	−0.55146	−0.64878
Dental school	47.500	46.685	0.81504	1.7159
Home for aged	34.300	36.175	−1.8753	−5.4673
Office building	32.000	34.363	−3.3633	−7.3852
Office building	31.900	32.240	−0.34011	−1.0662
Office building	40.000	39.514	0.48638	1.2159
Office building	49.500	47.478	2.0213	4.0844
Medical school	36.200	36.012	0.18837	0.52036
Union hall	24.000	23.538	0.46236	1.9265
Hospital addition	38.800	41.315	−2.5146	−6.4810
Office addition	20.000	20.718	−0.71753	−3.5876
College building	18.800	19.337	−0.53691	−2.3550
Office building	34.700	34.365	0.33503	0.96552
Office building	15.10	14.604	0.49633	3.2370
School, high	18.100	16.983	1.1165	6.1688
County correctional center	39.000	38.225	0.77473	1.9865
County jail	36.000	35.588	0.41230	1.1453
College dormitory	21.100	20.875	0.22502	1.0664
College dormitory	24.300	23.836	0.46433	1.9108
College building	30.000	30.536	−0.53583	−1.7361
Hospital addition	27.500	26.446	1.0542	3.8333
Foundry	11.300	10.963	0.33703	2.9836
Factory	14.500	14.900	−0.40001	−2.7537
Factory	10.000	10.353	−0.35312	−3.5312
Factory	14.750	14.810	−0.0600	−0.4046

residuals. Accordingly, compute

$$
\left.
\begin{aligned}
e_i^1 &= \frac{100(C_i - \hat{C}_i)}{C_i} \\[2mm]
\bar{e} &= \frac{1}{n} \sum_{i=1}^{n} e_i^1 \\[2mm]
\sigma_e^2 &= \frac{1}{n} \sum_{i} \left(e_i^1 - e\right)^2
\end{aligned}
\right\}
\tag{17}
$$

Now, define the function $p(x)$ by the relations

$$
\left.
\begin{aligned}
& & \min e_i^1 &< x_k < \max e_i^1 \\[2mm]
p(x_k) &= \frac{1}{2n\Delta x} \sum_{j} e_j^1 \quad \text{for} \quad & x_k - \Delta x &\leq e_j^1 \leq x_k + \Delta x \\[2mm]
& & x_k &= (2k-1)\,\Delta x; \quad k = 1, \ldots, m \\[2mm]
& & \Delta x &= \frac{\max e_i^1 - \min e_i^1}{2m}
\end{aligned}
\right\}
\tag{18}
$$

Clearly, $p(x_k)$ is a probability density function and gives

$$
\begin{aligned}
\sum_{k=1}^{n} 2p(x_k)\,\Delta x &= 1 \\[2mm]
P(x_1 < e^1 < x_2) &= \sum_{i=1}^{ } p(x_i)\,\Delta x; \quad x_1 \leq x_i \leq x_2
\end{aligned}
\tag{19}
$$

in which the second relation specifies the probability that the error, e^1, in a predesign cost estimate assumes a value between any two x_1 and x_2.

From Eq. 19 one finds the probability that the percentage of error, e^1, of a predesign cost estimate will be within a desired range $(x_2 - x_1)$. This is equivalent to the proportion of the estimates that have a percentage of error within the specified range. On the other hand, the question regarding the range of a percentage of error that contains a specified percentage of estimates has no unique answer and cannot be computed unless the distribution, $p(x)$, is specified or at least one limit of the range is defined. For the sample of data in Table 2.2.6, on the basis of which Eq. 8 was derived, the values, C_i, $\hat{C}_i$, $e_i = C_i - \hat{C}_i$, and $e_i^1 = 100\,e_i/C_i$ have been computed and are shown in Table 2.2.7. These data along with Eq. 17 gave $\bar{e} = 0.05\%$ and $\sigma_e = 3.06\%$. Furthermore, the same data along with Eqs. 18 and 19 led to the conclusion that 68.3% and 95.5% of the residuals had values in the ranges, $-0.05\% \pm 3.06\%$ and -0.05%

□ 6.12%, of the actual costs, respectively. This accuracy is well within that required for predesign estimating programs.

SUMMARY AND CONCLUSIONS

The results of this work consist of the derivation of a predesign cost-estimation function for buildings and of the development of criteria for evaluating the function. Given the location and time of construction, the height, type, quality, and technology of a building, one is able to find a cost, in dollars per square foot, and to comment with qualifications on the probability that the difference of the estimated cost from the actual cost will be within a certain percentage of the actual cost.

The function was tested with an apartment-office building in Los Angeles, Calif., 11 stories high, built early in 1972, and an office building in Detroit, Mich., 39 stories high, built late in 1973. The results were amazing with $e = 34.40 - 34.30 = \$0.10/\text{sq ft}$ and $e = 50.24 - 50.00 = \$0.24/\text{sq ft}$, respectively. But of particular interest is the fact that if the data buildings are classified on the basis of type with a description of their technology or extras and consulted in each application with regard to the variable values assigned, then the function provides better than expected estimates.

As a defect of the predesign cost-estimation function one may point out the limited sample used for derivation. This is undoubtedly a limitation. However, the value emphasis of this work is not on the construction of the function itself but on the general methodology including the use of mathematical tools. Utilizing the expounded methodology and the same mathematical tools, one is able to derive a truly global predesign cost-estimation function for buildings by the proper selection and classification of a quite larger data sample. It is needless to say that the scope of the function, with regard to generality and accuracy, dictates the size and the diversity of the data.

An examination of the mathematical tools employed during the development reveals a quite painful process. It utilizes the ideas of regression, correlation, and probability analysis in a lucid way that requires nothing more in the part of those who want to use the methodology than a knowledge of: (1) Simple differentiation; (2) solving systems of linear algebraic equations; and (3) understanding the concept of probability. This deliberate effort reflects the philosophy which advocates that in estimating there is no room for faith implicit in the use of methods or formulas not rigorously understood by the estimator. The implementation of this philosophy is judged as one of the contributions of the present work.

A review of the estimation function (see Eq. 8) along with the nature and structure of its variables shows that the most subjective variables, quality Q and technology T, are also the ones with the highest correlation coefficients. This reiterates the previously made assertion that the present work does not aim at replacing but rather at systematizing the estimator's judgment and knowledge. A specific answer concerning the effects of these variables is provided by the predesign cost-estimation functions

$$\hat{C} = -48.74 - 0.78\,H + 8.07\,Q + 7.57\,F + 15.0\,P + 8.05\,L;\ r_c^2 = 0.89$$

$$\hat{C} = 47.39 + 0.40\,H + 15.56\,F + 20.55\,P - 3.87\,L;\ r_c^2 = 0.75$$

(20)

which ignore the variables, T and Q, T, respectively.

Either of these expressions may be used to determine the predesign estimates. However, since they neglect variables that have high degree of correlation with the dependent variable, their accuracy decreases. Elimination of T reduces the original correlation from $r_c^2 = 0.998$ to $r_c^2 = 0.89$ while elimination of T and Q reduces it further to $r_c^2 = 0.75$, which cannot be acceptable except in cases where simply a general idea of the project cost is required.

A word of caution and of practical value is necessary before closing the analysis. The emphasis of this work is on the methodology and estimation ideas, not on the computational procedures. Accordingly, one may program the mathematical relations and perform the cost function derivation and analysis if one wishes to use a larger and better data sample or even change the variables. However, this is not imperative. As long as the estimation ideas behind those relations are understood, one may utilize one of the many available statistical analysis computer packages to perform the regression analysis. Indeed, this was the choice made by the writers in order to save time in computational work. Furthermore, the variables selected should not be construed as unique or the best; the same holds for their measurement. But the fact remains that if the methodology is applied to classes of buildings instead of to the whole population of buildings, one is bound to get very good results. Indeed, this work should be viewed in this context. It does not pretend to present results on the basis of the best available data nor does it concern itself with estimation techniques for projects where plans, specifications, or more detailed information are available. Refs. 2, 6, 10, 11 and 14 deal with the subject of cost estimation in general and those who wish to utilize more comprehensive or specified data along with the methodology developed in this work are encouraged to do so by consulting Refs. 3 and 5.

The methodology developed requires that a complete record of the sample building data is maintained with a concise definition of the variables and their values used in deriving the cost function. In estimating the cost of a new building project, it is instructive that the new project is compared with similar projects in the data for consistency in the assignment of the variables. It is also instructive that data are updated and the cost function is rederived periodically.

REFERENCES

1 'Average Cost of Buildings in $ per c.f. since 1915,' The Department of Buildings and Safety Engineering of the City of Detroit, Detroit, MI, USA.

2 Chilton, C. H., ed., Cost Engineering in the Process Industries, McGraw Hill, New York, NY, 1960.

3 Construction Information Sources and Reference Guide, Construction Publications, Phoenix, AR, 1974.

4 'Cost Roundup & Outlook,' Engineering News Record, McGraw Hill, New York, NY, June, Sept., Dec., 1961 and Mar., June, Sept., 1962.

5 Dodge Manual for Building Construction, Pricing and Scheduling, Dodge Building Cost Services, McGraw Hill Information Systems, New York, NY, 1974.

6 Jelen, F. C., Cost and Optimization Engineering, McGraw Hill, New York, NY, 1970.

7 Means, R. S., Building Construction Cost Data, Robert Snow Means, Dunbury, Mass., 1972.

8 Miller, I., and Freund, J. E., Probability and Statistics for Engineers, Prentice Hall, Englewood Cliffs, NJ, 1965.

9 Park, W. R., 'Pre-Design Estimates in Civil Engineering Projects,' Journal of the Construction Division, ASCE, Vol. 89, No. CO2, Proc. Paper 3627, Sept., 1963, pp. 11–23.

10 Park, W. R., Cost Engineering Analysis, John Wiley, New York, NY, 1973.

11 Popper, H., ed., Modern Cost Engineering Techniques, McGraw Hill, New York, NY, 1970.

12 Rowan, P. P., et al., 'Sewage Treatment Construction Costs,' Journal of the Water Pollution Control Federation, Vol. 32, June, 1960.

13 Rubey, H. and Milner, W. W., Construction and Professional Management, MacMillan, New York, NY, 1966.

14 Seiter, K., III, Introduction to Systems Cost Analysis, John Wiley and Sons, New York, NY, 1969.

15 'Steam-Electric Plant Construction Cost Annual Production Expenses,' Federal Power Commission, Annual Supplement 6–13, 1953–1960.

APPENDIX: NOTATION

The following symbols are used in this paper:

A, B = constants;
A_k = unknown constants, $k = 0, 1, \ldots, 6$;
C = unit cost;
$\hat{C}$ = estimated unit cost;
C_i = data unit costs, $i = 1, \ldots, n$;
$\hat{C}_i$ = data values of kth variable;
e = sum of squares of $(C_i - \hat{C}_i)$;

$\bar{e}$ = mean value of percentage error of estimate;
$p(x)$ = probability density function;
r_c = coefficient of determination;
t = time variable;
V_k = independent variables, $k = 1, \ldots, 6$; and
$\square_c$ = standard error of estimate.

2.3 Some examples of the use of regression analysis as an estimating tool[*]

R. McCaffer

Estimating based on statistical methods has the appeal of condensing historical records into a useable form. One of these statistical methods is multiple Linear Regression Analysis and an example of its use was described in the January/February 1974 edition of *The Quantity Surveyor* by G. A. Hughes. Regression Analysis as a cost modelling tool is a technique which enables the cost of project to be expressed in very few items. For example the cost model say residential apartment blocks may be

Cost (at model's base date)
= 36001.75 + (169.84 □ Area of Single Units)
+ (137.38 □ Area of Double Units)
+ (2553.8 □ No. of Storeys)
+ (3049.53 □ No. of Lifts □ No. of Storeys)
+ (139.85 □ Total Access Area)
+ (395.88 □ Common room Areas)
+ (13335.43 □ No. of Garages)

and if the model variables had the following values in one particular case

Area of single units   = 1044
Area of double units   = 281
No. of Storeys   =   3
No. of Lifts   =   1
Total Access Area   = 550
Common Room Areas =   87
No. of Garages   =   1

the estimator could quickly show that cost (at models base date)

[*]1975, *Quantity Surveyor*, Dec., 81–6.

$$
\begin{aligned}
= 36001.75 \ &+ \ (169.84 \ \square \ 1044) \\
&+ \ (137.38 \ \square \ 281) \\
&+ \ (2553.8 \ \square \ 3) \\
&+ \ (3049.53 \ \square \ 1 \ \square \ 3) \\
&+ \ (139.85 \ \square \ 550) \\
&+ \ (395.88 \ \square \ 87) \\
&+ \ (13335.43 \ \square \ 1) \\
= \pounds 393{,}422.97 &
\end{aligned}
$$

(Note the above formula is only applicable for the same type of buildings used in the analysis and should not be applied to other types of buildings. For further information contact the author.)

The technique also enables the estimator to say how accurate the estimate is, in numerical rather than subjective terms. If his assessed accuracy was $\square 9\%$ this would imply that for most cases the cost is expected to be within £358,014 to £428,831.

Obviously it is much quicker to calculate likely costs in this way rather than by approximate design, approximate quantities and approximate rates. In several cases we have found these cost prediction models were significantly more accurate than designers' early estimates. Thus in appropriate cases cost models based on regression analysis can provide a substantial improvement on existing methods of estimating.

All this doesn't come about by accident, it is necessary to assemble data relating to past projects and to do some fairly extensive analytical work to produce the cost prediction models. In practice it is valuable to update the models periodically by including recent data and discarding the oldest data. There seems a good case for assembling the relevant data centrally and providing the models for estimating purposes to subscribing members.

This paper briefly describes, in non-statistical terms, multiple regression analysis as it is used as a cost modelling technique. The results of some studies done within the Department of Civil Engineering at Loughborough University of Technology are given. Most of the examples are aimed at the estimating as done by the construction industry's professional advisers and not, with one exception, by contractors.

REGRESSION ANALYSIS – THE TECHNIQUE

Regression Analysis is a technique that will find an expression or formula or mathematical model which best describes the data collected. For example, if you were attempting to predict the cost of say residential apartment schemes the data that might be available to you from past schemes may be –

1   The cost of the scheme.
2   The total area of the single units.

3   The total area of the double units.
4   The number of storeys.
5   The number of lifts.
6   The total area of common rooms.
7   The total area of corridors and access space etc . . . etc. . . .

The list given is indicative of the data that could be obtained from drawings and contract documents. To undertake a regression modelling or cost modelling exercise data of this type would be required for a number of such schemes. The exact number required will be discussed later.

The first stage in finding a 'cost model' which best describes the data collected would be to find the one variable or item of data that explains most of the variation in costs between the different schemes. This is likely to be the variable that has the highest correlation with cost. For example, Fig. 2.3.1 shows cost plotted against one variable.

By finding a straight line which 'best fits' this plotted data we have a 'simple regression model'.

This cost model is

COST = a + B (total area of single units)

where 'a' is the intercept on the vertical axis and 'B' is the slope of the line in relation to the horizontal axis.

For example, the cost in pounds may turn out to be 325 + 7 □ total area of single units in sq. metres.

How well this 'model' fits the data can be determined by examining the residuals. That is the difference between the cost predicted by the model (i.e.

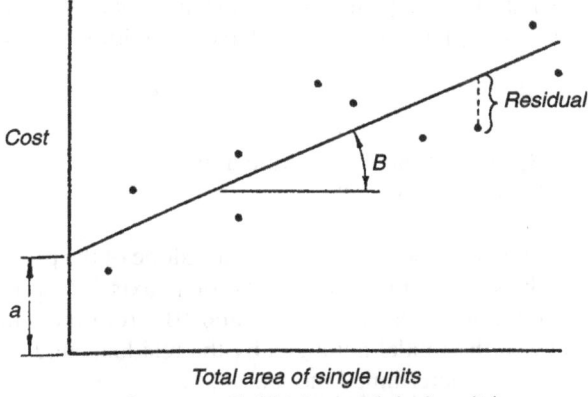

Total area of single units
Cost = a + B (Total area of single units)

Figure 2.3.1 A simple regression model.

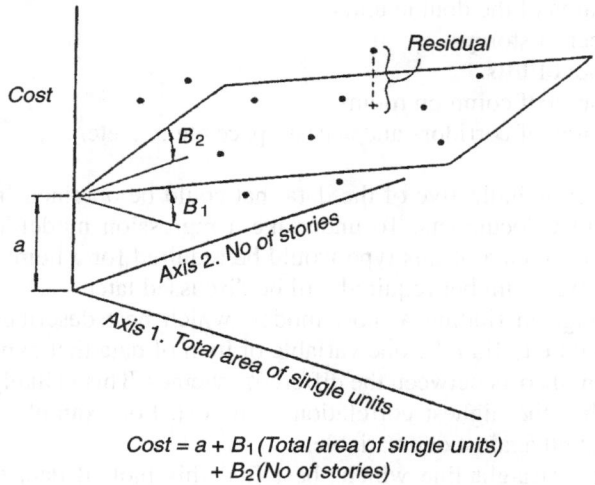

$$Cost = a + B_1 \text{ (Total area of single units)}$$
$$+ B_2 \text{ (No of stories)}$$

Figure 2.3.2 Multiple regression model.

a point on the line) and the actual cost recorded in the data (one of the plotted points). Examining the residuals is in effect examining the model's in-accuracy. The greater the residuals the less accurate is the model.

The second stage in finding the best model is to determine the second variable that explains most of the variation in costs after taking account of the first variable. For example, Fig. 2.3.2 shows cost plotted against variable (B) the total area of single unit as in Fig. 2.3.1 on one axis and against variable (4) the number of storeys on a second axis. The data points being plotted are no longer on a sheet of paper but in space and a sloping plane now fits the data not a single straight line.

When the best fitting plane is found we have a 'multiple linear regression model'.

The cost model is

$$\text{COST} = a + B_1, \text{ (total area of the single units)}$$
$$+ B_2 \text{ (no. of storeys)}$$

where 'a' is the intercept on the cost axis, B_1 is the slope of the plane in relation to axis 1 and B_2 is the slope of the plane in relation to axis 2. It should be noted that 'a' and 'B_1' will not be the same as 'a' and 'B' from the simple model. Again the accuracy of the model is judged by the residuals and the scatter of these residuals about the determined model.

The process can be extended to include the third most significant variable and the cost model could become

COST = a + B$_1$ (Total area of single units)
            + B$_2$ (No. of Storeys)
            + B$_3$ (Total area of corridors and access space).

However, it is impossible to present this graphically.

The model need not contain only three variables but can be extended to include any number of variables until the best model possible is found, that is until the residuals are made as small as possible. However, the more variables included in the model the greater is the amount of data required to construct the model. A very rough guide is that the minimum number of past schemes required for model building is two or three times the number of variables included in the final model.

Although the technique is based on items of data which are assumed to vary

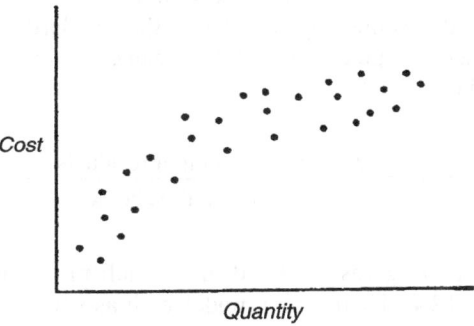

Figure 2.3.3 A non-linear relationship. In this case cost does not vary linearly with quantity.

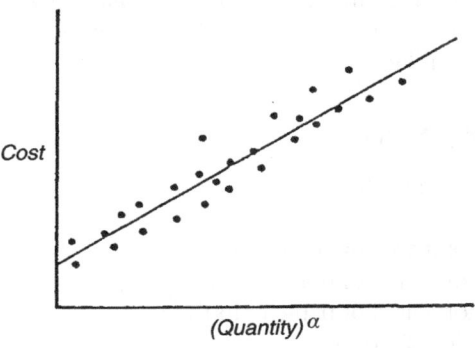

Figure 2.3.4 Transforming to linear. By transforming quantity to (quantity)$^{\square}$ a linear relationship with cost could be obtained.

linearly with cost this can be overcome by transforming the data before fitting a model. For example, if the cost of a particular item falls off with increasing quantity, see Fig. 2.3.3, the quantity item could be transformed by raising to a power, see Fig. 2.3.4.

Transformations of data are also used to re-arrange the collected data. For example, if you had cost as one variable and floor area as another this could be transformed into cost/unit area.

THE MODEL'S ACCURACY

The accuracy of the model is judged by the scatter of the residuals, the difference between the model's predicted cost and the actual cost. The scatter of these residuals can be examined in two ways, one is to plot them as in Figures 5 to 10, the other way is to calculate the 'coefficient of variation'. The scatter of the residuals can be determined by calculating the standard deviation of the residuals, this expressed as a percentage of the mean cost of all schemes gives the coefficient of variation

$$\text{coefficient of variation} \quad = \frac{\text{standard deviation of residuals}}{\text{mean cost of all schemes}} \quad \square \, 100$$

This coefficient, say 12%, gives the band into which most cases, say 65%, would fall, i.e. + or − 12%. Therefore a model with as small a coefficient of variation as possible is desired.

The residuals plotted or the coefficient of variation describe the accuracy of the model as it fits the data used to determine the model. The real test is how accurately does the model predict cases that are not included in its own data base. Experience indicates the coefficient of variation (c.v.) will increase by 25% to 50% when the derived model is applied to data outwith its own data base, that is a model with a c.v. of 10% will deteriorate to 15% − 20% when used on other cases of similar type.

THE MODEL'S MEANING

The model derived may be of the form

$$
\begin{aligned}
\text{COST} = a \quad &+ B_1 \,(\text{total area of single units}) \\
&+ B_2 \,(\text{total area of double units}) \\
&+ B_3 \,(\text{total area of triple units}) \\
&+ B_4 \,(\text{No. of storeys}) \\
&- B_5 \,(\text{total area of circulation space}) \\
&\text{etc.}
\end{aligned}
$$

This model as a whole predicts costs within its accuracy limits. It cannot be used other than as a whole and observing the negative sign on B_5 say if we increase circulation space the cost will decrease. Because by increasing the circulation space will change other variables and so alter the model's prediction. Each variable in the model is explaining some of the variability in cost, if one or more variable share too much of this explanation they will be compensated by a negative variable in the model. Therefore the model can only be taken as a whole.

BUILDING A MODEL

To construct a regression model access to the appropriate data is obviously needed.

The steps described above are only practicable using computers. There are several computer packages available that permit the user, say a quantity surveyor, to present his data in a prescribed manner to the computer package and all the calculations are done automatically. The inexperienced user of the technique must at this stage seek the advice of either an experienced user or of a statistician before drawing conclusions.

At Loughborough the computer package used is 'Nimbus' (Ref. 1) supported with programs for storing data and selecting different combinations of variables which were developed within the Department of Civil Engineering. Having developed our own computer systems regression analysis and cost model building have been reduced to a routine exercise given access to data.

For a full statistical explanation of regression analysis Draper and Smith (Ref. 2) is recommended and for further background 'Regression Analysis – New Uses for an Old Technique' by Trimble (Ref. 3) is recommended.

EXAMPLES OF COST MODELS

Examples of eight regression based cost models developed at Loughborough are given. The models are not described in full but their accuracy is given either in terms of the coefficient of variation or as scatter diagram of residuals.

(1) Reinforced concrete frames

An exploratory study by J. S. Buchanan (Ref. 4) produced a cost model for reinforced concrete frames in buildings. The model contained variables such as gross floor area, average load, shortest span, longest span, number of floors, height between floors, slab concrete thickness and number of lifts.

The residuals shown in Fig. 2.3.5 demonstrate that the model is more accurate for medium and high cost schemes rather than low cost schemes.

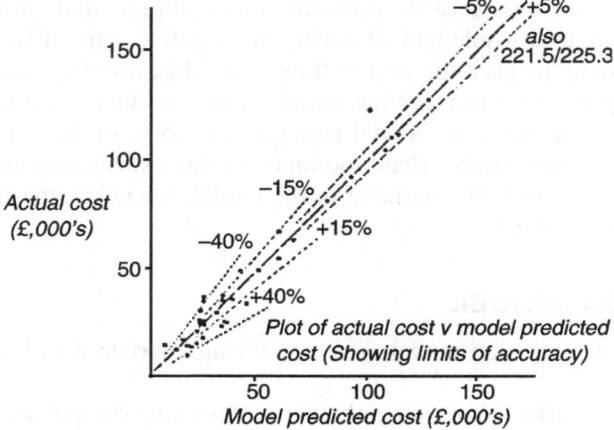

Figure 2.3.5 Actual vs predicted costs: R.C. structural frames.

(2) Services in buildings

P. R. Gould (Ref. 5) produced a cost model for heating, ventilating and air conditioning services in buildings. The variables included functions which described the heat and air flow through the building, the heat source and distance which it has to be ducted and shape variables.

The residuals are shown in Fig. 2.3.6 and demonstrate higher accuracies at higher costs.

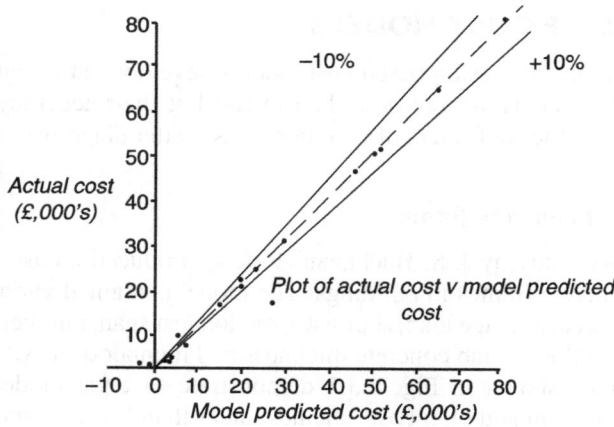

Figure 2.3.6 Actual vs predicted costs: Heating and ventilating services in building.

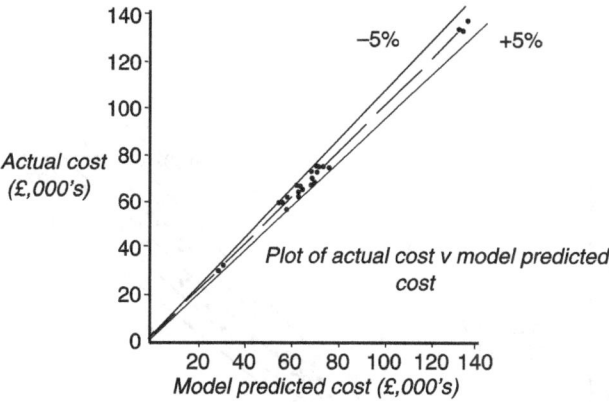

Figure 2.3.7 Actual vs predicted costs: Schools.

As a result of this further extensive work was undertaken as part of a research contract for H.M. Treasury. The results are not available for publication.

(3) School buildings

B. F. Moyles (Ref. 6) produced a cost model for system built school buildings and included variables such as floor area, area of external and internal walls, number of rooms and functional units, area of corridors, storey height and number of sanitary fittings.

The residuals shown in Fig. 2.3.7 demonstrate high accuracy.

(b) Houses

R. H. Neale (Ref. 7) produced a cost model for houses built by a private contractor for private sale. The model included such variables as floor area, area of roof and of garage, number of storeys, slope of site, unit cost of external finishes and cost of sanitary fittings, area and volume of kitchen units, site densities, regional factors, number of doors, area of walls, number of angles on plan, construction date and duration of development and type of central heating. The residuals shown in Fig. 2.3.8 show that only two cases fell outside the □ 10% band.

Homes for old people

J. Baker (Ref. 8) produced a cost model for residential apartment schemes for old people. His model included such variables as the area of single units, double units, triple units, common rooms, Warden's Flat, laundry, access corridors, number of lifts and garages and duration of contract. The model had

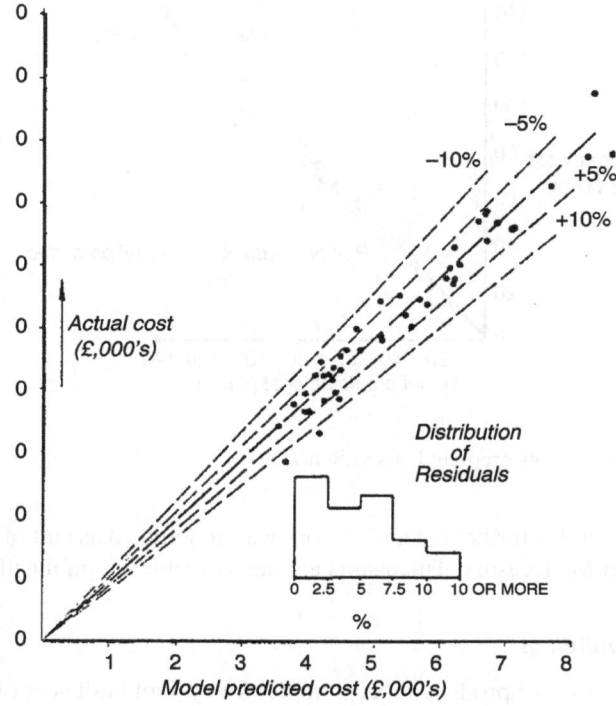

Figure 2.3.8 Actual vs predicted costs: Houses.

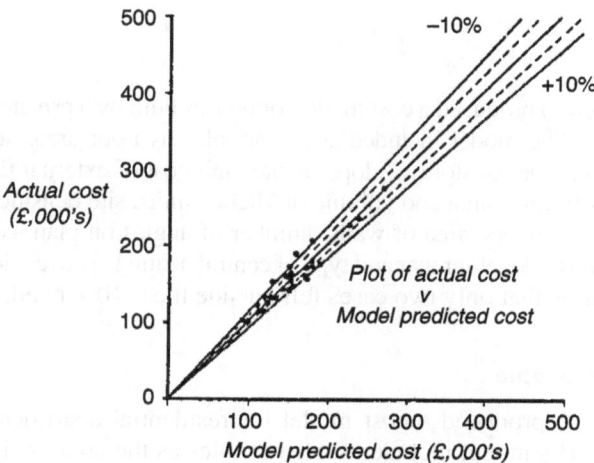

Figure 2.3.9 Actual vs predicted costs: Houses for old people.

a coefficient of variation of 9.16% and the residuals are shown in Fig. 2.3.9. Baker adjusted his basic cost data for regional variations, contract size and date before conducting his analysis.

Passenger lifts

J. D. Blackhall (Ref. 9) produced a cost model for passenger lifts in office buildings. The model's variables included the contract date, the dimensions of the car, the number of landings, the length of travel, the operating speed, and variables referring to the type of control system and the location of the installation. The coefficient of variation was 20.9%.

Electrical services in buildings

Blackhall (Ref. 9) also produced a cost model for electrical services in office buildings. The model's variables included number of distribution boards, fuşed load, number of active ways, number of socket and other outlets, voltage, contract date and a differentiation whether the building was for commercial or domestic use. The coefficient of variation was 20.0% and the residuals plotted in Figure 2.3.10 again show better accuracies for more costly schemes.

Blackhall also had the opportunity of comparing the accuracy of his model with the accuracy of traditional estimating. Blackhall calculated the accuracy of the traditionally produced estimates in a way that could be directly compared with the model's coefficient of variation. The result was

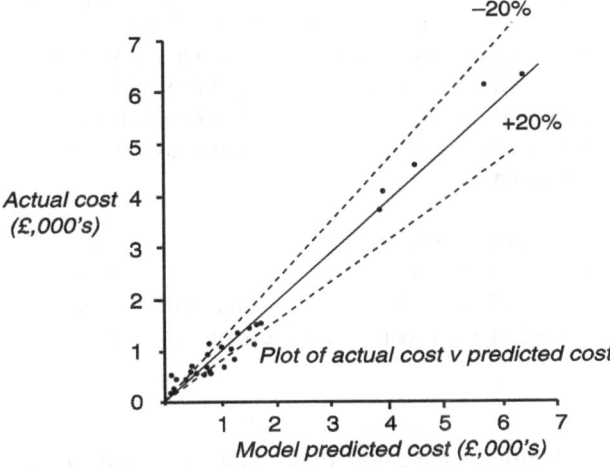

Figure 2.3.10 Actual vs predicted costs: Electrical services.

| Traditional estimates | 34% |
| Model's c.v. | 20% |

Even allowing for a deterioration in the model's performance when applied to data outwith its own data base, which will leave the model with a c.v. of 25% – 30% the model is still more accurate than traditional methods. Also and perhaps more important the time taken by the user of the model is much less than the traditional estimator.

Although the coefficient of variation of this model is high the performance of traditional estimating is also poor. In a highly variable situation neither can hope to be very accurate.

Motorway drainage

D. Coates (Ref. 10) produced an estimating model for a contractor's unit rate for laying motorway drainage using porous pipes, Hepline pipes and asbestos pipes. The three models one for each pipe type contained the variables, internal diameter, average depths and the cost of the pipe.

The coefficients of variation were, for porous pipes 12.8%, for helpline 9.2% and for asbestos pipes 6.9%.

SOME FURTHER COMMENTS

Models' accuracy

The models presented show a range of accuracies from well under 10% to over 20%. Before these accuracies are dismissed as a limitation of the technique the performance of traditional estimating should be examined. From studies done at Loughborough some measure of the accuracy of estimating by professionals, that is designers, architects, quantity surveyors, etc., has been obtained. For example the following standard deviations give figures for traditional estimating which could be compared to regression models' coefficients of variation.

Estimating cost of electrical services	34%
Estimating cost of H & V services	26%
Estimating cost of office buildings, in excess of	15%
Estimating cost of road works, in excess of	20%

Inflation

The cost model builders have not yet reached a consensus in how to deal with inflation. One approach has been to include the date in the model which has some merit provided the model itself is updated regularly. Another approach has been to reduce the cost information to a base date using cost indices before analysing the data. The model's prediction is then at the base date and needs to be corrected to time now again using indices. A third approach is a variation of the previous one and that is to construct a model from data that belongs to a limited time span and any prediction is for the cost at the time of the model data. Correction to time now is again needed.

Model's life

The life of any particular model will be limited not only because of inflation, but because of changing building designs. For example, there may be a shift from waterborne heat to air ducted heat or the insulation standards of buildings may change. A model constructed before these changes take place will be inadequate in predicting the newer buildings. Therefore regular updating of a model is necessary if it is to be used on a regular basis.

NOTES:

Figures 5 to 10 are scatter diagrams of residuals for six different regression based cost models. If the model predicted the cost exactly all points would lie on a 45□line through the origin. However, since the model does not predict the actual costs exactly the error shows as the distance between the plotted point and the 45□line.

References 4 to 10 are MSc project reports by post-graduate students at Loughborough University of Technology and held in the Department of Civil Engineering, or MSc Thesis, copies of which are in the University Library.

REFERENCES

1 Nimbus, B. M. D. *Biomedical Computer Programs*. University of California Press, 1971.
2 Draper, N. and Smith, H. '*Applied Regression Analysis*.' Wiley-Interscience, 1966.
3 Trimble, E. G. '*Regression Analysis – New Uses for an old Technique*.' Internal Paper, Department of Civil Engineering, Loughborough University of Technology.
4 Buchanan, J. S. 'Development of a Cost Model for the reinforced concrete frame of a Building.' MSc Project Report, 1969. Supervisor: Professor E. G. Trimble.
5 Gould, P. R. 'The Development of a cost model for H & V and air conditioning installations in buildings.' MSc Project Report, 1970. Supervisor: Professor E. G. Trimble.
6 Moyles, B. F. 'An Analysis of The Contractors' Estimating Process.' MSc Thesis, 1973. Supervisor: Professor E. G. Trimble.

7 Neale, R. H. 'The Use of Regression Analysis as a Contractor's Estimating Tool.' MSc Project Report, 1973. Supervisor: R. McCaffer.

8 Baker, M. J. 'Cost of Houses for the Aged.' MSc Project Report, 1974. Supervisor: Professor E. G. Trimble.

9 Blackhall, J. D. 'The Application of Regression Modelling to the Production of a Price Index for Electrical Services.' MSc Project Report, 1974. Supervisor: R. McCaffer.

10 Coates, D. 'Estimating for French Drains – A Computer Based Method.' MSc Project Report, 1974. Supervisor: R. McCaffer.

2.4 Cost optimisation of buildings*

A. D. Russell and K. T. Choudhary

INTRODUCTION

Today's building facilities are becoming more complex and costly. The capital and operating costs of such facilities directly affect the profitability of the processes carried on within them. To assist in designing efficient buildings, considerable work had been focussed to date on developing sophisticated analyses and design tools for the various subsystems which comprise a total building system (e.g., foundation, structure, mechanical, electrical, etc.). Some attention has also been directed toward the optimal design of these systems. For example, a great deal of work has been performed in the area of structural optimization, with weight or weight related cost being selected as the performance criterion.

Issues yet to be resolved before practical optimization based design aids become commonplace in the arsenal of tools available to building designers include the development of: (1) Objective functions which reflect the viewpoint of the owner or ultimate user, or both, and which account for labor, equipment, management, and capital inputs as well as material inputs; (2) the data base essential for formulation of the objective functions just described; (3) effective methods for coordinating the designs of the various systems comprising a building so that physical interactions between them are properly accounted for and the appropriate objective function is minimized; (4) efficient methods for investigating various alternative systems for each subproblem and for handling discrete variables; (5) computer codes which are efficient in terms of cost, are easy to use and which produce practical designs, e.g., yield actual member sizes which satisfy all relevant code provisions; and (6) solutions to a series of typical design problems encountered in practice which demonstrate the extent of the benefits to be derived by designers and owners from the application of optimization methods.

*1980, *American Society of Civil Engineers, Journal of the Structural Division*, January, 283–300.

Recent work directed at one more more of the foregoing issues has been described in the literature. Examples and their relationship to these issues include Bradley, et al. (2) for items 4 and 6, Lee and Knapton (5) for items 1, 2, 4, and part of 5, Lipson and Gwin (7), for items 1, 5, 6, and part of 4, Lipson and Russell (8) for items 1, 5, 6, and Miller, et al. (9) for items 3, 4, and 5.

In this paper, what are felt to be useful methodologies and results with respect to items 1, 3, 5, and 6 are described by way of considering the cost optimization of a particular type of light industrial buildings. Light industrial structures have been selected because of their relative simplicity and economic importance. Attention is first directed at a detailed statement of the problem in which the design of the roofing, bent, foundation, and cladding systems is treated. Following this, in order of appearance, are the formulation of cost models, consideration of the optimization method and computer model used, results from an example, and finally, conclusions.

DESCRIPTION OF PROBLEM

Light industrial facilities have column-free spaces as depicted in Fig. 2.4.1 provide the focus for the study described herein. Attention is directed at those systems which have a high degree of interaction with one another, namely the foundation, structural bent, and roof systems with allowance being made for the cost of the cladding system. Not considered at present are the mechanical and lighting systems, the designs of which can have important implications for both the enclosure and structural bent systems.

Only one alternative is considered for each of the systems treated. Pedestal footings comprise the foundation system, and steel bents consisting of wide flange columns and parallel chord Pratt trusses form the structural system. The

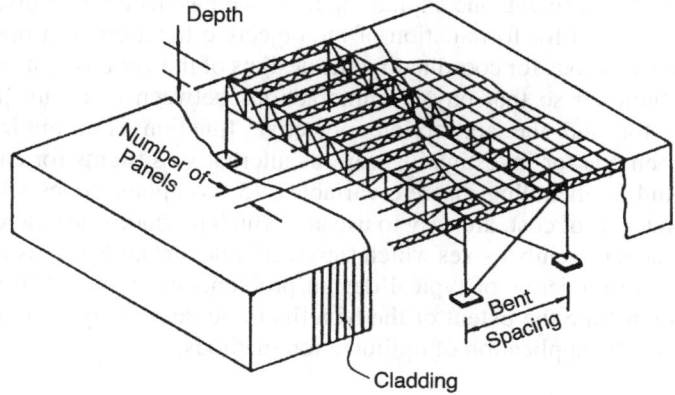

Figure 2.4.1 Schematic of design problem.

roof system consists of manufactured joists and steel Q decking and metal cladding is used for the walls.

Three key decision variables describe the overall configuration of the building. They are bent spacing, number of panels per truss, and truss depth to span ratio. Given values for these variables, the number of bents, foundations, purlins and joists (if required) can be readily determined.

Because of code provisions relating to frost penetration (10), a pedestal footing has been selected for this study. The footing is assumed to be square and the dimensions of the pedestal are taken to be equal to the column base plate plus 2in. (51 mm). The base plate is designed in accordance with Ref. 10. For analysis and design purposes, the column to footing connection is assumed to be pin connected.

The bent system is assumed to be fabricated from G40.21 grade 44 W steel. Chords are assumed to be continuous and of constant area and are fabricated from Tees. Even panel spacing is assumed with loads being applied to the panel points. Back to back double angles are used as web members. symmetry in terms of web member sizes is imposed with web member sizes being allowed to vary from panel to panel. All connections are assumed to be welded except for field chord splices and column connections which are assumed to be bolted. All members are sized in accordance with CSA Standard S16–1969 so as to satisfy the forces arising from consideration of six load combinations as specified by the National Building Code of Canada. These combinations are: (1) DL; (2) DL + LL; (3) DL + WL; (4) DL + EQ; (5) 0.75 (DL + LL + WL); and (6) 0.75 (DL + LL + EQ) in which DL = dead load, LL = live load, WL = wind load, and EQ = earthquake load.

Depending on panel length and decking gage, the roof system is comprised either of purlins and steel decking or of purlins, joists, and decking. Joists and purlins are selected from tables of short and long span manufactured joists and trusses. Twenty-two, 20, and 18 gage decking are considered. All roof system components are designed according to CSA-S16–1975, CSA–A136, and CSA–W95–1. An allowance of 8 lb/sq ft is made for insulation, roofing, felt, tar, and gravel.

Because the optimum depth to span ratio of the truss is influenced by the cost of the cladding system, allowance is made for the cost of a metal cladding system costing $2.00/sq ft of wall area ($21.50/m^2). A more detailed design algorithm would be required if enclosure costs were to be minimized or if design of the mechanical system were considered.

COST MODELS

Capital cost has been selected as the most important single performance criterion or objective function for the systems considered herein. If the mechanical and lighting systems and detailed design of the enclosure system were included in the design problem, life cycle cost would form a more appropriate objective function.

Normal design office practice is to compute system cost by multiplying the various material quantities required by their respective unit costs and then summing up to obtain total system cost. These unit costs are usually derived from published cost guides (3, 4, 11) and they reflect the inputs of labor, equipment, and materials. Such an approach can be useful for estimating an overall preliminary budget. It does not reflect, however, the methods of costing employed by contractors and fabricators which invariably account for costs which are a function of the number of components to be installed or erected. Thus, in order for the designer to be confident that his decision pertaining to both overall building configuration and individual system configuration yield minimum total capital cost, the cost models adopted must reflect a costing procedure similar to that employed by contractors. It is shown in this study that different design configurations are selected as being optimal in terms of capital cost, depending on the form of the cost model used.

In general, fabricators and contractors price out separately the components contained in the expression for the capital cost of the ith system, i.e.:

$$C_i = PR_i + OH_i + CL_i + CE_i + CM_i \qquad (1)$$

in which PR_i = profit; OH_i = overhead; CL_i = cost of labor; CE_i = cost of equipment and CM_i = cost of material, all for the ith system. For this study, profit and overhead are computed as a fraction of the direct costs, yielding the following expression for C_i:

$$C_i = (1 + M_i)(1 + O_i)(CL_i + CE_i + CM_i) \qquad (2)$$

in which M_i = the fraction of direct cost and overhead charged for profit and office overhead; and O_i = the fraction of direct cost charged for field or shop overhead, or both. Of all the components in Eq. 2, determination of the material input is the most straightforward, requiring a simple quantity takeoff. For example, for the foundation system ($i = 1$) the cost of material FOB site may be written as:

$$CM_i = \text{cost of formwork} + \text{cost of reinforcing steel} + \text{cost of concrete}$$

$$+ \text{cost of anchor bolts} + \text{cost of backfill} \qquad (3)$$

Accurate assessment of the inputs of labor and equipment, however, requires an understanding of the elemental operations involved in the fabrication or construction or both, of each system.

For each of the systems considered herein, the sequence of operations required for construction was determined in consultation with a contractor and a steel fabricator and erector. For example, Fig. 2.4.2 depicts a model of the sequence of construction operations required for the foundation system, and as

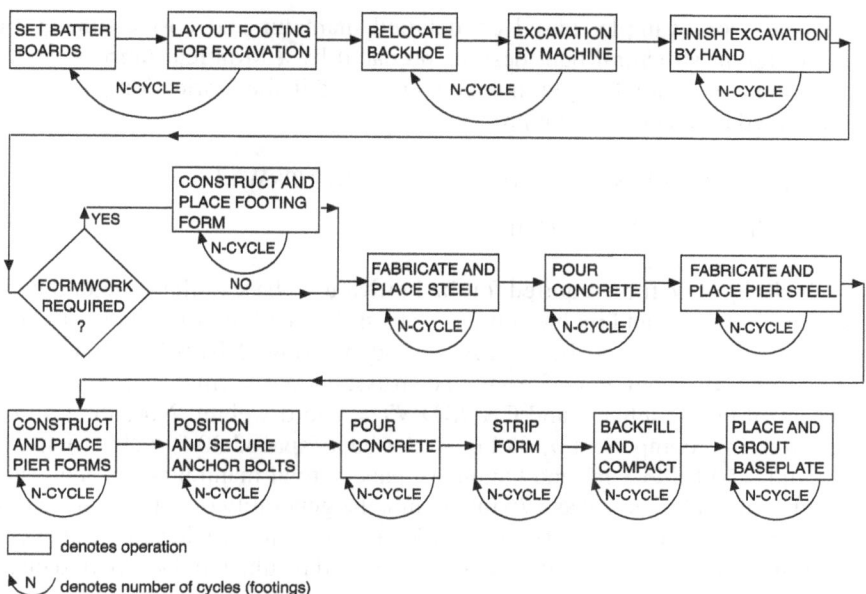

Figure 2.4.2 Sequence of operations for footing construction.

such directly accounts for the number of footings N to be constructed.

Depending on the system being constructed and on the specific nature of each elemental construction operation, different models are appropriate for predicting the time consumed by each operation. For example, after consultation with a contractor, a model of the form:

$$T_{ij} = \dot{a}_{ij} + b_{ij}Q_{ij} \qquad (4)$$

was selected as being appropriate for the operations associated with construction of the footings in which T_{ij} = the same required for the jth elemental operation; a_{ij} = the time required independent of the amount of material to be placed due to relocation of men and equipment, set-up, etc. for the jth operation; b_{ij} = the production rate for the jth operation (units = time/unit quantity); and Q_{ij} = the quantity of work performed or material placed for the jth operation. Values for a_{ij} and b_{ij} are a function of crew composition, equipment used, size conditions, etc. Despite the sizeable contribution that the a_{ij} component can make to the total time T_{ij} required for the ith system, jth operation, most cost models derived from published cost guides invariably exclude this component.

Time models of the simple form expressed by Eq. 4 are not always adequate. For example, based on detailed manhour and cost estimates of nine different structural bent designs prepared by a Montreal fabricator, the following expression for total manhours required in the fabrication phase was determined by a curve fitting procedure:

$$T_{3F} = \text{NBENT}* [87.2 + 2.35\ \text{NPANL} + 0.019\ \text{TRWT}^3$$

$$+ \text{NSP}\ (2.20 + 0.034\ \text{CHORDWT}^2)] \tag{5}$$

in which T_{3F} = the time required for the fabrication phase of the structural bent system (system 3) and includes time for truss fabrication, truss chord splices, column preparation and bent cleaning and painting; NBENT = number of bents; NPANL = number of panels per truss; TRWT = truss weight; NSP = number of chord splices; and CHORDWT = chord weight. Note the size of the fixed time component a_{ij} in Eq. 5. For bents spaced 30 ft (9.15 m) apart, spanning 150 ft (45.8 m) and having 16 panels, total manhours per bent estimated by the fabricator was 176 manhours. the generality of Eq. 5 as it relates to the type of structural system considered herein for a wide range of bent spans and spacings is currently being studied and results will be reported elsewhere.

Given the labor and equipment inputs required for each operation, a rate table containing costs per hour, CL_{ij} and equipment costs per hour, CE_{ij} can be developed. Then, continuing with the footing example, expressions for labor and equipment cost may be written as:

$$CL_i = N \sum_{j=1}^{n} T_{ij}\ CL_{ij} \tag{6}$$

$$CE_i = N \sum_{j=1}^{n} T_{ij}\ CE_{ij} \tag{7}$$

and finally, total system cost is obtained using Eq. 2 and summing all C_i, $i = 1$, . . . 4.

The approach to cost modeling described herein has the advantage that it accounts for the number of components to be constructed for each system and for both the fixed and variable costs associated with each component. Such an approach is essential for the accurate determination of the minimum cost building configuration. The drawback to this approach, however, is the requirement that design firms have access to contractors cost data and have knowledge of construction operations. This is possible within design-build firms – it is difficult at best for conventional design firms.

OPTIMIZATION PROCEDURE AND COMPUTER PROGRAM

Adopting the minimization of building cost per square foot as the objective function, the design problem may be stated as:

$$\text{Minimize } \frac{1}{SL}(C_1 + C_2 + C_3 + C_4)$$

$$= \text{Minimize } f(\text{x}) = \frac{1}{SL}[2(x_1 + 1)C_F + (x_1 + 1)C_B$$

$$+ x_1 C_R + 2(H + Sx_3)(S+L)C_C] \tag{8}$$

in which $f(x)$ = total cost per square foot; S = building width; L = building length; H = clear height of building; C_1 = cost of foundation system; C_2 = cost of bent system; C_3 = cost of roofing system; C_4 = cost of cladding; C_F = cost per footing; C_B = cost per bent; C_R = cost of roof system per bay; C_C = cost of cladding per square foot; x_1 = number of bays; x_2 = number of panels; and x_3 = depth to span ratio. Subject to:

$$\frac{L}{x_1^U} \square \frac{L}{x_1} \square \frac{L}{x_1^L} ; \quad x_1 \text{ an integer} \tag{9}$$

$$x_2^L \square x_2 \square x_2^U; \quad x_2 \text{ an even integer} \tag{10}$$

$$x_3^L \square x_3 \square x_3^U \tag{11}$$

in which x_2^L = lower bound for number of bays; x_1^U = upper bound for number of bays; x_2^L = lower bound for number of panels; x_2^U = upper bound for number of panels; x_3^L = lower bound for depth to span ratio; x_3^U = upper bound for depth to span ratio; and subject to the design decision variables pertaining to member sizes being selected so as to satisfy all binding codes such that C_F, C_B, C_R, and C_C, are optimized for each (x_1, x_2, x_3) tuple. In particular: (1) The foundation design must satisfy ACI 318–71 Building Code and National Building Code of Canada, 1975; (2) bent number sizing must be done in accordance with CSA Standard S16–1969; and (3) roof system components must be sized according to CSA Standard S16–1975, CSA–A 136–1974 and CSA–W95–1.

The problem, as stated, is a nonlinear mixed integer mathematical programming problem for which no all-inclusive optimization algorithm exists. The complexity of the problem is such that neither the objective function nor the constraints may be written explicitly in terms of the independent variables.

One way to solve such a problem is to break it up into a series of problems and find a method of coordination the solution of each of the subproblems so as to optimize an overall objective function.

In this study, the problem is decomposed and solved as follows. A master program is defined which has the three interface variables as its decision variables and the system cost (or weight) as the coordinating function. Values for the interface variables are determined using Box's Complex Method (1). Subproblems are then defined for the foundation, bent, and roofing subsystems (no detailed design procedure is used for the cladding system). These subproblems have as their goal the optimization of subsystem design variables, given specific values for the interface variables. Different methods are used to select optimum values for the subsystem variables. The foundation design function is a 'satisfying' one, i.e., find a feasible solution; bent member sizes are determined on the basis of a fully stressed design criterion while roof system design is determined on the basis of an exhaustive search with subsystem cost (or weight) as the merit criterion. Fig. 2.4.3 depicts the hierarchy of design problems for the system considered.

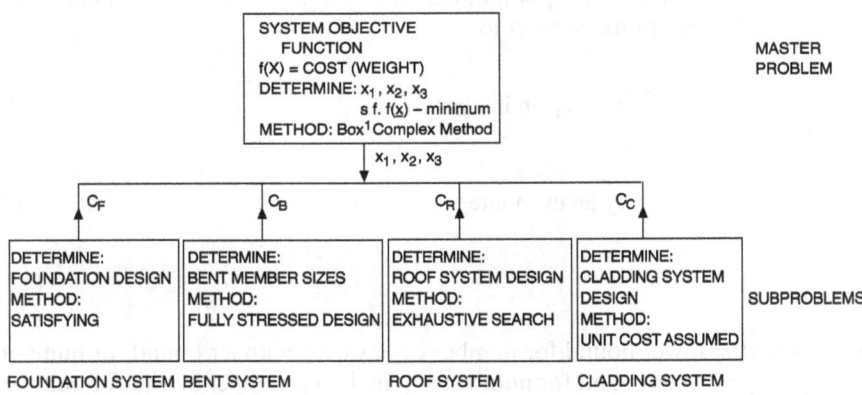

Figure 2.4.3 Hierarchy of optimization problems for system design.

A modified version of Box's Complex Method, as described herein, is used to determine the optimal values of the interface variables x_1, x_2, and x_3. This technique, in various forms, has been successfully applied to other complex structural design problems (6, 7, 8). Box's Method, as used for this study, may be described by the following sequence of steps:

1. Set random number base, reflection factor a, and stopping criteria. Specify whether a one or two cycle search is to be used and specify the number

of points in the complex for each cycle (k_1, k_2). Specify an initial feasible starting point.

2. For the first cycle, generate a complex of k_1 points, with $k_1 \square n + 1$, in which n = the number of interface variables. The first point in the complex corresponds to the initial feasible starting point determined previously. The remaining $k_1 - 1$ vertices of the complex are generated randomly as follows:

$$x_j = x^L + r_j(x^U - x^L)^T; \quad j = 1, 2, \ldots k_1 - 1 \tag{12}$$

in which $x_j = j$th point in the initial complex; r_j = uniformly distributed random number over the interval [0, 1] generated for the jth point; x^L = vector of lower limits for the independent variables; and x^U = vector of upper limits for the independent variables. This process ensures that explicit constraints of the form:

$$x_i^L \square x_i \square x_i^U; \quad i = 1, \ldots n \tag{13}$$

are satisfied but does not ensure that implicit constraints of the form:

$$x_i^L \square x_i \square x_i^U; \quad i = n + 1, \ldots m \tag{14}$$

in which $x_i = g_i(x_1, x_2, \ldots x_n); \quad i = n + 1 \ldots m$ (15)

are satisfied.

3. Except for the case of integer variables, treat the violation of implicit constraints by moving the trial point halfway towards the centroid of those points already selected (where the given initial point is included). This procedure will ensure that a satisfactory point will be found if the design space is convex.

4. Evaluate the objective function for each point in the complex.

5. Replace the point having the worst objective function value by a point reflected through the centroid of the remaining points $\square \square 1$ ($\square = 1.3$ for this study) times the distance between the centroid and the worst point on the lien connecting the centroid with the worst point. If this new trial point violates an explicit constraint, set the value for the offending variable equal to its lower or upper limit, as the case may be. If an implicit constraint is violated, move the trial point halfway toward the centroid of the remaining points. If the objective function value for this new point is inferior to that of the worst point, set $\square = \square/2$, repeat the feasibility check and reevaluate the objective function. If this new point is still inferior, set $\square = 0$, i.e., use the centroid. If no improvement results, evaluate a new point halfway between the centroid and the best point in the complex. Terminate the search for this cycle if no improvement is found. Otherwise, go to step 6.

6. Check stopping criteria. See if the complex has collapsed into a region such that:

$$\frac{f(x_w) \Box f(x_B)}{f(x_B)} \Box \text{TOL} \tag{16}$$

in which $f(x_w)$ = objective function value for the worst point in the complex, $f(x_B)$ = objective function value for the best point in the complex, TOL = tolerance value set by the user (0.005 for this study). See if the maximum number of iterations NIT_i for the ith cycle has been reached. If the stopping criteria are not attained, go to step 5. If one or more stopping criteria are attained, terminate the search if all cycles have been completed. Otherwise, use the best point achieved so far as the initial feasible starting point and go to step 2.

In step 5, the integer variables x_1 and x_2 are treated as follows. In determining a new trial point, x_1 is first treated as continuous and then rounded off to the nearest integer for purposes of checking the constraints and for evaluation of the objective function. A slightly different procedure is required for x_2, number of panels, because of it being constrained to assume even integer values. Rather than examine only one trial point corresponding to the even integer value of x_2 closest to the continuous approximation, two trial points (x_2^1, x_2^2) are examined and the point having best objective function value selected. The other point considered corresponds to $x_2^2 = x_2^1 - 2$, in which x_2^1 is the even integer value closest to the continuous value of x_2 generated by Box's Method. This approach of examining two values for x_2 keeps the complex from collapsing onto one value of x_2 for all points in the complex. This tendency to collapse was noted in preliminary runs, and thus the procedure described.

There is no guarantee that the preceding will yield a global optimum. The validity of the optimum achieved may be checked by conducting the search several times, each time using a different starting complex. In this regard, the use of a two cycle search procedure automatically provides a check, albeit limited, on whether a global optimum has been reached. As well, the two cycle search helps avoid the problem of the search procedure collapsing into a subspace of the total design space.

Several runs were made to study the behavior of the search method with respect to varying complex size, k, reflection factor, a, number of cycles and initial starting point. The criterion of weight optimization (bent plus roof system weight per square foot) was arbitrarily selected for these runs. So that a graphical display of the method's behavior could be made, the variable x_2, number of panels, was set at 14, which corresponds to a roofing system having no joists and 22 gage decking [$S = 100$ ft (30.5 m), $L = 520$ ft (159 m)]. Fig. 2.4.4 depicts the results for a two cycle search procedure. Trial points are primed in order of sequence for those cases where several points had to be examined before improvement could be made for a given iteration. When the centroid is used as a trial point, it is indicated by a triple prime. If the centroid still results in no improvement, a new point generated between the centroid

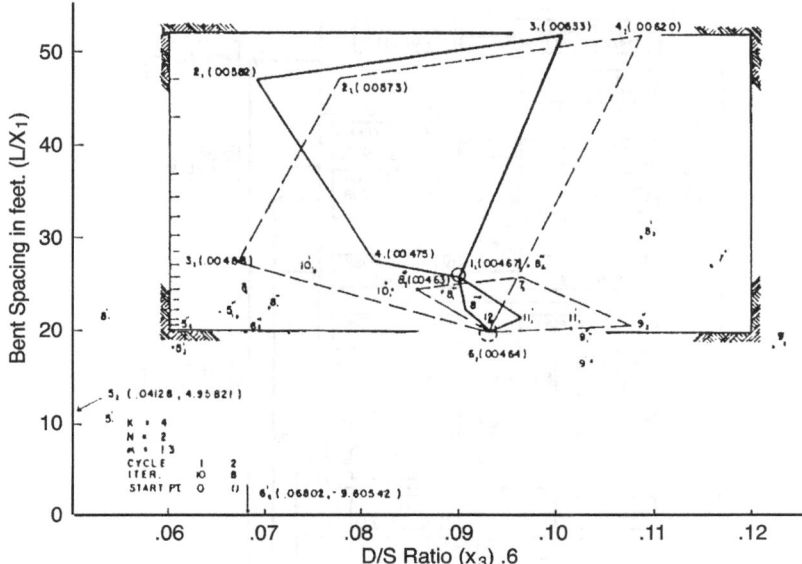

Figure 2.4.4 Behavior of Box's complex method (1 ft = 0.305 m).

and the best point in the complex is examined and is denoted by the quadruple prime. Subscripts on trial points refer to cycle number. The initial and final complexes for each cycle are shown, with the solid lines denoting the first cycle and the dotted lines the second. Objective function values for points in the initial complex are given in parentheses, in tons per square foot.

Based on these runs, it was concluded that k, the complex size should be equal or greater than $2n$ to avoid premature collapse of the complex, the impact of initial starting point on the final optimum is insignificant, small values of a, the reflection factor lead to premature collapse of the complex inside the initial complex and that a two cycle procedure, having the same total number of iterations as one cycle procedure, is more effective than a single cycle one in locating the optimum or avoiding local optima and premature collapse of the complex, or both.

The optimization algorithm used for subsystem design coordination along with the algorithms required for the anlaysis and design of the four subsystems considered were computerized and integrated into one package. Fig. 2.4.5 depicts the components of this package. The various modules required are sequenced in a manner which reflects the need to optimize the design of each subsystem subject to the same overall system configuration and which reflects the way in which the subsystems physically interact in terms of the load transferred from one to another.

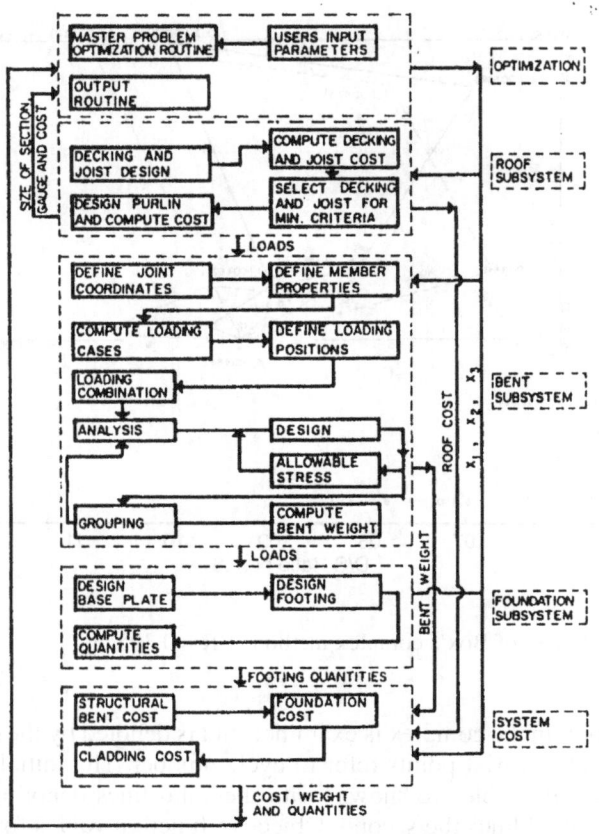

Figure 2.4.5 Structure of computerized design aid.

At present, one of three objective functions may be used to determine optimal performance. They are system weight, material quantity related cost, and total cost including the inputs of labor, equipment, and materials.

Example Problem. – An example problem is examined for purposes of illustrating the methodology described previously and to demonstrate the importance of careful formulation of the objective function. The example consists of a building whose plan dimensions are 100 ft by 520 ft (30.5 m by 159 m) and whose clear height to the underside of the trusses is 25 ft (7.6m). The allowable bearing capacity of the soil is taken as 4 ksf (192 kPa) and the strengths of the footing concrete and steel are taken as f'_c = 3 ksi (21 mPa) and f_y = 60 ksi (414 mPa), respectively. Steel used for the columns and trusses is assumed to be G40.21 grade 44 W steel. Loads are set in accordance with the National Building Code of Canada for the Montreal area. All prices reflect 1977 rates.

Before examining overall system performance, it is instructive to examine briefly individual subsystem performance in terms of material quantities required and costs as a function of the interface variables, number of bays, number of panels, and depth to span ratio. This examination will highlight the nature of the trade-offs being made between fixed and variable costs within a subsystem and between subsystems.

Material quantities and cost for the foundation are almost solely a function of the variable, number of bays, because the number of footings as well as the load carried by each footing is determined by this variable. Figs 2.4.6 and 2.4.7 depict the variation of footing material quantities and cost with x_1. As seen from these figures, it is more cost effective to put more material into a smaller number of footings than vice versa. In percentage terms, the variation in foundation cost over the range of truss spacings examined is 41.4% and the difference in cost between the minimum material configuration [$x_1 = 26$ ft (7.9 m)] and minimum cost configuration [$x_1 = 52$ ft (15.9 m)] is some 3–1 2¢/sq ft of plan area ($0.38/m²).

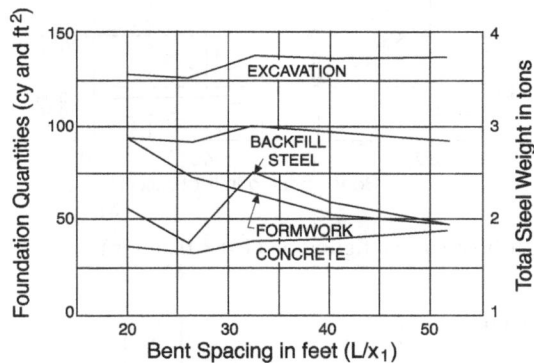

Figure 2.4.6 Variation of footing material inputs with bay spacing (1 ft = 0.305 m; 1 ton = 907.2 kg).

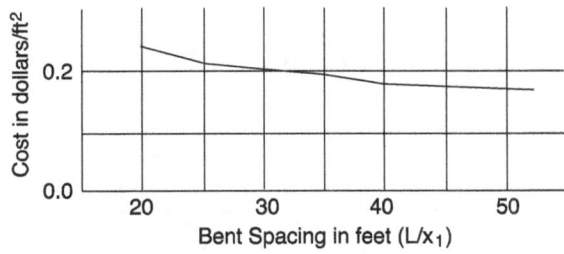

Figure 2.4.7 Variation of foundation cost with bent spacing (1 ft = 0.305 m).

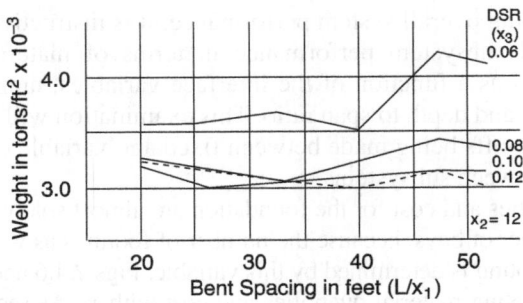

Figure 2.4.8 Bent weight as function of configuration (1 ft = 0.305 m).

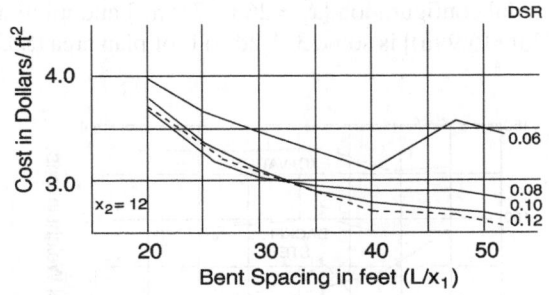

Figure 2.4.9 Bent cost versus configuration (1 ft = 0.305 m).

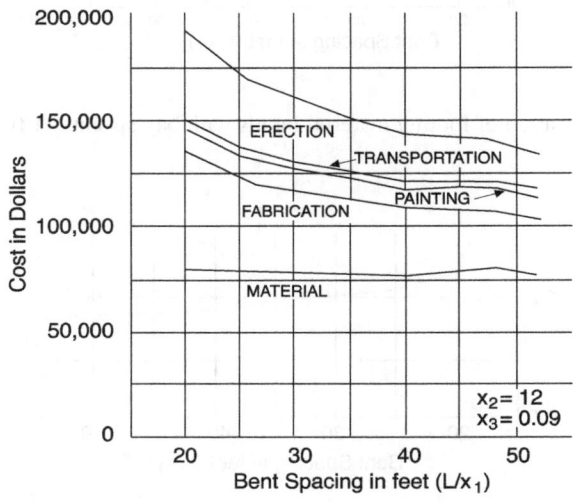

Figure 2.4.10 Composition of bent cost (1 ft = 0.305 m).

Material quantities and costs for the bent systems and functions of all three interface variables. Fig. 2.4.8 depicts variation of bent system weight with changes in configuration while Fig. 2.4.9 depicts the variation of bent system cost with configuration. Fig. 2.4.10 shows the composition of bent cost. The lack of smoothness in these plots is accounted for by the discrete member

Table 2.4.1 Design and cost details for various roof system configurations

Truss spacing in feet	Number of bays[a]	Number of panels	Number of purlins	Number of joists	Purlin size	Decking gage	Total weight in tons per square foot	Total cost in dollars per square foot
(1)	(2)	(3)	(4)	(5)	(6)	(7)	(8)	(9)
20.00	26	8	234	624	20CA8.5	22	0.00170	1.71
	26	10	286	780	20CA7	22	0.00170	1.82
	26	12	338	–	18CA6.5	20	0.00161	1.41
	26	14	390	–	16CA6	22	0.00146	1.37
	26	16	442	–	16CA6	22	0.00162	1.45
26.00	20	8	180	640	26CA12	22	0.00186	1.75
	20	10	220	800	26CA10.5	22	0.00190	1.88
	20	12	260	–	26CA9.5	20	0.00181	1.50
	20	14	300	–	26CA8	22	0.00161	1.43
	20	16	340	–	24CA8	22	0.00169	1.48
32.50	16	8	144	512	28CA15	20	0.00211	1.86
	16	10	176	640	28CA13	20	0.00215	1.96
	16	12	208	–	28CA13	20	0.00203	1.63
	16	14	240	–	28CA12	22	0.00191	1.62
	16	16	272	–	28CA12	22	0.00203	1.70
40.00	13	8	117	624	34CA17.5	22	0.00210	1.93
	13	10	143	780	34CA15.5	22	0.00217	2.08
	13	12	169	–	34CA13.5	20	0.00206	1.62
	13	14	195	–	34CA13.5	22	0.00203	1.66
	13	16	221	–	34CA13.5	22	0.00216	1.76
47.27	11	8	99	616	34CA23.5	22	0.00237	2.08
	11	10	121	770	34CA18	22	0.00230	2.11
	11	12	143	–	34CA15	20	0.00220	1.71
	11	14	165	–	34CA15.5	22	0.00218	1.73
	11	16	187	–	34CA15.5	22	0.00233	1.87
52.00	10	8	90	560	57CA28	22	0.00254	2.52
	10	10	110	700	57CA28	22	0.00282	2.87
	10	12	130	–	57CA28	20	0.00301	2.74
	10	14	150	–	57CA28	22	0.00311	2.93
	10	16	170	–	57CA28	22	0.00339	3.20

[a] L = 520 ft. (30.5 m).

Note: 1 ft = 0.305 m; 1 ton = 907.2 kg.

spectrum used for design. In order of importance, bent weight is sensitive to depth to span ratio, truss spacing and number of panels, with the last variable being a distant third. However, cost is sensitive to number of panels, because of the fixed costs associated with each additional web member. Fig. 2.4.10 emphasizes the importance of considering costs associated with the number of elements treated in the fabrication and erection phases as well as material costs. Based on the results obtained for a lattice of the three interface variables, the minimum cost configuration was found to be $x_1 = 10$, $x_2 = 8$, and $x_3 = 0.12$. This eight panel configuration requires a roof system comprised of purlins and joists.

Of the three interface variables, number of bays and number of panels have an important influence on the weight and cost of the roof system. Table 1 contains the minimum cost roofing system for various configurations. Joist sizes of 4CA12 and 4CA8 ($xCAy$, x = joist depth in inches, y = joist weight in pounds per lineal foot) are required for the eight and 10 panel configurations, respectively. The elimination of joists for configurations of 12 panels and more leads to substantial cost savings. The minimum cost solution occurs for a configuration of x_1 equal to 26 and x_2 equal to 14. The minimum weight solution occurs for the same configuration. For large bent spacings [e.g., 52 ft (15.9 m)], manufactured trusses must be used for purlins. As bent spacing increases (x_1 decreases), the number of purlins decreases substantially while the weight of purlins increases substantially. Thus, a tradeoff between

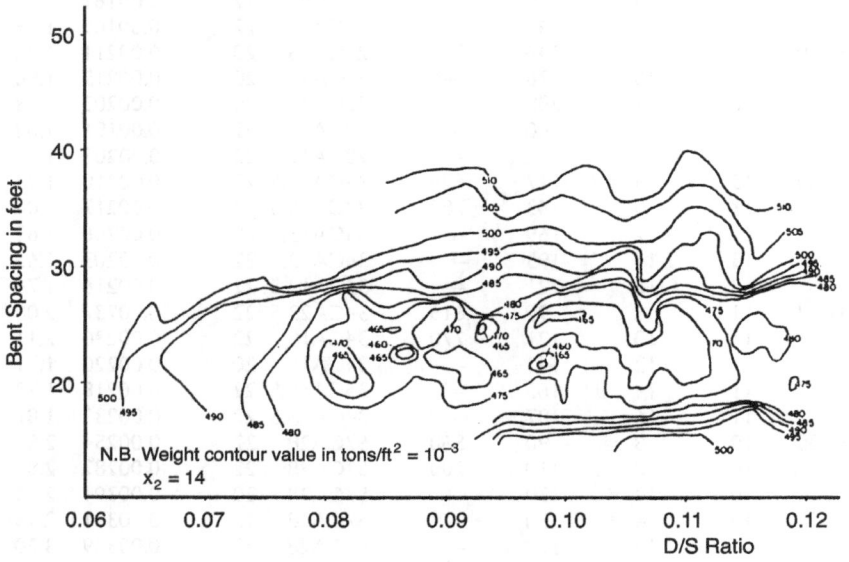

Figure 2.4.11 Contours for bent and roofing system weight (1 ft = 0.305 m).

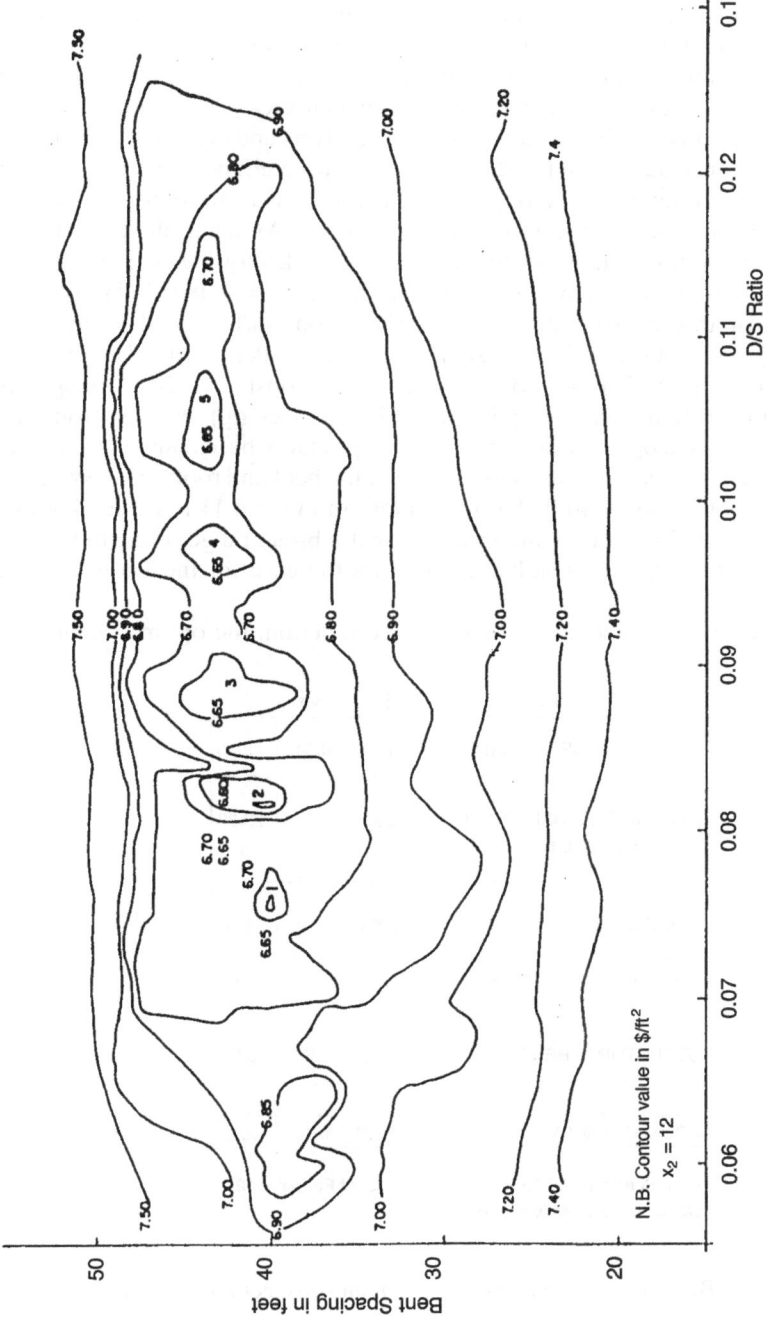

Figure 2.4.12 Contours for total system cost (1 ft = 0.305 m).

handling charges and material cost exists. For the costs considered herein, the additional material cost associated with the larger purlin spans outweighed the savings in handling costs associated with fewer purlins.

By using an optimization routine, such as described previously, to set values for the interface variables in a way which seeks to maximize overall performance, it is possible to coordinate the analysis and design of the various subsystems previously described. First, consider weight per square foot (bent plus roof subsystems) as the objective function, a measure typically used as a proxy or surrogate for cost by civil engineers. As determined by Box's Method, the optimal values for the interface variables for the cost parameters considered are $x_1 = 24$ bays, $x_2 = 14$ panels, and $x_3 = 0.08917$. The weight and cost of this configuration are 0.004545 tons/sq/ft (44.28 kg/m²) and $7.26/sq/ft ($18.06/m²) (all subsystems). Fig. 2.4.11 depicts the design space for the weight optimization problem with x_2 held constant at 14 (corresponds to a roofing system having no joists and 22 gage decking). As seen from this figure, many local optima exist and they may probably be accounted for by the discrete member spectrum used for design of the bent and roofing subsystems. The 0.00475 tons/sq/ft (46.27 kg/m²) contour in Fig. 2.4.11 is within 4.5% of the optimal weight solution and indicates that a broad range of depth to span ratio yields near optimal results. Such is not the case for the variable, bent spacing.

When cost is considered as the objective function, the optimal values for

DESCRIPTION	$	$/ft²	%
CLADDING FOR D/S RATIO	28200	0.54	8.33
CLADDING TO UNDERSIDE OF TRUSS	79980	1.54	23.33
ROOF SYSTEM	86255	1.66	25.16
STRUCTURAL BENT	138830	2.67	40.50
FOUNDATION SYSTEM	9530	0.18	2.78

Bent Spacing = 43.33 ft   Number of Panels = 12
Depth to Span Ratio = 0.088154

Figure 2.4.13 Breakdown of total cost for minimum cost configuration (1 ft = 0.305 m).

the interface variables are $x_1 = 12$ bays, $x_2 = 12$ panels, and $x_3 = 0.08820$. This configuration has a weight of 0.00512 tons/sq/ft (49.88 kg/m^2) and a cost (all subsystems) of $6/59$/sq/ft (70.86/m^2). Thus, a difference of 0.67/sq/ft (7.20/m^2) or $34,480, separates the minimum weight from the minimum cost design. Fig. 2.4.12 depicts the design space for the cost optimization problem with x_2 held constant at 12 (corresponds to a roofing system having no joists and 20 gage decking). Again, the design space exhibits several local optima, creating difficulties for any search procedure that may be used. The 6.90/sq/ft (74.19/m^2) cost contour which is approximately within 5% of the optimal cost encloses a broad range of configurations, both in terms of depth to span ratio and bent spacing. Fig. 2.4.13 depicts the cost contribution of each subsystem to the total cost for the optimal configuration. As seen from this figure, for the subsystem alternatives and conditions assumed, the bent, roofing, and cladding subsystems are by far the most important in terms of cost.

Twenty-five cost optimization runs were made. For these runs, the objective function value ranged from a low of 6.59/sq/ft (70.86/m^2) to a high of 6.81/sq/ft (73.23/m^2) or a 3.2% variation. The objective function value for 21 of these runs was within 1.6% of the best run.

CONCLUSIONS

A methodology for the cost optimization of buildings has been described and demonstrated. It consists of the decomposition of the overall building design problem into a series of problems, one for each building subsystem, e.g., foundation, structure, roofing. The solution of these problems is sequenced in a manner which accounts for the physical interaction between each subsystem. Their solution is coordinated by way of a master optimization problem, the function of which is to determine values for the variables which describe how the various subsystems are interfaced.

If cost is used as the objective function, it is important to include both quantity (variable) and nonquantity (fixed) related costs in order to achieve the minimum cost design, as viewed by fabricators and contractors. One way of estimating fixed and variable costs is to consider the elemental operations required for the fabrication or construction of a specific component or subsystem.

Not considered in the present work are ways in which the selection of alternative concepts for each subsystem could be incorporated into the optimization process. This topic should be treated in future work. As well, consideration should be given to including the design of other subsystems into the problem (e.g., mechanical and enclosure subsystems) and of using life cycle cost as an objective function.

ACKNOWLEDGEMENTS

The writers gratefully acknowledge the financial support for this work by La Formation de Chercheurs d'Action Concertée du Québec and by National Research Council operating grant A8952.

REFERENCES

1   Box, M. J. 'A New Method of Constrained Optimization and Comparison with other Methods,' *Computer Journal*, Vol. 8, 1965, pp. 42–52.
2   Bradley, J., Brown, L. H., and Feeney, M., 'Cost Optimization in Relation to Factory Structures,' *Engineering Optimization*, Vol. 1, 1974, pp. 125–138.
3   *Building Construction Cost Data*, Robert Snow Means, Duxbury, Mass., 35th ed., 1977.
4   *Dodge Construction Systems Costs*, McGraw Hill Information Systems, 1977.
5   Lee, B. S. and Knapton, J. 'Optimum Cost Design of a Steel-Framed Building,' *Engineering Optimization*, Vol. 1, 1975, pp. 139–153.
6   Lipson, S. L. and Agrawal, K. M. 'Weight Optimization of Plane Trusses,' *Journal of the Structural Division, ASCE*, Vol. 100, No. ST5, Proc. Paper 10521, May, 1974, pp. 865–879.
7   Lipson, L. L. and Gwin, L. B. 'Discrete Sizing of Trusses for Optimal Geometry,' *Journal of the Structural Division, ASCE*, Vol. 103, No. ST5, Proc. Paper 12936, May, 1977, pp. 1031–1046.
8   Lipson, S. L. and Russell, A. D. 'Cost Optimization of Structural Roof System,' *Journal of the Structural Division, ASCE*, Vol. 97, No. ST8, Proc. Paper 8291, Aug., 1971, pp. 2057–2071.
9   Miller, C. J., Moses, F. and Yeung, J. 'Optimum Design of Building Structures using Staged Optimization,' *Computers in Structural Engineering Practice*, Proceedings of the CSCE Speciality Conference, Montreal, Canada, Oct. 6 and 7, 1977, pp. 265–387.
10   National Building Code of Canada, 1975.
11   Yardstocks for Costing, a supplement to the *Canadian Architect*, Construction Data Systems and Southam Business Publication, Canada, 1977.

2.5 Predicting the tender price of buildings in the early design stage: method and validation*

R. McCaffer, M. J. McCaffrey and A. Thorpe

ABSTRACT

Predictions of the tender price of proposed buildings are needed during early design. Conventionally, these are prepared manually and based on limited data.

A computer system was produced which prepares estimates using a library of data containing rate, quantity and date for the constituent elements of previous buildings, inflation indices and statistical models.

Because the scatter of the rate vs quantity data for each element in a sample of buildings was so variable between samples, no one model was adequate. Thirty two different models were included, together with a criterion for selecting the most appropriate. This ensured the most precise prediction possible.

The co-efficient of variation of the ratio forecast to actual price measured for a set of system-produced estimates grouped by building type ranged from 10 to 19%. Current manual practice ranges from 6 to 21%, but this is only achieved much later when design is complete and more reliable data is available.

INTRODUCTION

The initial estimate of the cost of a building often carries the burden of being the cost limit for the project. The establishment of the cost limit is important in determining a project's viability, and so this initial estimate frequently carries a burden disproportionate to its reliability. The reliability of this estimate is questioned because it is frequently produced within a restricted time limit ranging from several days to a few hours, and because in most cases there is a shortage of relevant and recent historical price data on which it can be based. Furthermore, it is rare for the accuracy of these estimates to be monitored and/

*1984, *J. Opl Res. Soc.*, 35(5), 415–24.

or published. This means that the achieved accuracy of initial estimates of building cost is often believed to be somewhat better than it actually is.[1]

In the U.K. priced bills of quantities for previous buildings are the primary source of the data used to predict the tender price of a proposed building during design. Bills of quantities are lists giving the specification, descriptions and quantities of the various items of work comprised in a building contract. Contractors insert rates against each item quantity and thereby compute their tender price for the work. The priced bill of quantities of the successful contractor becomes a contract document.

The data used to make a tender price prediction is obtained by analyzing the priced bill of quantities of a previous similar contract to give rates at a coarser level of detail than that of the bill and more appropriate to the amount and type of information available during design. One form of analysis frequently used to provide the rates needed to price early designs is the Standard Form of Cost Analysis.[2] This combines the various item prices from the bill to give the price of each of 39 defined elements (e.g. external walls, roof, doors, etc.) which make up a building and which are common to all building types. Units of quantity appropriate to each element are defined, quantities measured and rates calculated. Other methods of analysis used during later design are less formally defined but are similar. They comprise the grouping of bill items at a coarser level of detail than that of the bill but at a finer level of detail than elements. This work used elements as the most common level of detail used during early design.

Currently it is common practice to produce tender-price predictions manually. The process comprises:

1   the measurement of the quantities of the elements which make up the proposed building;
2   the selection of one cost analysis of a similar completed building and the abstraction of the rates on which to base the prediction;
3   the synthesis of the prediction by updating and adjusting the rates from (2) by a published index and applying them to the quantities from (1).

The selection of the cost analysis of only one previous building means that the unreliability of the prediction will be that of the original data combined with that of the index (Beeston[3]).

However, by taking the average of several appropriately adjusted prices, Beeston[3] argues that the variability in the data can be reduced and submits that more reliable estimates would be obtained by using the updated rates from several buildings for each element. Enough data exists for this to be done, but to do so using current manual methods is impractical because of the data management problems and because the calculations take too long.

The use of computer-based methods allows estimates to be based on rates from many previous buildings but introduces problems of analysis.

This paper describes a computer-based tender-price prediction system, the analyses it performs, the statistical modelling methods used, the criteria for automatically selecting the most appropriate statistical model, the computation procedures employed and the results obtained using test data.

The tender-price prediction system

The tender-price prediction system originally developed by McCaffer[4] and later refined by McCaffrey[5] and Thorpe[6] is described in Table 2.5.1.
Figure 2.5.1 adds further explanation to this.

Table 2.5.1 Features of the tender-price prediction system

Files holding data on which estimates are based
(1)   Master library of rates, quantities and dates for the various items that make up a building stored project by project and sub-divided by building type.
(2)   Library of published building tender-price and cost indices.

User options available in constructing a data base for a specific estimate
(3)   Selection of the rate vs quantity data on which the estimate is to be based from the master library.
(4)   Selection of the indices to be used to adjust rates for inflation and market conditions.

Statistical modelling options
(5)   Automatic selection of the statistical treatment of the historical data used for *each* individual item in the estimate.
(6)   Manual pre-selection of the statistical treatment for the historical data used for each item.

Calculations
(7)   For each item in the estimate the rates are updated using the selected inflation index to the date of the estimate and the updated rates analysed in any one of 12 diffferent ways to produce an item estimate. The sum of the item estimates is the estimate for the whole building.

Calculation options
(8)   Single estimate for each time and single total estimate.
(9)   Simulation of the estimate 'n' times (n being specified by the user) to determine the variability of the total estimate. The mean and standard deviation of the simulated estimates are calculated.

System output
(10) A print-out of the estimate prepared from the input items and quantities provided, listing the items ranked in order of contribution and dispersion and the expected total price.
(11) Similar output for each simulation together with the mean of the 'n' simulated total estimates and their standard deviation.
(12) An optional listing item by item, of rates from any part of the master library adjusted for inflation to a common date.

Test facility
(13) The facility to test the system by preparing predictions for a nominated building in the master library using the price quantity data of the older buildings.

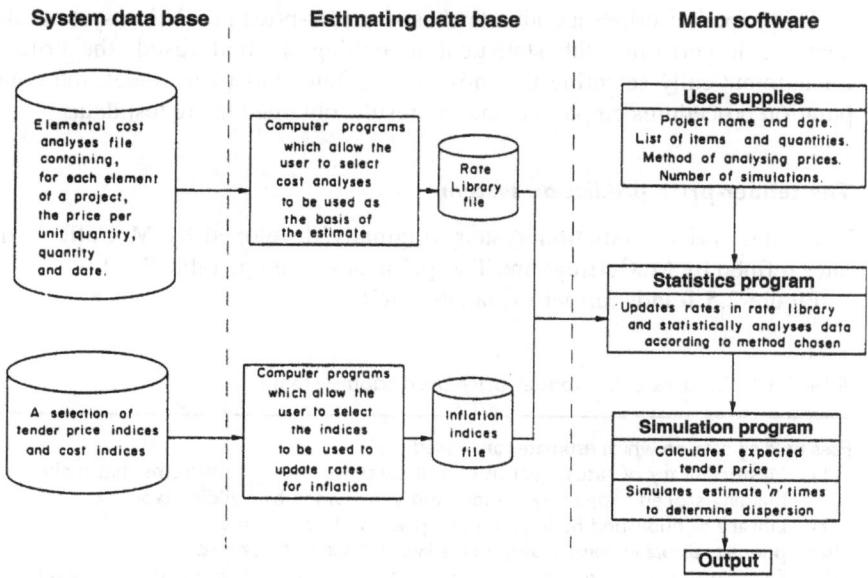

Figure 2.5.1 Flow chart of the tender–price prediction system.

DATA USED AND ANALYSIS

The data held in the 'cost analyses' file comprises a list of 39 elements used to describe a building[2] and, against each element, the price per unit quantity (rate), the quantity of the element (area of external walls, number of sanitary appliances, etc.) and the date. Rates are not adjusted to price-levels ruling at a particular time, so that the choice of the indices used to correct rates for inflation can be made by the user. Each rate, quantity and date entry refers to a particular building. Thus one historical cost analysis can provide a potential total of 39 sets of rate, quantity and date, that is one entry for each element.

The filed data is grouped by type of building, e.g. schools, factories, offices, etc. The user selects which type of building he requires. Then, from this group of similar (by type) buildings, he selects the particular ones required to form the data on which the estimate of the proposed building will be based.

Thus, for the estimate of any proposed building, the user assembles, for each of the 39 elements, as many sets of data from the file as are relevant. Using the date and the indices to update the rate for each element, this will result in a number of (quantity, rate) pairs. The problem is then to define a method of modelling the rate vs quantity data for each element that is appropriate for each case. The purpose of this model is to obtain an estimate of the rate for the element in the proposed building and a measure of its dispersion.

The estimate of the rate for each element combined with the quantity of the element and summed gives the estimate of the total price of the proposed building. Assuming independence between rates, the dispersion of the expected total price for the proposed building can be calculated algebraically from the dispersions of the residuals of the rates for each element and the element quantities. In practice, however, users prefer to see the effect on the total price of the variability of the element rates. 'n' different estimates of the total price are therefore produced using Monte Carlo simulation, and the mean and standard deviation of the 'n' total prices calculated. This approach has two other advantages. Firstly, different assumptions about the underlying distributions of the residuals of the element rates can be easily accommodated and secondly, it is possible to avoid any such assumption if a rate is calculated using a residual chosen randomly from the observed residuals.

Whether the dispersion of the total price is calculated algebraically or by Monte Carlo methods, the major problem is the validity of the assumption of independence between rates. Beeston[7] questions this assumption on the grounds that there are correlations between the rates for different elements. These correlations have not yet been quantified but it is possible that they are reduced to acceptable levels when the rates are updated. If this is not so, then a resampling (of the data) method[8] would produce a more realistic measure of the variability of the estimated total price, and this is the approach that was used to validate the system.

The two factors that were identified as having the greatest effect on the 'accuracy' of the estimated price were the index used to update rates held on file from date of rate to date of estimate and the method of modelling the rate vs quantity data.

If estimates of price are prepared for a group of buildings whose actual prices are known, the choice of the index used to correct price level determines the mean of the ratio 'prediction/actual' with respect to unity. The method of modelling the rate vs quantity data determines the dispersion of the ratios predicted/actual around the mean.

There exists several possible indices for price level adjustment, and the problem with respect to indices was seen as the selection of the most appropriate rather than the creation of a new index. The statistical modelling of the rate vs quantity data seemed to offer scope for improving the reliability of estimates.

Modelling methods

Two types of analyses were adopted:

1   univariate analyses, and
2   bivariate analyses.

Table 2.5.2 The statistical treatments available
(A) Univariate analysis

Type of analysis	Transformation of variables	
	Raw data y = rate	Log data y = log rate
Normal distribution $$\hat{y} = \frac{1}{n} \sum_{i=1}^{n} y_i$$	(1) Normal	(2) Log-normal (corrected for bias)
Triangular distribution $$\hat{y} = \frac{y_{min} + y_{mode} + y_{max}}{3}$$	(3) Triangular	(4) Log-triangular

(B) Bivariate analysis. Models of the Form $\hat{y} = a + bx$

Type of analysis	Transformation of variables		
	Raw data y = rate x = quantity	Single-log data y = $\log_e$ rate x = quantity	Double-log data y = $\log_e$ rate x = $\log_e$ quantity
Least squares regression Minimize $$\sum_{i=1}^{n} \left| y_i - a - bx_i \right|^2$$	(5) Regression	(6) Single-log regression	(7) Double-log regression
Absolute error regression Minimize $$\sum_{i=1}^{n} \left| y_i - a - bx_i \right|^1$$	(8) Error regression	(9) Single-log error regression	(10) Double-log error regression
$L_{1.5}$ regression Minimize $$\sum_{i=1}^{n} \left| y_i - a - bx_i \right|^{1.5}$$	(11) $L_{1.5}$ regression	(12) Single-log $L_{1.5}$ regression	(13) Double-log $L_{1.5}$ regression
Weighted regression Minimize $$\sum_{i=1}^{n} \frac{\left| y_i - a - bx_i \right|^2}{\left| y_i \right|}$$	(14) Weighted regression	(15) Single-log weighted regression	(16) Double-log weighted regression

Note 1:
$$n = \text{number of rates}$$
According to the transformation used:
$$\hat{y} = \text{the estimate of the item's rate or its logarithm}$$
(in which case estimated rate = $e^{\hat{y}}$)

each y_i = a recorded rate for the item or its logarithm;
$\hat{x}$ = the quantity of the item being estimated or its logarithm;
each x_i = a recorded quantity for the item or its logarithm;
y_{min} = the minimum of the recorded rates or its logarithm;
y_{max} = the maximum of the recorded rates or its logarithm;
y_{mode} = the mid-point of the modal class of recorded rates or its logarithm;
a and b = constants determined by minimizing the functions stated.

Recorded rates and recorded quantities are those held in the estimate's base data.

Note 2:
The 16 treatments are doubled to 32 by segmenting the data into different quantity ranges and applying the 16 treatments to both segmented and unsegmented data. The estimate's base data is automatically segmented to restructure the analysis to a specific range in which the estimate will fall.

Note 3:
The method of determining the most appropriate statistical model is to minimize the ratio:

$$\frac{\text{Median of the absolute values of the relative residuals}}{1 + \text{median of the relative residuals}}$$

where the relative residuals are defined in terms of the ratio R (= forecast – actual) as follows:

$$\text{Relative residual} = 1 - \frac{1}{R}, \text{ if } R < 1$$
$$= R - 1, \text{ if } R > 1$$

Using this transformation, an estimate of 200 of an actual price of 100 is quantitatively the same as an estimate of 50 except for change of sign. The distribution of the relative residuals is therefore symmetric if the overestimate ($R > 1$) occurs as often as the underestimate ($1 > R$).

The numerator of the ratio minimized is a measure of dispersion and the denominator a measure of location. The ratio is thus similar to the co-efficient of variation of the ratio forecast to actual but has the advantage that it is independent of the fitted models because it is calculated using median values rather than a standard deviation and mean.

The method was arrived at on an *ad hoc* basis following a number of trials.

(1) The univariate analyses were introduced for modelling rate vs quantity data sets that were small and for users who expressed a preference for 'simpler' analyses. These were mainly used when manually selected. The four methods of determining an expected value are as shown in Table 2.5.2, the log transformation being included to reflect a general acceptance that there is a positive skew in data of this type.[3]

For an estimate based on the triangular distribution, the data is partitioned into classes and the mid-point, the 'modal' value, of the class with the highest frequency of points calculated. The data is assumed to belong to a triangular distribution with a mode equal to this 'modal' value and with maximum and minimum equal to the maximum and minimum respectively of the observed values. The expected rate and its standard deviation are calculated using the usual formulae for the triangular distribution[9]

The calculated means and standard deviations are used to predict prices and their associated dispersions. The log transformation is similar except that the analyses are performed on logged data.

(2) Four bivariate treatments were used to model the data. These were all regression models of rate on quantity and differed only in the calculation of the regression parameters. The four treatments were:

a   simple least-squares regression,
b   absolute error regression,
c   L1.5 regression,
d   weighted regression (also termed 'chi-squared' regression in this paper).

The use of regression analysis for estimating in the construction industry has been described by McCaffer.[10] In this paper the authors address the problem of cost modelling by use of both simple and multiple regression techniques based on the work of several research students at Loughborough University of Technology.[11–17]

These regression models give an expression $y = a + bx$ where y is rate, x is quantity, and a and b are given by the analysis. For a given quantity, the value 'y' can be calculated and is taken to be the estimate of the rate for that element and the mean of a normal distribution whose standard deviation is set equal to the standard deviation of the residuals from the regression model. This normal distribution is subsequently used in the simulation of the whole building price. When the quantity of the element being estimated is outside the range of the data being analyzed, the regression model can break down and, for example, yield a negative rate. In these circumstances the system defaults to an appropriate univariate analysis to calculate the rate.

(a) For data whose parent population is assumed to be normal, simple least-squares regression is appropriate. The transformation of log price was introduced because rate vs quantity scatter diagrams showed that, for many elements, rates tended to change exponentially and not linearly with quantity. The transformation of taking logs of both rate (y) and quantity (x) was introduced as a scaling effect to reduce the quantity range in the analysis and so partially overcome the disproportionate effect of large, one-off projects.

(b) The second bivariate technique, Absolute error regression (termed in this paper 'error regression'), was introduced for the following reasons.

The least-squares technique is based on the assumption of normality, as already stated. However, there is rarely sufficient data for a true estimation of the form of the distribution of the data-set. The mean is very sensitive to deviations from normality due to outliers, longtails, etc., which appear in our data. To overcome these problems, a robust method is required which yields similar estimates to the least-squares estimates when the data is normally distributed but yields estimates little changed when the data contains outliers.[8]

To do this, Forsythe[18] recommended minimizing the pth power deviations for a power of p between one and two:

$$F = \sum_{i=1}^{n} \left| y_i \square a \square bx_i \right|^p ,$$ (1)

where a and b are the estimated regression parameters; for absolute error regression $p = 1.0$, and for least-squares regression $p = 2.0$.

The parameters a and b are not linear combinations of x and y (unless $p = 2$), and so it is more difficult to calculate their values, and iterative routines have to be used.

Several iterative routines were tested and compared by Forsythe and others, and these are described by Fletcher and Powell,[19] Fletcher and Reeves[20] and Davidson.[21]

Further work in this area by Vassilacopoulos[22] suggested the use of a standard N.A.G. routine.[23] This proved satisfactory for our problem, being robust but rather slow. The method is described by Mosteller and Tukey.[8]

From the absolute error regression line defined, the mean and standard deviation of the points around this line can be found in the usual way, as for the simple least-squares regression case. These are then used to predict rates and dispersions as before.

Transformation of log quantity and of both log quantity and log rate was used as with least-squares regression.

(c) The third bivariate technique introduced was L1.5 regression. In his paper 'Robust estimation of straight line regression co-efficients by minimizing pth power deviations', Forsythe[18] discussed test values for p of 1.25, 1.50 and 1.75. Vassilacopoulos[22] examined values of p of 1.0, 1.5 and various trimmed data sets. Both found that $p = 1.5$ was a fast and reasonably robust estimator for moderately contaminated data, giving values between the simple least-squares technique and the absolute error technique. Thus $p = 1.5$ was used as a 'middle of the road' technique, a compromise between the two extremes.

Forsythe[18] demonstrated that a value of $p = 1.5$, was 95% as efficient (in terms of root mean square error) as the least-squares method when the data was not contaminated (i.e. normally distributed) and up to 60% more efficient when the dat was contaminated or non-normal.

To implement and test this treatment, the same technique was adopted as used for the absolute error regression model, with the exception that the value of p was changed to 1.5 in the functional equation to be minimized.

The same N.A.G. routines[23] and iterative procedures were used and the means and standard deviations calculated from this model in the same way. The single-log and double-log transformations were included as before.

(d) The fourth and final simple bivariate technique to be introduced was weighted or chi-squared regression.

The chi-square distribution[24] is positively skewed and takes into account the size of the quantity being estimated – i.e. **relative** error. Estimates of the regression parameters are obtained by minimizing the function:

$$F \sum_{i=1}^{n} \frac{\left(y_i \square a \square bx_i \right)^2}{\left| y_i \right|}$$

The denominator is the absolute value of the actual y value, and this introduces relative size into the function.

The regression parameters, found from the usual formulae for weighted least-squares regression, are used to calculate estimates and standard deviations as for the least-squares regression model.

The logged transformations were included as before.

In addition to the foregoing models, a method of segmenting the data was introduced because, for some elements, the plot of rave v quantity was 'U' shaped. For this type of data it is more appropriate when using univariate and bivariate techniques to model the rate v quantity relationship locally to particular quantity levels. The method involved:

i   specifying a minimum number of class boundaries to be used to partition the data into quantity ranges;
ii inserting this number of class boundaries in the data by means of a N.A.G. routine;[23]
iii checking to see if there was at least the specified minimum number of points in the partitioned class where the estimate was to be made.

If enough points did exist in the particular class, the analysis (least-squares regression, absolute error regression, etc.) was performed on those points.

If, however, the required number of data points was not present, the class width was increased, new class boundaries defined and the process repeated.

The introduction of segmenting the data doubled the number of modelling methods available. This gave a potential total of 32 different statistical treatments.

Selections of modelling method

A data library of 107 buildings was assembled to predict a further 35 buildings (i.e. buildings not in the data library) whose actual prices were known. This data was subdivided by building type, giving four test samples of buildings to be estimated and four corresponding data libraries. By preparing estimates for each element in each of a sample of test buildings using each of the 32 modelling methods and by knowing the actual price for each element, the modelling method that give the smallest difference between the estimated price of the element and the actual price of that element in the building being estimated could be selected. This sample set of buildings could then be used to determine the appropriateness of the various modelling methods. Figure 2.5.2

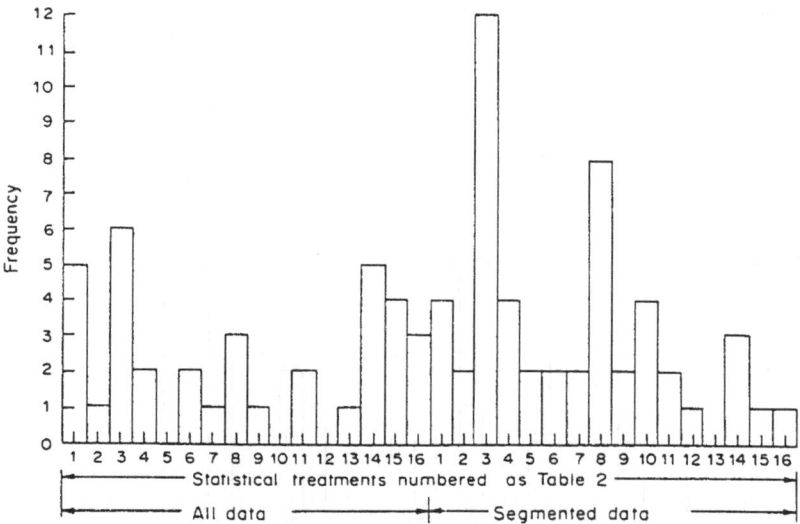

Figure 2.5.2 Frequency of use of the modelling methods producing the most accurate results.

presents the frequency of the modelling methods that were most accurate for one of the sample of buildings studied (i.e. the frequency of the modelling methods that produced element estimates closest to the actual price of the element). The observation was that no particular modelling method was favoured but there was a scatter across the range of modelling methods.

The requirement was to find a criterion which could be used to select the modelling method in a working situation (i.e. without knowing the actual price of the element of the building being estimated). Eight criteria were examined; most were rejected as inappropriate, and the three subjected to further tests were:

(i)   to minimize the mean square error of the residuals;
(ii)   to minimize the absolute error of the residuals; and
(iii)   to minimize the ratio of

$$\frac{\text{median of absolute values of relative residuals}}{1 + \text{median of relative residuals}}$$

residuals here being the residuals of the models when fitted to the base data.

Selection criteria (i) and (ii) were rejected as displaying some model dependency, and criterion (iii) was adopted as appropriate in that it showed no

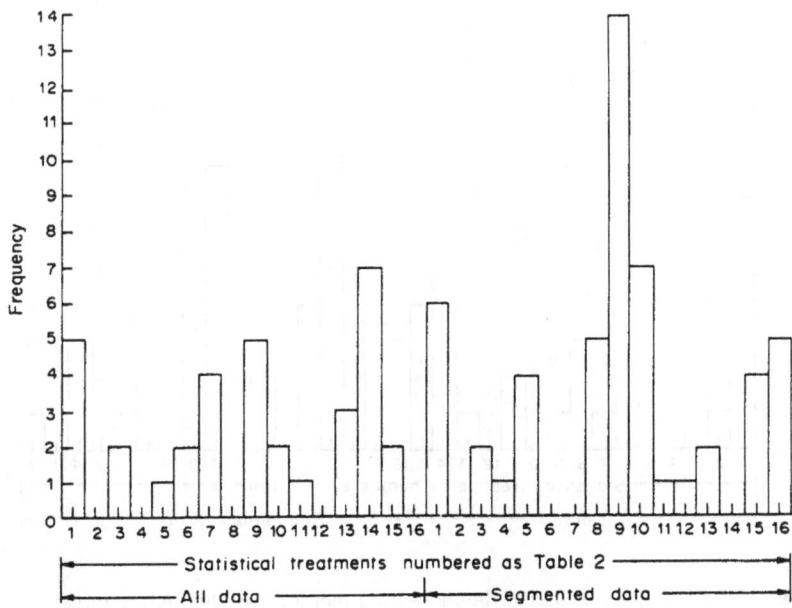

Figure 2.5.3 Frequency of selection of modelling method using the automatic selection criterion.

Table 2.5.3 Precision achieved in current practice at *pre-tender* stage

Building type	Co-efficient of variation of ratio $\frac{forecast}{actual}$	Data source	Sample use
Schools	6%	Local authority Q.S. section	15
Schools	11%	Morrison and Stevens[25]	256
Railway buildings	21%	Kennaway[16]	47
All building types	13%	Morrison and Stevens	860

model dependency and produced the most precise estimates as measured by the co-efficient of variation of the ratio forecast total price to actual total price for each of the samples. The frequency of selection of modelling methods using selection criterion (iii), and the same test sample of buildings used for Figure 2.5.2 is shown in Figure 2.5.3. Figure 2.5.3 has a scatter across

Table 2.5.4 Summary of precision of estimates by the tender-price prediction system using automatic selection of statistical models and data of the type used during the early design years

Building type	All results		Results with 10% of outlier results removed	
	Co-efficient of variation of ratio forecast actual	Sample size	Co-efficient of variation of ratio forecast actual	Sample size
Middle and secondary schools	10%	8	10%	8
Primary, middle and secondary schools	11%	13	9%	12
Health centres	14%	8	6%	7
Offices	19%	6	11%	5
All the above tests	15%	35	12%	32

the range of modelling methods comparable to that of Figure 2.5.2, which was created by selecting models which produced element estimates closest to the actual element price. This comparability was not evident when other selection criteria were used, and tests using other samples of buildings confirmed these observations.

Consistency of predictions

Data was collected from published research reports and from the quantity-surveying sections of a local authority and a nationalized industry to establish the achieved precision of current practice. This is given in Table 2.5.3.

The precision achieved using automatic selection of statistical models based on criterion (iii) are summarized in Table 2.5.4.

CONCLUSIONS

No one particular modelling method was appropriate for use in all cases. The automatic selection of statistical models proved to be the most consistent in that the co-efficient of variation of the ratios forecast to tender in a group of predictions was the smallest.

The tests using the system were conducted by including all the available data for a given building type prior to the date of the test estimate. No selection of the base data was exercised and no attempt was made to exclude from the tests items such as external works that are inadequately represented by elemental data. This meant that some of the test estimates had large percentage errors. Results with 10% of the outlier results removed are therefore also shown in Table 2.5.4.

The comments relevant to these results are that, compared to current manual methods, the tender price prediction system improved estimating precision (i.e. reduced the scatter of any group of predictions).

An improvement in estimating precision during the **early design stages** is claimed for the tender price prediction system. The precision given in Table 2.5.4 is that achieved using elemental cost analysis data (i.e. the type of data that can be used during the early stages of a project). This precision compares favourably with that in Table 2.5.3, which gives the precision of pre-tender estimates, i.e. estimates prepared at pre-tender stage, when the design is complete and more reliable data is available.

ACKNOWLEDGEMENTS

The authors are grateful to the Building Cost Information Service of the Royal Institution of Chartered Surveyors and other organizations for permission to use their data, to the Science and Engineering Research Council for funding the work and to Dr A. N. Pettitt, Department of Mathematics, Loughborough University of Technology for statistical advice, guidance and encouragement.

REFERENCES

1 B. C. Jupp (1980) Cost data estimating processes in strategic design of building. Report describing research undertaken for the Building Research Establishment.
2 The Building Cost Information Service of the Royal Institution of Chartered Surveyors (1969) *Standard Form of Cost Analysis* (re-printed 1973), London.
3 D. T. Beeston (1973) *One Statistician's View of Estimating*. Directorate of Quantity Surveying Development, Property Service Agency, D.O.E., London.
4 R. McCaffer (1976) *Contractors' bidding behaviour and tender–price prediction*. Ph.D. Thesis, Loughborough University of Technology.
5 M. J. McCaffrey (1978) Tender-price prediction for U.K. buildings – a feasibility study. M.Sc. Project Report, Department of Civil Engineering, Loughborough University of Technology.
6 A. Thorpe (1982) Stability tests on a tender–price prediction model. M.Sc., Thesis, Loughborough University of Technology.
7 D. T. Beeston (1983) *Statistical Methods for Building Price Data*. E. & F. N. Spon, London.
8 F. Mosteller and J. W. Tukey (1977) *Data Analysis and Regression*. Addison–Wesley, Reading, Massachusetts.
9 A. H-S. Ang and W. H. Tang *Probability Concepts in Engineering Planning and Design, Vol. 1, Basic Principles*. Wiley, New York.

10 R. McCaffer (1975) Some examples of the use of regression analysis as an estimating tool. *Quant. Surv.* **31**, 81–86.

11 J. S. Buchanan (1969) Development of a cost model for the reinforced concrete frame of a building M.Sc. Project Report, Department of Civil Engineering, Loughborough University of Technology.

12 P. R. Gould (1970) The development of a cost model for heating and ventilating and air conditioning installations in buildings. M.Sc. Project Report, Department of Civil Engineering, Loughborough University of Technology.

13 B. F. Moyles (1973) An analysis of the contractor's estimating process. M.Sc. Thesis, Loughborough University of Technology.

14 R. H. Neale (1973) The use of regression analysis as a contractor's estimating tool. M.Sc. Project Report, Department of Civil Engineering, Loughborough University of Technology.

15 M. J. Baker (1974) Cost of houses for the aged. M.Sc. Project Report, Department of Civil Engineering, Loughborough University of Technology.

16 J. D. Blackhall (1974) The application of regression modelling to the production of a price index for electrical services. M.Sc. Project Report, Department of Civil Engineering, Loughborough University of Technology.

17 D. Coates (1974) Estimating for French drains – a computer based method. M.Sc. Project Report, Department of Civil Engineering, Loughborough University of Technology.

18 A. B. Forsythe (1972) Robust estimation of straight line regression co-efficients by minimizing pth power deviations. *Technometrics* **14**, 159–166.

19 R. Fletcher and M. J. D. Powell (1963) A rapidly convergent decent method for minimization. *Comput. J.* **6**, 163–168.

20 R. Fletcher and C. M. Reeves (1964) Function minimization by conjugate gradients. *Comput. J.* **7**, 149–154.

21 W. C. Davidson (1959) *Variable metric method for minimization.* A.E.C. Research and Development Report ANL–5990.

22 G. X. Vassilacopoulos (1973) Robust polynomial regression. M.Sc. Project Report, Department of Mathematics, Loughborough University of Technology.

23 Numerical Algorithms Group (1981) *FORTRAN Library Manual Mark 8*, Vol. 1.

24 G. W. Snedecor and W. G. Cochran (1967) *Statistical Methods*, 6th edn. Iowa State University Press.

25 N. Morrison and S. Stevens (1981) *Construction Cost Data Base – Second Annual Report.* Property Services Agency, D.O.E., London.

26 W. G. Kennaway (1980) Estimating for abnormal conditions M.Sc. Thesis, Department of Building, Heriot-Watt University.

Part 3 Cost-product modelling

This part contains only a very small proportion of the extensive work that has been published on this topic world-wide. All these papers examine by various means the cost implications of design alternatives through the use of product models. Outside the market economies, where product element values (prices) are fixed by institutional means, the task is relatively simple and straightforward, the main requirement being a large computer to hold all the data. In a competitive environment however the large variability involved has resulted in a statistical approach to modelling in all except the earliest work. In contrast with the papers concerned with new models in Part Two, these papers fall well short of the standards of rigour. The main reason for this seems to be that useful product elements are difficult to identify and the relationships sought are extremely weak. The pity is that this is seldom made clear in the papers themselves.

The Wilderness **An investigation into building cost relationships of the following design variables: storey height, floor loading, column spacing, number of storeys** (ch 3.1) is the famous product of The Wilderness Dining Club for chartered quantity surveyors and other members of the RICS. It represents the product of 'several years research' and endeavours to show 'the effect on cost of certain fundamental design variables'. It aims to explicate what is the judgemental aspect in the use of single price models so that the cost implications, floor loadings etc. can be adduced. A series of beautifully produced charts resulted from the work, which involved extensive measurement and pricing (element and functional valuation) (using Spon's 1956/7, 82nd ed.), showing the likely effect of incremental changes in the design variables of storey heights, floor loadings, column spacing and number of storeys. The most impressive aspect of this work now is the great care taken and vast amount of time spent by (unfunded) practitioners in computing the results without a computer.

Flanagan and Norman's **The relationship between construction price and height** (ch 3.2) makes the distinction between static models, where the elements are assumed to be independent, and dynamic models, where the elements are interdependent, i.e. dynamic models consider simultaneously the effect of changing one element and the induced effects on other elements. The main purpose of the paper is to show that, in contrast with numerous other studies, the price per square metre of gross floor area (functional value) does not rise lineally with the number of storeys (element value). This is done firstly in a discursive manner by reference to other similar work in the field together with some personal observations relating to the building technology implications. The second part of the paper describes an empirical analysis to test the U shape hypothesis by regression analysis with significant results.

Maver's **Cost performance modelling** (ch 3.3) is a notable early reference to automatic cost forecasting and describes, from a designer's viewpoint the benefits of automatic cost forecasts when linked to CAD. The paper is concerned with what was then 'the new generation of computer-based design models' which aimed to make 'the trade-off between construction investment and return [cost value trade-off] substantively more explicit.' The philosophy for achieving this is that the designer generates a design hypothesis which is automatically costed by the computer program (task 2). The designer then evaluates the cost/performance profile and modifies the design accordingly. This process then continues iteratively until the designer converges on the design which gives a desired cost/performance profile representing an appropriate cost value balance. BCIS type product elements are used, the element values (quantities) being generated by the CAD system with the functional values (element unit rates) being held as constants in the system's costing files. This paper is probably the first to clearly establish the role and goal of cost evaluation (and therefore cost models) as an aid to the design process task 2.

Pegg's **The effect of location and other measurable parameters on tender levels** (ch 3.4) describes an important piece of work that has received too little attention by researchers. The main gain of this work is the empirical identification of nine variables (useful elements) that (statistically) significantly effect building prices date, location, selection of contractor, contract sum, building function, measurement of structural steelwork, building height, form of contract and site conditions over a very large sample (1188) of contracts. The disappointing aspect is that the method of analysis is not clearly described and levels of accuracy not given. To say that '. . . a considerable amount of variability remained' (after what appears to be a series of bivariate analyses) is not very helpful for other researchers in the field. A further, and even more serious, problem with this work is that access to Pegg's data is severely restricted so it is impossible to replicate his findings even with the same data.

3.1 An investigation into building cost relationships of the following design variables: storey height, floor loading, column spacing, number of storeys*

Wilderness Group

INTRODUCTION

Building productivity

This is National Productivity Year and it is perhaps a happy coincidence that the Wilderness Cost of Building Study Group is able at this time to submit this report. It is the result of several years of research and it is hoped that, by showing the effect on cost of certain fundamental design variables, it may lead to a greater awareness of the cost importance of the basic design concept and thus make a substantial contribution to productivity in the building industry.

How this study arose

'How much will it cost?' is a question often asked of a quantity surveyor at an early stage of a building project before the design is crystallised. Historical data of the cost of previous buildings will give some guide but how reliable are they when applied to a 'one off' building on a particular site with its own special limitations on building shape and height and the client's special requirements? How then, knowing that important policy decisions by client and architect may rest upon this first estimate of cost, does he give a considered and reliable answer?

In the early design stage before details of construction are known, which would enable approximate quantities to be prepared and priced, the estimate of cost will probably be based upon historical data modified by experience. The answer will be a compromise between what some other building did cost, what the proposed one should cost and what it might cost when the design has been worked out in detail.

*1964, Report to the Royal Institution of Chartered Surveyors. Copyright 1964: reproduced by permission of The Royal Institution of Chartered Surveyors. The original report contains 28 charts for a detailed investigation of cost relationships. This chapter uses just a few of the charts, for illustration.

Approximate estimates of cost

The 'cube' and 'floor area' methods of unit rate approximate estimating suffer from certain disadvantages in that they take no real account of building shape or the number of storeys over which the accommodation is provided. The 'cube' method makes quite false allowance for varied storey height and the 'floor area' method makes no allowance at all. When assessing the unit rate to allow for these factors the quantity surveyor uses his experience or where practicable prepares theoretical quantitative estimates for the particular project.

Storey Enclosure System of Approximate Estimating

To overcome, or at least ease, the difficulty of assessing the unit rate the Group devised the 'Storey Enclosure System of Approximate Estimating' which was presented in April, 1954 to a general meeting of quantity surveyors of The Royal Institution of Chartered Surveyors.

In principle the system aimed at achieving better results by narrowing the range of unit rates, according to building type and specification and compiled from historical cost data, to be applied to the areas of floors, roofs and enclosing walls multiplied by certain factors.

Although considerable interest was displayed at the time and a number of quantity surveyors use the system, little data has been forthcoming other than from members of the Group. Whilst this has been sufficient to test the system and to show that it is an improvement on the 'cube' and 'floor area' methods it is insufficient to provide the same extensive historical data as are available with them. Also the multiplication factors adopted, being based upon experience in use and not upon proven data, are open to question as to their reliability for universal application. Even so this system has resulted in unit rates being less variable within buildings of similar type than with the 'cube' and 'floor area' systems.

Like the 'cube' and 'floor area' systems the 'storey enclosure' system depends upon a price rate selected by judgment and little direct guidance is given as to ways in which economy of design might be achieved or the measure of possible economies. Such guidance is a long felt need of the building industry.

Limitations of historical data

In practice, the differences between buildings are so numerous and occur in such variety that it is almost impossible to find actual buildings which are sufficiently similar for their differences in cost to be related to particular factors and, furthermore, such cases yield information regarding the effect of these factors only insofar as they are conditioned by the other characteristics of the buildings.

It was found impracticable merely with historical data to isolate with any certainty the effect upon buildings cost of certain design factors individual to those buildings of which the costs were examined: in particular column spacings and floor loadings.

It was felt that the relative cost effects of these factors would have to be ascertained on a theoretical basis before historical data could be used to verify results of the study.

Reasons for this study

Having arrived at these conclusions the Group decided to go right back to the first principles upon which the cost of a building arises: in fact that it should embark upon an investigation into the economics of design.

THE WORK OF THE STUDY GROUP

Basis of study

The original intention was to investigate the design cost relationships of a large number of hypothetical steel framed buildings of equal total floor area and similar specification but with the accommodation arranged on one or more storeys in buildings of varying shapes with varying bay sizes, column spacings, storey heights and superimposed floor loadings. It was hoped that later the costs of varying methods of construction and specification and buildings of other sizes might also be investigated.

It was decided to limit the study to the structure and basic finishes of a building and to exclude internal partition layout, staircases, lifts, special features, fittings, services, external works and site clearance.

Functional components

The structure and basic finishes of the buildings studied were treated as an assemblage of a comparatively few basic types of 'functional components' i.e. parts of a building that can be separately designed and thus the costs isolated.

The main 'functional components' chosen were suspended slabs, beams, ties and columns assembled to form a single bay of suspended floor or flat roof (see Fig. 3.1.1). Each bay includes the cost of only one column (i.e. a quarter of a column at each of the four corners) and only half the cost of the beams or ties between such columns. Other 'functional components' include the ground slab to one bay, the foundation to one column and the various perimeter items e.g. external walling. Each includes its finishings and other ancillary items.

Volume of work limits extent of study

Each 'functional component' was separately designed to cover the selected variables of column spacing, storey height and loading and then measured and priced.

This work, involving many thousands of separate calculations, was done by members of the Group in their spare time and, in spite of mass production techniques, took several years to complete. That this report is limited in its present scope is largely due to the extremely time consuming nature of the processes required to provide data in sufficient variety for cost relationships to be established.

Although the study as originally planned is incomplete certain conclusions can now be drawn relating to the 'core' of a building, namely the floors, roofs and their supports, and it is the purpose of this report to bring them to the attention of the building industry and the allied professions in particular and to stimulate interest in further research.

Core

For the purpose of this study and report the 'core' of a building is deemed to be the structure and finishes excluding:

a. The extra cost of perimeter items i.e. work pertinent to the external plan perimeter.
b. Internal partition layout.
c. Staircases and lifts.
d. Roof structures (e.g. tank and lift motor rooms), roof lights, external doors, canopies and other special features.
e. Fittings and equipment.
f. Aesthetic treatment and embellishment.
g. Services (e.g. heating, hot and cold water supplies, rainwater disposal, sanitary work and drainage, mechanical ventilation and electrical).
h. External works and site clearance.

It must not be thought that the 'core' merely comprises the internal bays. Briefly it is the whole basic structure and finishings of a building (whether or not it contains internal bays) but excluding the net extra cost of the external walls.

In the case of the 'core' the particular 'functional components' (Figs. 3.1.1 and 3.1.2) considered consist of:

Roof:
$\begin{cases} \text{Slab} \\ \text{Beams and ties} \\ \text{Columns one storey high.} \end{cases}$

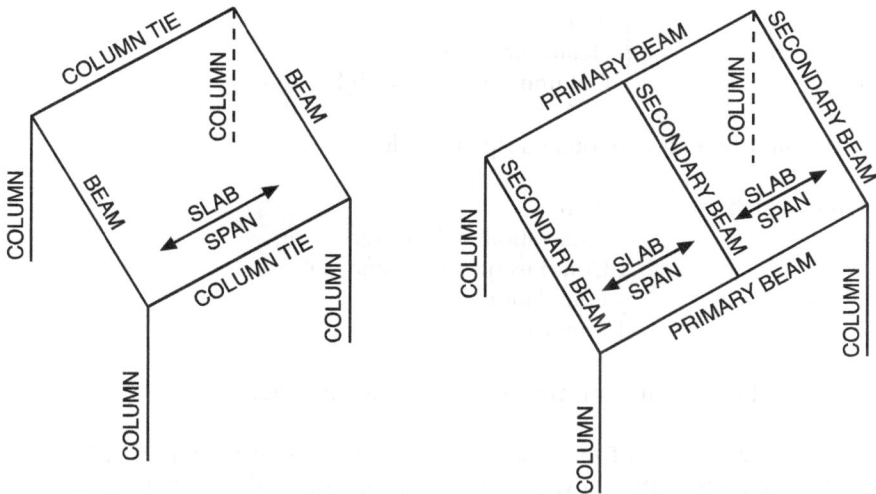

Figure 3.1.1 Beam and column arrangements.

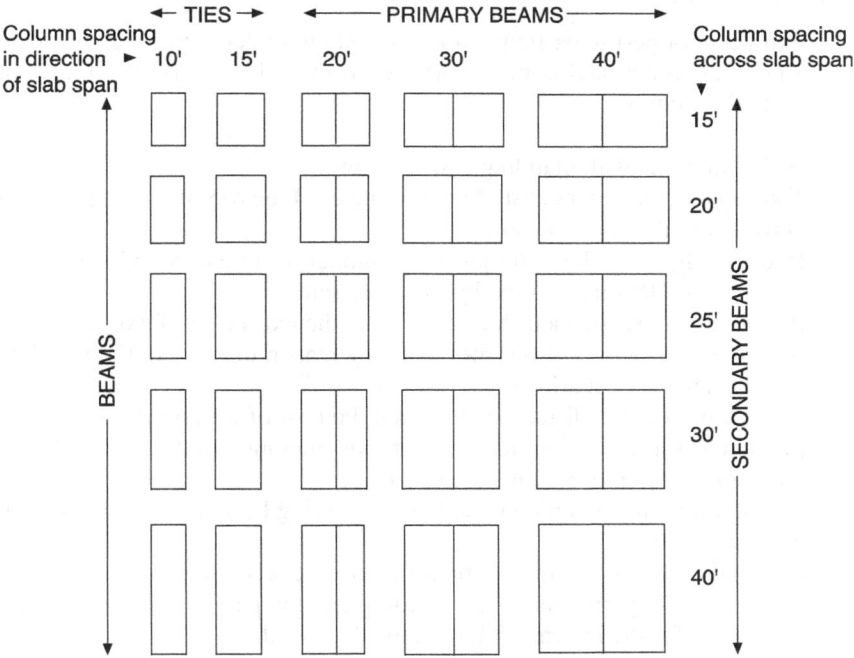

Figure 3.1.2 Roof and suspended floor panels.

Topmost   ⎧ Slab
suspended   ⎨ Beams and ties
floor: ⎩ Columns one storey high.

and similarly with each other suspended floor.

Ground slab: ⎧ Slab
                         ⎨ Excavation and filling.
                         ⎩ Columns below ground slab
Foundations: ⎧ Foundations
                         ⎩ Excavation and filling.

(Each inclusive of its finishings and other ancillary items).

The cost of each of the 'cores' studied has been arrived at by adding together the costs of the relevant 'functional components' and reducing the total to a cost per foot super of total floor area which is then described as the 'core rate'.

Perimeter items

The extra cost of perimeter items (not covered by the 'core rate') comes from the following 'functional components' (each inclusive of its finishings and other ancillary items):

1   Wall panel, i.e. wall cladding and windows.
2   Top edge, i.e. the extra cost of the perimeter of the roof including all work pertinent to the style of roof.
3   Bottom edge, i.e. the external wall foundation in excess of the cost of column foundations covered by the 'core rate'.
4   Roof slab and suspended floor edges, i.e. the extra cost of external edges of suspended floor and flat roof slabs by reason of non-continuity of reinforcement over bearings.
5   Wall beam, i.e. the difference between the cost of a beam in the external perimeter of the building and the part cost included in the 'core rate' for the internal beam or tie thus displaced.
6   Wall column tie, i.e. an unloaded member tying back a wall column to an internal beam.
7   External or wall column, i.e. the difference between the cost of a column in the external perimeter of the building and the part cost included in the 'core rate' for the internal column thus displaced.

Considerable data has been accumulated for these perimeter items but this aspect of the study is not sufficiently advanced for conclusions to be reached.

Specification

The specification common to all buildings adopted for study is given in the Appendix. It is of a fairy minimum standard and this was intentional.

For record purposes the specification includes external walling and other perimeter items for which the study is incomplete and not covered by this report.

Beam and column arrangements

The beam and column arrangements shown in Fig. 3.1.2 were those adopted for the worked examples of 'cores' and have been calculated for the three alternative floor loadings stated in the Appendix.

Storey heights

The storey height is measured from the top of a floor slab to the top of the next slab above.

The storey height variants of 10ʹ0ʺ, 12ʹ6ʺ, 15ʹ0ʺ and 20ʹ0ʺ were used; that of 20ʹ0ʺ being restricted to single storey buildings.

Pricing

Methods used by contractors' estimators and quantity surveyors for the build-up of prices vary and for any particular project the data are related to local conditions. For a comparative study of this nature it is essential to have a uniform basis of pricing and Spon's 'Architects' and Builders' Price Book' 1956-57 (82nd Edition) was chosen. No addition has been made to the unit prices therein for 'preliminaries' nor any allowance for extra hoisting in connection with multi-storey buildings. These are factors, therefore, that must be taken into account when using the charts.

This pricing is based upon building materials' prices as at March 1956 and wage rates as at May 1956. As it is cost relationships and not actual costs that have been studied these dates are considered to be relatively unimportant and to remain so unless and until there is a significant change in the relative costs of, for instance, concrete, steel and formwork.

Throughout this report the terms 'price' and 'cost' are used synonymously in the sense of the cost to the building owner.

Charts

The summarised data relating to the 'core' are presented graphically in the form of charts appended to this report.

These charts, which appear relatively simple, do contain an immense amount of factual information being derived from 1195 worked examples of 'cores'.

Purpose of this report

When the quantity surveyor is asked to advise upon the cost implications of fundamental design alternatives he should find the charts appended to this report a useful aid. This is felt to be especially so as, before a design is sufficiently advanced for approximate quantities to be prepared, it is design variables in the 'core' for which it has erstwhile been difficult to assess a cost.

These charts will not give, and it is not intended that they should give, ready-made answers to the problems that arise. The quantity surveyor should, however, have the knowledge and experience to interpret the data in the context of the overall concept of the particular project.

What the charts show

Charts Nos. 1 to 16 are all parts of one graph and enable comparisons to be made by superimposing two or more charts and 'looking through'. These charts show the relative cost of the 'core' (the vertical scale) depending upon:

a.   Number of storeys
b.   Storey height
c.   Superimposed loading of suspended floors
d.   Column spacing in direction of slab span
e.   Column spacing across slab span.

The vertical scale of the charts is relative cost of the 'core' per unit of floor area and the position of the datum of 100 is of no particular significance. There is no horizontal scale but by projecting a straight line through the appropriate points on the two column spacing scales (the diagonal scales bottom left) the relative costs can be read off from the points where the line cuts the various graph curves.

Charts Nos. 17 to 28 are examples illustrating the costs as calculated for the panels given in Fig. 3.1.2 and showing the distribution of these as between:

1   Flat roof slab
2   Beams for ditto
3   Suspended floor slabs   Each complete with finishings
4   Beams for ditto   and other ancillary items
5   Ground floor slab   as applicable.
6   Columns
7   Foundations

The accuracy with which the curves on Charts Nos. 1 to 16 reflect the calculated data can be seen from the specific examples illustrated on Charts Nos. 17 to 28. Such discrepancies as there are probably mainly arise because where any steel section is not quite large enough to fulfil the design requirements the next larger section has been allowed for. Sometimes this does not average out and tends mainly one way or the other.

How the charts can be used

The charts are intended primarily as a tool for the quantity surveyor who, as a member of the design team, may be asked at a preliminary stage to advise upon the cost aspects of such design variables as lie within the selection or influence of the architect and are not dictated by site limitation or users' requirements.

The charts show cost relationships. Their practical value comes from equating with particular points on the appropriate graphs the current 'core rate' of known examples, either actual buildings of which the cost is known or hypothetical or proposed buildings of which the cost can be estimated in other ways, e.g. approximate quantities. The effect of the varied design of the building under review and possible alternative design variables can thereby be assessed. Seldom can readings be used directly and assessment will generally be necessary.

They are intended for use when considering the cost aspect of such alternatives as:

a. Single or multi-storey design taking into account the stipulated floor loadings and minimum column spacings.
b. Increase in column spacing and/or storey heights above the stipulated minimum where this may be of some present or future value.
c. The grouping of accommodation where there are varied requirements of floor loadings and minimum column spacings and storey heights.
d. The users' proposed mechanical handling arrangements where these require increased floor loadings and/or column spacings and/or storey heights.
e. The direction of suspended slab spans.

Charts Nos. 17 to 28, which show the relative costs of the main groupings of 'functional components' in the 'core' for certain selected examples, will probably be found of most value when the quantity surveyor comes to consider varied forms of construction and specification such as:

a. Other forms of foundations and/or differing soil bearing capabilities.
b. Beams of limited depth.
c. Varied specification for one or more of the main groupings of 'functional components.'

Such studies will require approximate quantities to be prepared for the actual examples required which can then be compared with the data given and the overall result assessed.

Limitations to data shown on charts

Some caution should be exercised in taking readings from towards the top left hand ends of the curves (i.e. for the larger column spacings) particularly with the higher floor loadings. It is here that plain or compound steel beams cease to be acceptable and recourse has to be made to built up plated beams; also the degree of permissible deflection becomes an important consideration.

No attempt should therefore be made to guess the costs of co-ordinates outside the boundaries of the charts: with the larger column spacings for the reason given above and at the other end of the curves because with still smaller column spacings costs again tend to rise.

It is emphasised that the charts show cost data limited to:

a. The 'core' of a building, and do not take into account the two other important design variables of the size of the building and its plan shape.
b. Steel framed structures of simple design, rectangular in plan, without abnormal eccentricity of loading, without offset, or set back of upper storeys and without specific provision for wind bracing.
c. Solid in-situ reinforced concrete slabs with the beam arrangements shown in Fig. 3.1.2.
d. A specification as given in the Appendix.

It must be appreciated that concrete cased steelwork is not normally used for single storey buildings nor is structural steel framing normally economic for buildings of one or two storeys where load bearing brickwork could be used, and that the cost of extra hoisting in connection with multi-storey buildings has not been taken into account.

It cannot be too strongly stressed that the charts must not be used out of context since they deal only with part of the cost of the building and are intended primarily for use by quantity surveyors. other factors which must be taken into account include external wall column spacing (which may differ from the column spacing of the 'core'), loss of usable floor area because of staircases and lifts and the costs of these and of services which will vary with the disposition of the accommodation, as well as the effects on cost of the shape of the building and varied specification.

Conclusions

This study is believed to be the first serious attempt within the building industry to isolate the cost effects of fundamental design variables taking into account the interacting cost effects of each upon the others.

The charts speak for themselves and little would be served by explaining the results in detail. Suffice it to say that many of the conclusions reached from their examination will only confirm the 'hunches' of experienced practitioners. What will, however, probably be quite new is the individual effect in terms of cost of the design variables investigated.

The charts appended to this report, contain the information now available which is considered of immediate practical value to the practitioner: the working documents and data which cannot readily be summarised for publication with this report could be made available for the use of others who may wish to continue this study.

FUTURE STUDY

Whilst a great amount of design, measurement and cost data is available relating to a considerable number of whole buildings it is not in finalised or graphical form that can be published. The charts appended to this report relate only to the 'core' which, although a major factor in building costs, is not the whole answer. The Group is keenly aware of this but the fact remains that the time needed for compilation by manual processes of sufficient data relating to whole buildings is beyond that which can be given by volunteers in the evenings and at weekends. This is the reason why the cost of external walls and other perimeter items (upon which considerable basic research has been done by the Group) is excluded from this report so precluding consideration of the effect upon cost of shape and size of whole buildings. What has however been done demonstrates the practicability of the methods adopted and provides a sound basis on which future study can be pursued.

Furthermore the simple multi-dimensional charts appended to this report have been prepared by a method which, although laborious and empirical may now be capable of mathematical formulation. It may be of interest that the curves within the boundaries of the charts are drawn as arcs of circles of varying radii and varying centres and that the scales for column spacing are graduated relative to the squares of the distances between columns. Whilst these expedients have produced charts summarising with reasonable accuracy the data obtained by calculation it may well be that further study on a pure mathematical basis would disclose a formula that would fit these graphs and provide a ready means for preparing further graphs with other design and pricing data.

The Group considers that further study requires financial backing and would best be conducted by experienced quantity surveyors and structural engineers within or under the auspices of an independent research or educational organisation and not relying entirely on spare time work of volunteers. If such organisation had access to an electronic computer it should accelerate the arithmetical processes involved and facilitate revision of the results to take account of changed data e.g. specification and pricing.

RECOMMENDATIONS

The Group considers that the charts provide a source of new and useful information and advocates their use by quantity surveyors. It therefore recommends the publication by The Royal Institution of Chartered Surveyors of this report and charts with a view to their wide circulation.

The Group also recommends that the research it has begun be widened and pursued: such further research under the auspices of an independent research or educational organisation would, it feels, be a proper contribution by the quantity surveying profession to the field of building design economics.

APPENDIX: SPECIFICATION ADOPTED FOR THE STUDY

This specification includes external walling and other perimeter items which do not form part of the 'core'.

1   Ground floor slab.
   (a) Vegetable soil excavated and carted off site.
   (b) 9″ Hardcore blinded.
   (c) 6″ Portland cement concrete (1:2:4) reinforced with steel mesh (4.32 lb/sq.yd.) in bays of 200 sq. ft. with ³/₈″ bitumen joint filler.
   (d) 1″ Granolithic (1:2¹/₂) paving laid whilst concrete is green.
2   Column foundations.
   (a) Column foundations Portland cement concrete (1:2:4) on 3″ blinding (1:3:6), top of foundation concrete at a depth of 2′0″ below Ground Floor level and calculated for 2 tons/sq. ft. bearing capacity of soil.
   (b) Steel base plates to stanchions or steel joist grillages as requisite with same and steel stanchions to underside of ground floor slab encased in concrete.
   (c) No holding down bolts.
   (d) Hardcore filling around bases of stanchions.
3   Suspended floor panels.
   (a) Portland cement insitu concrete (1:2:4) slabs reinforced with mild steel rods and designed to C.P. 114:1948 for 40, 100 and 200 lb/sq. ft. superimposed loading plus 40 lb/sq.ft. for floor and ceiling finishes and partitions.
   (b) Steel framework of beams and stanchions (see Figs. 3.1.1 and 3.1.2) to B.S. 449:1948 of riveted and bolted construction with no restriction on the depth of any beam. (Universal sections have not been used.)
   (c) Beams and stanchions encased with 1:2:4 concrete of rectangular section to give 1″ cover and with tops of steel beams level with underside of slab. Steel fabric wrapping.
   (d) Soffit of slab, beams and columns plastered two coats gypsum and twice distempered.

(e) 1□ Granolithic (1:2$^1/_2$) paving on 1$^1/_2$□ Portland cement and sand screed.

(f) 6□Granolithic coved skirting to columns.

4   Flat roof panels.

(a)
to   As for 40 lb/sq.ft. suspended floor panels 3(a) to (d) above.
(d)

(e) Roof screed of 2$^1/_2$□(average) Portland cement and sand (1:3).

(f) $^3/_4$□Mastic asphalt with felt underlay.

(g) 6□Granolithic coved skirting to columns.

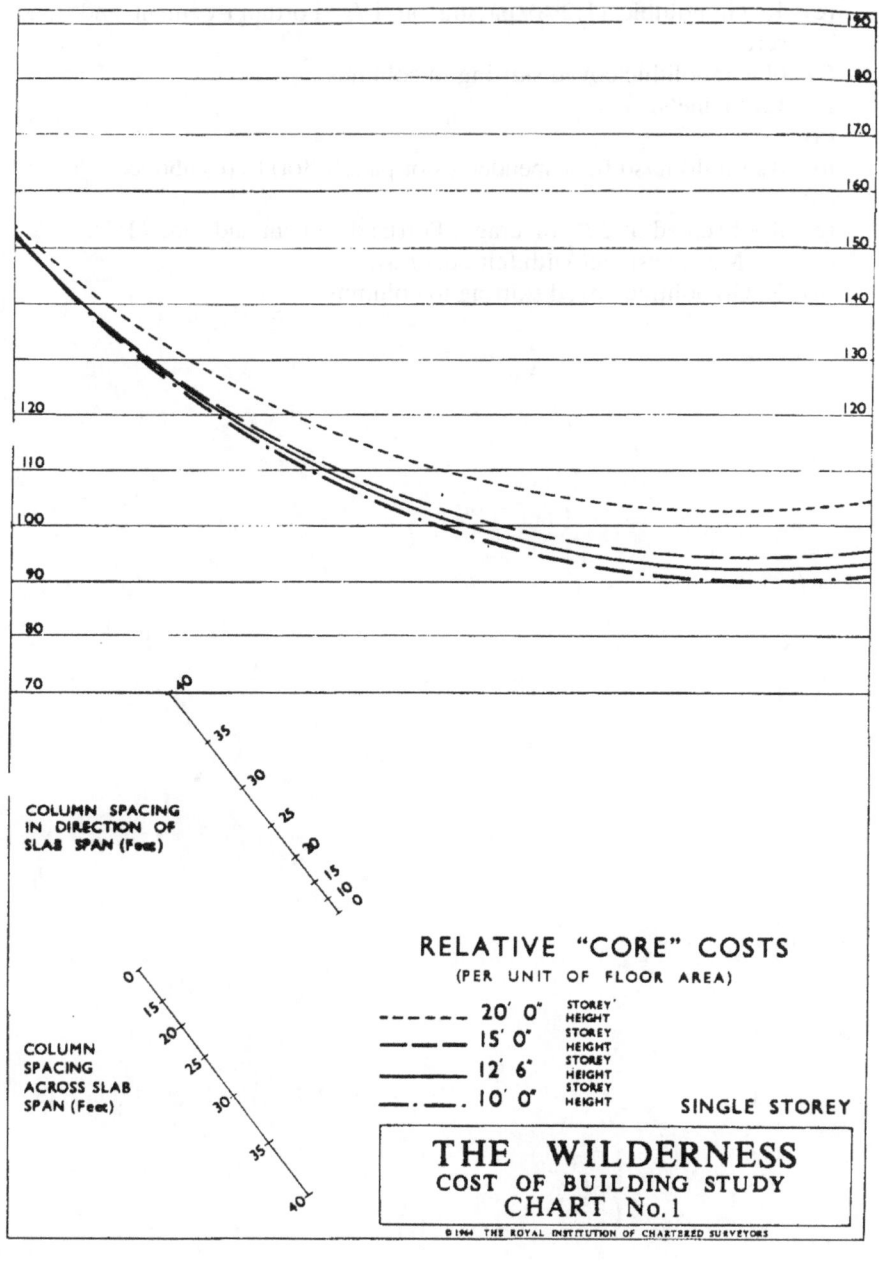

RELATIVE "CORE" COSTS
(PER UNIT OF FLOOR AREA)

------- 20' 0" STOREY HEIGHT
— — — 15' 0" STOREY HEIGHT
———— 12' 6" STOREY HEIGHT
—·—·— 10' 0" STOREY HEIGHT

SINGLE STOREY

COLUMN SPACING IN DIRECTION OF SLAB SPAN (Feet)

COLUMN SPACING ACROSS SLAB SPAN (Feet)

THE WILDERNESS
COST OF BUILDING STUDY
CHART No. 1

© 1964 THE ROYAL INSTITUTION OF CHARTERED SURVEYORS

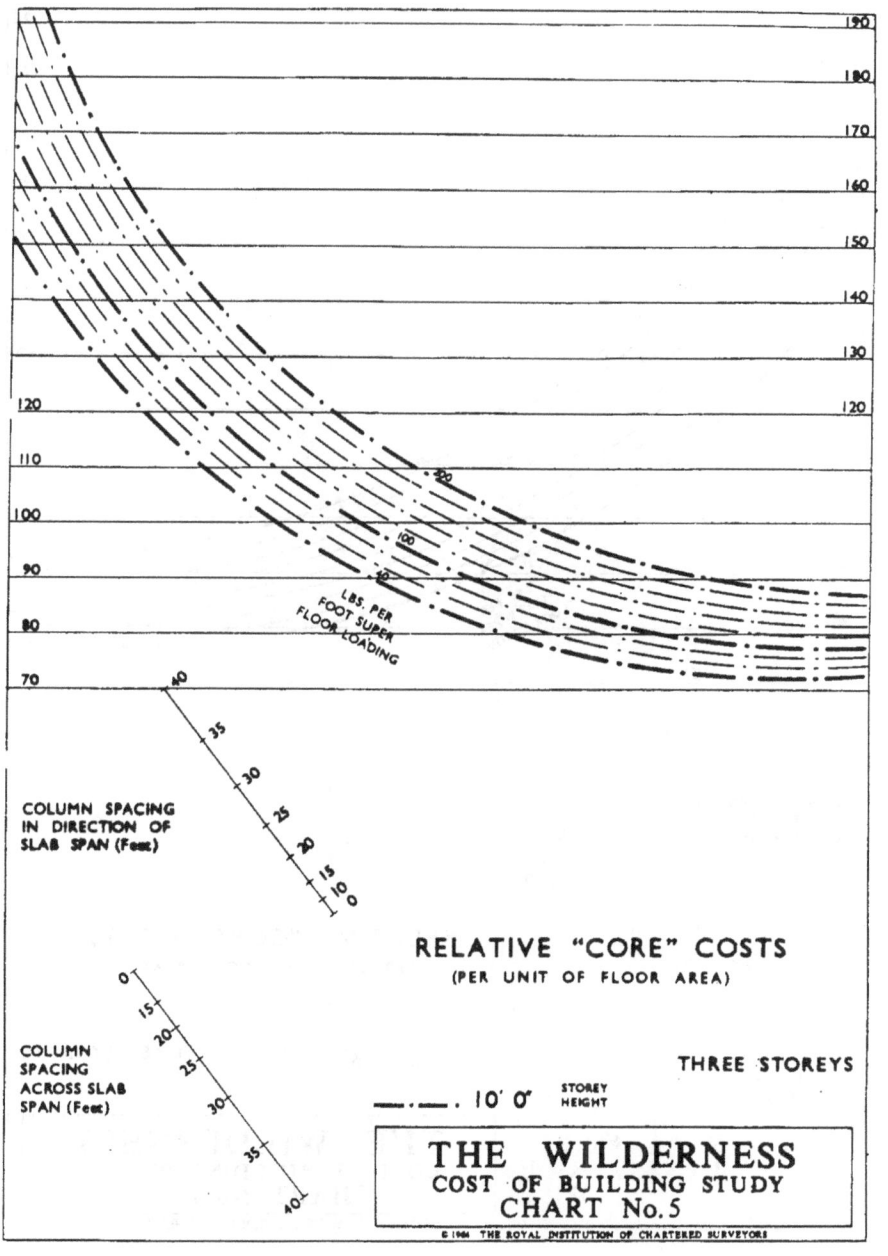

190
180
170
160
150
140
130
120   120
110
100
90
LBS. PER
FOOT SUPER
FLOOR LOADING
80
70

120
110
100
90
80
70

COLUMN SPACING
IN DIRECTION OF
SLAB SPAN (Feet)

40
35
30
25
20
15
10
0

RELATIVE "CORE" COSTS
(PER UNIT OF FLOOR AREA)

COLUMN
SPACING
ACROSS SLAB
SPAN (Feet)

0
15
20
25
30
35
40

THREE STOREYS

— · — · 10' 0"   STOREY
                 HEIGHT

THE WILDERNESS
COST OF BUILDING STUDY
CHART No. 5

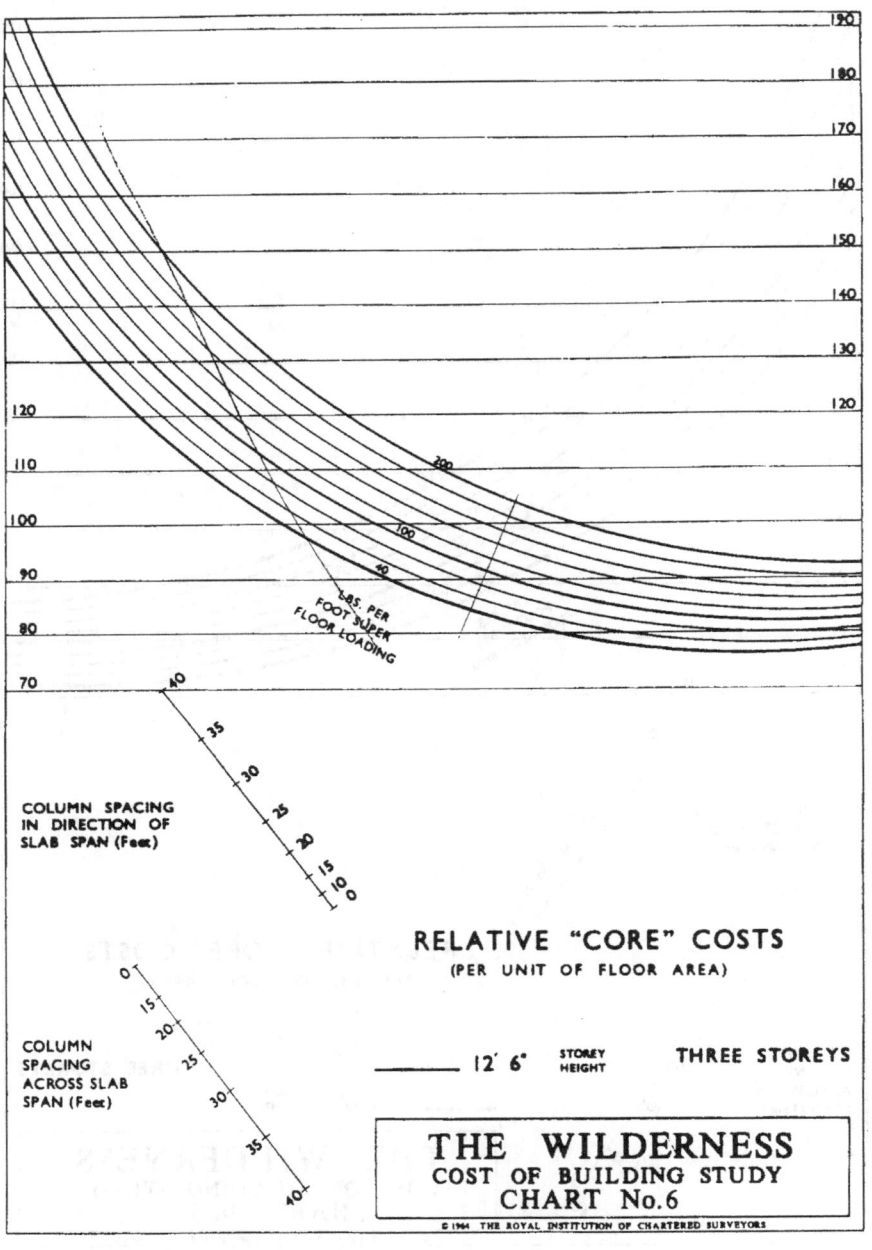

COLUMN SPACING
IN DIRECTION OF
SLAB SPAN (Feet)

RELATIVE "CORE" COSTS
(PER UNIT OF FLOOR AREA)

COLUMN
SPACING
ACROSS SLAB
SPAN (Feet)

_____ 12' 6" STOREY HEIGHT   THREE STOREYS

THE WILDERNESS
COST OF BUILDING STUDY
CHART No.6

© 1964 THE ROYAL INSTITUTION OF CHARTERED SURVEYORS

LBS. PER
FOOT SUPER
FLOOR LOADING

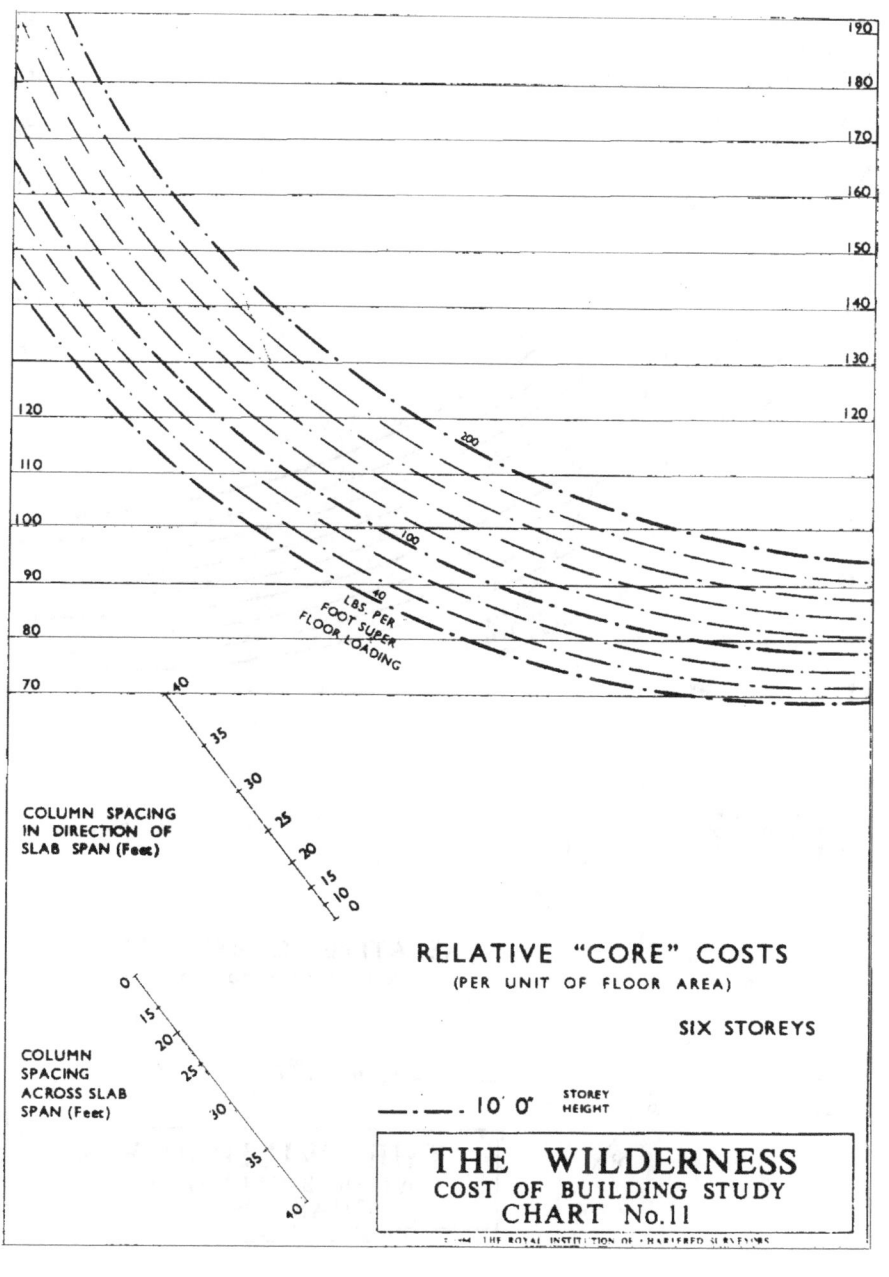

COLUMN SPACING
IN DIRECTION OF
SLAB SPAN (Feet)

LBS. PER
FOOT SUPER
FLOOR LOADING

RELATIVE "CORE" COSTS
(PER UNIT OF FLOOR AREA)

SIX STOREYS

COLUMN
SPACING
ACROSS SLAB
SPAN (Feet)

—·—·— 10' 0"   STOREY
                HEIGHT

THE WILDERNESS
COST OF BUILDING STUDY
CHART No.11

THE ROYAL INSTITUTION OF CHARTERED SURVEYORS

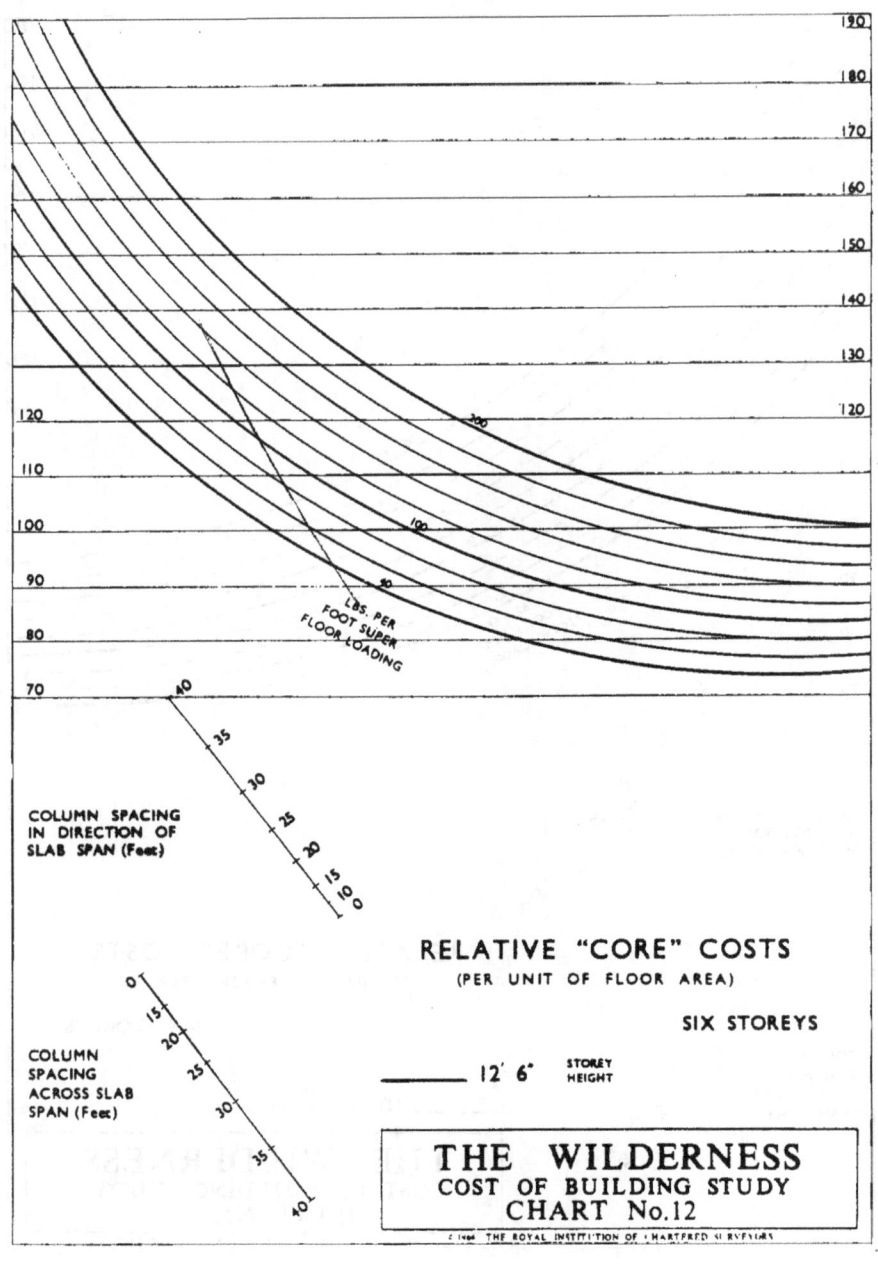

RELATIVE "CORE" COSTS
(PER UNIT OF FLOOR AREA)

SIX STOREYS

———— 12' 6" STOREY HEIGHT

THE WILDERNESS
COST OF BUILDING STUDY
CHART No.12

© 1964 THE ROYAL INSTITUTION OF CHARTERED SURVEYORS

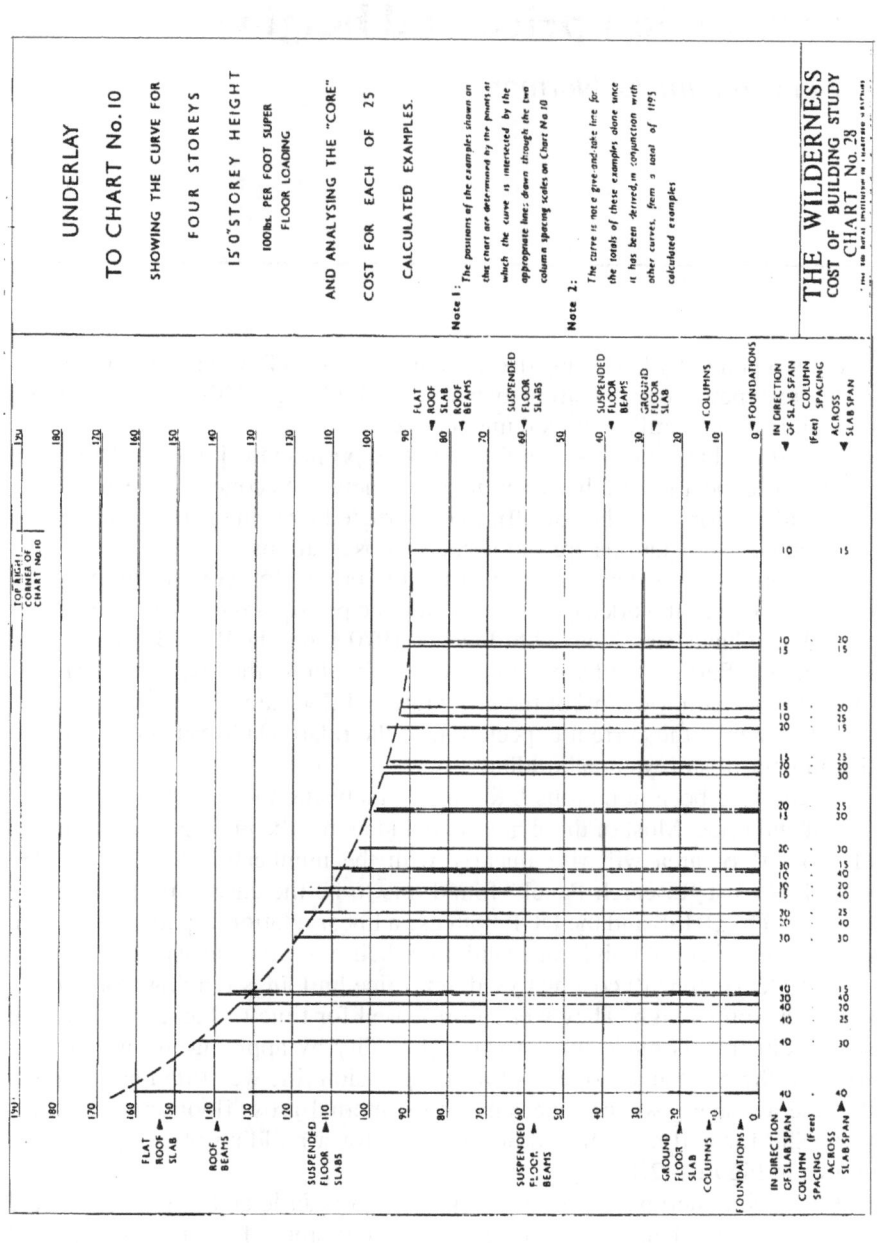

UNDERLAY

TO CHART No. 10

SHOWING THE CURVE FOR

FOUR STOREYS

15'0"STOREY HEIGHT

100lbs. PER FOOT SUPER
FLOOR LOADING

AND ANALYSING THE "CORE"

COST FOR EACH OF 25

CALCULATED EXAMPLES.

Note 1: *The pressures of the examples shown on that chart are determined by the points at which the curve is intersected by the appropriate lines drawn through the two column spacing scales on Chart No.10*

Note 2: *The curve is not a give-and-take line for the totals of these examples alone since it has been derived, in conjunction with other curves, from a total of 1195 calculated examples*

THE WILDERNESS
COST OF BUILDING STUDY
CHART No. 28

3.2 The relationship between construction price and height*

R. Flanagan and G. Norman

There is a generally held view that for the same gross floor area, tall buildings are more expensive to construct than low-rise buildings. This facet of building economics is generally not well understood.

This study has two aims therefore. First, to propose the kind of relationship we would expect to find between price and height. Secondly, to examine the cost analysis for a number of offices constructed over the period 1964–75 to see whether expectations are supported by observations.

The prices considered are construction prices for completed projects, excluding external works, expressed as a price per square metre of gross floor area. These have been indexed to January 1970 using the RICS Building Cost Information Service tender price index. For simplicity they will be referred to in the text as £/m² (ie £/m² of the gross internal floor area).

What then is the current expectation of the relationship between £/m² and the number of storeys in a building?

There have been numerous UK studies analysing the association between height and price. Most of these have concluded that the price per square metre of gross floor area will rise linearly with the number of storeys. In 1958 the RICS Cost Research Panel[20] found this to be the case with multi-storey housing. The DHSS and the DOE both use a linear relationship to take account of the price effect of height in calculating their cost allowances.

In 1972 Tregenza[21] cost analysed ten office buildings ranging from one to eighteen storeys high. They were all re-based for time to January, 1971, and some adjustments made for the varying quality by applying mean values to some of the elemental prices. A linear regression line was fitted and implied that the average cost per square metre of internal gross floor area increased from £88 for a 10m high building to £112 for a building 60m high; this is shown in Figure 3.2.1.

Research undertaken in Finland by Professors Jarle and Pöyhönen looked separately at the height effect on price of multi-storey housing. Figure 3.2.2

*1978, *Chartered Surveyor B. and QS Quarterly*, Summer, 69–71.

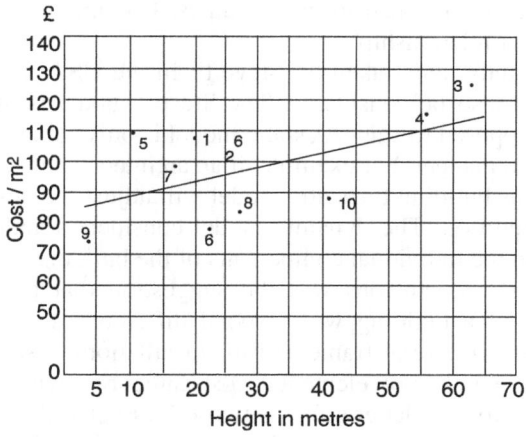

Figure 3.2.1 Height–cost relationship.

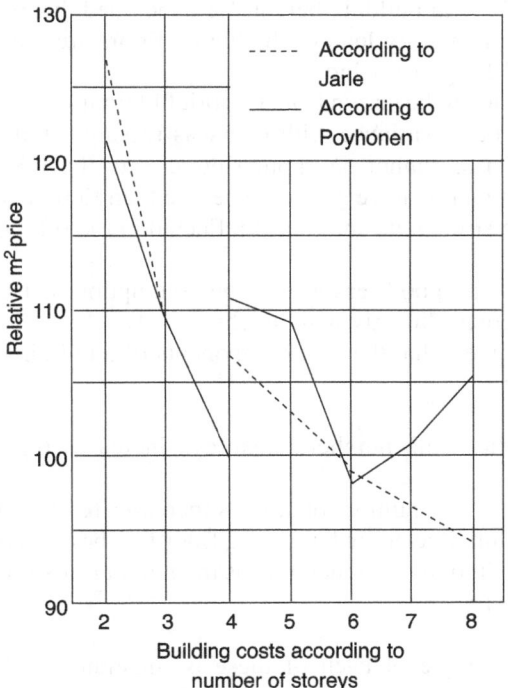

Figure 3.2.2 Number of storeys–cost relationship.

shows the conclusions reached in these studies. It is interesting to note the rejection of a linear relationship.

A theoretical study undertaken by Steyert[22] in the USA using modelling techniques should also be considered. The Steyert model is dynamic in its methodology as opposed to static. A static model is only capable of adjusting variables in a fixed manner. For example, if an architect wants to increase the storey height of his building the static model will adjust the design variables in arithmetic proportion. The dynamic model considers simultaneously the induced effects on the individual components of the building.

One of Steyert's conclusions was the suggestion that the costs of the various elements of a building will respond differently to changes in the number of storeys. Structural frame and lift installations costs will increase with height but a number of elemental cost categories will decrease with height. The reason for this latter effect is two-fold. Firstly, there is a learning curve effect, or improvement curve as it is sometimes called, with respect to labour output.

Secondly, the total cost of such items as the roof and to some extent foundations will increase less than proportionately with gross floor area.[23]

Steyert further showed that column costs and the cost of withstanding wind load both rise rapidly with building height. He concluded that the cost implications of carrying the gravity load of the higher floors are more significant than the problem of design for wind.

The main conclusion to be drawn from this brief literature survey is that UK practice appears to be at variance with overseas research findings. Clearly, international experience cannot be applied directly to the UK situation eg North American contractors have greater expertise than European contractors at building high rise structures, and this is reflected in the pricing strategy of bids.

There seems to be no good reasons for the assumption that £/m² for UK buildings should increase linearly with height.

Theory would suggest that the cost components of a building can be split into four categories:

(a) those which fall as the number of storeys increases (eg roofs, foundations);
(b) those which rise as the number of storeys increases (eg lift installations):
(c) those which are unaffected by height (eg floor finishes, interior doors);
(d) those which fall initially and then rise as the number of storeys increases (eg exterior closure).

A hypothetical example of each of these is illustrated in Figure 3.2.4. Adding these curves together would imply that the relationship between £/m² and the number of storeys would be U-shaped as indicated by the (hypothetical) total cost curve in Figure 3.2.4.

Learning curves have been mentioned previously and these can be considered in more detail to lend further support to our theoretical expectations.

Repetition of site operations will lead to a decrease in operational costs on the one hand, and to indirect cost savings caused by the reduction of construction time on the other. The direct savings in operational costs depend on the gradual decrease in operational time attained in repetitive work.[24] The learning curve theorem states that every time the number of repetitions doubles the output time declines by a fixed percentage.[25] The fixed percentage identifies the improvement achieved. A typical learning curve is illustrated in Figure 3.2.3.

If we consider a two storey and a five storey building with the same gross floor area, there is some reason for thinking that the learning curve effect would introduce cost savings to the taller building. In the context of an individual building there is a limit to this process, however, in that savings in some elemental prices brought about by the learning curve will not be sufficient to offset increases in prices of other elements of the construction process.

The hypothesis is that £/m² will decline initially as the number of storeys increases, but will eventually rise.

To test this hypothesis we considered data on fifteen offices with two or

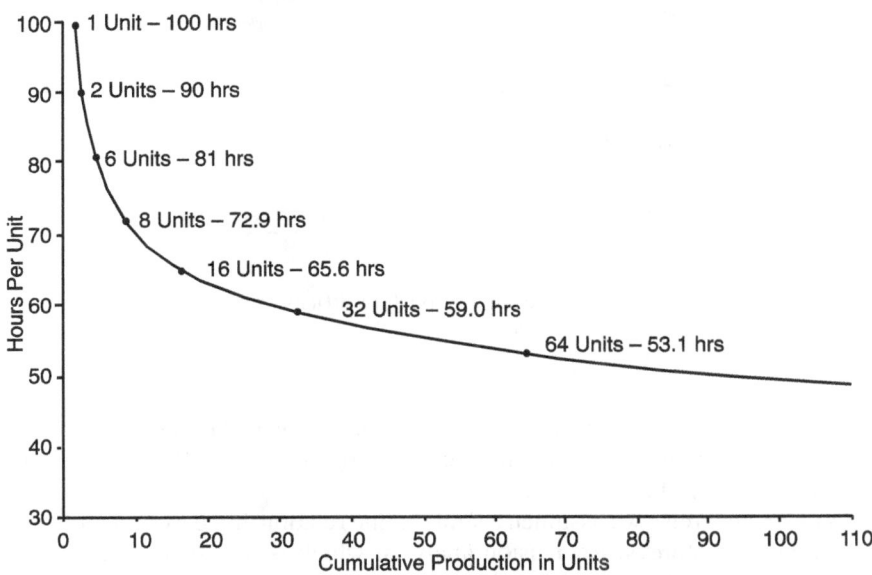

Figure 3.2.3 Illustration of a 90% learning curve.

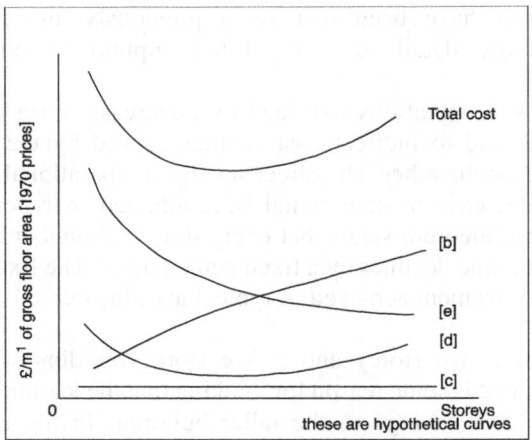

Figure 3.2.4 Total cost curve.

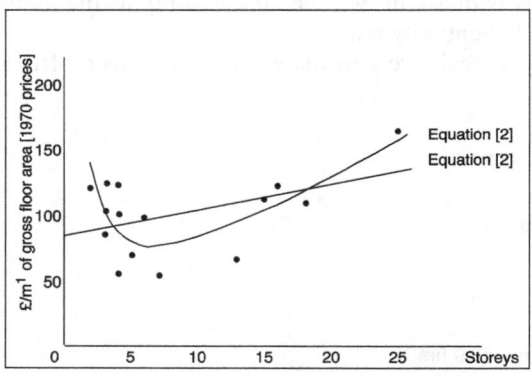

Figure 3.2.5 Number of storeys–cost relationship: offices.

more storeys[26], reported in the *Architects Journal* over the period 1964-75. They include the ten observations used by Tregenza in his 1972 study. The data are plotted in Figure 3.2.5.

As a first exercise we assumed a straight line relationship between £/m² and the number of storeys. Regression analysis, which is a statistical technique where historical data is analysed in a structural format, was used to find the relationship between height and price. A commercially available computer program package was employed in performing the statistical analysis.

The relationship obtained from the data is:

$$£/m^2 = 84.29 + 1.97 \square \text{ Number of storeys} \qquad (1)$$

This equation is graphed as the straight line in Figure 3.2.5.

Equation (1) states that $£/m^2$ is £84.29 plus £1.97 for each floor (including the ground floor). In other words, adding a storey will increase $£/m^2$ by £1.97.

While the relationship in equation (1) is the best straight line approximation to the data, it is only just statistically significant in that it explains less than 20 per cent of the variation in construction prices.[27] Looking at the equation in more detail, it can be seen that the straight line does not reflect at all the pattern of variation of $£/m^2$ for buildings under 10 storeys, and underestimates the increase in $£/m^2$ for the high rise buildings.

The next step therefore, was to try an equation which would give an initial decline in $£/m^2$ as the number of storeys increased, followed by a subsequent rise. The equation estimated was:

$$£/m2 = 5.94 \square \text{ Number of storeys} + \frac{248.59}{\text{No of storeys}} \qquad (2)$$

This equation is shown in Figure 3.2.5 as the U-shaped curve. The interpretation of the equation is illustrated in Table 3.2.1.

As can be seen from this table, while the second term in equation (2) declines with the number of storeys, the firmest term increases, and eventually becomes dominant above 6 storeys.

Equation (2) is statistically significant, explaining 54 per cent of the

Table 3.2.1 Illustration of equation (2)

No. of storeys (A)	First term of (2) 5.94 × number of storeys (B)	Second term of (2) 248.59 ÷ number of storeys (C)	Predicted £/m¹ (B) + (C) = (D)
2	11.88	124.30	136.18
4	23.76	62.15	85.91
6	35.64	41.43	77.07
8	47.52	31.07	78.59
10	59.40	24.86	84.26
15	89.10	16.57	105.67
20	118.80	12.43	131.23
25	148.50	9.94	158.44

(January 1970 prices)

variation in construction costs of the various offices. It approximates the data in Figure 3.2.5 much more closely than equation (1). The data would therefore appear to support our hypothesis as illustrated by the total cost curve in Figure 3.2.4.

There are many other influences, eg. quality, geographical location, size of project, site characteristics, etc., in addition to height which will affect construction prices. Further, equation (2) is one of many possible equations which

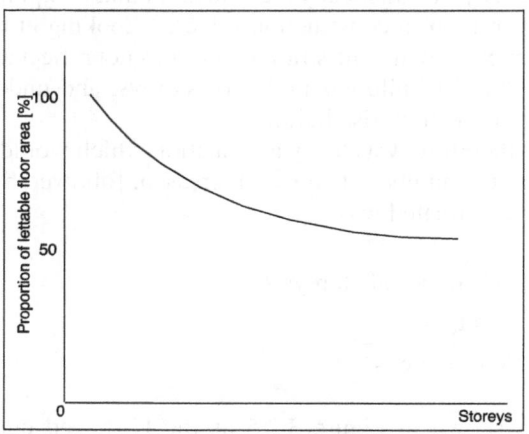

Figure 3.2.6 Hypothetical lettable floor area model.

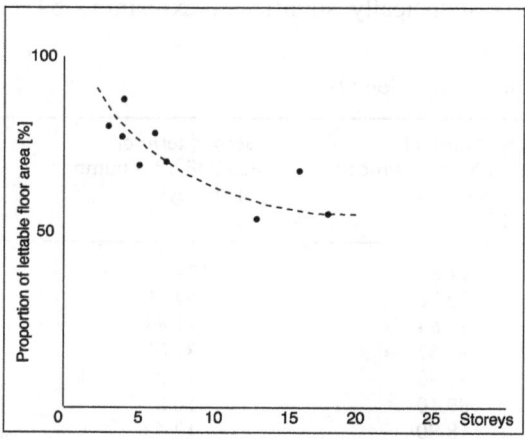

Figure 3.2.7 Empirical model.

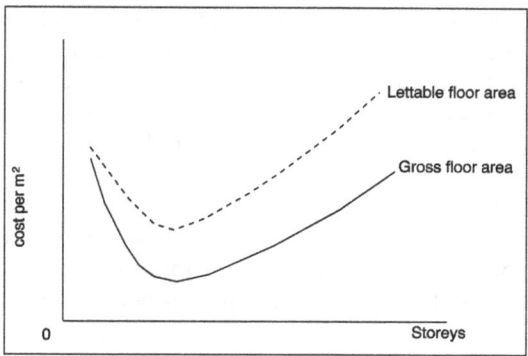

Figure 3.2.8 Lettable floor area and gross floor area.

would give rise to a U-shaped curve. Its main attraction is its simplicity, but this is obtained at some loss of accuracy in its approximation to the data. These considerations result in 46 per cent of the variation in construction costs being unexplained by equation (2). Looking at figure 3.2.5, it can be seen that there is wide variation in costs of construction for the three 4 storey buildings in the sample. A more detailed breakdown of the costs for these buildings indicates that the differences in costs arise mainly from the preliminaries and super-structure elements, reflecting the heterogeneous nature of buildings.

A further point is worthy of note. We have concentrated on the relationship between £/m² of gross floor area and the number of storeys. In commercial office buildings there is some suggestion that the amount of lettable floor area will decrease as additional storeys are added because of the need for circulation space, lift shafts, etc. This is illustrated in Figure 3.2.6. After an initial steep fall the rate of reduction in the proportion of lettable floor area is shown to decline. It should be emphasised that the curve as drawn is hypothetical but it is supported by the data used by Tregenza. These are plotted in Figure 3.2.7 with a suggested correlation between lettable floor area and the number of storeys (this has been fitted by eye).

If consideration is given to the price per square metre of the lettable floor area as opposed to the gross floor area, a different cost curve will be obtained. This has significant implications for commercial office building developers. The effect will be to shift vertically the curve relating £/m² and the number of storeys, the move being greater the more storeys there are. Figure 3.2.8 illustrates this point. The solid curve relates £/m² and number of storeys and the dotted curve relates £/m² of lettable floor area to number of storeys. While the dotted curve is flatter than the solid curve, there is no reason to believe that the U-shape will disappear completely.

In conclusion, while equation (2) must be treated with some caution if used as a predictive equation, we feel the initial hypothesis has been justified. The relationship between £/m² and the number of storeys in an office structure can be expected to be U-shaped. it will exhibit an initial decrease in £/m² as the number of storeys is increased, before eventually rising.

It must be stressed that an equation can only be as good as the data from which it is generated.

The precise nature of the impact of height on construction prices needs further investigation. This study has outlined a basic theoretical structure and has adduced supportive evidence. It is clear that further detailed research is required. We are currently undertaking some further theoretical work.

NOTES AND REFERENCES

1   RICS Cost Research Panel, 'Factors Affecting Relative Costs in Multi Story Housing. *Chartered Surveyor* March 1958.

2   P. Tregenza, 'Association between building height and cost.' *Architects Journal*, November 1 1972 p 1031–1032.

3   R. S. Steyert (1972). *The Economics of High Rise Apartment Buildings of Alternate Design Construction*. construction Research Council, American Society of Civil Engineers.

4   This has to be qualified with respect to foundation costs; these will be influenced by total dead and live load to be carried, the soil bearing capacity of the site and the foundation perimeter to ground floor area ratio.

5   Report of an enquiry undertaken by the committee on housing, building and planning, *Effects of Repetition on Building Operations and Processes on Site*. United Nations, new York 1965, Reference ST/ECE/H03/14.

6   Jack K. Chellow. 'The use of the learning curve.' *American Association of Cost Engineers Conference Transactions* 1971, p 163–168.

7   Single storey buildings have been excluded since these are liable not to exhibit the characteristics of multi-storey buildings.

8   With three or more observations there is a potentially infinite number of straight lines which could be used to approximate to the data. Regression analysis is a method for choosing the line with the best fit but there remains the possibility that the data are random or do not follow a straight line.

    A test of significance is a method for testing the appropriateness of the suggested relationship. It indicates the extent to which we have explained the variation in £/m² exhibited by the observations. The percentage stated in the text are an indication of the proportion of this variation actually explained; the greater the percentage the more confident we can be in the equation.

3.3 Cost performance modelling*

T. Maver

In this edited version of a paper given at a meeting of the QS research and development forum on December 4, Professor Maver discusses how the incoming generation of computer-based design models promises to transform the practice of architectural design

To Bronowski[1], the single most signal step in the ascent of man was the creation of the cave paintings – the first bold attempt to model, for the young men of the community, a future reality as yet outwith their experience. In the course of the ascent of man model building, as a means of understanding and predicting future reality, has become staggeringly sophisticated.

The success of the first lunar landing is due primarily to the power and validity of simulation models developed by design engineers working on the space programme. Curiously, however, in the field of architectural design, our modelling capability has scarcely advanced in the 5000 years since plans and elevations first appeared.

This paper is concerned with an emerging generation of computer-based models relevant to architectural design which, in comparison to the ubiquitous plan and elevation, are predictive rather than descriptive and dynamic rather than static. Their impact on the practice of architecture promises to be enormous.

Figure 3.3.1 is a simple and somewhat dated representation of the building design problem as four sub-systems[2]. At the extreme right we have the **client/ user objectives**, in order to satisfy which the client/user body engages in a set of *activities*.

Looking now at the extreme left, we have the **building hardware**; this exists to create and maintain an **environment**. the design problem is focused at the interface between the environment and the activities; it may be said that the designer is concerned to assemble the *hardware* which will maintain an *environment* which promotes rather than inhibits the **activities** necessary to fulfil the client/user **objectives**.

*1979, *Chartered Quantity Surveyor*, 2(5), Dec., 111–15.

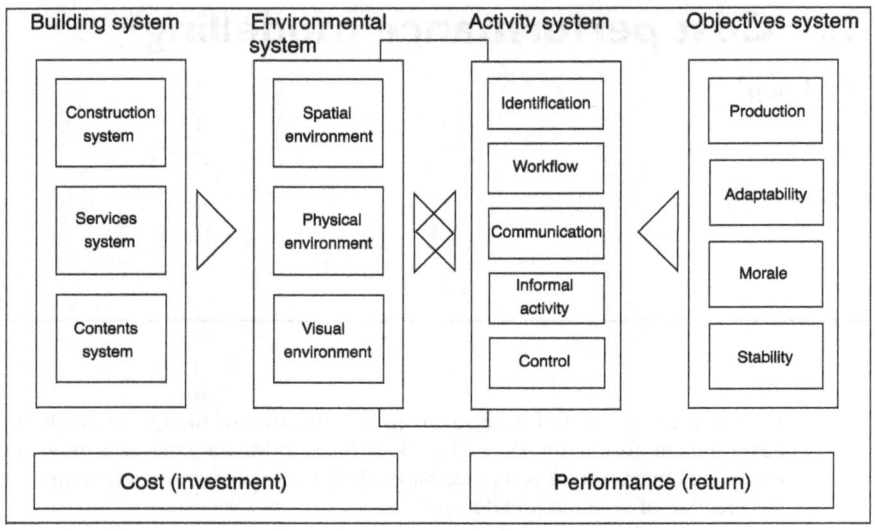

Figure 3.3.1 A representation of the design problem.

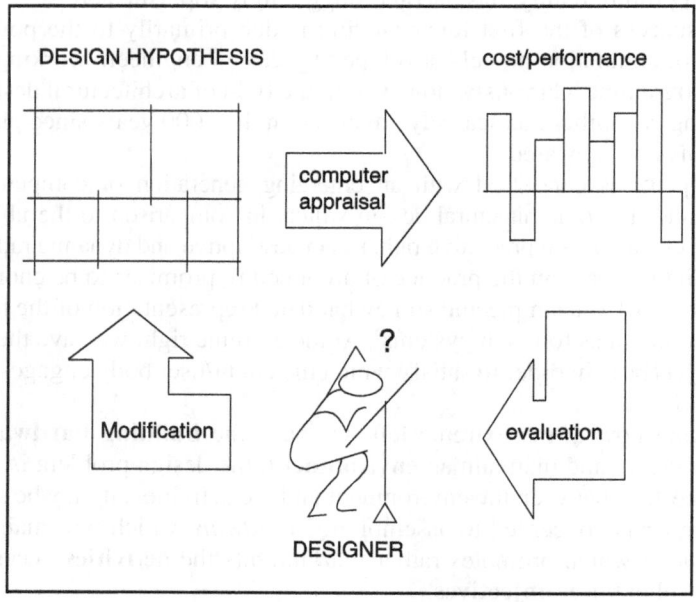

Figure 3.3.2 Concept underlying new generation models.

The value of this view of the design problem is that we can summarise the hardware and environment in cost-in-use terms (ie *investment*) and we can summarise the degree of promotion of activities and fulfilment of objectives in performance terms (ie *return*). We can then say that the designer is concerned to maximise the return on the client's investment.

The simple concept underlying the new generation of computer based design models is to make the trade-off between investment and return substantively more explicit. Figure 3.3.2 represents the models' operation.

The designer (or design team) generates a design hypothesis which is fed into the computer; the program models the scheme and predicts the future reality in cost and performance terms; and the designer evaluates the cost/performance profile and modifies the design hypothesis accordingly. Iteratively, he/she then converges on that design for which the cost/performance profile represents the appropriate balance of investment and return.

FORM AND OPERATION OF MODELS

The modelling concept embodied in figure 3.3.2 is as applicable to the early stages in design as to the detailed stages; to the re-design of existing buildings as to the design of new buildings; and to the design of the whole building as to the design of some sub-system within the building.

The ABACUS group within the Department of Architecture and Building Science at the University of Strathclyde has developed, over the last decade, a range of computer-based models which nest one within the other or may be deployed in specific design contexts. These include BILD[3] for detailed design of whole buildings, PHASE[4] specific to hospital design, HELP[5] specific to housing layouts, ESP[6] focussing on energy issues, AIR-Q[7] focussing on movement simulation and PARTIAL[8] intended to promote user participation in design.

All share the same characteristics: they deal with cost and performance; they are cheap, readily accessible and easy to use; and they operate interactively with the design team.

The program GOAL[9] (= General Outline Appraisal of Layout) will serve as an example. It is intended for use at the early stages in design when broad decisions are being taken on the form and fabric of the building. The design hypothesis is entered into the program by drawing the geometry on the graphics screen of the computer terminal, floor by floor, and by selecting constructional and other choices from a set of data files.

Geometry input is achieved by means of a cursor and a set of graphical commands (3). Additional graphical representations of the design hypothesis – eg axonometrics, perspectives (4) and sections – can be generated automatically. At any stage, the basic form of the design can be readily modified.

The standard data files hold functional constructional, costing, environmental and climatic data. The capital cost file is exemplified in (5). In

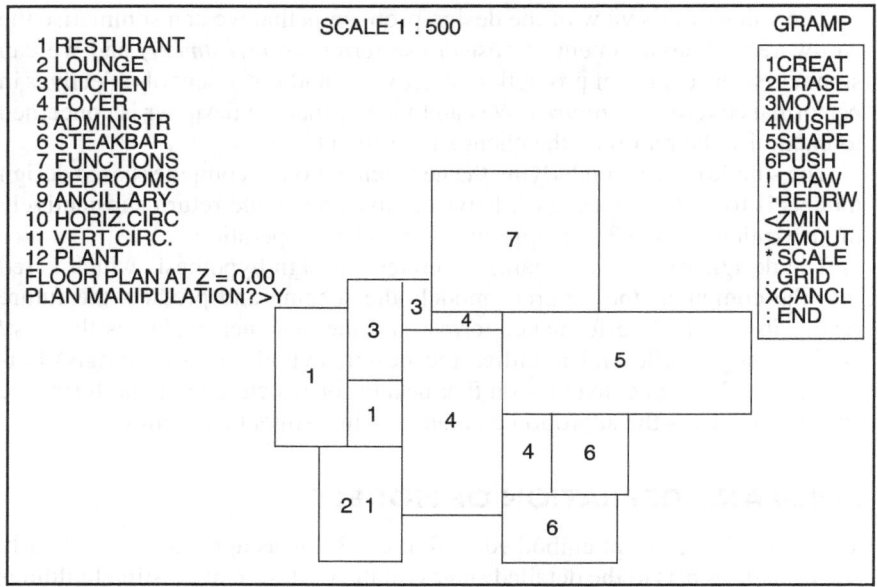

Figure 3.3.3 Geometry of the design hypothesis being input.

conjunction with the construction file, members of the design team can create constructional alternatives by adding to or changing categories and data within the constructional heading. The basis of capital cost estimation can be changed by altering the quantity basis (eg from wall area – TWAC – to floor area – TFA) by which unit costs rates are multiplied up.

Further cost parameters held in file include energy tariffs, additional running costs and amortisation determinants (6). Output from the program is extensive and includes spatial descriptions (7), planning efficiency (8), thermal behaviour (9) and, of course, capital and recurring costs (10). As each section of output is displayed the opportunity is provided to modify any or all of the decisions which make up the design hypothesis – building shape, orientation, construction, fenestration, unit costs, and so on. The operation, then, is as in 2.

USE OF MODELS IN PRACTICE

Notwithstanding a dynamic conservation within the rank and file of the professionals making up the building design team, acceptance of the new generation of design models is accelerating. PHASE is now used regularly in the early stages of design of all district general hospitals in Scotland; ESP has

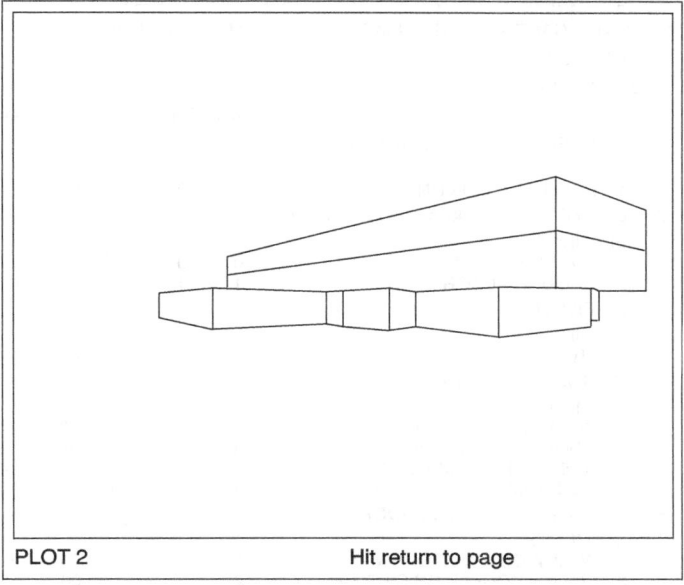

PLOT 2   Hit return to page

Figure 3.3.4 Perspective view of output on screen.

been acquired by the DHSS, Eastern Electricity, Building Design Partnership and others; AIR-Q provides the core of a California-based airport design consultancy; BIBLE features regularly in design presentations; and PARTIAL has been used in two modest but ultimately very important urban renewal proposals.

GOAL is assuming increasing importance in practice and is currently being developed as the main instrument in a European cost advisory service operating from Brussels. Its impact can, perhaps, be extrapolated from its role in the education of architects and quantity surveyors.

Figure 3.3.11 shows a few discrete points in the search sequence adopted by one architectural student in his pursuit of an 'optimal' solution to an hotel brief: first he explored a range of building forms, then a range of constructional alternatives.

In the course of his 'crit' discussion focused, appropriately, on the balance between **cost** and **performance** rather than on unsubstantiated guesses of the technical and economic merit of the scheme.

Experience gained in observing the application of such computer-based models in practice suggests a number of implications.

First, a great deal of design time is lost as design hypotheses are passed to and fro between the architect and specialist members of the design team. Quite

ITEM	GP		CONSTR		RATE	QUANTITY	
\multicolumn{8}{	l	}{STANDARD DATA : CAPITAL COST ELEMENTS 30-Aug-79 17:54}
\multicolumn{8}{	l	}{HOTEL DESIGN DATA}
\multicolumn{8}{	l	}{HOTEL COST DATA 1977}
101	1	A PRELIMS & CONTINGENCIES					
102	1	ELEMENTS					
103		1 PRELIMS & CONTINGENCIES	0	0	6.93	11	TFA
201	2	BC FOUNDATIONS & GROUND FLOOR					
202	2	ELEMENTS					
203		1 FOUNDATION	0	0	3.40	12	GFA
204		2 GROUND FLOOR SLAB	1	1	20.00	12	GFA
301	3	D FRAME					
302	1	ELEMENTS					
303		1 FRAME	0	0	18.00	11	TFA
401	4	E ROOF & ROOFLIGHTS					
402	4	ELEMENTS					
403		1 ROOF-HIGH U-VALUE	4	1	21.72	42	TRAC
404		2 ROOF-LOW U-VALUE	4	2	24.99	42	TRAC
405		3 ROOFLIGHTS (SINGLE GLAZ.)	3	1	105.90	34	TLAC
406		4 ROOFLIGHTS (DOUBLE GLAZ.)	3	2	105.90	34	TLAC
501	5	F WALLS, WINDOWS, DOORS					
502	7	ELEMENTS					
503		1 WALL A:CONCRETE	2	1	42.60	22	TWAC
504		2 WALL B:DOUBLE BRICK & CAVITY	2	2	38.00	22	TWAC
505		3 WALL C:COMPOSITE PANEL 1	2	3	60.00	22	TWAC
506		4 WALL D:COMPOSITE PANEL 2	7	4	67.00	22	TWAC
507		5 WINDOW:SINGLE GLAZ.	3	1	60.72	32	TGAC
508		6 WINDOW:DOUBLE GLAZ.	3	2	180.00	32	TGAC
509		7 DOORS	0	0	3.00	11	TFA
601	6	G INTERNAL WALLS					
602	1	ELEMENTS					
603		1 INTERNAL WALLS	0	0	18.81	11	TFA
701	7	H UPPER FLOORS					
702	2	ELEMENTS					
703		1 UPPER FLOOR INT.	0	0	10.00	14	IUFA
704		2 UPPER FLOOR EXT.	1	2	10.00	15	EUFA
801	8	I STAIRCASES					
802	1	ELEMENTS					
803		1 STAIRS	0	0	5.00	11	TFA
901	9	J FIXTURES					
902	1	ELEMENTS					
903		1 FIXTURES	0	0	20.00	11	TFA
1001	10	KM INSTALLATIONS					
1002	2	ELEMENTS					
1003		1 WATER, SANITARY	0	0	10.35	11	TFA
1004		2 HEATING INSTAL.	0	0	23.16	11	TFA
1101	11	NOR SERVICES					
1102	3	ELEMENTS					
1103		1 MECHANICAL	0	0	0.60	11	TFA
1104		2 ELECTRICAL	0	0	15.00	11	TFA
1105		3 GAS	0	0	1.00	11	TFA
1201	12	ST EXTERNAL SERVICES					
1202	2	ELEMENTS					
1203		1 EXT. SERVICES	0	0	7.00	11	TFA
1204		2 DRAINAGE	0	0	7.05	11	TFA

Figure 3.3.5 The user-modifiable capital cost file.

```
FUELS
ITEM                    J/UNIT        EFFFIX+RATE/UNIT
1      ELECTRICIT       0.4E+07       1.00      7.60      0.0340
2      COAL             0.3E+08       0.75      0.00      0.0477
3      OIL              0.4E+08       0.75      0.00      0.1257
4      GAS              0.1E+09       0.75      0.00      0.3194
LIGHT FITTINGS

STANDARD DATA : ADDITIONAL RUNNING COSTS
                                              30-Aug-79 17:55

HOTEL DESIGN DATA
ITEM                              RATE  QUANTITY
1      HOT WATER                  12.00   1 NOCC
2      MAINTENANCE                 9.36  11 TFA
3      ADMINISTRATION             23.00  11 TFA
4      RATES                       8.32  11 TFA

STANDARD DATA : FINANCIAL                     30-Aug-79 17:56

HOTEL DESIGN DATA
1      INFLATION RATE    % PA      6.00
2      INTEREST RATE     % PA     10.00
3      BUILDING LIFE YEARS        60
```

Figure 3.3.6 User-modifiable energy tariff and additional costs' files and amortisation parameters.

frequently the scheme on which the architect has lavished time and effort is found by one or other of the specialists to be infeasible.

With access to explicit appraisal techniques, however, it is possible to check a wide range of criteria simultaneously from the outset of the design activity. Moreover, it is entirely practical for each member of the design team to have access to, and operate on, the common design model – whether or not they share a design office. The models, then, provide a strong integrating force.

Second, the extent of the search for a solution which balances investment and return (ie cost and performance) is significantly enlarged. Experiences from the use of PHASE in hospital design indicates a ten fold increase in the design alternatives explored.

The significant factor is not simply that the search sequence is **extended** (by an order of magnitude); it is **directed** by the cost/performance profile.

Third, contrary to the fears of many design practitioners (particularly architects) the use of computer-based models focusses *increased* attention on subjective value judgements, rather than less. As the cost/performance attributes of design alternatives are made more explicit, confidence grows in

```
OUTPUT : GENERAL          4-Sep-79 10:01
AREAS (M2)
```

COMPONENT	EXT. FLR	INT. FLR	TOT. FLR	EXT. WALL	ROOF	VOL (M3)
1 1 RESTAURANT	103.0G	0.0	108.0	78.0	0.0	324.0
1 2 RESTAURANT	39.0G	0.0	39.0	0.0	0.0	117.0
2 1 LOUNGE	109.3G	0.0	109.3	78.0	109.3	387.8
3 1 KITCHEN	60.0G	0.0	60.0	27.0	18.0	180.0
3 2 KITCHEN	15.0G	0.0	15.0	9.0	9.0	45.0
4 1 FOYER	198.0G	0.0	198.0	39.0	71.5	594.0
4 2 FOYER	16.0G	0.0	16.0	0.0	0.0	48.0
4 3 FOYER	44.0G	0.0	44.0	16.5	24.8	132.0
5 1 ADMINISTR'	270.0G	0.0	270.0	181.5	0.0	810.0
6 1 STEAK BAR	91.0G	0.0	91.0	180.0	91.0	273.0
6 2 STEAK BAR	64.0G	0.0	64.0	48.0	36.0	192.0
7 1 FUNCTIONS	252.00G	0.0	252.0	129.0	252.0	756.0
8 1 BEDROOMS	101.3U	654.8	756.0	417.0	0.0	2268.0
8 2 BEDROOMS	0.0	756.0	756.0	417.0	756.0	2268.0
TOTAL	101.3U 1266.3G	1410.8	2778.3	1500.0	1367.5	8334.8

```
WALL TO FLOOR RATIO       0.54
VOLUME COMPACTNESS        0.53
```

Figure 3.3.7 Tableau of spatial descriptors output by the computer.

```
OUTPUT: ACTIVITY PERFORMANCE
STANDARD DEVIATION = 0.773
TOO FAR APART
```

	COMPONENTS			DISTANCE	FACTOR	
6	STEAKBAR	3		KITCHEN	36.4	2.66
7	FUNCTIONS	1		RESTAURANT	33.9	2.23
6	STEAKBAR	1		RESTAURANT	35.2	2.06
5	ADMINISTR'	2		LOUNGE	43.5	1.59
7	FUNCTIONS	3		KITCHEN	21.7	1.42
5	ADMINISTR'	4		FOYER	22.8	1.17
7	FUNCTIONS	4		FOYER	22.3	0.98
3	KITCHEN	2		LOUNGE	21.3	0.93

```
TOO CLOSE TOGETHER
```

	COMPONENTS			DISTANCE	FACTOR	
2	LOUNGE	1		RESTAURANT	13.6	0.00
8	BEDROOMS	7		FUNCTIONS	24.8	0.18
8	BEDROOMS	3		KITCHEN	26.1	0.19
8	BEDROOMS	1		RESTAURANT	27.1	0.20
8	BEDROOMS	6		STEAKBAR	28.3	0.21
8	BEDROOMS	4		FOYER	15.6	0.23
6	STEAKBAR	5		ADMINISTR'	18.4	0.54
7	FUNCTIONS	5		ADMINISTR'	24.5	0.54

Figure 3.3.8 Tableau of planning efficiency output by the computer.

ENVIRONMENT : WALL HEAT LOSS, WINTER 4-Sep-79 10:10

BUILDING ORIENTATION 210 DEG

COMPONENT		WALL		W/M 2	AREA	KW
1	1	RESTAURANT	NW	75.6	13.5	1.020
4	1	FOYER	NW	69.4	33.0	2.291
1	1	RESTAURANT	NE	63.3	40.5	2.562
6	1	STEAKBAR	NW	63.3	39.0	2.468
2	1	LOUNGE	NW	57.1	28.5	1.628
2	1	LOUNGE	NE	57.1	34.5	1.971
5	1	ADMINISTR'	NW	57.1	40.5	2.313
6	2	STEAKBAR	NW	57.1	1.5	0.086
8	1	BEDROOMS	NW	57.1	168.0	9.596
8	2	BEDROOMS	NW	57.1	168.0	9.596
5	1	ADMINISTR'	SE	44.6	51.0	2.275
6	1	STEAKBAR	SW	44.6	21.0	0.937
6	2	STEAKBAR	SW	44.6	24.0	1.071
7	1	FUNCTIONS	SE	44.6	54.0	2.409
7	1	FUNCTIONS	SW	44.6	42.0	1.874
8	1	BEDROOMS	SE	44.6	168.0	7.496
8	2	BEDROOMS	SE	44.6	168.0	7.496
3	1	KITCHEN	SE	33.7	18.0	0.606
3	2	KITCHEN	SE	33.7	9.0	0.303
8	1	BEDROOMS	NE	31.3	40.5	1.267
		AVERAGE		22.6		
		TOTAL			4145.0	93.540

Figure 3.3.9 Tableau of heat losses output by the computer.

the formulation of the value judgements needed in the exercise of design decision-making.

The future implications are altogether more exciting; let us consider for a moment the dilemma facing the designer in figure 3.3.2. From the program output he identifies a poor performance in terms of thermal environment but the modifications he can make to the design hypothesis are manifold.

Should he, for example, change the shape, change the orientation, change the thermal resistance, change the thermal mass, decrease fenestration, or what? It has to be conceded that the causal relationships between *design* and cost/ performance *consequences* are, to put it at its most polite, ill-understood.

So what do the new generation of models have to offer? Quite simply, the opportunity to explore, parametrically, just such relationships. By keeping all other design variables constant, members of the design team can determine the effect of changing a single design variable (eg shape or construction) on any one of the range of cost/performance variables (eg capital cost, energy consumption, or movement efficiency).

Moreover, the models lend themselves to progressive refinement. By modelling existing buildings, the model user/developer can compare the

```
COST ANALYSIS : RUNNING COSTS4-Sep-79 10:20

HEATING ELECTRICITY                          23494.        16.6        8.5
LIGHTING                                      4136.         2.9        1.5
HOT WATER                                     1248.         0.9        0.4
MAINTENANCE                                  26004.        18.3        9.4
ADMINISTRATION                               63900.        45.0       23.0
RATES                                        23115.        16.3        8.3

TOTAL                141898.         100.0              51.1

COST ANALYSIS : CAPITAL COST GROUPS                            4-Sep-79 10:21

A     PRELIMS & CONTINGENCIES        19253.        3.7         6.9
BC    FOUNDATIONS & GROUND FLOOR     29630.        5.7        10.7
D     FRAME   50009.                  9.6         18.0
E     ROOF & ROOFLIGHTS              29702.        5.7        10.7
F     WALLS, WINDOWS, DOORS          75194.       14.5        27.1
G     INTERNAL WALLS                 52259.       10.1        18.8
H     UPPER FLOORS                   15120.        2.9         5.4
I     STAIRCASES                     13891.        2.7         5.0
J     FIXTURES                       55565.       10.7        20.0
KM    INSTALLATIONS                  93099.       17.9        33.5
NOR   SERVICES                       46119.        8.9        16.6
ST    EXTERNAL SERVICES              39034.        7.5        14.1

      TOTAL   518875.                100.0      186.8

COST ANALYSIS : PRESENT WORTH AND ANNUAL EQUIVALENT
                        PRES. WORTH      ANN. EQUIV.
            CAP. COST       518875           52009
            RUN. COST      1414297          141878

            TOT. COST      1933172          193887
```

Figure 3.3.10 Recurring and capital costs output.

prediction output by the model with reality. In the case of cost, for example, the base data and aggregating logic can be progressively refined until an acceptable level of correspondence is reached between the 'design' (the representation of future reality) and the 'building' (the full scale 3D manifestation of the design intent).

COST-PERFORMANCE SPECIFICATION

But issues remain; for example, against what criteria should a boost-performance profile be evaluated?

Building design is unique in the following respect. While, with other artefacs a 'performance specification' is quoted in the brief and those tendering for the commission seek to meet the specification at minimum cost, in building

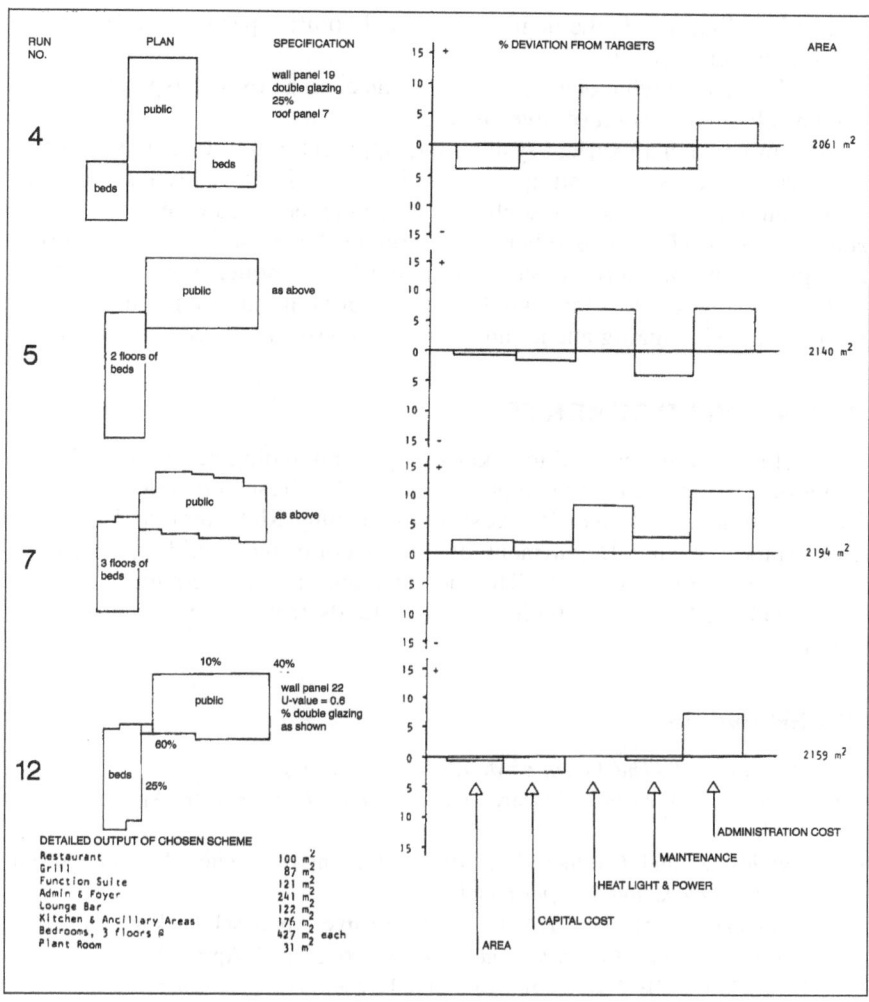

Figure 3.3.11 Discrete stages in the search for an 'optimal' solution.

design a 'cost limit' is set (based entirely on capital expenditure) and neither designer nor contractor is bound by performance criteria.

The explicit nature of computer-based models provides new opportunities for the introduction of performance specification. Let us assume for the moment an escape from the tyranny of the £1m² control by the expedient of inverting the metric.

How much more meaningful would it be, in balancing investment (cost-in-use) against return (performance), to state the amount of space per £, the level of daylight per £, the degree of thermal comfort per £, the efficiency of movement per £, etc, etc?

In other words: what degree of performance, on a host of disparate criteria, can be achieved per unit of investment?

Assuming such a metrical system, two approaches to the setting of criteria for performance specification are possible. One is an analytical approach to optimum targets (as set by sophisticated computer-based investment models such as INVEST[10]). The other is an appraisal approach based on norms emerging from long-term usage of computer-based models such as GOAL.

Either may seem less than wholly satisfactory but must be preferable to the shifting sands, dragging anchor and faulty compass of current design practice.

ACKNOWLEDGEMENTS

In conclusion, I am pleased to acknowledge, and apologise for any misinterpretation of, the intellectual inputs of ABACUS' research fellows (notably Harvey Sussock and Alan Bridges) and of Denny McGeorge and Professor Tom Markus of the Department of Architecture and Building Science at the University of Strathclyde. The models mentioned in the paper have been developed predominantly with the aid of funds from the Science Research Council.

REFERENCES

(1)   Bronowski, J., *The Ascent of Man*, BBC Publications.

(2)   Building Performance Research Unit, *Building Performance*, Applied Science Publishers.

(3)   Gentles, J. and Gardner, W., 'BILD Building Integrated Layout Design', *ABACUS Occasional Paper* no 64, 1978.

(4)   Kernohan, D. et al., 'PHASE: An Interactive Appraisal Package for Whole Hospital Design', *Computer Aided Design*, vol 5, no 2, April 1973.

(5)   Sussock, H. 'HELP Housing Evaluation Layout Package', *ABACUS Occasional paper* no 53, 1976.

(6)   Clarke, J., 'A Design Oriented Thermal Simulation Model', *Proceedings of CAD 78*, IPC, 1978.

(7)   Laing, L.W. W., PAIR-Q: A Flexible Computer Simulation Model for Airport Terminal Building Design', *DMG-DRS Journal*, vol 9, no 2.

(8)   Aish, R., 'Prospects for Design Participation', *Design methods and Theories*, vol 1, no 1.

(9)   Sussock, H. 'GOAL General Outline Appraisal of Layouts', *ABACUS Occasional Paper* no 62, 1978.

(10) Maver, T. W., Fleming, J., and McGeorge, D., 'INVEST A Program for Analysis in Hotel Design', *ABACUS Occasional Paper* no 63, 1978.

3.4 The effect of location and other measurable parameters on tender levels*

I. Pegg

INTRODUCTION

The tender price of a project is affected by many parameters. BCIS has been measuring tender price levels since the first quarter of 1974 and these are published as the BCIS Tender Price Index (TPI). Some of the factors affecting prices have been measured and papers describing the effect of 'date' and 'location' have been previously published in BCIS.

Now that the data are held on computer it has become feasible to measure the effect on tender price indices of other factors and this paper reports on the affect of: date; location; selection of contractor; contract sum; building function; measurement of structural steelwork; building height; form of contract; site conditions; type of work; fluctuations; client; contract period; form of construction; and method of measurement. It should be borne in mind that other factors not investigated in this study can produce substantial variations in tender prices.

This study updates and extends Cost Study F23, (BCIS Study of Regional Tender Levels).

The techniques used are capable of further improvement and these refinements will be made in the future. In the meanwhile these first results should prove useful to subscribers in the interpretation of project tender price indices and the adjustment of cost analyses.

METHOD

The principle behind these calculations is the comparison of the individual project index with the index expected for that project after taking account of six major variables. The resulting factors are grouped according to the criteria under study and the significant differences between groups are then identified.

*1984, *Cost Study F33*, BCIS 1983-4-219-5 to 17, © Building Cost Information Service, Royal Institution of Chartered Surveyors (extracts). Updated tables are published by BCIS.

Project indices

The individual project indices are calculated in the normal course of preparing the Tender Price Indices (see Cost Study F5). It should be remembered that the method used is based on sampling techniques and is therefore subject to measurement error. Towards the end of the sample period the base Schedule of Rates used for comparisons was changed together with some of the item sampling and calculation rules. This should not affect the general results apart from the study of structured steelwork.

Calculation of factors

In order to study the average pricing level of projects in different categories, the individual projects falling within the category are divided by the expected project index to produce a project factor. The expected project index is calculated taking into account the sixth major variables – date, location, contract sum, tendering procedures, building function and the measurement of structural steelwork. Details of these effects are given later in this report.

When trying to study the effect of a particular factor, the factor itself was of course, excluded from the calculations of the expected project indices. Once these six effects had been estimated, the process was repeated to see if the estimates changed and if so the process was repeated until a stable estimate was produced. This iterative procedure is necessary as some of the effects are not independent; for example, when looking at office buildings (which tend to be more expensive than the 'average') it is difficult to isolate the effect of the building function from the effect of the size of the contract sum.

The geometric mean of the project factors are given for the various categories. The geometric mean was chosen since it is used to calculate the Tender Price Index and because particular statistical inferences can be drawn from the particular nature of this data. The arithmetic (ordinary) standard deviation has been given since this is more readily interpreted.

Sample

The projects included in this study are those indexed for the BCIS Tender Price Index, dated first quarter 1978 onwards. Older projects have not been included in the study as resources have not yet been available to code and enter additional projects on the computer. The selection process for bills of quantities to be included in the Tender Price Index is based on random requests for bills from the BCIS membership and therefore the sample represents a reasonable cross section of building work within the following criteria:

• Accepted tenders;
• New buildings or horizontal extensions;

- Contract sums over £45,000 (currently)
- Tenders based on firm or approximate bills of quantities.

For technical reasons the sample is not as large as that used in the published TPI. 1188 projects from the TPI are included in this study.

GENERAL RESULTS

The following effects were investigated:–

1 Date
2 Location
3 Selection of Contractor
4 Contract sum
5 Building function
6 Measurement of structural steelwork
7 Building height
8 Form of contract
9 Site conditions
10 Type of work
11 Fluctuations
12 Client
13 Contract period
14 Form of construction
15 Method of measurement

Items 1–6 were used to provide an expected index for individual projects. Items 1–9 were found to be statistically significant and details are given on the following pages.

The analysis by type of work (item 10) was not significant because of the small effect that was present and the use of extraneous codes. The comparison between new build and horizontal extension did show a significant difference, the new build being on the mean with the horizontal extension being 2% above on average.

The study into Fluctuations (item 11) only looked at the difference between firm and fixed contracts and traditional and formula fluctuations. The 'date' factor was divided into projects with and without fluctuations (as with the tender price index) and so no comparison between contracts with and without fluctuations provisions was possible in general terms. Table 3.4.1 shows the difference between firm and fluctuating price contracts.

The client effect (item 12) was significant until the building function was taken into account. This is because certain clients are associated with particular building types – the effect of which is highly significant. Once the effect of building function had been removed there was no significant pattern detected.

The length of contract period (item 13) showed no measurable effect on the index, although it should be remembered that the contract sum will be highly correlated with the contract period. A future area for investigation may be the duration of contract compared with the duration predicted given the contract sum.

The form of construction (item 14) was not significant, which is reassuring given that the TPI is intended to measure pricing level independently from design. If any effect was present it would probably be removed by the measurement of structural steelwork adjustment.

There was no significant difference between jobs measured under different standard methods (item 15). This is another satisfactory test of the indexing process.

Residuals

After adjusting the projects in the manner described above, a considerable amount of variability remained. This is partly due to sampling error within the indexing process, but mostly attributable to other factors which have not been, or are too esoteric to measure.

Some tests were carried out on the residuals in order to determine the shape of the distribution so that the validity of some of the statistical techniques used could be verified. The tests were not conclusive, but suggested that the logarithm of the indices were approximately normally distributed and that assumption has been employed throughout this study.

The practical implications of this area that if the factors in this study are employed to estimate an individual project tender price index, the measured index figure has a 90% chance of falling within the range − 15% to +18%. There is therefore considerable scope for professional assessment in the light of particular circumstances.

NOTES

The tables in the main part of this study have common headings, and the terms employed are listed here:-

Factor

The geometric mean of the project factors as described in the method section.

90% CI

The 90% Confidence Interval is a measure of the reliability of the above factor and is influenced by the sample size and variability of the individual factors. In practical terms it says that there is a 90% chance that the 'true' factor for all projects in this category lies within this range. It does *not* say that 90% of project factors will fall within this range.

S.D.

The Standard Deviation is a measure of the dispersal of the figures around the mean. Note that the logarithm transformation has not been used in this calculation to make interpretation easier.

Range

The lowest and highest factors in the sample.

Sample

The number of buildings in this category used in the study.

USE OF THE FACTORS

When using the figures in this study to adjust average or individual project tender price indices the factors should be multiplied, even where several adjustments are to be made.

Example:

To forecast the project tender price index of a factory in Avon tendered for in competition with a base month of August, 1983.

Forecast for 3rd quarter 1983   =   227
Factor for Avon   =   0.93
Factor for Factories   =   0.97
Factor for competitive tender   =   0.99

Forecast of project index = 227 □ 0.93 □ 0.97 □ 0.99   =   203

Great caution should be employed when using these factors since the remaining variability is very large for individual project indices.

DATE

Table 3.4.1a shows the change in tender levels over time. The index figures are geometric means of the individual project tender price indices. Each of the index series is separately based with 1st quarter 1974 equalling 100.

Table 1b has been calculated using the individual project indices adjusted

Table 3.4.1 Tender price indices – comparison of adjusted and unadjusted indices

| Quarter | | a) Project indices (used in this study) | | | | | | b) Adjusted project indices (used in this study) | | | | | |
| | | All | | Firm price | | Fluctuating price | | All | | Firm price | | Fluctuating price | |
		Index	Sample	Index	Sample	Index	Sample	Index	Sample	Index	Sample	Index	Sample
1978	i	137	55	142	29	134	37	138	55	143	29	134	37
	ii	147	41	145	22	148	19	149	41	146	22	151	19
	iii	152	52	156	29	147	23	155	52	156	29	155	23
	iv	164	42	164	18	164	24	159	42	161	18	159	24
1979	i	173	89	179	34	169	54	171	83	176	34	169	54
	ii	178	55	182	26	174	29	181	55	187	26	175	29
	iii	200	59	196	32	203	37	201	69	200	32	203	37
	iv	214	58	219	22	212	36	210	58	215	22	207	36
1980	i	215	81	222	30	211	51	212	81	221	30	207	51
	ii	222	40	232	21	212	19	220	39	229	20	211	19
	iii	229	46	239	29	212	17	225	46	233	29	211	17
	iv	217	49	223	23	212	26	216	49	221	23	212	26
1981	i	212	52	214	23	210	29	211	52	209	23	212	29
	ii	211	37	216	19	205	18	211	37	214	19	208	18
	iii	209	57	206	36	214	21	211	57	211	36	210	21
	iv	204	71	202	39	205	32	208	70	208	39	207	31
1982	i	216	91	221	54	208	37	215	90	217	54	210	36
	ii	212	64	218	42	200	22	217	64	220	42	210	22
	iii	210	70	214	45	203	25	213	70	214	45	212	25
	iv	209	34	212	25	199	9	214	34	213	25	214	9
1983	i	219	13	223	11	193	2	220	13	232	11	198	2

for the main variables – location, contract sum, tendering procedures, building function and the measurement of structural steelwork. It is these adjusted figures in Table 3.4.1b which are used within this study.

LOCATION

This section supersedes the previous BCIS Study of Regional Tender Levels (F23).

The areas used in this analysis were counties in England and Wales and regions in Scotland. These areas have been grouped into Regions by the Department of the Environment Standard Statistical Regions (except for the South-East for which smaller groups are used) and split into Districts where several projects are available from the same district. Sample sizes are very small in some groups and the District, and some county, figures should be used with extreme caution. It should also be acknowledged that there may be considerable variation within counties or even districts. The factors are given in Table 3.4.2. A similar study was done based on the National Grid Reference which showed broadly the same pattern.

SELECTION OF CONTRACTOR

Each project was coded according to the method employed to select the contractor.

The factors relating to each code are given in Table 3.4.3. The factor for serial contracts should be treated with extreme caution since in addition to the small sample size they are not all for different clients. The high factor for negotiated contracts does not of course necessarily imply bad value, since the client may benefit in some other way (e.g. a faster programme).

CONTRACT SUM

In order to study the influence of contract sum on tender prices the contract sum was adjusted to 1st quarter 1974 prices and grouped according to the logarithm of this figure. When the factors for each group are plotted as in Figure 3.4.1 a strong downward trend can be seen with some disturbance at either end. There are fewer projects at the top of the range and so it is not possible to tell whether the trend continues downward or levels out. Small contracts (under £25,000 at 1974 prices) do appear significantly cheaper than the trend line would indicate.

A weighted least squares regression was run to compare the log of the factor with the mid point of the group interval. This produced a corrected R^2 of 85% when all observations were included or 91% when the first three groups were omitted. This indicates a good fit to the line. The line produced by the latter equation is shown in Figure 3.4.1 and the following equations can be used to give a factor for a particular contract sum over £25,000 at 1974 prices:–

Table 3.4.2 Regional, county and district factors

Area	Factor	90% CI	S.D.	Range	Sample
Northern Region	0.97	0.95–0.98	0.09	0.74–1.19	73
Cleveland	0.95	0.92–0.98	0.07	0.84–1.05	14
Cumbria	0.98	0.90–1.07	0.14	0.74–1.15	9
Durham	0.98	0.94–1.03	0.09	0.88–1.15	13
Northumberland	0.95	0.85–1.06	0.11	0.81–1.07	5
Tyne & Wear	0.96	0.94–0.99	0.07	0.84–1.19	32
Newcastle upon Tyne	0.93	0.89–0.97	0.05	0.88–0.99	6
North Tyneside	0.95	0.91–1.01	0.07	0.84–1.08	7
South Tyneside	1.01	0.93–1.09	0.10	0.90–1.19	6
Gateshead	0.97	0.94–1.00	0.04	0.92–1.02	8
Sunderland	0.97	0.88–1.06	0.10	0.91–1.15	5
Yorkshire and Humberside Region	0.92	0.90–0.93	0.09	0.74–1.19	80
Humberside	0.91	0.87–0.94	0.06	0.76–1.00	12
North Yorkshire	0.95	0.90–1.00	0.10	0.74–1.10	13
South Yorkshire	0.89	0.86–0.93	0.08	0.77–1.07	14
Rotherham	0.87	0.82–0.91	0.05	0.82–0.92	5
Sheffield	0.90	0.86–0.94	0.06	0.84–1.02	7
West Yorkshire	0.92	0.90–0.95	0.09	0.75–1.19	41
Bradford	0.94	0.89–0.99	0.11	0.81–1.19	14
Leeds	0.87	0.84–0.91	0.06	0.75–0.97	12
Wakefield	0.95	0.92–0.99	0.06	0.88–1.03	9
East Midlands Region	0.93	0.91–0.95	0.11	0.72–1.27	73
Derbyshire	0.89	0.86–0.92	0.09	0.72–1.08	21
Leicestershire	0.97	0.91–1.02	0.14	0.75–1.27	19
North West Leicestershire	0.98	0.82–1.15	0.20	0.75–1.21	6
Leicester	0.97	0.89–1.06	0.14	0.84–1.27	9
Lincolnshire	0.90	0.86–0.94	0.08	0.78–1.04	10
Northamptonshire	0.98	0.94–1.02	0.08	0.81–1.11	13
Northampton	0.99	0.93–1.05	0.09	0.81–1.11	8
Nottinghamshire	0.91	0.86–0.96	0.09	0.81–1.07	10
Scotland	0.96	0.95–0.98	0.10	0.74–1.20	120
Border Region	1.05	0.90–1.23	0.04	1.02–1.07	2
Central Region	1.01	0.87–1.17	0.09	0.93–1.11	3
Dumfries and Galloway	0.96	0.79–1.16	0.11	0.84–1.03	3
Fife Region	0.84	0.80–0.89	0.06	0.78–0.95	6
Grampian Region	0.93	0.88–0.98	0.09	0.77–1.08	13
Aberdeen	0.96	0.92–1.00	0.07	0.87–1.08	9
Highland Region	0.94	0.83–1.06	0.11	0.77–1.06	5
Lothian Region	0.98	0.95–1.01	0.09	0.77–1.19	25
West Lothian	0.97	0.90–1.04	0.10	0.77–1.10	3
Edinburgh	0.98	0.94–1.02	0.09	0.33–1.19	15
Orkney Islands Area	1.04				1
Strathclyde Region	0.97	0.95–1.00	0.11	0.74–1.20	53
Glasgow	1.02	0.95–1.08	0.13	0.78–1.20	13
Inverclyde	1.01	0.92–1.12	0.10	0.90–1.10	5
Cunninghame	0.91	0.84–0.98	0.08	0.77–0.99	6
Kilmarnock and Lowdon	0.96	0.86–1.08	0.12	0.84–1.12	5
Tayside Region	0.98	0.89–1.00	0.10	0.85–1.19	9

Northern Ireland	0.84	0.79–0.89	0.11	0.68–1.09	14
Antrim	0.93	0.87–1.00	0.08	0.86–1.09	6
Armagh	0.75	0.62–0.92	0.09	0.68–0.86	3
Down	0.73	0.65–0.83	0.05	0.68–0.75	3
Londonderry	0.86	0.67–1.10	0.05	0.83–0.90	2
Offshore Islands	1.10	0.92–1.31	0.11	0.97–1.17	3
Channel Islands	1.06	0.62–1.83	0.13	0.97–1.16	2
Isle of Man	1.17				1

Table 3.4.3 Selection of contractor

Group	Factor	90% CI	S.D.	Range	Sample
Not known	1.00	0.98–1.02	0.11	0.82–1.41	55
Competitive tender	0.99	0.99–1.00	0.10	0.69–1.33	1062
Negotiated	1.11	1.08–1.15	0.14	0.75–1.51	60
Extension contract	1.03	0.92–1.16	0.15	0.88–1.22	6
Serial contract	1.09	1.03–1.15	0.05	1.02–1.12	4

$$F \ = \ 1.533 \ \square \ C^{-0.03515}$$
$$\text{where } \ F \ = \ \text{Factor}$$
$$C \ = \ \text{Contract sum at 1074 prices}$$

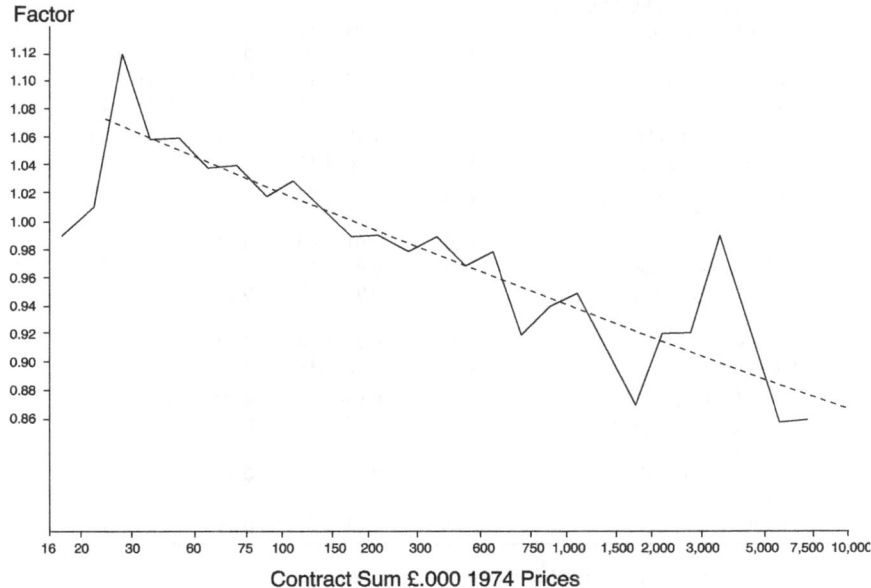

Figure 3.4.1 Influence of contract sum on tender price.

Table 3.4.4 Building function

	Group	Factor	90% CI	S.D.	Range	Sample
2	TRANSPORT & INDUSTRIAL BUILDINGS	0.99	0.97–1.00	0.12	0.70–1.32	215
21	Railway buildings	1.14	1.08–1.21	0.14	0.90–1.30	15
22	Road transport buildings	1.01	0.95–1.07	0.13	0.95–1.07	15
25	Agricultural buildings, farms	1.02	0.87–1.19	0.17	0.82–1.28	5
27	Factories	0.97	0.95–0.98	0.11	0.74–1.32	128
28	Warehouses, storage buildings	0.97	0.94–1.00	0.11	0.71–1.16	48
3	ADMINISTRATIVE, COMMERCIAL BUILDINGS	1.02	1.01–1.03	0.	0.72–1.46	201
31	Official administrative buildings	1.04	1.00–1.09	0.11	0.87–1.24	18
32	Office buildings	1.03	1.01–1.05	0.11	0.72–1.45	85
34	Shops etc	0.98	0.86–1.00	0.11	0.73–1.32	64
38	Public Service buildings	1.06	1.03–1.09	0.10	0.83–1.34	34
4	HEALTH & WELFARE BUILDINGS	1.00	0.99–1.02	0.09	0.80–1.30	112
41	Hospitals	1.03	1.00–1.06	0.10	0.84–1.30	33
43	Other health buildings	1.01	0.98–1.03	0.08	0.85–1.17	35
44	Homes	0.98	0.96–1.01	0.10	0.80–1.18	42
5	REFRESHMENT, ENTERTAINMENT RECREATION BUILDINGS	1.05	1.03–1.06	0.10	0.81–1.30	111
51	Refreshment building	1.07	1.03–1.11	0.10	0.92–1.23	19
53	Community buildings	1.05	1.03–1.07	0.09	0.84–1.30	48
54	Swimming pools	1.04	0.94–1.15	0.14	0.81–1.19	7
56	Sports buildings	1.02	0.99–1.05	0.10	0.86–1.26	36
6	RELIGIOUS BUILDINGS	1.09	1.05–1.14	0.12	0.94–1.33	20
63	Churches, chapels	1.10	1.06–1.15	0.10	0.99–1.33	15
7	EDUCATION, CULTURAL, SCIENTIFIC BUILDINGS	1.03	1.02–1.04	0.10	0.79–1.32	201
71	Schools	1.02	1.01–1.04	0.09	0.85–1.32	138
72	Universities, colleges	1.02	0.98–1.05	0.11	0.84–1.26	27
73	Research, scientific centres	1.02	0.97–1.07	0.13	0.79–1.20	22
76	Library buildings	1.02	0.92–1.14	0.09	0.93–1.16	4
78	Studies, etc.	1.08	1.02–1.14	0.08	0.96–1.20	7
8	RESIDENTIAL BUILDINGS	0.96	0.95–0.97	0.09	0.64–1.22	325
81	Housing, dwellings in general	0.95	0.94–0.96	0.09	0.64–1.22	199
84	Special residential buildings	0.96	0.95–0.97	0.09	0.78–1.22	118
85	Hotels, etc.	1.05	1.02–1.08	0.05	0.97–1.12	8

Note – Groups with a sample size of less than 4 have been excluded.

BUILDING FUNCTION

The first two digits of the CI/SFB code for the building function were used to classify the projects in the study.

The factors appear to take on a pattern if the building function is taken as a proxy for complexity of construction. Buildings that are straightforward or repetitive, like factories and housing, have low index figures while complex and unrepetitive buildings, such as churches and hospitals have higher tender levels. Table 3.4.4 gives the factors for the various building types.

MEASUREMENT OF STRUCTURAL STEELWORK

This variable was investigated not because of any expected difference in the actual pricing level of contractors on jobs with and without measured steel-work, but because of a known problem with the indexing process. The rates for measured steelwork used in conjunction with the 1973 Schedule of Rates were high compared with other trades. This resulted in projects indexed under the 1973 Schedule yielding artificially low index figures when the steelwork was included in the indexing process. To omit steelwork from the index entirely would give a false measure of tender prices, but individual project indices were not readily comparable with the overall mean and so a crude adjustment for this effect has been made to projects indexed under the 1973 schedule only. The factors derived for this purpose are given in Table 3.4.5.

Table 3.4.5 Measurement of structural steelwork

Group	Factor	90% CI	S.D.	Range	Sample
No steelwork	1.01	1.01–1.02	0.10	0.69–1.43	743
Measured steel frame	0.93	0.91–0.95	0.12	0.65–1.22	82
Some measured steelwork	1.00	0.98–1.02	0.07	0.85–1.22	39
P.C. Sum for steelwork	0.99	0.98–1.00	0.10	0.73–1.33	249

BUILDING HEIGHT

The results of this analysis are given in Table 3.4.6 and shown graphically in Figure 3.4.2. It is difficult to detect any underlying trend in these figures, but the peak at 4 storeys might be caused by the transition between site access methods. The measured effect could equally be due to some spurious correlations.

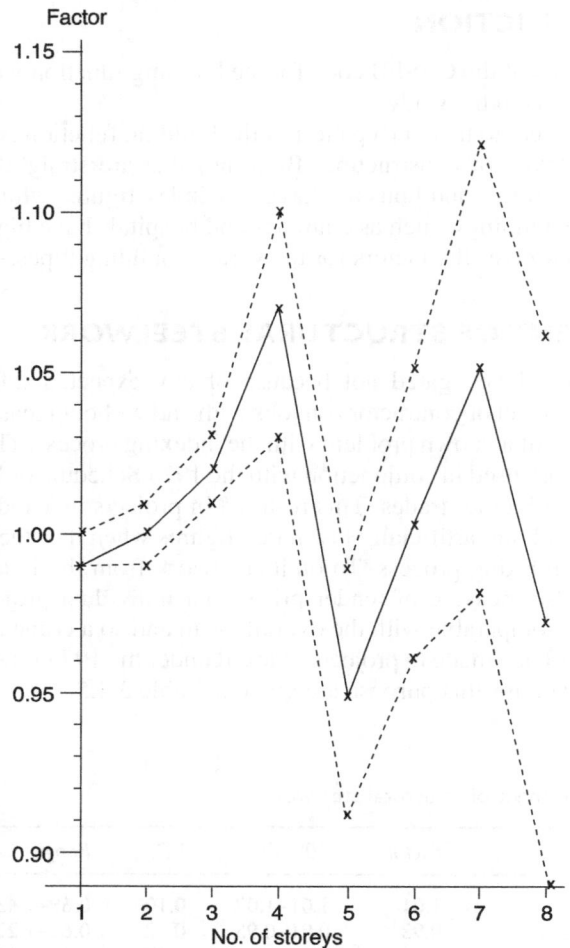

Figure 3.4.2 Influence of building height on tender price.

FORM OF CONTRACT

It is widely held that some forms of contract are more onerous than others, and this part of the study was intended to detect any influence the form of contract might have on pricing levels. Unfortunately the sample sizes for the less common contracts are too small to allow any firm conclusion, but the results obtained are given in Table 3.4.7.

Table 3.4.6 Building height

Primary number of storeys	Factor	90% CI	S.D.	Range	Sample
Not known	0.98	0.96–1.00	0.07	0.87–1.11	22
1	0.99	0.99–1.00	0.10	0.70–1.36	610
2	1.00	0.99–1.01	0.10	0.68–1.41	365
3	1.02	1.01–1.03	0.09	1.01–1.03	119
4	1.07	1.03–1.10	0.11	0.88–1.34	32
5	0.95	0.91–0.99	0.10	0.78–1.12	16
6	1.00	0.96–1.05	0.07	0.91–1.10	8
7	1.05	0.98–1.12	0.09	0.93–1.20	7
8	0.97	0.89–1.06	0.07	0.88–1.06	4
9	1.01				1
11	1.34				1

Table 3.4.7 Form of contract

	Factor	90% CI	S.D.	Range	Sample
Not known	0.97	0.87–1.08	0.06	0.91–1.02	3
CCC/Works 1 or GC/Works 1	1.06	0.92–1.38	0.06	1.02–1.11	2
JCT (edition unknown)	1.02	0.99–1.06	0.10	0.85–1.24	22
JCT 1963 edition	1.00	1.00–1.01	0.10	0.68–1.35	1035
JCT 1980 edition	0.98	0.97–0.99	0.09	0.74–1.20	119
Other	1.15	0.95–1.41	0.20	0.98–1.41	4

SITE CONDITIONS

In order to determine the influence of site conditions, two aspects of the construction site were taken and coded, namely working space and ground conditions.

If anything was known about these aspects, one of three codes could be chosen to indicate the absence or presence of difficulties. In the majority of cases very little was known about the site and coding was therefore difficult and consequently haphazard. The results given in Table 3.4.8 indicate that there is some consistent pricing differences between 'easy' and 'difficult' sites, but it is hoped that more specific information can be obtained in future to allow further investigation.

While the analysts by working space alone was statistically significant, the overall analysis was not – probably because of the smallness of the effect compared with the residual variations.

Table 3.4.8 Site conditions

8a Factors

Working space	Ground Conditions				
	All	Not known	Good	Moderate	Bad
Not known	1.00	1.01	0.94	0.98	0.98
Unrestricted	0.98	1.00	0.99	0.98	0.99
Restricted	1.01	1.01	0.96	1.01	1.03
Highly restricted	1.02	1.03	–	0.97	1.01

8b Sample Size

Working space	Ground Conditions				
	All	Not known	Good	Moderate	Bad
Not known	311	258	9	17	27
Unrestricted	585	258	200	76	51
Restricted	239	177	13	21	28
Highly restricted	50	36	–	3	11

8c Standard Deviations

Working space	Ground Conditions				
	All	Not known	Good	Moderate	Bad
Not known	0	0.10	0.11	0.10	0.11
Unrestricted	0	0.10	0.10	0.09	0.10
Restricted	0.1?	0.10	0.10	0.09	0.11
Highly restricted	0.15	0.15	–	0.11	0.11

Part 4 Cost-process modelling

In contrast to Part 3, this part contains most of the papers published on process modelling for this domain which reflects the difficulty of the task which largely depends on finding an aspect of the production process that is common to all contractors. The most recent work relies heavily on the stochastic simulation of the production process by the use of network planning models in combination with knowledge modelling of the planning process. This is perhaps the newest of all fields due to its dependence on the linking of latest developments in information technology for practical applications.

Beeston's **Cost consequences of design decisions: cost of contractors' operations** (ch 4.1) describes an early attempt at developing a cost forecasting system for designers based on a process model. The computer program COCO represents the decision processes of contractors' planners when they are determining the required plant, labour and construction time for design realisation. The major emphasis is in determining the type of tower crane to use on site, as this was found (after discussions with 'several major contractors') to be pivotal in resource allocation. COCO is interactive and poses a series of questions to obtain design information from the user. The answers to these questions are then passed through a decision tree structure to eliminate plant choices. COCO covers the basic superstructure of a reinforced concrete framed or loadbearing cladding, floors, roof, brick/blockwork, partitions, windows, etc. In many respects, COCO is a more flexible version of the Wilderness studies – instead of interpolating graphs, COCO calculates the costs directly. In retrospect, COCO's main problem would have been the relatively crude computer technology around at that time. Running interactively on a 10K machine in time sharing mode must have been a real challenge. Today's equivalents are certainly more technologically advanced. Of course, the COCO model is clearly amenable to structural validation and the modelling process itself is to some extent self-validating.

Holes and Thomas' **General purpose cost modelling** (ch 4.2) describes CIRCE (Construction Industry Resource Cost Estimating), a computer system which explicitly relates the dimensions of the spaces comprising the building, to production process requirements. The system purports to be 'smooth' as it enables the same computer software and basic production process data to be applied at any stage in the design process, permitting a smooth development of project costing as the design is developed. The approach exploits what are termed the 'regularities' of buildings in terms of their layout, construction and the site operations involved. The system comprises two components. The 'estimating' component centres around the determination, by the user, of 'resource collectives'. These collectives are developed in an hierarchical tree structure which runs up from raw resources to a complete building. Resources may therefore be collected to form an operation, operations to form an element and so on. The 'regularities' invoked here are such matters as work gang constituents, material content of common constructions etc. The second system component (termed the 'modelling' package in later work by these authors) generates the quantities of building elements from the overall dimensions of the building and its constituent spaces. It requires that the building is orthogonal, working on the assumption that the building is a cellular combination of rectangular forms, the enclosing areas and junction lengths of which are fully defined, given the cell and overall building dimensions. The work is important in that element values are explicitly modelled within the system to form product based units, the link to the process being formed when functional values for the product are derived, again explicitly, from resource costs.

Bennett and Ormerod's **Simulation applied to construction projects** (ch 4.3) describes the Construction Project Simulator developed at Reading University out of Brian Fine's provocative ideas of the 70s. With hindsight, this work was undoubtedly ahead of the technology available at the time, and could be done better and certainly much cheaper today. The idea is simple enough – to randomise activity times (element values) in a construction network. The existence of parallel activities makes the resulting problem (puzzle?) too difficult to be treated analytically and so is modelled for stochastic (Monte Carlo) simulation. Once the arduous task of inputting the data to the system is completed, the system builds up time and cost histograms by repeated trials. Of course, independence between activities is assumed and all except activity times are assumed to be constants. One of the most interesting features of the system is the use of graphics to represent input and output states, not to mention the rather long run times needed. A major difficulty with the system is in testing its accuracy as it produces a **frequency distribution** of likely costs and times (it is already structurally validated to some extent by the data collection procedure).

Bowen *et al.*'s **Cost modelling: a process modelling approach** (ch 4.4) addresses what they term the 3 weaknesses of product type models – inexplicability (lack of explanatory ability), unrelatedness (assume independence) and determinism (do not accommodate uncertainty). Assuming the **purpose** of models is to help provide an explanation of **why** costs change along with design (task 3), they conclude that process models are logically necessary.

They assume an 'adequate' understanding of interdependency and relatedness between components, elements and systems is necessary. They assume that the uncertain aspects of the system will somehow interact in a way that will give significantly different results to a similar system where certainty equivalents are used (or that some measure of uncertainty of the total system is needed) and this necessarily implies the use of stochastic simulation. Systems to satisfy these assumptions are to be based on what they term 'formal modelling'. As a result they propose the use of probabilistic (for uncertainty), networked (for relatedness) process (for explicability) models to satisfy all three criteria. They propose that the overall construction plan network is broken down into sub-networks, each representing a product 'element' of the building. The combination of the sub-networks then creates a problem similar to that addressed by Russell and Choudhary (2.4). Mathematical 'nebulae' theory is suggested as a possible approach in this task and knowledge based systems (heuristic rules) are proposed to overcome difficulties caused by lack of data. The paper is a welcome attempt at a formalised criteria-based approach to model structuring. Here we see the beginnings of a framework where we can start applying some real logic. Can the necessity for simulation, for example, be proved from the basic assumptions? What classes of systems satisfy the basic assumptions? The main drawback is the woolliness over assumptions – particularly concerning **purpose** and lack of any basic assumption concerning **accuracy**. Also, the important distinction between modelling contractors' estimators **perceptions** of site activities and the 'reality' of site activities needs to be made.

Marston and Skitmore's **Automatic resource based cost-time forecasts** (ch 4.5) extends Bennett and Ormerod's CPS (ch 4.3) by introducing (1) an 'intelligent' front end in the form of an automatic planner and (2) a management control aspect to the stochastic simulation. This ambitious development is counterbalanced by the restricted area of application (housing refurbishment) to avoid the need for an infinite amount of resources to develop the system!

4.1 Cost consequences of design decisions: cost of contractors' operations*

D. T. Beeston

INTRODUCTION

A new approach to the way in which the quantity surveyor can give cost guidance to the designer is being developed within the Directorate of Quantity Surveying Development of the PSA. The technique can be used at any stage in the design process but is of considerable value at the earliest stage of design – the period during which present methods are found by many to be less than adequate.

During the early design stage of a project the designer looks to the quantity surveyor for cost guidance on alternative design proposals. It is at this stage that the designer is formulating his ideas, crystallizing them into a form which will dictate and determine much of the eventual design development of the project. This is a stage when his decisions have far reaching consequences when it is important that the quantity surveyor is able to advise him and give him the comparative cost guidance which will enable and help him to understand the cost consequences of his design decisions.

A computer program 'Costs of Contractors' Operations' (COCO) has been written which represents the decision processes of contractors' planners when they are determining the required plant, labour and construction time for a project. Operating the program provides a means of calculating comparative cost and time consequences for alternative design decisions taking into account the way these decisions determine contractors' operations and planned use of resources.

Greater understanding of contractors' planning methods and of how constructional problems affect resource requirements and consequently cost should lead to more economical design which should in turn be reflected in bid prices. Better estimates can also be made of economical construction times.

PRESENT METHODS

Techniques and achievements of present method

The main means of exercising cost control during the design stage of a project is to ensure that the estimated cost of the project falls within a cost limit.

*1973, *Report*, Directorate of Quantity Surveying Development, Property Services Agency, Department of the Environment.

Present methods of assessing such costs are based largely upon making comparisons of the elemental distribution of costs against a yardstick of historic data derived from the cost analyses of similar work. These historic data are adjusted or updated for time, location, quality and other factors which the quantity surveyor deems appropriate. This method helps to identify elemental parts of the design which appear to be more or less expensive than might be expected, when gauged against similar completed projects. By this means the necessary information can be derived either to confirm that the estimated cost of the design satisfactorily falls within set cost limits, or, alternatively to identify and direct the designer to areas where economies might be possible or should be sought.

Limitations of present method

The use of costs related to measured work as a basis for elemental estimating provides a useful means of obtaining a single estimate, of cost checking or of making broad comparisons. It is a technique which has built up over a number of years and has been largely determined by the available sources of data. Increasingly, however, the price shown against a bill item fails to represent the true work content of the item as the contractor diverts costs in various directions and in particular towards 'preliminaries'. In this way, elemental cost data may become less useful than it once was and an attempt is now being made to go some way towards solving the problem. When considering, for example, alternative designs, a change which has little apparent effect on measured quantities or rates could cause an important change in the contractors' mode of operation, on the plant that he would use, on the time that he would take and consequently upon the amount of his tender. Conversely, for reasons of size of plant and or continuity of operations, the contractor may not be able to allocate fewer resources to a design change although the measured work content is less, consequently an anticipated reduction in cost may not be fully achieved.

Present estimating techniques are rarely capable of quantifying these effects and although we may be conscious of a need to make an appropriate adjustment, in practice this is seldom done in any precise way.

The argument put forward therefore is, that based on current methods, it is possible to deduce from historic data and from geometrical considerations that one design alternative appears to be cheaper than another whereas, if the specific constructional implications were taken into account the difference might be reduced, disappear or even be reversed. Conventional estimating methods might cause the designer to sacrifice a client benefit by selecting what seems to be a cheaper design alternative while in fact due to the resource implications of that alternative, the contractor's tender may not be significantly affected.

OBJECTIVES OF THE 'COCO' TECHNIQUES

The cost consequences of design decisions particularly those which affect the choice of major plant and the speed of construction are especially likely to be inadequately estimated by current methods of cost advice. The key is realistic representation of the ways in which costs arise.

COCO is a developing technique based on this premise and is aimed at helping the quantity surveyor give improved comparative cost guidance to the designer by taking account of contractors' planning and methods of resource allocation.

When considering, for example, dimensions of concrete frame components for alternative bay dimensions, the size of crane, the different labour and form-work requirements and consequent different erection time should all be taken into account. Constraints such as building shape and site access affect plant location which must be considered in relation to the length of reach required by the crane jib and weights to be lifted, especially when heavy components are involved.

The level of detail which COCO considers is that required to ensure that the contractor can exploit the economies incorporated in the design. It is that level which enters into the contractors' planning procedure at tender stage.

CONTRACTOR'S RESOURCE ALLOCATION

Since it is the contractor who establishes the price the client has to pay by means of his tender, several major contractors were interviewed to establish how they went about preparing their tender estimates. It was found that resource allo-cation formed a fundamental part of the tendering process. A computer pro-gram was therefore built up to represent the sequence of decisions which project planners make in setting resource and time requirements.

The level of detail considered by the contractor's planner has been found to be deep in some respects but not in others. From this it is immediately possible therefore to see what is most significant about designs from the point of view of the contractor and construction. Since detail which contractors planning staff ignore will not materially affect resource allocation, we can similarly disregard such matters in giving design guidance.

The approach adopted is to simulate on the computer what in practice is done by the contractor when tender planning. This can have the effect of bringing to the design team some of the benefits of having a contractor in the team without any of the attendant disadvantages.

The validity of this new method is strengthened by contractors confirming that they follow a similar basic approach to contract planning.

THE COMPUTER PROGRAM 'COCO'

Input and processing

The computer program is interactive and poses a series of questions required to obtain data from the operator. By this means the computer is able to thread its way through the various branches of a decision tree structure. A small part of the flow chart is reproduced at Appendix A.

It requests various building dimensions from which it eliminates certain plant choices, then by evaluating constraints such as height and reach, weight to be lifted etc it finds the size and type of crane necessary to satisfy these conditions. The program then asks for data about the structure – number of columns, beams, cladding units, quantity of formwork, concrete, bricks etc. It also requests the unit cost (supply and delivery only) of these components. From this and the application of contractors' productivity constants built into the program it can evaluate a basic workload based on crane lifts and labour for fixing etc.

Certain allowances are made to take account of wind, learning curves, density of working area, high rise factors and other matters of complexity. With this additional information the total crane cycle time is calculated in relation to the workload.

If the load is excessive the program will consider means of reduction such as pumping of concrete, use of hoists, or use of additional cranes until a solution is found which makes optimal use of contractors' resources. In some cases constraints may be such that an optimal solution cannot be found and the user will be asked to answer questions in an attempt to find the most economic outcome.

Typical questions which arise during a straightforward run appear on the example at Appendix C and a simplified outline of the main program steps is shown at Appendix B.

Output

The output gives details of the crane type/size, the construction time for the elements considered, the labour and material resources required with their costs, the number of formwork uses possible, the percentage crane utilisation and finally a comparative cost for the work covered is stated. Typical results appear at the end of the print-out (Appendix C).

A separate data file of productivity (or output) constants is maintained for each contractor interviewed and output can be obtained for any number of contractors specified. The program will recycle as required using each contractors data in turn to arrive at his particular resource allocations etc. The source and content of each data file is not divulged to users.

A summary of output (single contractor) is shown at Appendix D. This shows the results obtained for a range of possible design alternatives. The examples are necessarily simple for illustrative purposes and the financial differences between alternatives are relatively small in this case, however, there is a saving of almost 10% between the dearest and cheapest alternative.

It is assumed that each alternative will attract similar overhead and profit etc additions and this may therefore be ignored in making comparisons. Time dependent site costs should however be taken into account when comparing results showing different construction times.

Using the program

The program is written in Fortran IV and is run on a time sharing terminal connected to a PDP10 computer system. The response time is very rapid so that no delay after input is usually discernible. In its present stage of development, initial input of data is completed in about 20 minutes. To investigate the cost consequences of choosing design alternatives it is often necessary to change only a few inputs to the program and since only the new values need be input, each succeeding run is dealt with more quickly. These rapid repeat runs give both the quantity surveyor and the designer the opportunity to try many more alternatives than hitherto and to build up patterns or trends. The latter are particularly useful if firm information is not available in the first instance in which case inputting a range of values will allow the user to gauge the sensitivity of the output.

Some of the input data which the program seeks is in a form unfamiliar to quantity surveyors – eg weights of components, 'delivered to site but unfixed' material costs of precast components, made-up formwork etc. The User Manual (DQSD No 2.01) includes an Appendix which gives general guidance on design matters which are undefined at early cost advice stage. It gives rule of thumb methods and tables to enable dimensioning and costing of components – eg assessing the dimensions and cost of a beam if only span and loading are known. The Manual explains aspects of the questions and their interpretation and gives guidance on methods of assessing the dimensions or quantities called for by the program – eg it considers such matters as siting of cranes and the size of shutters that the program calls 'large'. In addition the manual is a complete guide to all aspects of using the program.

FURTHER DEVELOPMENT

The present program (MK I) covers the basic superstructure of a reinforced concrete framed or loadbearing brick building of three or more storeys having a total contract value of about £150000. This includes the frame, cladding, floors, roof, brick/blockwork, partitions, windows etc. The Mark II version of the program with incorporate minor improvements and add excavations and foundations and the means of dealing with substructures and basements. This is currently under final test and will be available shortly.

Further areas for development are steel frames (now under way) and mechanical and electrical engineering aspects. Ideally the whole of the building process would be covered with due allowance being made for interaction between the various elements. Opportunities will be taken to widen the scope of contractor representation and to maintain adequate liaison with contractors to ensure that the information upon which the program is based will continue to keep abreast of current methods and techniques.

It is believed that the growing use of this program will enable the Q.S. to give designers better and more appropriate information to enable selection of suitable design alternatives and in so doing help to ensure that the client receives better value for money.

APPENDIX A

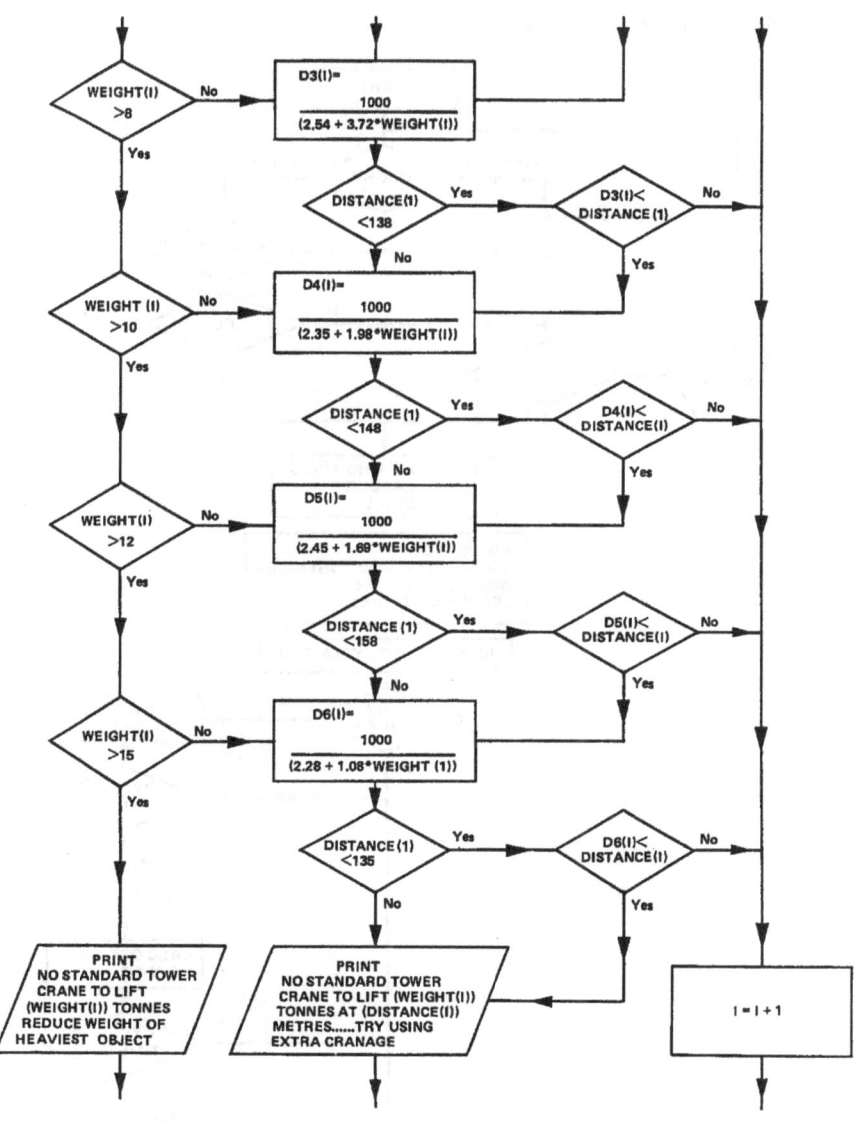

PART OF 'COCO' FLOW CHART (MARK 1)

APPENDIX B

COCO: Greatly simplified flow chart

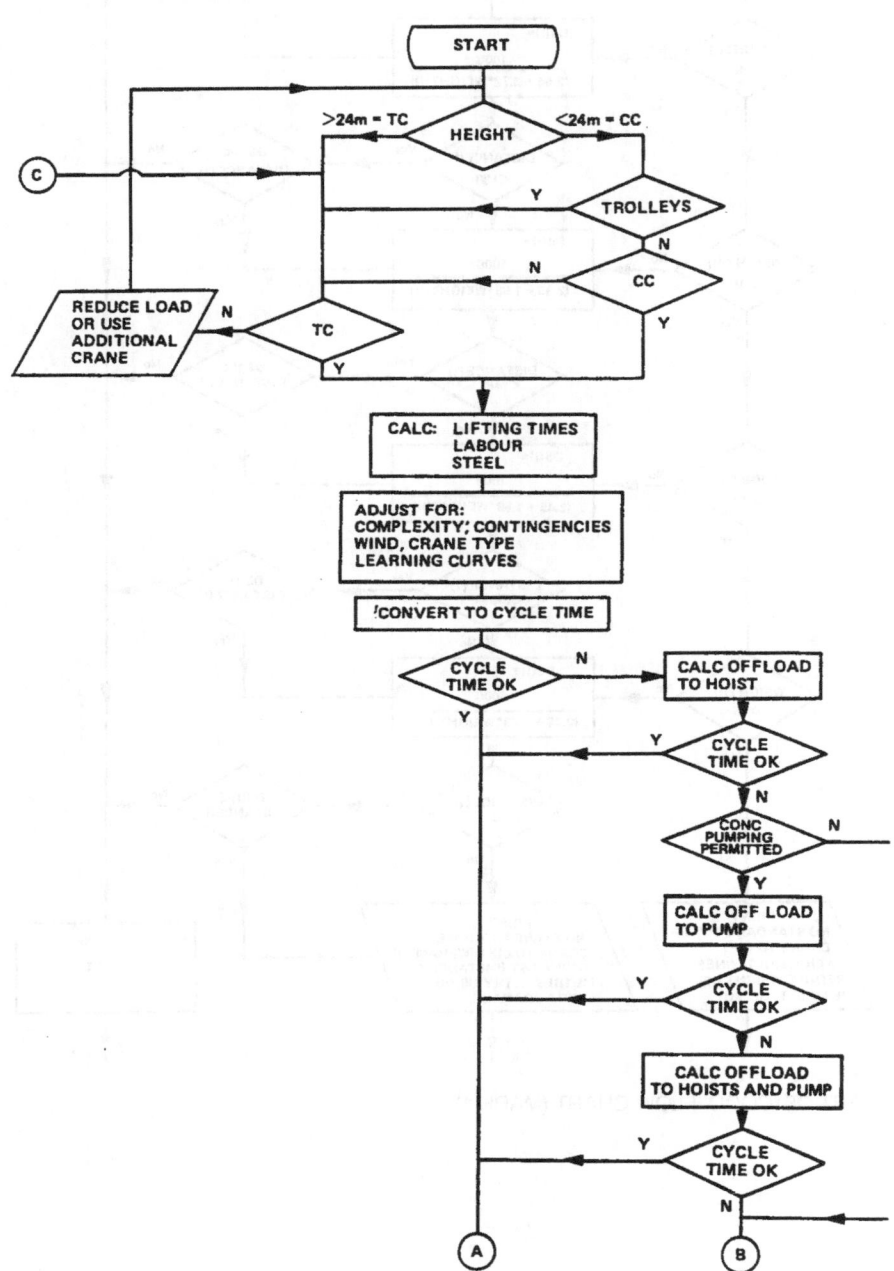

APPENDIX B (CONTD)

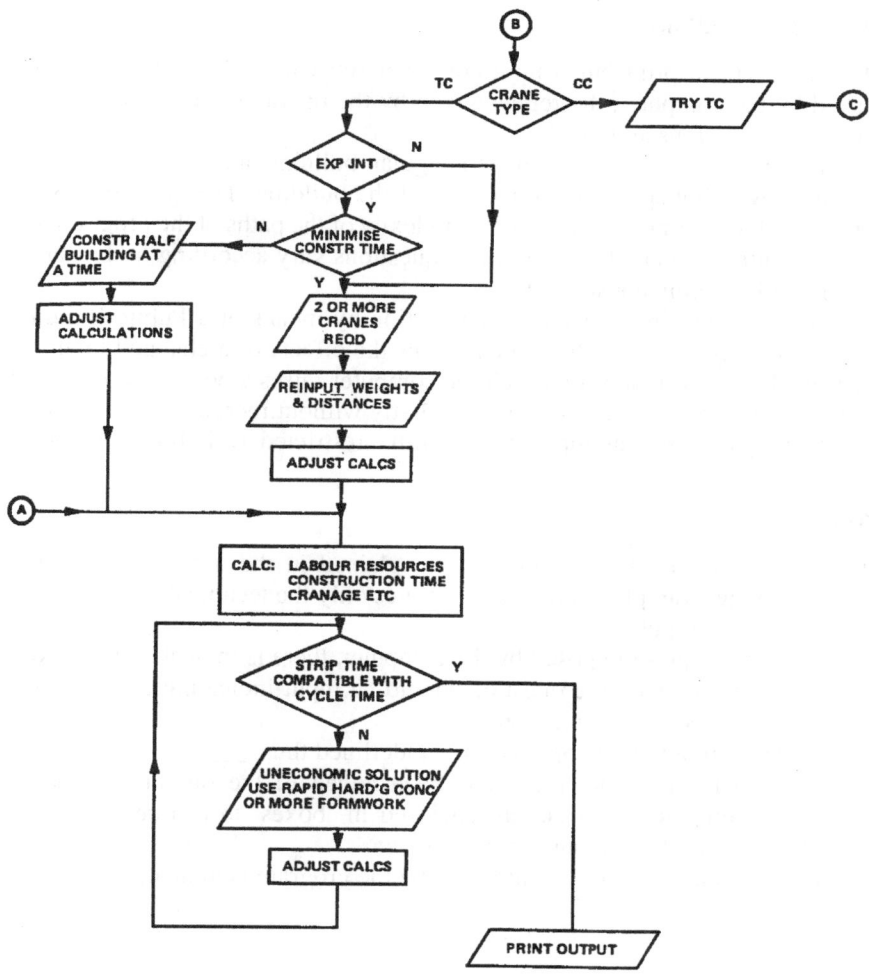

APPENDIX C

Typical 'COCO' print out

This is a typical print out from a computer run which shows the questions asked by the computer, the replies given by the operator and the subsequent computer decisions and results.

The project is a four storey rectangular building and a site constraint excludes working space along one face of the building. The questions asked are typical for a simple project, the complexity of the paths of the program that the computer uses and the consequent questions vary according to the nature of the project being considered.

This example first considers construction by means of a Public Building Frame (mostly pre-cast). It then examines the effects of a changed input for cast-in-situ construction for which the computer raises a possible problem of labour recruitment. This is considered firstly without recruitment restrictions and then by changing an input answer with a restricted availability of labour.

Key:

1   Login and logout procedures are explained in the users' manual. The following example, therefore, illustrates only the technical content of the input and output.
2   After each question posed by the computer, there is indicated in brackets the form in which the question should be input indicating the units and format (punctuation) required.
3   Answers input by the operator are underlined thus _____
4   A typewritten commentary has been added to the output by way of explanatory notes. These are enclosed in 'boxes' to differentiate them clearly from the computer print-out.
5   Final output of resources allocated by the program is indicated thus: ‖

99. RESULTS REQUIRED FOR 1, 2, 3, 4 OR 5 CONTRACTORS ◄────── Question asked by Computer
1 ◄

1. FLOORS EXCLUDING BASEMENTS (NR) ◄
4

> Answer input by user. (In a normal run, more than one contractor would be selected)

2. MAXIMUM HEIGHT OF BUILDING (M)
11

> Next question from computer

3. GROSS FLOOR AREA (M2)
2495

5. CAN TROLLEY/TABLE FORMWORK BE CONSIDERED
NO

QUESTIONS 6–10 REFER TO A CRAWLER CRANE

6. PERMISSIBLE WORKING SPACE (M)
10

7. HEAVIEST OBJECT TO BE LIFTED (TONNE)
4

8. MAXIMUM DISTANCE OF THIS OBJECT FROM EDGE OF BUILDING (M)
10

9. OBJECT FURTHEST FROM CRANE (TONNE)
2

10. DISTANCE OF THIS OBJECT FROM EDGE OF BUILDING (M)
13.5

> From answers to Questions 6–10 the computer calculates that a crawler crane would not be suitable. It next tries a tower crane.

QUESTIONS 11–14 REFER TO 1 TOWER CRANE

11. HEAVIEST OBJECT TO BE LIFTED (TONNE)
4

12. MAXIMUM DISTANCE OF THIS OBJECT FROM CRANE (M)
22 ◄
> User decides where crane should be sited.

> From these answers the computer provisionally selects a tower crane to suit required loads and distances.
> The next questions (30 onwards) ask for data to calculate its work-load.

13. OBJECT FURTHEST FROM CRANE (TONNE)
2

14. DISTANCE OF THIS OBJECT FROM CRANE (M)
27

> NOTE: Questions in the range 15–29 may arise in complex cases which need, for example, hoists or more than one crane.

30. PRECAST COLUMNS (TF:NR/COST EACH)
4:42/12 ◄
> This input format means there are 4 floors, each with 42 columns, each costing £12 delivered to site but not fixed.

31. PRECAST BEAMS (TF:NR/COST EACH)
4:24/22, 16/30

> The computer automatically checks that the number of floors in each 'per floor' input equals the number input in answer to Question 1. The number of staircase flights is normally one less than the number of floors; the top floor (with zero quantity) must therefore be entered before the computer will proceed.

32. PRECAST STAIR/LANDING UNITS (TF:NR/COST EACH)
3:6/75
1:0 ◄

33. PRECAST FLOOR UNITS (TF:NR/COST EACH)
4:200/18

34. PRECAST CLADDING UNITS (TF:NR/COST EACH)
4:27/40

35. PRECAST CROSS WALL UNITS (TF:NR/COST EACH)
0

> NOTE: All costs are supply and deliver to site only. Costs arising from labour and time fixing/erecting etc are automatically derived from constants held within the computer.

36. SPECIAL UNITS (TF:NR/COST EACH)
0

37. POURED CONCRETE IN WALLS, BEAMS AND COLUMNS (TF:M3/COST PER M3)
4:31/8.5

38. POURED CONCRETE IN FLOOR SLAB (TF:M3/COST PER M3)
4:31/8.5

39. SMALL SHUTTERS FOR WALL FORMWORK (IGNORING REUSES) (TF:NR/M2)
0

40. LARGE SHUTTERS FOR WALL FORMWORK (IGNORING REUSES) (TF:NR/M2)
0

> Definitions, assumed by the program for 'small' and 'large' shutters are explained in the users' manual.

44. TRADITIONAL BEAM AND SOFFIT FORMWORK (TF:M2/COST PER M2)
4:108/3.5

45. TRADITIONAL WALL AND COLUMN FORMWORK (TF:M2/COST PER M2)
4:244/4

46. STAIRCASE FORMWORK (TF: M2/COST PER M2)
0

47. BRICKS (TF:NR/COST PER 1000)
0

48. CONCRETE ETC. BLOCKS (TF:NR/COST PER 1000)
4:1100/65

49. PARTITION AND WINDOW UNITS (TF:NR/COST EACH)
4:92/55, 27/40 ◄

> This input means there are 4 floors, each with 92 partition units at £55 each and 27 windows at £40 each.

50. IS THERE BRICK/BLOCKWORK IN (1) CLADDING GENERALLY (2) BACKING TO PANELS OR (3) NEITHER (1, 2 OR 3)
2

> Backing blockwork assumed to concrete cladding panels.

51. AREA OF BUILDING REQUIRING SCAFFOLDING (M2)
400

> Scaffolding assumed for cast-in-situ and walls only.

61. ROOF HOUSINGS (NR/COST EACH)
1/500

62. TYPE OF ROOF: (1) FELT OR ASPHALT ON CONCRETE OR (2) METAL DECKING (1 OR 2, M2/COST PER M2)
1, 585/3

64. CRANE TRACK FOR 1 TOWER CRANE(S) (M)
18

66. VERTICAL CIRCULATION AREAS (NR)
2

67. FORMWORK STRIPPING TIME (DAYS)
14

> Output for 1 contractor for PB frame building

CONTRACTOR 1

TIME TO WATERTIGHT   28 WEEKS

LABOUR RESOURCES: 4 JOINERS
                    3 GENERAL LABOURERS
                    2 STEELFIXERS
                    2 SUPERVISORS
2/1 GANG OF BRICKLAYERS FOR   6 WEEKS

TYPE OF CRANE(S):
TOWER CRANE MAXIMUM CAPACITY 10 TONNES E.G. RECORD 764F

COSTS: LABOUR   8980.
        1 CRANE(S)   11634.   (UTILISATION 75%)
        FORMWORK   622.   (USES 4)
        MATERIALS   54562.
        ROOFING SUBCONTRACT   1755.
        SCAFFOLDING   556.

COST OF SUPERSTRUCTURE   78109.

> This answer enables an alternative design to be selected for comparison. The computer then asks which questions to change.

DO YOU WISH TO CHANGE ANY ANSWERS
YES

TYPE IN QUESTION NUMBERS TO BE CHANGED SEPARATED BY RETURNS
FINISHING WITH AN EXTRA RETURN
7
8
9
11
12
13
30
31
32
33
37
38
44
45
46
51

These are the questions whose answers must be changed if our next alternative for this example is cast-in-situ construction.

QUESTIONS 6–10 REFER TO A CRAWLER CRANE

A crawler crane is tried for the revised inputs.

7. HEAVIEST OBJECT TO BE LIFTED (TONNE)
2.25

8. MAXIMUM DISTANCE OF THIS OBJECT FROM EDGE OF BUILDING (M)
13.5

9. OBJECT FURTHEST FROM CRANE (TONNE)
2.25

QUESTIONS 11–14 REFER TO 1 TOWER CRANE

A crawler crane is again found unsuitable so a tower crane is tried and provisionally chosen.

11. HEAVIEST OBJECT TO BE LIFTED (TONNE)
2.25

12. MAXIMUM DISTANCE OF THIS OBJECT FROM CRANE (M)
27

13. OBJECT FURTHEST FROM CRANE (TONNE)
2.25

30. PRECAST COLUMNS (TF:NR/COST EACH)
0

31. PRECAST BEAMS (TF:NR/COST EACH)
0

32. PRECAST STAIR/LANDING UNITS (TF:NR/COST EACH)
0

33. PRECAST FLOOR UNITS (TF:NR/COST EACH)
0

37. POURED CONCRETE IN WALLS, BEAMS AND COLUMNS (TF:M3/COST PER M3)
4:94/8.5 ◀

This includes poured concrete to in-situ stairs.

38. POURED CONCRETE IN FLOOR SLAB (TF:M3/COST PER M3)
4:83/8.5

44. TRADITIONAL BEAM AND SOFFIT FORMWORK (TF:M2/COST PER M2)
4:949/3.5

45. TRADITIONAL WALL AND COLUMN FORMWORK (TF:M2/COST PER M2)
4:380/4

46. STAIRCASE FORMWORK (TF: M2/COST PER M2)
3:28/4.5
1:0

51. AREA OF BUILDING REQUIRING SCAFFOLDING (M2)
1400

Scaffolding assumed this time will be necessary for the whole face of the building.

69. WILL IT BE POSSIBLE TO RECRUIT 13 JOINERS FOR 24 WEEKS
YES

> This answer assumes that
> 13 joiners could be
> recruited.

> Output for 1 contractor for
> cast-in-situ building
> assuming 13 joiners
> available.

> Question 69 is not in the basic
> input data form.
> It should now be recorded as an
> 'additional question'.

CONTRACTOR 1

TIME TO WATERTIGHT   26 WEEKS

LABOUR RESOURCES: 13 JOINERS
                             4 GENERAL LABOURERS
                             5 STEELFIXERS
                             4 SUPERVISORS
2/1 GANG OF BRICKLAYERS FOR   6 WEEKS

TYPE OF CRANE(S):
TOWER CRANE MAXIMUM CAPACITY 8 TONNES E.G. RECORD 646E

COSTS: LABOUR   19150.
        1 CRANE(S)   8654.   (UTILISATION 62%)
        FORMWORK   3796.   (USES 4)
        MATERIALS   38432.
        ROOFING SUBCONTRACT   1755.
        SCAFFOLDING   1798.

COST OF SUPERSTRUCTURE   73585.

> If labour problems are
> likely, Question 69 can
> be recalled for
> changing.

DO YOU WISH TO CHANGE ANY ANSWERS
YES

TYPE IN QUESTION NUMBERS TO BE CHANGED SEPARATED BY RETURNS
FINISHING WITH AN EXTRA RETURN
69

69. WILL IT BE POSSIBLE TO RECRUIT 13 JOINERS FOR 24 WEEKS
NO

> According to the labour problems
> expected, the user should assess
> how many joiners would be
> available.

70. MAXIMUM NUMBER OF JOINERS (NR)
9

> Output for 1 contractor for
> cast-in-situ building with
> restricted labour available.

              CONTRACTOR 1

TIME TO WATERTIGHT   32 WEEKS
LABOUR RESOURCES: 9 JOINERS
                             3 GENERAL LABOURERS
                             4 STEELFIXERS
                             3 SUPERVISORS
2/1 GANG OF BRICKLAYERS FOR   6 WEEKS

TYPE OF CRANE(S):
TOWER CRANE MAXIMUM CAPACITY 8 TONNES E.G. RECORD 646E

COSTS: LABOUR   17584.
        1 CRANE(S)   10454.   (UTILISATION 52%)
        FORMWORK   3720.   (USES 4)
        MATERIALS   38432.
        ROOFING SUBCONTRACT   1755.
        SCAFFOLDING   2241.

> It is instructive to
> compare the three
> outputs with special
> reference to:
> TIME to watertight
> LABOUR (numbers and costs)
> CRANE (Type, size
> cost and
> percentage
> utilisation)
> FORMWORK and
> SCAFFOLDING

COST OF SUPERSTRUCTURE   74186.

DO YOU WISH TO CHANGE ANY ANSWERS
NO

> This answer terminates
> the run.

CPU TIME: 4.60 ELAPSED TIME: 32:28.80
NO EXECUTION ERRORS DETECTED

EXIT

'Coco' – Comparison of detailed outputs for one contractor and alternative designs

DESIGN VARIANTS – REF NO		1	2	3	4	5	6	7	8	9
PLAN SHAPE AND HEIGHT		RECTANGULAR – 4-STOREY					SQUARE 4-STOREY		RECTANGULAR 7-STOREY 'TOWER'	
TYPE OF CONSTRUCTION		PB FRAME				CAST IN SITU	PB FRAME	CAST IN SITU	PB FRAME	CAST IN SITU
MATERIAL ALTERNATIVE		–	–	–	SMALL CLADDING PANELS	–	–	–	–	–
RESOURCES	CRANE ACCESS	ONE SIDE	–	FIXED	as 1	as 1	FIXED	FIXED	FIXED	FIXED
TIME TO WATERTIGHT – WEEKS		28	28	28	29	26	28	27	26	25
LABOUR – JOINERS		4	4	4	4	13	5	13	6	15
GEN LABOURERS		3	3	3	3	4	3	4	3	4
STEEL FIXERS		2	2	2	2	5	2	5	2	5
SUPERVISORS		2	2	2	2	4	2	4	2	4
2/1 GANG BRICKLAYER – WEEKS		6	6	6	6	6	6	6	6	6
CRANE TYPE		10 T Tower	30 T Crawler	10 T Tower	10 T Tower	8 T Tower	10 T Tower	10 T Tower	8 T Tower	8 T Tower
COSTS – LABOUR £		8,980	8,980	8,980	9,312	19,150	9,781	19,944	9,794	19,772
CRANE(S) £		11,634	5,030	13,155	12,015	8,654	13,155	12,775	9,200	8,900
FORMWORK £		622	622	622	622	3,796	676	3,920	966	4,477
MATERIALS £		54,562	54,562	54,562	55,242	38,432	53,422	36,804	57,488	40,419
ROOF SUB/CONTRACT £		1,755	1,755	1,755	1,755	1,755	1,710	1,710	1,053	1,0530
HOIST(S)		–	3,650	–	–	–	–	–	–	–
SCAFFOLDING £		556	556	556	2,020	1,789	695	1,604	939	2,224
TOTAL SUPERSTRUCTURE (ADD OVERHEADS)		78,108	75,154	79,629	80,965	73,585	79,439	76,756	79,440	76,844
CRANE UTILISATION		75%	75%	75%	75%	62%	75%	58%	75%	56%
FORMWORK USES		4	4	4	4	4	4	4	4	4

4.2 General purpose cost modelling*

L. G. Holes and R. Thomas

INTRODUCTION

Conventional building cost predictions are based on calculations whereby one or more variable quantities, which are derived from facts known about the project at the time, are multiplied by suitable money rates, representing predictions of facts, as yet uncertain.

As the designing proceeds and more decisions are made, a greater variety of variables becomes available for incorporation into the calculations. Hence the use of, firstly, a functional unit rate, then a rate per square metre of gross floor area or various element unit rates and, ultimately, the unit price rates entered against the quantities of measured items in the bill of quantities. These last are intended to cover the cost of the resources likely to be used during the construction of each item, and rates used at earlier stages will probably be based on combinations of such items from other projects. Thus, it would seem that all data on money rates implicitly represents, or 'models', the consumption of materials and components and the work of operatives with plant, as well as the expenditure that will motivate these events.

Our approach is to relate building measurements to likely resource consumption rates, and then to apply unit resource price rates that reflect appropriate market conditions; project and general overheads are separately considered. That is to say, we concentrate on the need for resources and regard cost as largely derivative.

Obviously, a great deal of data is involved, and our purpose in this paper is to outline how a computer can be used to deal with it. Essentially, we set out to exploit the regularities to be found in the individual facts about buildings and their construction and also in the relations between them. Whatever the stage in the design process, we use much the same resource data and the same computer programs, storing within the computer the portable data that represent these regularities.

*1982, in Brandon P. S. (ed.) *Building Cost Techniques – New Directions*, E & F N Spon, 220–7.

COMPUTER SYSTEMS

There are two complementary computer-aided systems:

1   a system originally developed for general contractors' estimators and in use, as such, commercially. This can be used for cost modelling either on its own, provided some approximations of measurements are acceptable, or in conjunction with
2   a system for measurement and the production of a bill of quantities, which can also be used on its own.

Both computer systems are content-free; that is, they provide a structure for project data, whatever it is, which the computer will relate and process under the control of the user. Data in frequent use can be stored and recalled using simple references. The systems are set up by creating computer files of such data, representing the kinds of activities that are being modelled. Project input data is prepared, using natural language for the descriptors and making reference to entries on the files; the computer amplifies this initial input by copying repeated data and by recalling data from the files. This amplified input is processed in accordance with programmed instructions and directions included in the data.

RESOURCE AND COST ESTIMATING

The Construction Industry Resource and Cost Estimating system that we call 'CIRCE' concentrates on relating item quantities to resources, treating the resource cost rates as simple facts which can be stored and changed at will. Its principal characteristic is that it will combine data on basic resources in ways that match the ways in which the resources themselves will actually be combined on the site; for example, to constitute a workgang, to make concrete, to construct a concrete foundation or even the whole of a foundation wall. These we refer to as 'collectives of resources'. The computer simply responds to the data in the input, which will consist of the details of basic resources and the specifications of the relations between these, the collectives and the item sizes and quantities. We have found that a ratio of no more than four constituents to one collective works well in practice. Since each constituent can, itself, be a collective, it is possible to relate an item quantity to up to four, or sixteen, or sixty-four, or whatever number of basic resources is appropriate. Each reference to a resource or collective can be associated with at least two (and up to five) numerical factors. The ability of the computer to store and recall data and to implode and explode it as required presents both a challenge and an opportunity, and CIRCE has already been used in a variety of ways.

REGULARITIES

To start with, one must consider what regularities there are in the facts about buildings and building processes and their relationships, and seek to reflect these in project data.

Such regularities can include the following:

1   the occupation of operatives and plant employed in the enterprise and the materials and components they habitually use,
2   collectives of operatives and plant when working on particular processes (we use a file with about one hundred such collectives in order that all labour hours and costs can be allocated to production cost control centres),
3   collectives of materials such as mortars and formwork,
4   collectives of the foregoing, representing resource requirements apportioned to each measured unit of a bill item,
5   collectives of (4) that will have the same item quantities; here it becomes necessary to consider what regularities there are in the buildings themselves.

As a result of the brief and the thought and calculations of the designer, the first quantitative information on the building will probably be in the form of location or sketch drawings. These will simultaneously indicate both the spaces to be provided for the occupiers and the physical structure that will enclose them.

At the risk of being obvious, it must be said that, generally, there will be horizontal constructions above and below the rooms, etc., and a vertical structure between one and the other and next to the environment. Also, that the junctions between these constructions will probably be at right-angles; that is, the building will be orthogonal. Thus, the various rooms and other 'cells' nest within the building envelope rather like a box of boxes.

We know that the finishes to the ceilings, walls and floors of the cells are made from different materials and that, although there are likely to be skirtings, and sometimes cornices, there are seldom any extra treatments at the junctions between the finishes to adjoining walls. Thus, for most projects, we shall be concerned with the surfaces at the top, at the sides, and at the bottom of each cell, and also with the junctions between the top and the sides, and between the sides and the bottom.

These three kinds of surface and two kinds of junction are also present in the building envelope in the form of the roof, the external walls, the ground floor, the eaves or verge, and the foundations to the external walls. The constructions at such surfaces have two variable dimensions and are likely to be regarded intuitively at areas. In the case of junctions, only their lengths vary, whilst the storey height is likely to be constant for each floor. These variable dimensions are, essentially, also those of the spaces they help to enclose.

Here, we begin to diverge from both CI/SfB (1) and the Standard Form of Cost Analysis (SFCA) (2). Although in the former, elements are regarded as being 'parts with particular functions', in practice, they appear to act more like classifiers of items such as those which are implied or specified by name in equivalent sections of the SFCA. However, if constructions at the surfaces of spaces are regarded as the elements proper, they become more easily recognised as entities with functions. This allows junctions to be similarly distinguished, their role being to maintain the continuity of functions between neighbouring elements.

SPACE-RELATED DIMENSIONS

In view of the foregoing, we hypothesise that, in the case of orthogonal buildings, we should be able to generate the areas of elements and the lengths of junctions by copying their scalar dimensions (or 'scalars') from those of the spaces. More-over, we should be able to generate the dimensions of measured items, repre-senting the parts of elements and junctions, by modifying such copied dimensions if necessary, as is done manually. The computer could then generate the girth of the internal dividing walls on any one floor, as this will be, in effect:

*   the sum of the lengths and widths of the cells,
                  minus
*   the length and width of the building envelope,
                  plus
*   (n − 1) □ the thickness of the internal walls
    (where n = the number of cells).

We also hypothesise that doors, windows and other openings are analogous, in that these are a combination of elements proper and junctions with other elements.

If the plan shapes of cells and the building envelopes were always simple rectangles, the dimensions could be copied directly from those of the spaces, the areas of ceilings, roofs, and floors being the product of the lengths and widths, and the girths of walls being twice the sums of lengths and widths. By regarding complex orthogonal plan shapes as consisting of simple rectangular components, and applying the same rules, the total areas can be correctly obtained, but the sum of individual girths will exceed the actual perimeter by twice the dimensions of the 'inner boundaries' between the component spaces.

One apparent solution to this difficulty is to regard compensatory spaces of, initially, zero thickness as lying between component spaces. These are given, a timesing factor of −1.00 as their initial purpose will be to remove the dimen-sions of these inner boundaries (as they have no physical existence) whilst having no effect on the areas.

When all the scalars, including the ones with negative timesing, are modified by the addition or subtraction of, say, a wall thickness, in order to obtain actual dimensions, these 'negative scalars' are found to continue their compensatory role, maintaining the correctness of both the areas and the perimeters. It would seem that 'negative spaces' can be in any plane, and the techniques has been developed so that the computer can generate the measurements of individual cuboids.

PROCEDURE AT SCHEME DESIGN STAGE

At this stage, only the size and arrangement of the spaces, and the general quality of the building will have been settled. Because the details of the technical solutions which will meet the functional requirements of the elements have yet to be decided, these can only be represented in some way, and related to the areas of elements and lengths of junctions, generated as described above.

An interesting possibility is to relate the various quantities to data on the cheapest possible technical solutions, and to include factors to represent the ratios between the likely cost of work of an acceptable quality and these minima. Such ratios are similar to the 'cost/worth ratios' used in Value Engineering procedures. Suitable data models can be held on computer files, and can include those of installations and external works. Where the file prices of resources have become outdated, current ones can be included in the job input.

All this can be carried out on the CIRCE systems alone, although the modification of scalars to allow for overlaps at corners, etc. is not possible. Even so, being a single system for use within a single organisation, it could also be a satisfactory way of preparing tenders for design-and-build contracts, particularly as budgets would be computed for production management.

However, if the scalars are to be modified, it is necessary to use the bill production system to compute the quantities. In such cases, the scalars could be retained for use later.

PROCEDURE AFTER THE DESIGN IS COMPLETE

The final stage of designing is to make decisions on the technical solutions, that is, on the parts that will constitute the elements and junctions, including their sizes, the materials from which they will be made, and any constraints on the quality of the result.

There are other regularities here, as wall leaves, screeds and other continuous parts of elements with two space-related dimensions will only need decisions on their thicknesses, whereas joists and other skeleton parts, and fascias, architraves and other parts of junctions will need widths and either thicknesses or heights.

The computer-aided system we are about to outline provides for the copying of dimensions from those of spaces, and, if necessary, their subsequent modification by the addition or deduction of such widths or thicknesses, so that the actual dimensions of the planes in each of the parts are generated. No attempt is made to aggregate into girths, as the computer can deal with practically any number of individual dimension sets. Indeed, there are advantages in this naive approach, as each set can be individually retrieved for operational purposes.

This is made possible by two more regularities as follows:

1   the three Cartesian (scalar) dimensions of objects and spaces indicate how much there is each. We also need to be able to state how many there are of each size,
2   in the case of parts that are constructed *in-situ*, their quantities will be the product of the number and one or more dimensions.

These regularities are given effect by entering the timesing, the three dimensions and the other facts in rows in a straightforward way and by indicating to the computer which of the dimensions are to be used in calculating the quantity. Where regularities exist between spaces and parts of elements or junctions, the scalar dimensions of the spaces are entered first, and followed by the items data, but with instructions to copy and modify particular columns of the preceding scalars. In the case of regular technical solutions, these copying instructions will also be regular (e.g. a floor finish will always require lengths and widths). Both these and the quantification indicator can be attached to their part descriptors in the computer file, and recalled with the part name in project data, so that no knowledge of this aspect of the Method of Measurement would be required by the user.

We have found it beneficial to make use of the regularities of elements and junctions in order to schedule the data on technical solutions in a methodical way. This schedule is regarded as the source document and the various references to be used in the input are added to it. Given a stock of descriptors, etc., we do not see why a computer should not be programmed to produce these schedules during an orderly detail design procedure and, at the same time, generate much of its own input for bill production.

BILL PRODUCTION

We have hypothesised that the items in a bill of quantities are sorted chiefly on the basis of the materials being used, even though they are arranged in worksections which may be called by the names of elements or operative occupations. In our bill production system, the descriptors of these materials are placed in separate lists and numbered sequentially to indicate to the computer where their respective items are to be placed, thus giving general control over the arrangement of the output.

The computer places the items in the conventional order of 'cubes, supers, runs and numbers', by reference to the quantification indicator, and uses the widths and thicknesses to provide almost every one with a unique position.

During the first process, the computer amplifies the input by copying descriptors and items from its files. The dimensions are then copied and modified and an 'abstract' is prepared, consisting of the items in bill order, with individual dimension sets, quantities and retrieval references. This is retained in the computer and a bill of quantities is produced by copying selected portions. Where the quantity of the material is different from the item quantity, this can be calculated by the computer under a set of controls that make use of brick sizes, etc.

Within each worksection, items are arranged under headings, and the information required by estimators when dealing with an item is contained either in the item description or in the first heading above it.

Partial bills are possible, and the dimension sets can be re-expressed to suit operational needs by redefining the component spaces from which they will be copied when the program is re-run.

ESTIMATING

Either the unit rate or the operational estimating approach is possible. Quantities from the bill are input to the CIRCE system, with references to basic resources and collectives, appropriate factors and updating details. An operation is regarded as a collective of resources, the costs of which will be apportioned to one or more bill items. In the output, item unit rates are analysed and totalled and the quantities and costs of resources are given.

Further developments are to be expected in the CIRCE system as the resources analysis can be arranged to suit the need for production information.

CIRCE seems particularly suited to the larger micro-computers and has also run on main-frame machines. A scaled-down version is available for use on a PET.

REFERENCES

1 Ray-Jones, Alan and Clegg, David. *CI/SfB Construction indexing manual.* London: Royal Institute of British Architects Publications, Ltd., 1976.
2 Building Cost Information Service. *Standard form of cost analysis.* London: The Royal Institution of Chartered Surveyors, 1969.

4.3 Simulation applied to construction projects*

J. Bennett and R. N. Ormerod

ABSTRACT

This paper describes the uncertain environment in which construction activity occurs and in particular the variability in productivity and the occurrence of external interferences. The paper then describes a suite of computer programs known as the Construction Project Simulator (CPS) whose features include: a hierarchy of linked bar charts, 'preliminaries schedule', weather data, direct costs, and resources for costing and/or resource restraint. The results of an actual project are presented to demonstrate the type and range of output available, including duration and cost prediction and cash flow curves. The results of four varied construction projects subjected to stimulation are presented, with the results demonstrating an improvement in accuracy over the common deterministic estimating procedures of the UK construction industry.

Keywords:

Uncertainty, risk, simulation, interference, variability

INTRODUCTION

This paper describes the results of a research project supported by the UK Science and Engineering Research Council under the Specially Promoted Programme in Construction Management. The research project culminated in the development of a suite of computer programs for a common business micro-computer known as the *Construction Projects Simulator* (CPS). The suite of programs is based upon the theoretical and practical hypothesis that much of construction activity is dominated by the uncertain environment in which it takes place. This general uncertainty is subdivided into two specific elements, namely, productivity variability and external interferences to the

*1984, *Construction Management and Economics*, 2, 225–63.

construction process on site. Data describing a particular project – bar charts, direct and indirect costs, resources, weather, productivity data – is input and fed to a series of stochastic simulation programs which produce cost and time predictions for the project. The data input is designed to be simple and suitable for computer illiterates.

The simulator was validated against four actual construction projects, drawn from typical market sectors in the UK building market, and the results of one are used to illustrate the type and style of the output available. The results of all the case studies are presented and the simulator is shown to provide better predictions than normal methods.

THE CONSTRUCTION PROBLEM

A construction project presents a unique problem to those involved in managing the construction process. Each project is different from all others, and must be carried out at a different location each time. The project must be formulated and executed by integrating the efforts of a large number of different organizations and individuals, all of whom have different and often conflicting priorities and objectives (Feiler, 1972). The manager of the process must consider and assess different technologies and alternative combinations of labour and equipment. While the manager knows what must be done he has considerable latitude in how it is to be executed. Furthermore, he must consider the effects of imponderables such as weather, material shortages, labour problems, unknown subsurface conditions, and inaccurate estimates of duration and cost. All these considerations combine to form a complex, dynamic problem (Riggs, 1979).

The construction manager must assess this dynamic problem in the context of future events and performance. Estimates and predictions must be made which attempt to forecast the future. Inevitably the forecasted value will deviate from the actual outcome, due to a lack of complete information about future events.

Another major effect common in construction projects is that the initial definition of the project is changed in scope due to modifications in the basic plan to incorporate changes. Quite often it is these unknown factors which create surprises for the client who commissions the construction project (Traylor et al., 1978).

All managers face a certain amount of unknown possibilities, which can affect the process they are attempting to control. However, the construction manager is likely to have to confront and resolve problems of a type and magnitude not found in other industries. Every manager engaged in construction, at whatever level, encounters these problems in the course of every working day. The problems are readily identifiable and are accepted in the industry at large – a few typical examples are listed below (Lichtenberg, 1981).

the actions of external agencies (e.g. government)
unknown or unassessed elements of the work to be performed
errors and omissions in working drawings
delays in obtaining management approvals
the effects of the weather on construction operations
late delivery of purchased material and equipment
unknown results of testing and commissioning complex services systems
unknown rates of inflation
variable performance work rates
mechanical breakdown or malfunction of equipment
rejection of poor quality work and re-work

All these factors and the complexity of the construction problem have long been recognized. This recognition led to the establishment of the Project Management discipline and the adoption in recent years of computerized, critical-path-network-based systems to provide assistance in planning, budgeting, scheduling and controlling projects. Yet, the availability of such sophisticated management tools and the establishment of a construction management profession has not generally provided the hoped-for improvement in project performance. Total project cost and time overruns are still commonly reported and clearly the use of sophisticated management techniques has not eliminated project estimating or performance problems. In view of this track record the client, who must use the facility provided by the construction process, has become increasingly critical of the ability of construction managers and has come to view the industry with mistrust – as highlighted by the recent British Property Federation proposals (British Property Federation, 1983).

A major reason for this perception has been the many changes forced on projects by the external, uncontrollable forces described before. However, this perception has been unwittingly nurtured by the tendency to characterize projects by single-value measures – single cost, single duration – which give illusions of certainty. The economist Ken Boulding, has said, 'an important source of bad decisions are illusions of certainty'. It is clear that uncertainty is endemic in construction and needs to be explicitly recognized by construction managers.

SIMULATION

As well as an awareness of uncertainty a manager requires a management tool to use in quantifying and assessing the impact this uncertainty may have on the project. The manager must be aware of the effects uncertainty may have so that he may make a rational decision as to what level of risk to accept in the light of the circumstances at the time. A management tool incorporating uncertainty will not make decision-making easier, but it will present more and better information, than currently available, on which to base a decision and therefore arguably improve the possibility of a good decision.

The techniques which is employed to form a management tool in an uncertain environment, to augment the decision maker's intuition and experience, is *simulation*. Simulation is also referred to as the Monte Carlo technique – due to the gambling aspect of the process – or a stochastic technique – due to the presence of random processes. Stochastic simulation typically generates durations and costs for each activity in a plan by randomly calculating a feasible value for each from a statistical probability frequency distribution – which represents the range and pattern of possible outcomes for an activity. To ensure that the chosen values are representative of the pattern of possible outcomes, a large number of repetitive deterministic calculations – known as iterations – are made. The result is typically presented as a cumulative distribution plot and a frequency histogram.

Due to the repetitiveness of the process and the handling of large amounts of numerical information it has only been feasible to implement this technique since the advent of computers. The current practices of industry, in using single-value, deterministic methods are a legacy from the era before computing liberated people from laborious calculations. Also it allows a deeper investigation of problems – such as construction – which do not have a single-value solution that can be represented by a set formula, nor operate in a totally random environment which can be represented by statistics, but has a limited random component (i.e. stochastic) which can be investigated most effectively through simulation.

It has been claimed that the introduction of simulation methods for construction management is likely to have as great an impact on the construction industry as did the introduction of network planning and scheduling methods some two decades ago (Nunnally, 1981). Some of the advantages claimed for the technique are summarized below.

1   The major advantage is that the results of such a simulation, given the validity of input assumptions, provide an unbiased estimate of the project completion distribution. This is particularly important in the light of evidence of inherent bias in deterministic network techniques.
2   Simulation provides an almost unlimited capacity to model construction operations, and they permit the construction manager to quickly evaluate many different combinations of equipment and methods under varying conditions of operation at modern cost.
3   Simulation can give the manager an insight into which factors are important – and hence where to concentrate his effort – and how they interact.
4   Simulation allows the user to experiment with different strategies without the risk of disturbing the real project and incurring costs. Also simulation enables one to study dynamic systems in much less time than is needed to study the actual system.

5   Additionally, if a person can interact with the computer simulation in a gaming environment, experience can be gained under realistic conditions before the work is started. This should lead to better management through a deeper knowledge of the problem.

6   Finally, and most importantly, simulation models often predict things which are not specifically incorporated into the model. Simulation of repetitive processes has shown that when uncertainty exists there are large penalties rather than benefits of scale.

There is some evidence that in the construction industry it is the larger projects which go wrong most frequently. Also, most work-study experts talk about the benefits of specialization. Simulation shows that for large projects subject to uncertainty there are penalties of specialization. Further, most models of construction processes assume that the cost of a project is the sum of the costs of the activities. Simulations of repetitive processes show that costs are largely generated by the uncertainties that exist, and that simple additive models like the Bill of Quantities seriously under-estimate cost. Finally, the traditional approach to construction would expect financial benefit to accrue from productivity increases and would direct effort towards increasing the speed of production. Simulations incorporating uncertainty direct attention towards obtaining benefits from reductions in interferences. The benefits to be gained by these reductions in interference can be shown to be very much larger in magnitude than gains made possible by productivity increases (Fine, 1982)

The only disadvantage of simulation techniques is that they are time-consuming and expensive in computer time, due to the requirement to perform many iterations of the same calculation. This has been true in the past when the only computers with the required capacity were large main-frame computers with high operating costs. Presently, however, due to the pace of technological development relatively cheap micro-computers with sufficient capacity are readily available. The objection that the process is time-consuming is also overcome by allowing a fully automatic operation and carrying out the processing at times when the micro-computer would not normally be used (i.e. at lunchtime or at night). Looking into the near future, if even some of the promises of the 'fifth generation' computers are realized then even these drawbacks will be nullified.

The use of computer simulations for construction applications is a recent development dating from the early 1970s. However, in this time nearly 30 different simulation models have been developed. All have their inherent strengths and weaknesses and they were reviewed during the course of the research. The objective in developing a new computer program was to overcome many of the shortcomings of previous techniques. Specifically the CPS includes processes to model the two components of uncertainty: variability and interference, it operates on a relatively cheap, easily available microcomputer, it has an increased capacity over earlier models in being able to accommodate more facilities such as more activities, costs, weather, resources and preliminaries in one package; and above all it is a model which is simple to operate, being designed for use in industry by managers who are not computer literate.

UNCERTAINTY

The many contributing factors to the construction problem – some of which are outlined above – are referred to collectively as **uncertainty**. An underlying hypothesis is that this global uncertainty can be subdivided into two major components, namely, **variability** in the performance of a task, and **interference** from outside the task which frustrates its progress. This categorization is supported by evidence from the direct observation of the construction process and by evidence from other construction academics (Bishop, 1966; Crandall, 1977; Woolery and Crandall, 1983).

When making decisions managers cannot be certain of good outcomes, because they cannot completely control external events or have total foresight. However, management can try to increase the probability of good outcomes by taking into account all the major factors. In particular where there is uncertainty as to which events might occur, the logic of the decision process should include that information.

There is a basic need to be able to quantify and assess the impact that uncertainty can have on a project, and to incorporate this knowledge in the project brief and the management of the project. This expanded awareness of the project gives executive management and the client a more complete view of the project and a basis for decision making. It provides a much higher quality of information on which to base a decision, and allows an assessment of both uncertainty and risk while incorporating the manager's own value judgement into the decision. The inclusion of uncertainty further supports the efforts of management in concentrating on the essential elements of the project. If construction project prediction and performance are to be improved, uncertainty and its sources must be dealt with seriously and specifically. Not as a final rudimentary 'contingency item' (whether explicit or not), but detailed systematically as an integrated part of the management process (Lichtenberg, 1981).

VARIABILITY

Introduction and definition

The use in the simulator model of variability as a component of uncertainty is based on the factual observation of the construction process, as the following discussion will illustrate. However, neglect of this variable is still proposed by texts of 'classical' network analysis, with such statements as 'only an occasional construction project will have variances in activity durations' (Harris, 1982). While observation of construction operation leads to the conclusion that it is pervasive and very large (Hall and Stevens, 1982). The importance of its inclusion in any realistic model of construction is illustrated by the fact that when only one activity was allowed to be variable (by □45%) in an otherwise deterministic 28 activity network, the result was an increase in project duration of three days from the deterministic result of 76 days (+3.9%) (Feiler, 1976).

Variability can be defined as the range and frequency distribution of possible durations in the execution of a particular task.

This arises as performances are subject to wide variation since in any given discipline there is a wide range of capabilities. Different skill levels, degrees of motivation and fatigue, and other behavioural factors will affect worker performance and produce the normal variations found in practice. In addition, certain construction tasks are inherently uncertain due to the nature of the material, e.g. excavation with variable ground conditions or the refurbishment of boiler linings (Feiler, 1972). Also in practice an appreciable and uncertain amount of repeat work is carried out due to poor workmanship or design, and substandard materials.

It is important to realize that the variability referred to here is specific to a particular task. Thus in gathering data on variability it is necessary to compare like with like. The quantity and quality of work carried out, and the method of execution should be identical or at least very similar. Thus, comparing the data for the input to two-and three-storey houses with different areas is misleading; while it is proper to compare the variability for the same task for each of these projects. Similarly it would be misleading to compare the same task using different methods of execution, e.g. with placing concrete by tower crane or by hand and barrow. The degree of variability for the same task with different quantities may well be of the same magnitude, but the density function will be about a different mean. The objection to different quantities of work could be overcome by expressing the duration for a unit of work, e.g. square metres of half-brick walling. One further step which would aid the standardization of data and the ease with which data from different sources could be amalgamated, would be to express the data in a unit of time which reflects the resource input. The best measure would be man hours, as this would not require a knowledge of gang sizes. The shape and range of a distribution would be the same whatever units were used, but the adoption of man hours would allow the production of distributions where the data is drawn from more than one source. This will have the effect of increasing the size of the data sample and hence the reliability of the distribution.

It is also important to consider only the effort expended in completing the task at the workplace and not to account for any time when the activity was suspended through inclement weather, lack of information, etc. There is considerable evidence that, in building work typically, many separate visits are needed to complete a task (McLeish, 1981, Roderick, 1977); and so it is important to only account for the time actually spent on the task. The data used should include the following: all work done, setting out and checking, preparation, handling materials, and nonproductive time at the workplace.

Learning and experience effects

It is also worth briefly discussing learning curves, as their effect is often assumed to affect variability. A learning effect is observable while an unfamiliar task is mastered, and an experience effect is present through the gradual improvement of experience gained while repeating the familiar task. The categories are not mutually exclusive and the effects overlap.

The learning effect is apparent in the building industry and studies of site operations note the improvement in productivity during the first few repetitions of a task (Clapp, 1965, McLeish, 1981). Once the task is mastered a further improvement is observable for the same gang under controlled conditions. However, although this effect is theoretically possible it is not common as a number of factors relevant to the construction process negate its effect.

Firstly, in building, the nature of the product imposes certain limits on the learning effect, as the workers and not the product have to be moved from one area to another (Relf, 1974). Also the unfavourable influence of interruptions to work render the problems of coordination and continuity of work more difficult to solve in construction than in other production work (United Nations, 1965). Continuity of work is essential, as any interruption to the process dramatically reduces any improvement. This is illustrated in Fig. 4.3.1. From actual observation of construction operations it is apparent that interruptions, for whatever reason, are frequent (Bishop, 1966, McLeish, 1981, Roderick, 1977). Interruptions are mostly caused by external influences and can completely neutralize any advantage gained through repetition (United Nations, 1965).

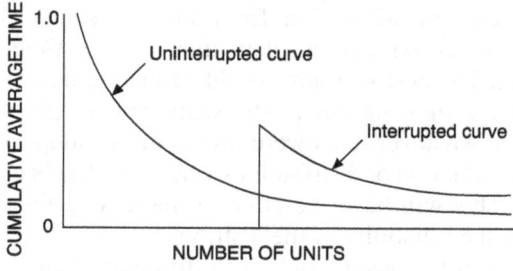

Figure 4.3.1 Cumulative experience curve with interruptions.

The series of repetitions in the construction cycle must be large enough to sustain the experience effect (United Nations, 1965). Also the operations carried out have to be identical, as even small differences can seriously affect improvements (United Nations, 1965). Further, any experience effect is lost when members of the gang change. Changing gang members is the rule rather than the exception in construction where the labour turnover is high.

In practice the combined effect of all those factors usually result in a severe limitation of the learning effect. Improvements are seen on the first few repetitions of a project, but there is much evidence to show that due to adverse pressures continuing improvement is not carried through the rest of the work (McLeish, 1981). However, where circumstances are better controlled or more favourable then the learning effect can be experienced throughout the project (United Nations, 1965). Even in these circumstances variations in output are still the norm.

In the context of productivity variability the learning effect is apparent normally for the first few repetitions of a project, and should be accounted for by modelling these earlier activities with an appropriately increased duration. However, it is likely that in all but the most advantageous circumstances this learning curve is not continued into the rest of the project, due to the influence of the factors discussed before. It is for this reason that a further learning effect, after the first few repetitions, can be discounted as contributing to the usual production variation in the construction industry.

SOURCES OF DATA

Any attempt to quantify variability and differentiate between different activities is plagued by a dearth of information. All the published sources draw on data collected by the Building Research Establishment (BRE) and are for traditional house building only. As well as a lack of information, the accuracy of the data is also questionable. Some studies include time related to interruptions in the variability measure. The source of earlier data is often from site records and its accuracy dubious. The most accurate data was collected by direct observation, with the BRE site activity analysis package (Stevens, 1969) whose accuracy can be calculated (see Table 4.3.1) although rarely specified, but is often of the order □5 to 35%.

The size of the sample also has a bearing on the accuracy; and the largest sample used is 52, but is more often in the range 20 to 30. This is exacerbated by the need to compare tasks on identical buildings, which are limited to one site. An improvement in sample size could be achieved by using a common unit of expression for the same task on different sites. It is also necessary to limit the observations to reasonably precise tasks (e.g. secondary work packages), as the variability of an operation (e.g. brickwork superstructure) is greater than that of the whole (e.g. house completions) (Clapp, 1965). Comparison of results is also complicated by different methods of

Table 4.3.1 Accuracy of activity sampling observations

Estimated manhours from observations	95% confidence limits
10 000	□2%
1000	□6%
100	□20%
10	□64%
5	□87%

presentation. The best method would be a density function, as this has an influence on which duration is chosen on each iteration in the model, but three methods are used in the literature and are given for all data, where available: the range expressed as a ratio of the largest data value to the smallest; the range expressed as a □ percentage about the centre; and the coefficient of variation, i.e. the standard deviation, expressed as a percentage of the mean.

Several studies have drawn on the data collected by the BRE, but these are limited to traditional house building on three separate sites. Limiting the discussion to those studies which conform with the definition of variability it is possible to draw tentative conclusions, although considerably more research is needed in this area.

A typical histogram representing the variability in output for bricklayers is presented in Fig. 4.3.2. From this source and several others relating to

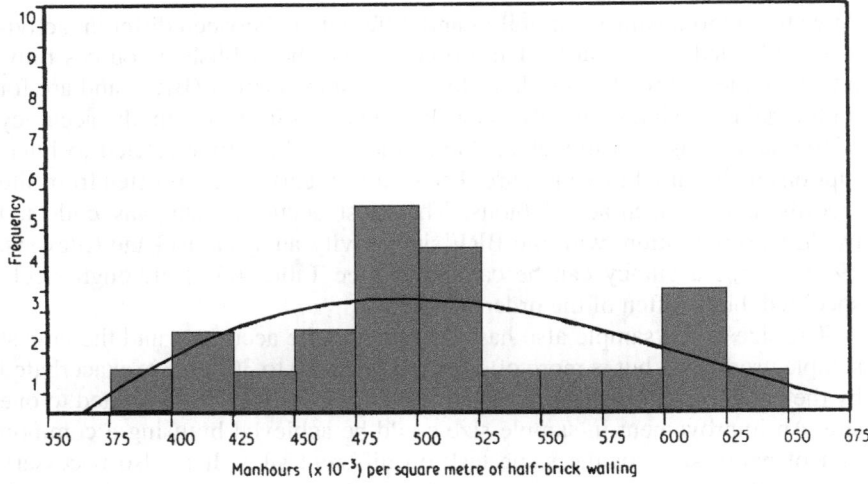

Manhours (x 10⁻³) per square metre of half-brick walling

Figure 4.3.2 Brickwork superstructure (half-brick walls only), variability of output for all gangs, one site (McLeish, 1981). Sample size = 22; range = 1.6:1 (□24%); coefficient of variability = 14%.

brickwork it is apparent that brickwork has a variability in the order of □30%, with a coefficient of variability of 15%. This is considerably lower than the other trades and is possibly more reliable being based on a larger sample size.

Brickwork superstructure was the only operational activity which exhibited any significant difference from the norm in the pattern of performance when the standard *T*-test was applied.

To explain this difference it has been suggested that historically bricklaying has always been organized into relatively large and relatively independent operations, with lower non-productive times. Support for this view, with particular regard to housing, can be found in the housing project used as a case study. In this case the only operation in the sub-structure primary work package (PWP) which had planned, and actual, continuity of operation was the foundation brickwork to damp proof course (DPC) level. This arrangement prevented the earliest start on subsequent operations of the first housing blocks, i.e. if the brickwork item had been discontinuous in the early stages then a start on following operations could have been made sooner. In the superstructure PWP the dominance of brickwork required that the sequence of locational areas for work to progress through was changed from those operations before and after the brickwork. This change was necessary to give continuity of work to bricklayers at a constant gang size. For all other building operations it is only possible to state that the range is in the order of □50 to 60%, with a coefficient of variability from 30 to 40%.

Data from industry

The previous section describes the currently available best sources of data on variability of building work, but the search for relevant data was extended to organizations with links to industry and to commercial contracting companies.

Many organizations were approached and none within the UK were able to supply data in a relevant format. However, two large progressive contracting organizations were interviewed as to the state of their productivity data base. In both cases the results were similar. Extensive libraries of productivity output rates exist for very detailed tasks. These are used for planning and bonus targetting. Where the source of the data is recorded it is based upon work study exercises carried out when this technique was in favour, in the 1960s and early 1970s. Limited exercises have been carried out since then, but mainly to update data for operations involving increased mechanical aids (e.g. concrete pumping) or on sites where progress was slow as part of other measures to improve productivity.

However, to use the raw data in a deterministic commercial environment, the variability of the data has been reduced by the use of work study relaxation allowances for such things as the conditions of work, and the physical effort expended (i.e. a subjective judgement of individual productivity against some standard). Further treatment of the data, by the exclusion of certain 'non-productive' times which contribute to variability (e.g. idle/inefficient, early tea/lunch, talking to supervisor), by the omission of extreme values, and the averaging of the remaining data values to produce a single figure as an output rate, render the libraries of information unsuitable for use with the simulator. Sufficient raw data was collected during the work study exercises to form the basis of suitable frequency distributions, but this raw data has been lost or dispersed through the company to an extent that any re-analysis is practically frustrated.

The discussions with industry tended to confirm the existence of variability, but did not provide any better data than that described above. The data obtained from published sources was therefore used in subsequent simulations of live projects.

INTERFERENCE

Introduction and definition

That interference exists in the construction process, as external influences on the progress of site works which cause stops to production, is readily apparent to all practitioners in the building industry. It stems from a wide variety of sources such as: the structure and work of the industry, acts of God, or human and social factors (Lichtenberg, 1981), risks outside the contractor's control (Ahuja, 1982), integrating the effort of a number of different organizations and individuals, many not under direct control (Feiler, 1976), legal (Woolery and Crandall, 1983) and environmental factors exhibiting random, seasonal or periodic behaviour.

These give rise to specific events which prevent work starting or continuing. The most frequently mentioned are: the weather, planning permission and building codes, lack of design details, non-availability of materials, non-attendance by subcontractors, labour strikes and shortages, equipment breakdown, and theft and vandalism. Occurrences of interferences on any one project are to be found in abundance.

The recognition of interferences as a separate category of uncertainty is an important step for practical as well as theoretical application. By focusing attention on the external influences it redefines the strategic and tactical management of a project to encompass the control of external events and the mitigation of their effects. The concentration of management effort on reducing the impact of interferences may have a greater beneficial effect than the control of production variability, which is essentially a site, operational, management task.

Classification of interferences

Previous attention in this area has been limited. However, a few studies have attempted a subdivision of the general interference class.

A classification by source defines: interruptions from those not directly involved in the building process (government departments, planning procedures, the general public, etc.); interruptions from resources (labour, equipment and materials); and interruptions from within the building team (Tavistock Institute, 1966). Another classification has been between interferences due to nature and 'acts of God', and other human and social interferences (Lichtenberg, 1981). Fig. 4.3.3 illustrates the effect these interferences have over and above the productivity variability.

A further classification is between those risks within, and those outside, a contractor's control (Ahuja, 1982). This source is from the USA and includes factors such as design which are traditionally outside a contractor's control in the UK.

Previous work in this area has thus been sparse, and a need exists to further refine and classify the components of interference, to gain a wider acceptance of the concept and to enable measurement of the magnitude of each element. It is postulated that the general classification of interference is made up of four major components, each of which affects to some degree the progress and cost of actual construction activities. These components are: interferences originating from the design process, interferences originating from the procurement process, interferences from the weather and those other external human and social factors remaining.

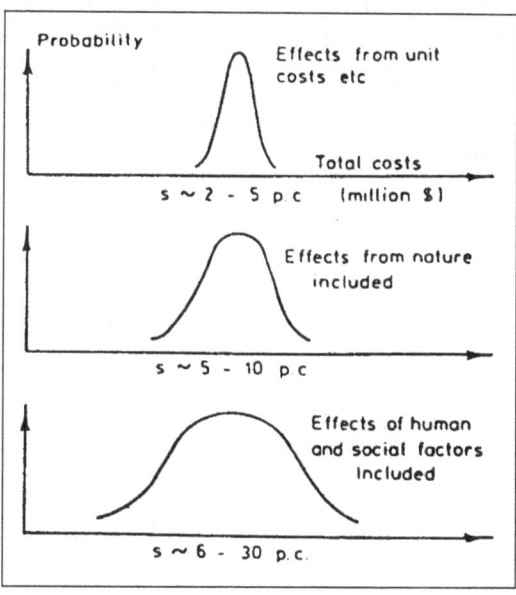

Figure 4.3.3 Illustrating the effect of interferences from various sources on task uncertainty (Lichtenberg, 1981).

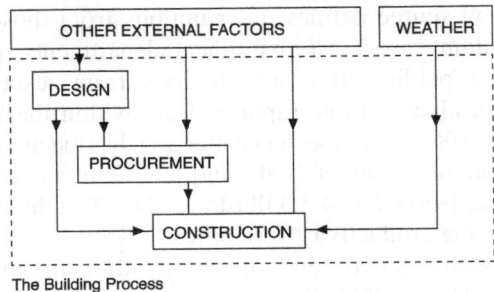

The Building Process

Figure 4.3.4 The components of interference and their interaction.

This classification is illustrated in Fig. 4.3.4. One type of interference which is sometimes encountered is termed 'internal interference' and results from a conflict between on-going operations. It can be explained with reference to a line of balance diagram, illustrated in Fig. 4.3.5, where due to a difference in productivity one gang prevents another from working through not completing their task and not vacating the locational work area. This is internal interference, as opposed to an external interference where work stops for other reasons. This internal interference in a project is automatically taken into account by the simulation process and it is important, particularly for data collection, not to include any amount for this in values used. Its automatic inclusion is generated through the choice of different durations through the influence of variability and the logical connections between activities – this leads to a different gradient in the line-of-balance for each operation each iteration and produces internal interferences.

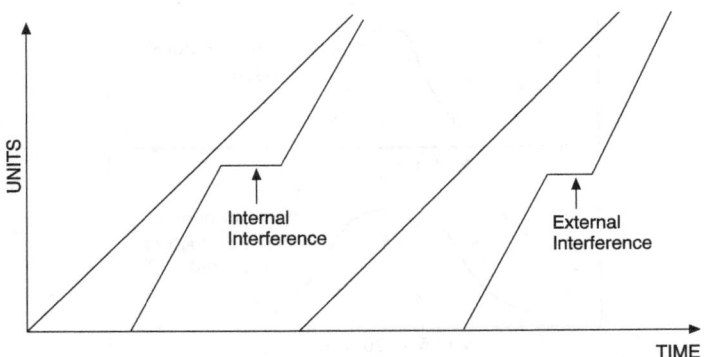

Figure 4.3.5 Internal and external interference illustrated with a line-of-balance diagram.

Interference data

Though interferences are readily recognized and accepted in industry and academia, the present research project and the previous one (Bennett and Fine, 1980) were the first to identify it as a distinct and separate classification within the general uncertainty of the building industry. For this reason, and the general lack of data collection practices by industry and research bodies, after an extensive search no data in an entirely suitable form has been identified.

Some information on the frequency with which different interferences occur could easily be recorded. This, however, would not give any indication as to the duration of the interferences. It is anticipated one day that reliable data may become available, and for this reason a facility has been provided in the simulator for the entry of a histogram plotting the relative frequency and duration of interferences experienced by any operation.

Data on the frequency with which different interferences occurred on the projects forming the case studies were collected from formal reports, where available. These, however, may not be comprehensive in that every interference which occurred may not have been recorded. It is also probable that the type of interference recorded has been influenced by the contractual responsibilities of a project. For example, it is clear that the extent of sub-contract non-attendance or labour supply problems has not been accurately recorded for those subcontractors appointed by the contractor, and for whom he has a contractual responsibility. However, the interferences from subcontractors appointed by the professional team, by nomination, and for whom the contractor can contractually claim time and cost extensions to the contract, have been more diligently recorded. With these reservations borne in mind, Table 4.3.2 lists the identified sources of interferences and their relative frequency of occurrence for the case studies.

One further technique which is applicable to industry has also been identified. This is the use of Foreman-Delay Surveys (Rogge and Tucker, 1982) which were originated in the USA, but which have been used at least once in the UK. The delay forms are completed by trade foremen at the same time they complete their daily time sheets. Work study exercises carried out in conjunction with the delay surveys indicate that after an initial 'settling down' period there is a good correlation between the results of a formal work study and the results of the delay survey. An important point stressed in the practical application of this technique is the active involvement of the foremen completing the forms. This is achieved by using the results of the surveys as the regular basis of two-way communication between the workforce and management. This brings delays to light, for the action of management, and ensures that foremen are not penalized for the causes of delay thus helping to ensure their accurate reporting. This method of highlighting delays and the focusing of management effort on them consistently resulted in a reduction in time lost due to delays when the method was adopted. Productivity improvement is accomplished primarily through the identification of specific

Table 4.3.2 Frequency of formally reported interferences on the case studies as a percentage of the total reported interferences

Interferences		Case studies			
Source	Description	7[a]	8	9	10
Design	Late of inaccurate design information, from any source	44%	47.8%	58%
	Additional work through design changes or unforeseen work	26%	21.5%	16%
Procurement	Subcontract supply problems of labour and materials	29%	21.5%	20%
Other	Rework through accident or bad workmanship (an element of variability)	NA	9.2%	6%
	Other unspecified	1%	NA	NA

[a]No data available

Table 4.3.3 Typical foreman-delay survey values, site averages (Tucker, Rogge and Hendrickson, 1982)

Foreman-delay survey category percentage (1)	Average, as a percentage (2)	Maximum, as a percentage (3)
Design rework	4.4	12.6
Construction rework	1.6	5.4
Pre-fabrication rework	1.2	6.2
Total rework	6.1[a]	17.1
Crew interference	0.9	5.8
Waiting for construction equipment	0.8	2.5
Waiting for materials	0.7	6.8
Moving to new work site	0.6	1.4
Waiting for information	0.3	1.9
Waiting for tools	0.2	1.3
Crowded work areas	0.2	1.6
Other delays	1.4	–
Total delays (non-rework)	4.9	20.5

[a]This average includes all sites, some of which did not subdivide rework categories. Therefore, this value is not equal to the sum of separate rework category averages.

problems causing delay and the resulting corrective action which is taken. This management of interferences, on the two sites studied in the USA, resulted in a general increase in performance as foremen reported a reduction in delays (Rogge and Tucker, 1982).

Some results from this technique are available from eight industrial construction sites in the USA. These are included to give an impression of the type and quality of data which could be expected using this method to quantify interferences, the specific values are not directly applicable as they refer to a different country with a different method of managing building construction. Average and maximum weekly values for each category of delay for all eight sites are presented in Table 4.3.3 in order of decreasing severity.

Table 4.3.4 presents similar information for six commonly occurring crafts at these sites. Foreman-delay surveys performed on the above industrial construction projects reported delays from near 0% to more than 20% of the working week. Values varied with size of project and craft mix. Reported delays generally increased as the size of the manual work force increased (Tucker, Rogge and Hendrickson, 1982). Given the absence of suitable data part of the research became calculating allowances for interference in the case studies which modelled reality. The way in which this was done and the results are described later in this paper.

Table 4.3.4 Typical foreman-delay survey values for common crafts in the USA, as a percentage of the working week lost (Tucker, Rogge and Hendrickson, 1982)

Foreman-delay survey category	Carpenter		Electrician		Ironworker		Labourer		Pipefitter	
	Average	Maximum	Average	Maximum	Average	Maximum	Average	Maximum	Average	Maximum
(1)	(2)	(3)	(4)	(5)	(6)	(7)	(8)	(9)	(10)	(11)
Design rework	0.4	3.2	3.3	17.2	3.2	14.8	0.5	7.7	8.3	23.0
Pre-fabrication rework	0.2	3.8	0.0	0.1	0.9	9.2	0.0	0.0	3.2	11.7
Construction rework	0.8	7.8	1.4	10.4	1.7	6.5	−0.1	0.5	2.9	12.6
Total rework	1.9[a]	11.3	4.7[a]	23.8	3.1[a]	14.4	0.5[a]	7.8	13.5[a]	38.5
Waiting for materials	1.4	10.9	0.3	1.1	0.9	4.9	0.0	1.0	1.1	8.7
Waiting for tools	0.1	0.6	0.1	0.4	0.3	9.3	0.1	0.5	0.5	5.3
Waiting for construction equipment	1.4	10.9	0.3	1.1	1.7	6.6	0.4	3.5	1.5	5.3
Waiting for information	0.3	5.9	0.2	3.8	0.1	0.5	0.1	1.0	0.7	6.3
Crew interference	0.1	1.5	0.6	3.8	0.4	1.8	0.0	0.8	1.8	9.9
Crowded work areas	0.1	0.6	0.5	5.4	0.1	0.3	0.0	0.3	0.4	8.8
Move to other work areas	1.0	11.5	0.5	2.3	0.9	2.7	0.3	1.3	1.0	5.1
Other delays	0.7	–	0.5	–	1.7	–	0.3	–	2.2	–
Total delays	5.1	46.1	3.0	11.0	6.1	63.5	1.2	4.4	9.2	26.0

[a]This average includes all sites, some of which did not subdivide rework categories. Therefore, this value is not equal to the sum of separate rework category averages.

COMPUTER PROGRAM

The development of the computer program known as the *Construction Project Simulator* (CPS) was undertaken with the aid of a research grant from the Science and Engineering Research Council and is based on the ideas and theories outlined above.

The software philosophy has been to make the operation of the programs as easy and simple as possible. This was accomplished by the extensive use of computer graphics, the inclusion of many checks for errors and error messages, through minimizing the use of the keyboard, and the inclusion where practicable of 'help' screens to prompt the user in the correct operating procedure. Also the structure of the programs is based on a hierarchical approach to allow ease of use and present a choice of levels for answers to increasingly complex questions, as schemes and strategies for a project develop.

The suite of programs is driven by a menu where all program operation must return to before another function can be chosen. The suite of programs are basically configured into two different categories; the series of programs allowing the entry of data describing a project, and two programs where this data is used to perform simulations and produce the end result as output.

The program have been coded in compiled Basic, and operate on a 512 Kbyte ACT Sirius I micro-computer with twin floppy disc drives providing 1.2 Mbyte of data storage.

DATA INPUT PROGRAMS

The following programs and facilities require input from the user and all provide data in one form or another to the simulation programs described later. Some programs can be used only after another program has been successfully completed and saved on disc, as they draw on information produced earlier – program operation automatically checks for these conditions.

Commencing a project

Two floppy discs are required, one containing the CPS suite of programs and one data disc.

The programs are then initiated by entering the three characters 'cps'. This starts an automatic process which loads the graphics software and checks the data disc for existing data. If the data disc is blank a program is entered which allows the entry of the project title, start date, anticipated end date, and a time unit to be used in the primary bar chart. If the data disc already has project data present then the menu programs is run – through which the user moves from one program to another.

Bar charts

The bar chart programs represent the heart of the CPS data input routine. The bar charts are arranged as a hierarchy, which is reflected in the structure of most other programs, and also models the way effective managers plan. The first bar chart to be presented is the primary level bar chart, in which the user defines separate activities known as primary work packages (PWPs). Each one of these PWPs may subsequently be 'exploded', or magnified, to define a secondary level bar chart which allows the constituents of the PWP to be planned in detail. Within the secondary level bar chart the user defines separate activities known as secondary work packages (SWPs).

The primary bar chart is presented from the data entered previously – the start and end dates, and the time unit (weeks or months) – and a calendar of dates and time unit numbers are calculated and presented on the screen along with the project title. The time unit and the length of the project defines how large the steps are between cursor movements on the screen – as the cursor moves horizontally in steps or jumps of one time unit. Thus activity durations have to be in whole time unit lengths. All of this process is carried out automatically and the user is presented with a titled, dated and scaled blank bar chart screen, ready for the entry of PWPs.

The maximum number of PWP bars permitted is 39, however, only 15 are displayed at one time on the screen, but the screen 'scrolls' up or down, one bar at a time, so that all bars may be viewed. The user enters a bar by typing its title in the title panel, and by defining the start and end positions, using one function key and two direction keys, the bar is automatically presented. Only one bar per line is permitted. Once a bar has been set up the start and end dates may be changed, and the bar redrawn in the desired position. Changes to bar charts can thus be rapidly executed. Bars may be inserted or deleted from the bar chart, and bar titles may be retyped.

The PWPs are interconnected by the device of logical links, entered by the user and drawn on the screen as thin lines. The maximum number of links is 250. The links act as logical restraints to progress and are entered by setting the cursor at the desired position on one bar, pressing one function key and moving the cursor, by four direction keys (or one day by the use of a mouse!), to the desired position on another bar and pressing the same function key again. Certain strict rules govern link arrangements, as this is the basis of the logic employed during simulation. Basically all links must pass down the screen when drawn, and are sorted by 'rescheduling' so that at least one link between any two bars is vertical and none slope from right to left. Links may be erased by placing the cursor on their origin and pressing one function key on the key board.

The combination of rapid data entry, graphic display, and the ease with which changes can be made provides a powerful tool to investigate the effect of logic or duration changes, as at the throw of a switch the logical effect of any change is made – much as the non-graphical and untimescaled critical path analysis performs. An example of a primary bar chart is illustrated in Fig. 4.3.6.

Holiday periods may be entered and deleted in much the same fashion as bars are entered. The maximum number of holiday periods permitted is nine. As the cursor moves in time unit steps the duration of a holiday period must be a whole number of time units.

After a session of data input to the bar chart, whether initial input or as the result of some change to the plan, the user must press the 'reschedule' function key before he is permitted to save the bar chart details permanently on disc. The reschedule checks, and if necessary rearranges, the logical links to ensure the plan is logically correct. The requirement to always reschedule before the save function key is enabled is to ensure that all plans contained on disc are correct ones.

Once a correct primary bar chart has been created and saved each PWP of this bar chart can be expanded to form the basis for a secondary level bar chart, when the more detailed constituent SWPs can be entered by the user.

The secondary level bar chart program takes the PWP title and calendar (and week number) start and end dates, and the presence of any holiday periods from the primary bar chart program. The user then has to select a time unit, equal to or less than the primary bar chart time unit – in weeks,

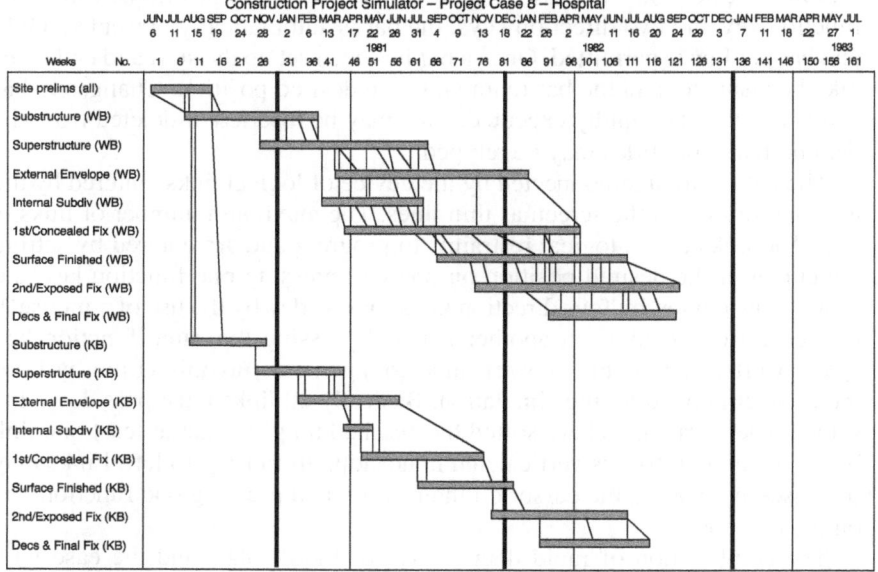

Figure 4.3.6 Typical primary bar chart.

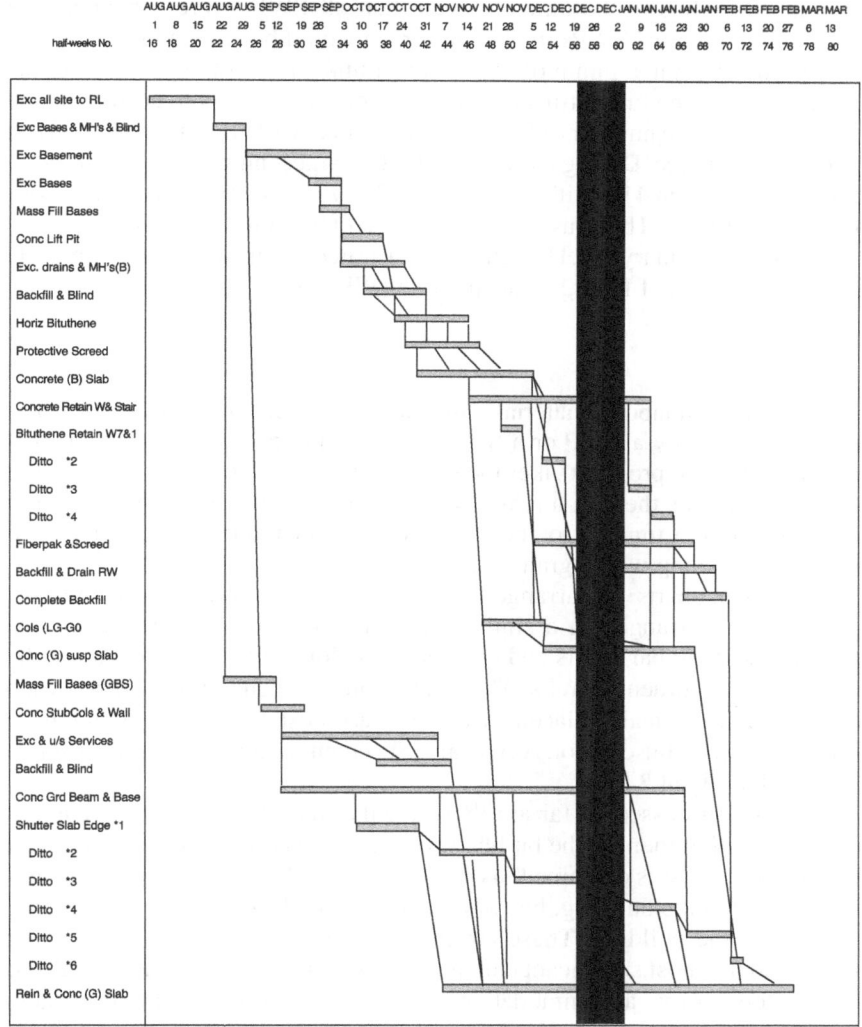

Figure 4.3.7 Typical secondary level bar chart – expanded plan of PWP 2 in Figure 4.3.6.

half-weeks, or days – and the computer presents a title, dated and scaled blank bar chart screen with any holiday periods in the right place, ready for the entry of SWPs.

The processes used to enter SWPs and logical links are exactly the same as those used in the primary bar chart, and described above. The maximum number of SWP bars permitted is 39, but the maximum number of logical links is set at 200. Thus in the unlikely event that all PWPs were expanded, the maximum number of SWPs in the entire plan would be 1521 and they would be restrained by a maximum of 7800 logical links. It is not expected that this facility will ever be fully utilized, but the upper limit was set to accommodate the largest possible number of SWPs and links in one PWP, which are likely to be found in practice. During use of the CPS the largest number of SWPs found in practice has been 410, with just less than 2000 secondary logical links, contained in 28 PWPs. This was for a complicated £9 million, 1.8 year project.

A typical secondary level bar chart is illustrated in Fig. 4.3.7 and represents the expanded plan of PWP2 – substructive (WB) in Fig. 4.3.6.

Cost entry

The direct cost in labour, materials and small plant need to carry out the operations described by a PWP or a SWP must be entered into the relevant cost program. The cost program takes the work package titles from the relevant bar chart and presents them in a table. Numbers up to eight characters in length may be entered. A running total of the labour and material costs are displayed on the screen along with a grand total.

As the bar charts are arranged as a hierarchy the structure of the cost information is arranged in a similar manner. Thus, there are as many cost tables as there are bar charts and the cost totals for a PWP are made up of the sum of the constituents SWPs. The totals from a secondary cost table when saved are automatically placed in the primary cost table, thus removing a potential source of user error. A typical cost screen, as it appears to the user, is illustrated in Fig. 4.3.8.

The costs discussed so far are those for the materials and effort required for incorporating them in the building. However, in construction a significant element of the cost is contained in cost centres which will not be permanently incorporated in the building, but which are nevertheless essential for the construction of the building. These are the 'preliminaries' costs for a building project and the most significant elements are: supervisory staff, plant, scaffold access, temporary accommodation, temporary services and materials-handling/cleaning (Gray, 1981). These costs are often shared by more than one activity on the plan, and they are specifically dependent on the duration and logic of certain elements of the plan. For example, the tower crane may be erected when a certain SWP in the substructure PWP has been completed and may be dismantled when a certain SWP has been completed in the superstructure PWP. The timing and hire period of the item of plant is linked to the logic of the plan, and if the durations of activities are extended in practice then the crane will be required for a longer period and the cost will increase.

This logic has been modelled in the CPS by the provision of a preliminary schedule where up to 15 separate 'preliminary categories' may be attached to

PWP No. 2	AMEND COST DATA		Substructure (WB)	
Activity	Labour Cost	Material Cost		
Exc all site to RL	<11610 >	<9395 >		
Exc Bases & MH's & B1 ind	<8645 >	<6680 >		
Exc Basement	<14925 >	<12075 >		
Exc Bases	<6635 >	<5365 >		
Mass Fill Bases	<520 >	<1300 >		
Conc Lift Pit	<1615 >	<3565 >		
Exc drains & MH's (B)	<7900 >	<11705 >	LABOUR TOTAL	
Backfill & Blind	<1065 >	<450 >	<	118520>
Horiz. Bituthene	<775 >	<1590 >		
Protective Screed	<440 >	<905 >	MATERIAL TOTAL	
Concrete (B) Slab	<3500 >	<7735 >	<	160300>
Conc Retain W & Stair	<4040 >	<8920 >		
Bituthene Retain W*1	<250 >	<455 >	GRAND TOTAL	
Ditto *2	<245 >	<455 >	<	278820>
Ditto *3	<245 >	<455 >		
Ditto *4	<245 >	<455 >		
Fibrepak & Screed	<1445 >	<2955 >		
Backfill & Drain RW	<12020 >	<10725 >		
Complete Backfill	<3510 >	<2845 >		
Cols (LG–G)	<2150 >	<4755 >		

Figure 4.3.8 Typical cost screen.

the logic of the primary bar chart. Once this program is engaged all the information entered into the primary bar chart is reproduced, except the majority of the logical link graphics is omitted for clarity and the link positions merely marked by short vertical lines. The user may then enter preliminary categories to define the start and end of a category, using the same procedure as drawing logical links in a bar chart. The only difference being that the cursor moves from link position to link position, rather than in time unit steps, and the start and end can thus only be attached to a logical link position. Heavy dashed lines are used to represent a preliminary category on the screen. Categories may also be deleted if required.

The effect of setting up a preliminary category, is that the logical links associated with the start and end positions are specifically identified, or 'flagged', to allow computation of the number of weeks between the start and end position. This information is then presented in a preliminaries cost table where a title, a weekly cost and a fixed cost can be entered for any preliminary category – to form the basis of the indirect cost calculation. A typical preliminaries schedule and cost table are presented in Fig. 4.3.9.

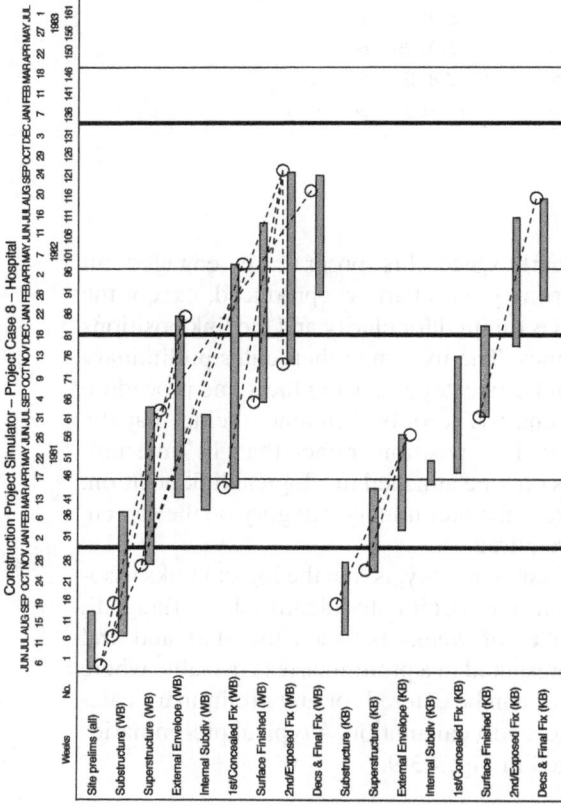

Construction Project Simulator – Project Case 8 – Hospital

Construction project simulator – Preliminaries description and costs

No.	Weeks start		Week end	Description	Labour		Materials
1.	Site prelims (all)	(0)	to 2nd/Exposed Fix (WB) clean, other, profit	(115)	Staff, hutting, ower,	2480	333 285
2.	Site prelims (all)	(0)	to External Envelope (WB)	(81)	Structural foreman 2 No.	380	0
3.	Substructure (WB)	(16)	to Superstructure (WB)	(60)	Tower Crane No. 1	655	5150
4.	Substructure (WB)	(8)	to 1st/Concealed Fix (WB)	(92)	Assistant engineer + chain boy	315	0
5.	Substructure (WB)	(8)	to Superstructure (WB)	(60)	Engineer + assistant	400	0
6.	Superstructure (WB)	(60)	to Decs & Final Fix (WB)	(110)	Hoists 2 No.	255	800
7.	Superstructure (WB)	(25)	to External Envelope (WB)	(81)	Scaffold	890	0
8.	1st/Concealed Fix (WB)	(42)	to 2nd/Exposed Fix (WB)	(115)	Services coordinator	230	0
9.	Surface Finishes (WB)	(62)	to 2nd/Exposed Fix (WB)	(115)	Finishing foremen 3 No.	485	0
10.	2nd/Exposed Fix (WB)	(71)	to 2nd/Exposed Fix (WB)	(115)	Completion agent	205	0
11.	Substructure (KB)	(16)	to External Envelope (KB)	(54)	Tower Crane No. 2	660	4650
12.	Superstructure (KB)	(24)	to External Envelope (KB)	(54)	Scaffold	155	0
13.	Surface Finishes (KB)	(58)	to Decs & Final Fix (KB)	(108)	Finishing foreman	165	0

Figure 4.3.9 Typical preliminaries schedule and cost table.

Resources

Where resources are desired to form the basis of resource restraint or cost calculation then three stages of data entry are required. However, the structure of the data entry has been designed to minimize the typing of resource titles, etc., many times. The first two stages are the creation of a library of standard trades and trade gangs, which once created need only be occasionally updated. Thus the entry of resource data is particularly rapid.

The first stage is to enter typical trades and their cost per time unit into a table of 100 available categories. This is accomplished in much the same way as direct costs are entered, with the user moving between defined fields to enter trade descriptions and costs. The number associated with a particular trade is then used as the trade reference number. This data entry takes place in the top half of the screen, and although there are only ten trades on display at any one time the top half of the screen scrolls independently.

This information is then used to build up typical trade gangs in the lower half of the screen. A gang title is entered an dup to five different trade members may be assigned to any one gang. It is common for there to be only one member of a gang, and the average number is for two or three. In this way a library of typical gangs is built up, although without any quantities of members, to allow a gang to be used in many different instances and then have different quantities added later. The number associated with a particular gang is then used as the gang reference number. The data entry takes place in the lower screen, and although only 10 gangs are displayed at one time the lower screen scrolls so that all the 100 available gangs may be viewed. A typical resources screen, as it appears to the user, is illustrated in Fig. 4.3.10.

Once the library of resource gangs has been assembled the business of applying them to particular activities can proceed. Resource restraint can only apply to the more detailed secondary level plans. Thus as many resource screens as there are secondary bar charts may be entered, although if resource restrain is desired in only a few PWPs, then only these require data entry.

For the selected PWP, expanded into a secondary level plan, another program presents the user again with a screen split horizontally, with each half again containing a table. In the lower screen the table of resource gangs is presented again for reference, and may not be changed. The upper screen displays the SWP titles from the relevant bar chart in a table. Each SWP title has twelve fields displayed against it. The second, fourth, sixth, eighth and tenth fields are displayed in reverse video, and the user is not permitted to enter these fields. The rest of the fields are defined by brackets and are allocated for user input. The upper screen is used to enter the number of a resource gang and to enter quantities against each constituent trade member.

In the first field of the upper screen the user enters a resource gang reference number. This has the effect of automatically displaying the trade members in the reverse video fields. This presents the user with details of gang members and an

CREATE RESOURCE SETS			
	TITLE		COST
23.	<Grdwk bricklyr	>	<150 >
24.	<Grdwrk plantdvr	>	<120 >
25.	<Drainlayer	>	<105 >
26.	<Formwork Carp	>	<125 >
27.	<Steelfixer	>	<120 >
28.	<Concretor	>	<100 >
29.	<Conc. finisher	>	<105 >
30.	<Scaffolder	>	<165 >
31.	<Bricklayer	>	<185 >
32.	<Bricklyr Labour	>	<190 >

RESOURCE SET		CODE REFERENCES								
11.	<RC Stairs : TC/P	>	<27>	<26>	<28>	<	>	<	>	
12.	<RC Stairs : H/D	>	<27>	<26>	<28>	<	>	<	>	
13.	<RC Foundations	>	<26>	<27>	<28>	<	>	<	>	
14.	<Formwork	>	<26>	<	>	<	>	<	>	< >
15.	<Steel fixing & concreting	>	<27>	<28>	<	>	<	>	<	>
16.	<Structural steel erection	>	<60>	<	>	<	>	<	>	< >
17.	<Asphalt roofing	>	<35>	<	>	<	>	<	>	< >
18.	<Tiled roofing	>	<36>	<	>	<	>	<	>	< >
19.	<Scaffolding	>	<30>	<	>	<	>	<	>	< >
20.	<Brickwork	>	<31>	<32>	<90>	<	>	<	>	

Figure 4.3.10 Typical resources library screen.

adjacent empty field in which the quantity of members required. Member quantities may only be placed in fields with an adjacent trade reference number. Thus the same gang structure can be used for all similar activities (e.g. brick-work), but the numbers varied to reflect relative resource usage. In the last field the cost per time unit of the whole gang is automatically computed from the data entered previously, and is displayed when the data is saved. A typical resource allocation screen, as it appears to the user is illustrated in Fig. 4.3.11.

Once all the resource data for the desired SWPs has been entered, one final data input procedure is necessary to set the upper limit of resource availability for this PWP. The program displays all the separate trade members used in all the gangs employed and this information is displayed as a table. The user can then enter the maximum number for each resource type and the computer validates this to ensure it is not less than that entered previously. These totals act as the upper limit for resource restraint during simulation, and the user by manipulating these figures can experiment with different combinations and quantities to determine the best mix.

PWP No. 2   **ALLOCATE RESOURCES AND COST**   Substructure (WB)

ACTIVITY   RESOURCE SET   [TRADES]+<NUMBERS>

	RESOURCE SET					
4. [Exc Bases]	<26>* 24 <1 >:	91 <1 >:	22 <1 >:	< >:	< >:<	298>
5. [Mass Fill Bases]	<24>* 28 <3 >:	< >:	< >:	< >:	< >:<	300>
6. [Conc Lift Pit]	<13>* 26 <2 >:	27 <2 >:	28 <3 >:	< >:	< >:<	790>
7. [Exc. drains & MH's (B)]	<26>* 24 <1 >:	91 <1 >:	22 <2 >:	< >:	< >:<	396>
8. [Backfill & Blind]	<27>* 12 <1 >:	13 <2 >:	< >:	< >:	< >:<	410>
9. [Horiz. Bituthene]	<27>* 12 <1 >:	13 <3 >:	< >:	< >:	< >:<	540>
10. [Protective Screed]	<27>* 12 <1 >:	13 <3 >:	< >:	< >:	< >:<	540>
11. [Concrete (B) Slab]	<2 >* 27 <4 >:	26 <2 >:	28 <3 >:	< >:	< >:<	1030>
12. [Conc Retain W & Stair]	<8 >* 27 <2 >:	26 <2 >:	28 <3 >:	< >:	< >:<	790>
13. [Bituthene Retain W*1]	<27>* 12 <1 >:	13 <2 >:	< >:	< >:	< >:<	410>

RESOURCE SET	**CODE REFERENCES**				
18. [Tiled roofing]	[36]	[ ]	[ ]	[ ]	[ ]
19. [Scaffolding]	[30]	[ ]	[ ]	[ ]	[ ]
20. [Brickwork]	[31]	[32]	[90]	[ ]	[ ]
21. [Window fixing]	[37]	[38]	[ ]	[ ]	[ ]
22. [Plumbing]	[43]	[ ]	[ ]	[ ]	[ ]
23. [Mastic pointing]	[39]	[ ]	[ ]	[ ]	[ ]
24. [Concreting only]	[28]	[ ]	[ ]	[ ]	[ ]
25. [Excavating [large scale]]	[24]	[92]	[ ]	[ ]	[ ]
26. [Excavation [small scale]]	[24]	[91]	[22]	[ ]	[ ]
27. [General labouring]	[12]	[13]	[ ]	[ ]	[ ]

RESOURCE TOTALS

24. Grdwrk plantdvr : <2 >
92. Excavator : <1 >
91. JCB : <2 >
22. Grdwk Labourer : <2 >
28. Concretor : <3 >
26. Formwork Carp : <4 >
27. Steelfixer : <4 >
12. Ganger : <1 >
13. Labourer : <3 >

Figure 4.3.11 Typical resources allocation screen.

Creating and allocating distributions

Frequency distributions, or histograms, must be created and assigned so that variability and interference routines in the simulation programs have the correct information. Certain areas of the distribution program are reserved for specialist purposes; 1 to 50 are available to the user to save created distributions, 51 to 62 are reserved for the weather, and 63 to 140 are reserved for secondary simulation results.

The four most common distributions used in simulation, namely, uniform, triangular, normal and beta, are permanently available. The user can also create skewed distributions for the triangular, normal and beta to model pessimistic or optimistic production rates. Bimodal distributions can also be created in a wide variety of shapes, and actual histograms of real data can be entered and saved as the histogram or as a 'best fit curve' (see Fig. 4.3.2). Some typical distributions are illustrated in Fig. 4.3.12).

The distributions created in this program are contained in a library or data base which will rarely require updating once compiled. The distributions in this data base should be regarded as a library to aid the manager in applying his intuition to the simulation processes. As each can be regarded in a similar manner to the familiar PERT three-point estimate, and the different shapes used to express feelings about an activity, e.g. it is optimistic, thus allocating a distribution with a pessimistic skew. Some familiarity with this concept is obviously required, however there is evidence that it coincides with how people view the real world.

All the activities, at whatever hierarchy level, must have certain parameters defined pertaining to a particular activity before a simulation may be

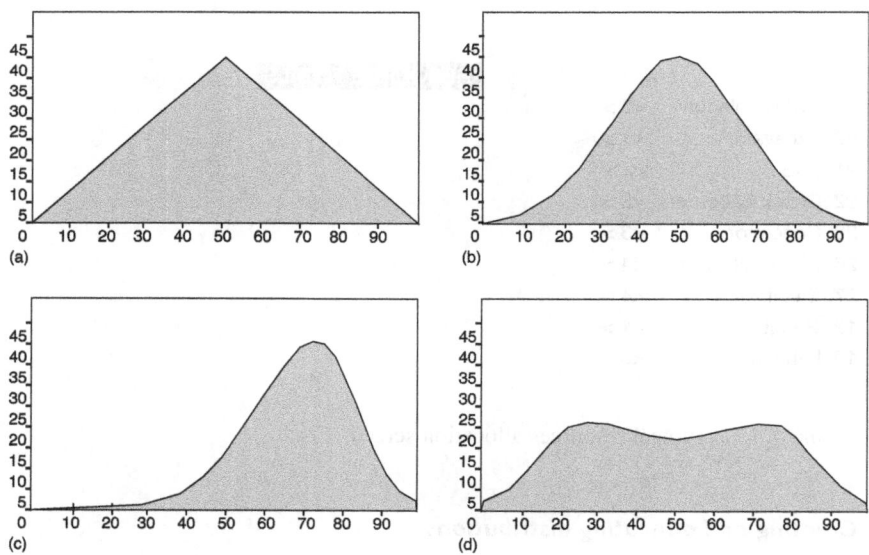

Figure 4.3.12 Typical frequency distributions. (a) Triangular distribution, (b) normal distribution, (c) skewed beta distribution and (d) bimodal distribution.

executed. Values for interferences, productivity variability and relevant distributions must be assigned to each activity by the user. The structure of these programs follow the same hierarchical structure as the bar charts, so there is one for the primary bar chart and one for each PWP. The structure of the screen display is similar to the cost programs with the activity title displayed along with fields to accept user input. In this program three fields are defined for use by the user.

The first field defines the interference factor acting upon the activity. A figure between 0 and 99 firstly defines the percentage chance that an interference will occur at any link position on the activity bar. This is separated by an oblique stroke '/' from an optional frequency distribution number between f1 and f50, which is held on file in the distribution program. This, if entered, is used to select a period of an interference from a user input distribution. If no distribution number is present the simulation programs will assume an interference period of one time unit.

The second field defines the range of the variability amount affecting an activity. A figure between 0 and 99 defines the plus and minus ($\square$) percentage variation about the midpoint of the distribution, which defines the range of possible activity durations that may be chosen during simulation. The third field defines the number of the frequency distribution which applies to an activity.

Weather data

The inclusion of weather processing in simulation is particularly suitable as the basic problem with weather data is that typical weather behaviour may be known historically, but the *actual* weather which will be encountered during construction is not known. The effect is qualitatively detectable, but not enough knowledge exists to assess the risk quantitatively. However, weather data is a classical stochastic problem (Dressler, 1974) where a frequency distribution can be obtained from reliable historic records and repeatedly sampled randomly to reflect the possible weather effects on a project. The treatment of weather information is thus an ideal candidate for simulation.

The availability of data is excellent and can be purchased from the Meteorological Office for a very modest fee. The data used in the CPS is based upon a combined plot when the rainfall during the working day was greater than 0.1 mm h^{-1} **and/or** the wind was gushing greater than 10ms^{-1} **and/or** the air temperature was less than 2$\square$C **and** the mean wind speed is greater than 4ms^{-1}. This combined plot is then available for each month of each year for the last 20 years to provide a frequency distribution for the hours per working month lost through the effect of the weather. Two such distributions are reproduced in Fig. 4.3.13 to illustrate the difference between February and July when the data is presented in this format.

The twelve histograms for the weather for each month of the year are

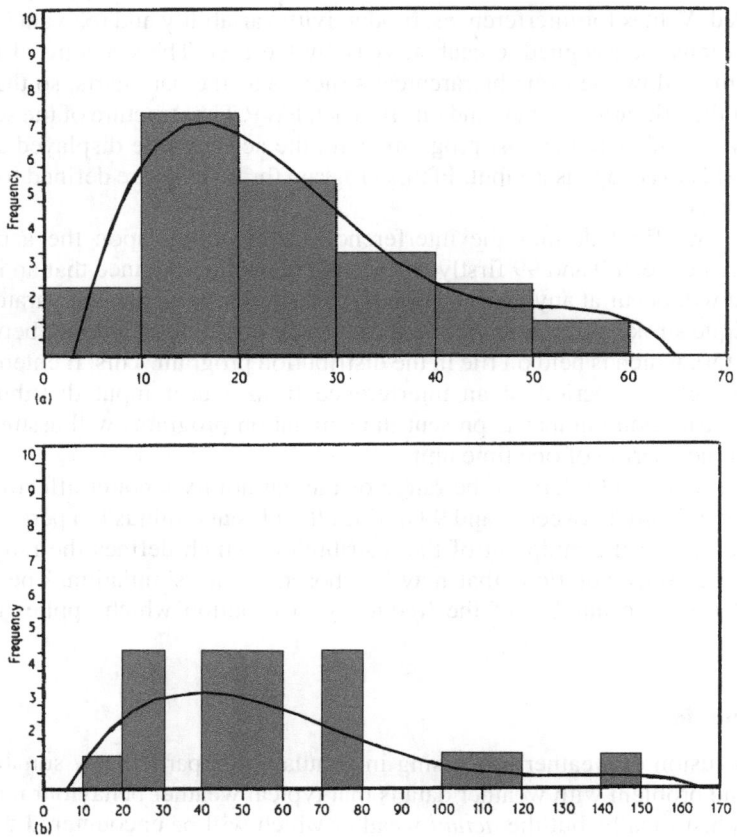

Figure 4.3.13 Typical weather frequency histograms. (a) Hours lost in the working month of February and (b) hours lost in the working month of July.

entered into a separate program. The weather data is specific to one location, so that new data should be collected for projects in different parts of the country as well as other countries. Also this program allows the identification of each PWP on the primary bar chart as being sensitive to weather effects or not. Thus the 'substructure' would be designated as weather sensitive while 'decorations and final fixes' would not. In this way the likely effect of the real weather can be included in a simulation.

SIMULATION PROGRAMS

Once the data input procedure has been completed, then the simulation of the project can be carried out. However, the CPS has been designed as a hierarchy to facilitate the simulation of schemes at various stages, so that data input can have different degrees of detail and simulations still performed.

A simulation of the primary level only may be carried out after the primary bar chart and weather data, and optionally the preliminary schedule and primary cost table, have been entered. This allows a quick assessment of schemes at an early stage in their development, or at a tendering stage.

A simulation of one or more secondary plans may be carried out after a secondary bar chart, and optionally the secondary cost table and resources details, has been entered. This allows an assessment of single PWPs when design details become available as the scheme develops, or allows re-assessment of a PWP after some change to the construction method.

A full scale simulation of all the secondary plans, followed by a simulation of the primary level can be carried out automatically. This allows an assessment of the whole scheme, at a level consistent with a good contract programme, to produce the most reliable simulation result. The duration of this most detailed simulation procedure is one and a half to three hours, depending on the size of the project plan.

Simulation method

The purpose of simulation is to imitate the conditions of a system with a model or simplified representation of the system, so that the model mimics important elements of the system under study. The end result is to produce a prediction of the likely range and pattern of contract duration and cost which are feasible under the conditions and constraints of a specific project.

This is carried out inside the computer in a mathematical process using the data provided by the user. The basic problem is to choose from the range of possible durations for one bar of the plan – represented by the assigned frequency distribution and variability percentage – in a manner which is representative of the original data.

The manner in which this is achieved is via the use of random numbers (RNs). Random numbers are generated by the computer and are applied to the assigned distribution to see if the chosen value is representative of the original data. If it is representative then this is the value which is used in the simulation. If it is not representative then the value is rejected and more RNs are generated. The whole process being repeated until a feasible value is chosen. The random numbers are used in conjunction with a frequency distribution which limits the randomness to that found in practice. Thus the process is not totally random, but has a random element, hence the term stochastic. The use of random numbers is really only a device to produce unbiased choices of duration, and although we cannot say precisely what the duration will be after any one choice we can say what the pattern and range will be after a sufficiently large number of choices. So it is necessary to make a large number of choices to produce the confidence that the resulting choices are representative of the original data. The number of choices are referred to as *iterations*, and the number of iterations performed is critical to the final accuracy of the whole process. The minimum number of iterations acceptable is 100, and this produces an accuracy of □2%. For an increase in the number of iterations to 200 the accuracy can be improved to □1%. However, there is a penalty to be paid for increasing numbers of iterations, as the simulation takes progressively longer. Thus the number of iterations chosen is a balance between accuracy and processing time. The CPS has the ability to perform any number of iterations between 100 and 200, the final decision being left to the user.

Briefly, the procedure on one iteration of a simulation is to take each bar segment defined by logical links in turn and to choose an actual duration via RNs, the assigned distribution and the variability amount. This actual duration will not be the same on every iteration, but will reflect the possible values and their frequency of occurrence found in practice. Each link position is then examined to see if an interference occurs there. This is governed by the assigned interference amount and the use of RNs. The duration of the interference is chosen by RN if an assigned interference distribution is present. The next bar is then taken and the same process followed until all bars have been considered. The whole bar chart is then 'rescheduled' by the logical restraints. If resource restraint is to be simulated then the resource usage is compared to the available resources to see if progress is impeded. The cost of the bar is then calculated from the ratio of the actual bar duration to the original bar duration multiplied by the labour cost and added to the material cost, or if resources are involved the labour cost is calculated from the resource costs. The final duration and cost are then recorded. The whole process is then repeated up to another 199 times to complete the simulation.

Each iteration of the simulation creates a result. Due to the presence of RNs and the duration of each bar being chosen from a feasible range, there are a bewildering combination of possible bar durations and interactions which lead to a range of results being obtained rather than one single result. The results are also presented as a frequency distribution giving the range and pattern of outcomes.

The process just described is true of the general simulation process at whatever hierarchy level. However, two further facilities are employed to feed the results of the secondary level simulation into the primary level and to include weather and preliminaries processing in the primary level simulation. Each secondary level bar chart can be simulated to produce a distribution of results for the duration and costs. These distributions are saved onto disc to be used as data input to the primary level simulation process, and to allow the fine detail to affect the overall result.

The effect of the weather is included in the primary level simulation, where the start and end dates of a bar are known after the actual duration has been chosen in a stochastic manner. The actual weather delay each month is chosen via RNs from a distribution of feasible weather delays (see Fig. 4.3.13), and this period is added to the bar to extend its duration. In this way a slightly different amount of weather delay is experienced on every iteration of the simulation. Thus the likely effect of the weather can be analysed, although nobody can predict what the actual weather will be except perhaps a day or two ahead.

The costs of the preliminary categories are also calculated at the primary level. As the start and end dates of each logical link is known it is an easy matter to calculate the period between the start and end of a preliminaries category and multiply the weekly hire rate by this figure. This is added to any fixed materials costs to arrive at a total preliminaries cost. As the position of the start and end links will rarely be the same in the simulation a range of preliminaries costs will be produced.

This then briefly represents the process which occurs during a simulation. However, the myriad choices of duration and the combination of these factors is almost inconceivable to the human mind, and it is only through the use of computers that problems of this nature can be assessed. Thus the computer has brought techniques to the aid of human affairs which would not be applicable without it.

Simulation results

Typical results produced by a simulation are illustrated in Fig. 4.3.14. The results are presented in two forms, as a frequency histogram (darker shading) which gives the user an impression of the pattern of results, and as a cumulative frequency curve (lighter shading) which allows the user to read off risk levels or confidence factors from the vertical scale. This vertical scale is labelled 'percentage' and represents the chance of a certain result occurring. This represents the 100 (or more) simulation results sorted into ascending order, so the lowest one had a small chance of occurring (or is a high risk), while the highest result had a very high chance of occurring (or is a small risk). The cumulative frequency curve is often the most useful as it is possible to read the risk associated with a particular result (or vice versa) directly from the graph. As there are a number of varied results there are also a number of varied patterns of possible expenditure over the life of the project. This gives rise to cumulative cash flow curves which may be examined at different risk levels to determine likely cash flow targets, and establish cash flow levels (Fig. 4.3.15) which are unlikely to be exceeded.

The CPS has been used with several projects and the results for a hospital project are presented to indicate the use the output may be employed for. The tender period for the project was 141 weeks and the tender cost (not including an allowance for inflation) was £4 856 115. The actual period for the project was 148 weeks and the final uninflated cost was £4 631 785 with the final cost including inflation being £5 769 600. The inflation rate affecting the project was 16%.

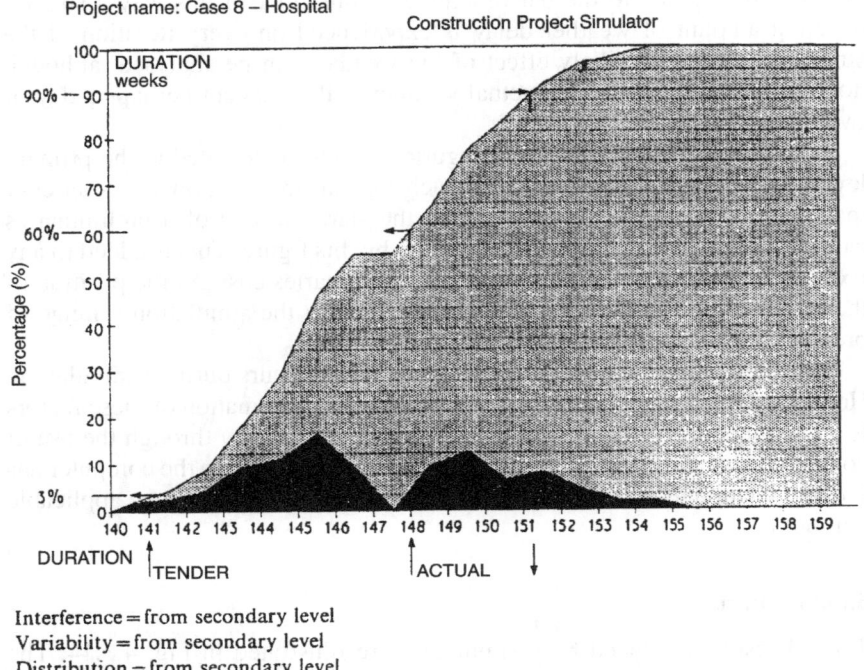

Interference = from secondary level
Variability = from secondary level
Distribution = from secondary level
Minimum = 139
Maximum = 156

Mean total duration = 147.76

Figure 4.3.14 Typical simulation result, including weather, for project completion.

The result of the most detailed simulation, including weather, for project completion is presented in Fig. 4.3.14 and the associated project uninflated final cost is presented in Fig. 4.3.16. From these figures it can be seen that the tender period only had a 3% chance of success, i.e. it was very risky, while the actual period had a 60% chance of success, i.e. it was very much less risky. This indicates that the contract period fixed in the tender documents was too optimistic, and if the client was serious about obtaining completion in 141 weeks then the design solution should have been reviewed or extra management effort expended. However, if the client could accept some later completion date he would have been aware from the output what the upper limits to his risk were, and so plan accordingly. If a risk judgement had been made based on a high confidence level of 90% then the client would have been aware that a possible time contingency of 10 weeks would be advisable, of which seven weeks was expended. The tender cost was not feasible and was

Project name: Case 8 – Hospital

Construction Project Simulator

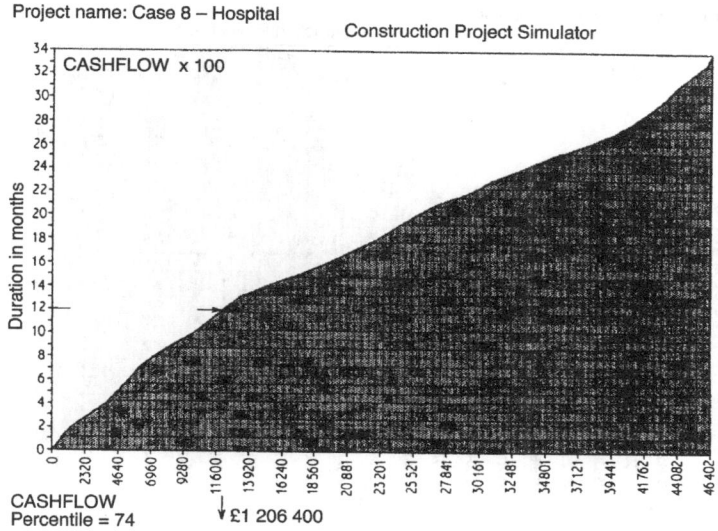

Figure 4.3.15 Cash flow curve associated with the 74% confidence level, and the likely expended sum at month 12.

£224 330 higher than the actual cost, even though the project was extended by seven weeks.

The professional adviser to the public client clearly overestimated his project contingency, which in this instance was included in the building budget. However, if a risk judgement had again been based on a high confidence level of 90% then the client or his professional adviser would have set the building budget at £4 637 700. This would have represented a reduction in the overall budget of £218 415 or 4.5%, which could have been crucial in the viability assessment of the scheme by the client. The contingency amount then built into the contract budget assuming a feasible tender cost at the 5% confidence level of £4 559 380, would have been £78 320 or 1.7%, of which £72 405 would have been expended.

The effect inflation can have on a project can be included in the simulation, and Fig. 4.3.17 illustrates the cost graph with the anticipated inflation amount included. No estimate of the inflated tender cost was available, but the actual value had an 80% chance of success.

As the weather routine in the simulator is an additional feature, to model a specific component of interference, it is possible to simulate the project behaviour as if it had not been subject to the effects of weather. The results of the simulation omitting the effects of weather are illustrated in Figs. 4.3.18 and 4.3.19. These show that the tender period had a 82% chance of success, while the actual period was not feasible. The increase in mean project duration

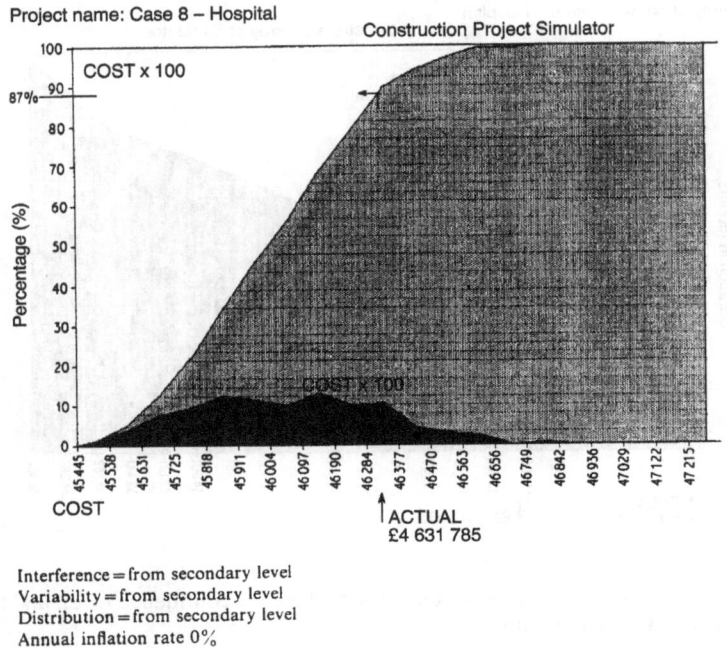

Project name: Case 8 – Hospital

Construction Project Simulator

COST x 100

Percentage (%)

COST x 100

COST

ACTUAL
£4 631 785

Interference = from secondary level
Variability = from secondary level
Distribution = from secondary level
Annual inflation rate 0%
Minimum = £4 553 529
Maximum = £4 678 396

Mean total cost = £4 617 358
Mean preliminaries cost = £1 001 172
Mean additional direct weather cost = £68 390.24

Figure 4.3.16 Typical simulation result, including weather and zero inflation, for project final cost.

solely attributable to the effects of the weather is 8.5 weeks or an increase of 6.1%. The increase in the mean total cost is £107 201 or an increase of 2.4%. The increase in the mean project preliminaries cost due to weather is £42 411 or an increase of 4.5%. This indicates that the preliminary costs are more sensitive to the weather than the total cost.

As well as judging a project's vulnerability to weather it is possible to judge a project's sensitivity through changing the project start date. To this end the actual start date of 9 June 1980 was changed to 9 November 1980. The holiday periods in the primary bar chart were adjusted, but otherwise everything else was unchanged. However, in this case the risk levels were not significantly changed through starting the project at a more unfavourable time of year for construction operations, and the conclusion in this case was that the project was not significantly sensitive in time or cost to a change in the project commencement date.

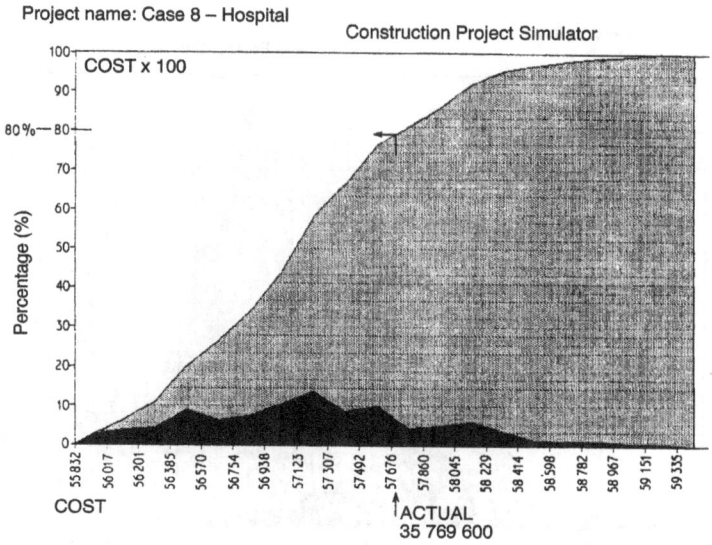

Project name: Case 8 – Hospital

Construction Project Simulator

Figure 4.3.17 Typical simulation result, including weather and 16% inflation, for project final cost.

The use of the cash flow facility has not been validated against actual expenditure, as it is a recent addition to the suite of programs. However, Fig. 4.3.15 is used to illustrate one of the 100 cashflow curves available.

CONCLUSIONS

One particular example has been used to illustrate the output of the simulator and how it can be applied to a specific project. However, during the development phase of the research four different projects were simulated to test the model's ability to accommodate a representative range of construction projects – new build office block, refurbished office block, new build hospital, and a housing estate – to establish the apparent level of interference affecting the projects and to compare the simulators' predictions with the deterministic predictions of the tender stage, and the final outcomes.

Project name: Case 8 – Hospital

Construction Project Simulator

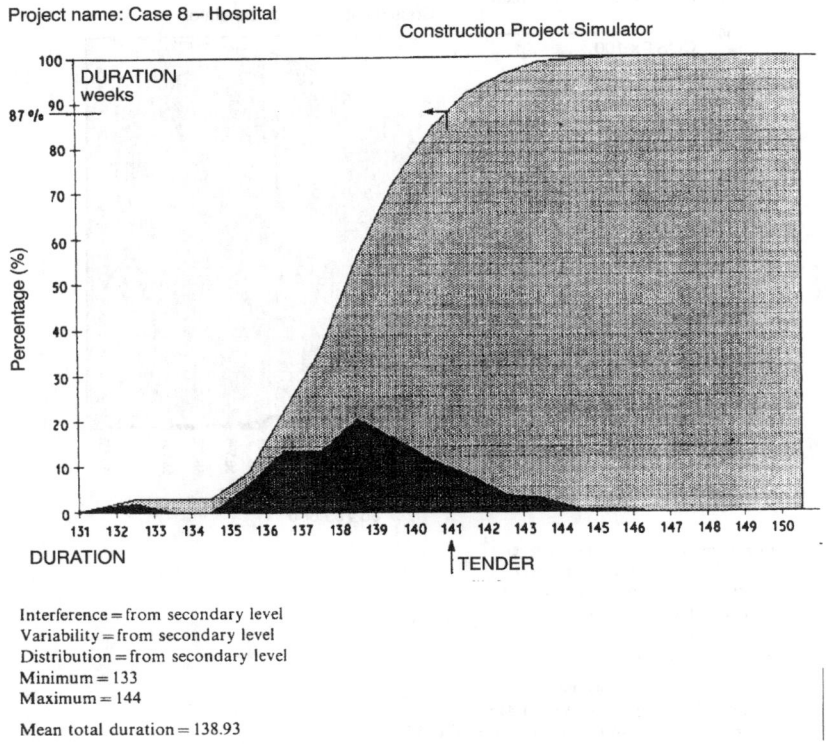

Interference = from secondary level
Variability = from secondary level
Distribution = from secondary level
Minimum = 133
Maximum = 144

Mean total duration = 138.93

Figure 4.3.18 Typical simulation result, excluding weather, for project completion.

The final construction period for all the cast studies was predictable using high confidence factors between 44 and 86%. However, the initial tender periods in all instances were too optimistic. In only one case study did the tender period have a possible confidence factor and then of only 26%, and for all the others the tender period was either unrealistically optimistic or had a very low, between 5 and 10%, chance of being attained. In all cases the tender period had been set by the client's advisers in the tender enquiry, and from the results of he simulation were too low and unrealistic. If the client had not been aware of the risk involved and had taken the tender period as a serious estimate then severe practical problems could have resulted. If realistic tender periods had been established by using reasonable confidence levels (30–50%) with the simulator predictions then this might have led to a less defensive and 'claims-conscious' attitude by the contractor and a more positive construction process. The client would also have been aware of the extent and risk of more unacceptable outcomes which could form the basis of time-contingency

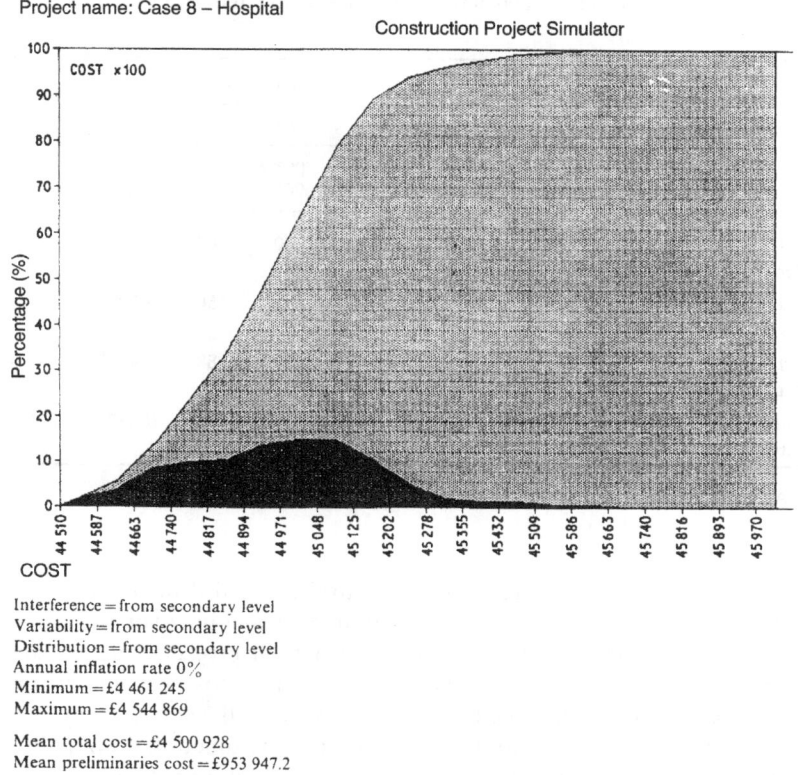

Project name: Case 8 – Hospital

Construction Project Simulator

Interference = from secondary level
Variability = from secondary level
Distribution = from secondary level
Annual inflation rate 0%
Minimum = £4 461 245
Maximum = £4 544 869

Mean total cost = £4 500 928
Mean preliminaries cost = £953 947.2

Figure 4.3.19 Typical simulation result, excluding weather and zero inflation, for project final cost.

management and been assessed during the scheme viability process.

The final construction cost for all case studies was predictable using high confidence factors between 56 and 88%. However, the initial tender cost in all instances was unrealistic, with one cost being unrealistically optimistic and three being unrealistically pessimistic (see Table 4.3.5). These results illustrate weaknesses in the current, deterministic practice of cost contingency management. For two projects undertaken for public clients the bill of quantities was structured in such a manner as to produce a tender figure which would not be exceeded – as this would be unacceptable to government accounting. The contingency amount built into the contract for both cases, however, was higher than necessary. A more realistic contingency budget and tender cost could have been predicted using the simulation results and a high confidence factor of 90%. The other two projects were undertaken for private clients and illustrate two current methods of contingency structuring. In one

Table 4.3.5 Case study tender and actual outcomes contrasted with the simulation predictions

Case study	Tender		Actual		Simulation				
					Time (weeks)		Cost (£)		
	Time (weeks)	Cost (£)	Time (weeks)	Cost (£)	90%	x	90%	
New build office block	73	1313950	82	1449890	80	83	1422422	1451020
New build hospital	141	4856115	148	4631785	146	150	4609558	4646000
Refurbished office block	75	4006595	79	4241000	79	83	4241089	4271000
Housing estate	130	3968425	135	3884000	133	136	3843319	3907400

N.B. All costs exclude inflation allowances.

case the tender cost was unrealistically optimistic, and if the client were not finally to be financially embarrassed a contingency amount over and above the tender cost must have been included in an overall project budget although not in the building fund. In the other case the tender cost was unrealistically pessimistic and higher than the final cost. In this case the contingency amount had been included in the building fund and it is assumed that no additional contingency was allocated in an overall project budget. In both cases an upper level of cost could have been realistically set at high confidence levels using the simulation results. The strategic decision would then have been necessary as to which section of the budget the contingency amount should have been allocated to. If the simulation results had been available to the client and his advisers then all parties would have been aware of the exposure to risk and contingency management placed on a more accurate, objective and explicit footing than occurred in practice.

In all cases, for both time and cost predictions, the simulators prediction of a contract duration and cost was better than that available through current US working practices. The further highlighting of other possible outcomes – both better and worse – and the quantification of these other values would add a new dimension, not previously available, to the management of the project.

A major element of the research was to demonstrate that the important characteristics of variability and interference, which dominate much construction activity, can be explicitly identified and incorporated in a model of the construction process. This effort has produced a reasonable amount of data for the variability, such that the magnitude of this effect could be identified and determined for use as an input parameter in the simulator.

Typical distribution shapes have also been identified for several trades and incorporated in a data bank in the CPS. Also through the distribution entry program it is possible to produce skewed distributions for use where activity durations are considered optimistic or pessimistic. This facility allows the experienced construction manager to include his important intuitive assessment into the simulation.

A similar effort was devoted to the definition and identification of interference. It soon became clear that the more general and inclusive category of interference would have to be refined for a greater theoretical and practical acceptance. This resulted in the identification of the major components of interference into: weather, design, procurement, and 'all other'. This has proved to be more acceptable. The data available for interferences, with the notable exception of weather, has been sparse despite an intensive search in both literature and practice Data on the weather has been found to be plentiful, in the correct format and cheap. The appropriate value for use as an input parameter to the simulator at the secondary level for the interference category (with the effect of weather modelled separately) has been determined by validating values through sensitivity analysis of real projects.

This process has identified the general interference magnitude to be 7%, with the weather modelled separately, for three of the four case studies. The odd result being for the housing case study where values of 2% had to be employed to produce sensible results. This result is not unexpected, as the design interference in a project with many identical repetitions is inevitably reduced after only a few repetitions of the task. In a similar manner once procedures have been established to procure materials they become a matter of routine and become less susceptible to interruption. In a similar vein it would be expected that the design element of a refurbishment project would be more open to interruption through the need to accommodate new work and existing work – the extent of which is not always known – than that for an entirely new building project. However, the refurbishment project chosen as a case study did not produce sensible results at higher levels of interference. The design effort was particularly responsive to the construction operation as up to six architects were based on site available for almost instant design decisions and inevitably forming a more cohesive and responsive design/construction team. Thus it is postulated that the interference magnitude on this project was beneficially affected by the project management structure. Also it has been demonstrated that interference can be adequately defined and typical values identified through sensitivity analysis.

The CPS now exists as a useful means of modelling important management characteristics of construction projects. It is being used at present to help control a live project and to form the basis of further research in construction management.

REFERENCES

Ahuja, H. N. (1982) A conceptual model for probabilitist forecast of final cost, in *Proceedings of the PMI/INTERNET Symposium*, Toronto, Canada, Project Management Institute, pp. 23–31.

Bennett, J. and Fine, B. (1980) Measurement of complexity in construction projects: SERC Research Project GR/A/1342.4: Final Report. University of Reading (Available as Occasional Paper No. 8.)

Bishop, D. (1966) Architects and productivity – 2. *Building* **272**, 533–63.

British Property Federation (1983) *The British Property Federation System for Building Design and Construction*, London.

Clapp, M. A. (1965) Labour Requirements for Conventional Houses: Current Paper 17. Building Research Establishment, UK.

Crandall, K. C. (1977) Scheduling under uncertainty, in *Proceedings of the PMI/INTERNET Symposium*, Chicago, USA, Project Management Institute, pp. 336–43.

Dressler, J. (1974) Stochastic scheduling of linear construction sites. *ASCE Journal of the Construction Division* **100**, 571–87.

Feiler, A. M. (1972) Project risk management, in *Proceedings of the Third International Congress on Project Planning by Network Techniques*, Stockholm, Sweden, INTERNET, Vol. 1, pp. 439–61.

Feiler, A. M. (1976) Project management through simulation. School of Engineering and Applied Science, Project TRANSIM, LA California, USA.

Fine, B. (1982) The use of simulation as a research tool, in *Building Cost Techniques: New Directions* (edited by P. S. Brandon). E. & F. N. Spon, London, pp. 61–9.

Gray, C. (1981) Analysis of the preliminary element of building production costs, MPhil thesis, University of Reading.

Hall, B. O. and Stevens, A. J. (1982) Variability in construction: Measuring it and handling it, in *Proceedings of the PMI INTERNET Symposium*, Toronto, Canada, Project Management Institute, pp. 559–66.

Harris, R. B. (1982) *Precedence and Arrow Networking Techniques for Construction*, J. Wiley & Sons, London.

Lichtenberg, S. (1981) Real world uncertainties in project budgets and schedules, in *Proceedings of the INTERNET Symposium*, Boston, USA, INTERNET, pp. 179–93.

McLeish, D. C. A. (1981) Manhours and interruptions in traditional house building. *Building and Environment* **16**, 59–67.

Nunnally, S. W. (1981) Simulation in construction management, in *Proceedings of the CIB Symposium on the Organization and Management of Construction*, Dublin, Eire, International Council for Building Research, Vol. 1, pp. 110–25.

Relf, C. T. (1974) Study of the Building Timetable: The Significance of Duration (Part 1). Building Economics Research Unit, University College London.

Riggs, L. S. (1979) *Sensitivity Analysis of Construction Operations*, PhD thesis, Georgia Institute of Technology, USA.

Roderick, I. F. (1977) Examination of the Critical Path Methods in Building: Current Paper 12. Building Research Establishment, UK.

Rogge, D. F. and Tucker, R. L. (1982) Foreman delay surveys: Work sampling and output, *ASCE Journal of the Construction Division* **108**, 592–604.

Stevens, A. J. (1969) Activity Sampling on Building Sites: Current Paper 16. Building Research Establishment, UK.

Tavistock Institute (1966) *Interdependence and Uncertainty: A Study of the Building Industry*. Tavistock Publications, London.

Traylor, R. C. *et al.* (1978) Project management under uncertainty, in *Proceedings of the PMI Symposium*, Los Angeles, USA, Project Management Institute, Vol. 2, pp. f1–f7.

Tucker, R. L., Rogge, D. F. and Hendrickson, F. P. (1982) Implementation of foreman-delay surveys, *ASCE Journal of the Construction Division* **108**, 577–91.

United Nations: Economic Commission for Europe: Committee on House Building and Planning (1965) *Effect on Repetition on Building Operations and Processes on Site*, Geneva.

Wookery, J. C. and Crandall, K. C. 91983) Stochastic network model for planning scheduling. *ASCE Journal of Construction Engineering and Management* **109**, 342–54.

4.4 Cost modelling: a process modelling approach*

P. A. Bowen, J. S. Wolvaardt and R. G. Taylor

ABSTRACT

Attempts at modelling the cost of buildings have hitherto attempted to explain cost as (simple) functions of different measurements of the finished building, element or component, and therefore by implication seeing building as a single discrete step. Such an approach ignores the fact that construction is a **process** consisting of separate but dependent physical activities occurring over a period of time and subject to uncertainty in terms of, *inter alia*, cost and duration. It does not explain when and how costs are incurred (inexplicability), disregards interrelatedness (unrelatedness), and ignores uncertainty (determinism). To overcome the above weaknesses, a process-based modelling approach is suggested. PERT-like networks can be used in a structured manner to model cost by representing expenditure as it occurs during construction activities. The proposed system will use A.I. (artificial intelligence) based on Nebula Theory to link sub-networks representing the construction of different elements (e.g., foundations, basement, etc.) to create a complete, representative network for the entire project. For each element, different sub-networks representing different designs (different in type, and not size or similar parameters) will be available, enabling the user to explore different design/construction alternatives. Being computer-based, the proposed system will use a data-base containing cost information, sets of sub-networks, different default values, and definitive functions.

Key words

Expert system, networks, Cost model, Construction.

*1987, in Brandon P. S. (ed.) *Building Cost Modelling and Computers*, E & F N Spon, 15–24, 387–95.

†This paper repeats in part material included in an earlier publication by Bowen, Taylor and Edwards incorporated in the proceedings of the C.I.B.86 Washington conference on Advancing Building Technology. The conceptual work discussed in the earlier paper is expanded and further developed here.

INTRODUCTION†

In response to the need for change in existing cost modelling methodologies which have been largely deterministic and non-explanatory, stress has bene laid upon the need for fresh thinking upon fundamental issues (Science and Engineering Research Council, 1984), rather than a modification of existing methods, and a prophetic call has been made for a 'paradigm shift' from the questionable norms of traditional approaches in building cost research to new methods derived from proven theories and founded upon an adequate knowledge-base (Brandon, 1982). In this paper the writers propose a network (activity) cost modelling system that represents the manner in which costs are generated, and which addresses the weaknesses considered to be inherent in traditional approaches to cost modelling, namely, inexplicability, unrelatedness and determinism. The conceptual framework advocated here is the subject of doctoral research at the University of Natal.

INEXPLICABILITY

Current cost models such as bills of quantities, elemental estimating methods, regression models, and optimisation models make no attempt to explain the systems they purport to represent. Such cost models are usually structured to represent 'finished' buildings (or building components) and are thus concerned more with ends than with means. Truly realistic **cost models** must, however, by definition, take into account, in a logically transparent manner, the cost implications of the way in which buildings are physically constructed, on the grounds that different construction methods significantly affect cost. Such models must **simulate** the construction process.

UNRELATEDNESS

Researchers have hitherto attempted to create global solutions or models, without an adequate understanding of the interdependency and relatedness existing between building components, elements and systems.

DETERMINISM

Traditional approaches to the provision of design/cost advice generally rely upon the use of historical (price) data to produce single-figure (i.e., deterministic) estimates of building cost, which are offered as predictions of future events without explicit qualification of their inherent variability and uncertainty. Such a deterministic understanding of the economics of buildings has been considerable criticised, e.g., by Wilson (1982) and Taylor (1984).

If a deterministic approach to techniques of cost modelling is no longer valid, then the direction of model development is clearly towards 'stochastic variability' (Mathur, 1982), where the duration and cost of activities or groups of activities are recognised as being uncertain, and are hence modelled as stochastic variables. Probabilistic approaches to the economics of buildings have been demonstrated through the research of Wilson (1982), Mathur (1982), Spooner (1974), Diekmann (1983) and others.

According to Hardcastle (1984), the fundamental weakness of traditional attempts to improve predictions in design evaluations has been the failure of existing communication documents to satisfy the criteria of an efficient information system. Following upon Hardcastle's recommendations, the major implementation problems facing building consultants are likely to be those of achieving adequate access to stochastically qualified information.

THE PARADIGM SHIFT

Although the nature of the paradigm shift in techniques of cost modelling can be explained in the sense of a shift from historical determinism to stochastic variability, from inexplicability to logical transparency, and from unrelated-ness to interdependence, a balanced approach to cost modelling requires a holistic consideration of the new paradigm.

The concept of the paradigm shift is illustrated diagrammatically in Figs. 4.4.1–4.4.2.

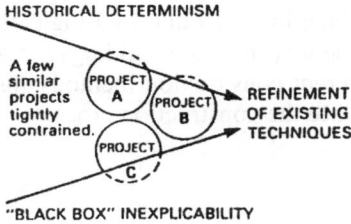

Figure 4.4.1 The existing paradigm, (*source*: Bowen and Edwards, 1985).

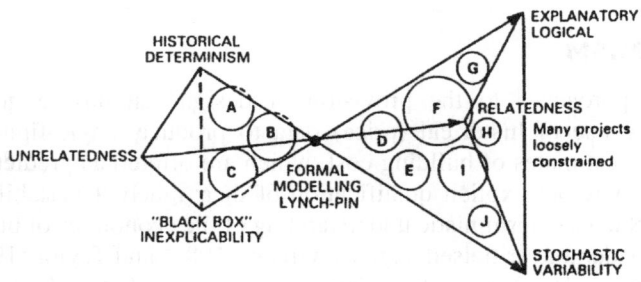

Figure 4.4.2 The paradigm shift (adapted from Bowen and Edwards, 1985).

The existing paradigm (Fig. 4.4.1) provides the framework for historical deterministic models which are statistically questionable in the inference they make, and which fail to explain the system they purport to represent. Such a paradigm adequately encompasses relatively few projects, similar in type, in a tight, over-constrained manner.

Through its emphasis on stochastic variability, explanatory understanding of construction processes and process interdependence, the new paradigm (Fig. 4.4.2) allows a far wider variety of often dissimilar projects to be embraced successfully in a more loosely constrained fashion. The lynch-pin of the new paradigm, providing the underlying theoretical foundation for its three thrusts, is identified as **formal modelling**. From the diagrammatic representation of the old paradigm it can be seen that continued refinement of existing techniques must eventually lead to a cul-de-sac in development possibilities (Fig. 4.4.1), whereas the adoption of formal modelling as the foundation for the new paradigm provides ever-widening opportunities for exploration (Fig. 4.4.2).

NETWORKS AND RELATED MODELS

Many attempts have been made in the past to model the process of construction in terms of networks. More recently, there has been growing interest in a probabilitistic approach to such models e.g., Bennett and Fine (1980). The underlying theme of these modelling attempts appears to be the desire to produce a modelling system which simulates reality as closely as possible, and permits the explicit treatment of the uncertainty inherent in the process. Traditionally these models have been derived as a means of enabling the more efficient and cost effective construction management of projects, with little or no attention being directed to the use of such models in the field of design economics and comparatives costing (Baxendale (1984), Legard (1983), White (1985) and Bennett and Ormerod (1984).

Earlier work by the Department of the Environment in Britain, which resulted in the COCO system (Department of the Environment, 1971), showed an appreciation of the need to consider and model the construction process in exploring the consequences of design decisions, but this system of the early 1970's failed to incorporate mechanisms to deal with data variability and uncertainty. Nevertheless, its significance for the development of process-based algorithmic modelling systems for design economics purposes should not be under-rated, particularly in its establishment of basic principles for exploring the cost consequences of design alternatives.

A PROCESS-BASED MODELLING SYSTEM

To diminish the three weaknesses discussed earlier, it is proposed that networks very similar to PERT/CPM networks be used as the basic modelling structure. The advantages of such an approach are immediately apparent. Inexplicability, for instance, is obviated since costs are introduced in the context (on the network arcs) in which they occur. Certain dependencies become apparent as a function of the network structure, while others (e.g., bulk buying) can be modelled. Lastly, stochastic variability can be introduced into the process and will be modelled in parallel to the network system.

Cost modelling using networks

For a given structure of type k, be it an office block or single-storey dwelling, it is possible to define the appropriate operational network for a given element or component as:

$$n \overset{\square}{=} (N_k, A_k)$$

By a network we mean a directed graph (N,A) in which N represents the set of nodes and A is an ordered binary relation over N. The appropriate network (N_k, A_k) is predetermined and can be chosen from a **set** of relevant networks via a selection procedure whereby various data, corresponding to a vector, $x^{(k)}$, of inputs and/or default values describing the element (or component) or structure type k are provided by the cost modeller.

Consider, as an example, the network depicted in Fig. 4.4.3 relating to numbered activities for a hypothetical piled foundation construction process.

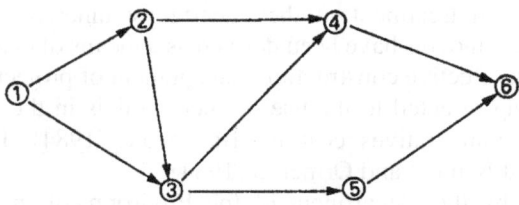

Figure 4.4.3 Piled-foundation construction process network.

In this instance, N = (1, 2, 3, 4, 5, 6)
and   A = ((1, 2), (1, 3), (2, 3), 2, 4), 3, 4), (3, 5), (4, 6), (5, 6))
            = $(a_1, a_2, a_3, a_4, a_5, a_6, a_7, a_8)$ i.e., the names of the activities associated with the arcs.

The number of possible network configurations for any discrete building element process is limited and, hence, the concept of a set of appropriate (say) foundation network processes is theoretically possible. The exact network chosen as being representative of the project element under consideration will be determined as a function of user inputs.

The vector, $x^{(k)}$, may be conceptualised as the physical and other parameters defining the particular construction process. For example, such a vector would include such information as the number of pile caps and piles, presence of column bases, length of strip footings, area on plan of foundations, etc. In the absence of such data, as is very likely in the early stages of design, default data relating to design/construction norms will be used. Such default data would be contained in an appropriate data-base. In reality, the vector $x^{(k)}$ will comprise two (or more) vectors, one relating to the input values, and another to data-base information such as relevant construction costs, etc.

Associated with the relevant network (N_k, A_k) and the vector $x^{(k)}$ will be vector functions $d^{(k)}(x^{(k)})$, $c^{(k)}(x^{(k)})$, etc. This relationship is depicted below as a mapping of vector functions. Vectors $d^{(k)}$ and $c^{(k)}$ refer to duration and cost vectors and are functions of user inputs, changing in a functional way as the values of $x_1, x_2, \ldots, x_n$ change. More specifically, $d_i^{(k)}(x^{(k)})$ and $c_i^{(k)}(x^{(k)})$ are the duration and cost associated with the ith arc of the network relating to project k. The exact calculation of duration and cost of each arc is possible because the relationships are expressed as predetermined functions defined by the designer (Fig. 4.4.4).

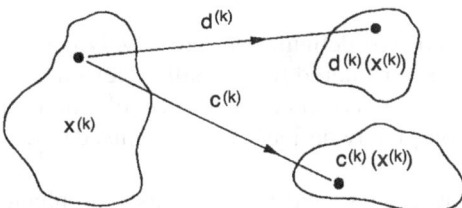

Figure 4.4.4 Mapping of vector $x^{(k)}$ onto cost and duration.

For example, consider activity 3 which may be (say) the process of excavating for foundation trenches. The time associated with this activity may be defined by the following functional relationship:

$$\text{duration}_3 = d_3(x) = \frac{x_1}{\square} \ \square \ \frac{1}{x_{17}} \ \text{days}$$

where x_1 = volume to be excavated = length $\square$ breadth $\square$ depth of footing

x_{17} = number of men in gang

$\square$ = volume of excavation/man-day (defined as a constant relating to that firm or the industry).

Similarly, the cost of the operation could be determined using the following costing algorithm:

$\text{cost}_3 = c_3 = c_3(x) = x_{17} \,\square\, d_3 \,\square\, x_{50}$
where x_{17} = number of men in excavation gang

d_3 = duration of activity (in days) (already computed)
x_{50} = cost/man-day

It follows from the above line of reasoning that the parameters originally defining the relevant network may now be expanded to include, *inter alia*, the duration and cost of each activity constituting the network. Relating this to the example illustrated in Fig. 4.4.3, the relevant network may be defined as:

$\text{n} = (N_k, A_k,$ $d^{(k)}(x), c^{(k)}(x), \ldots)$
        pre-determined   duration, cost, etc. –
        in terms of   function of user input
        operations. x.

where $N = (1, 2, 3, 4, 5, 6), A = (a_1, a_2, a_3, a_4, a_5, a_6, a_7, a_8)$
$d = (d_1, d_2, d_3, d_4, d_5, d_6, d_7, d_8)$
$c = (c_1, c_2, c_3, c_4, c_5, c_6, c_7, c_8)$

Naturally the parameters defining each arc or activity may further be expanded to include all factors considered functionally relevant.

Once the network and associated activities relating to the relevant element or component is completely defined, it may be used to determine the network duration and cost.

It is possible, at a theoretical level at least, to enhance the hypothesised model by employing Monte Carlo simulation to add a probabilistic dimension to he simulation model. By sampling from known or assumed distributions of the factors and costs involved in the building process, and with repeated iterations, it is possible to determine the expected value of the factor and/or cost under consideration, together with its associated probability distribution.

Such a technique is not without problems, however, the issue of dependency between asks, components, etc. ranking foremost amongst these (Brandon and Newton, 1986). Much research needs to be directed to problems such as this, and the issue of the nature of the probability distributions themselves, if meaningful models are to result.

Thus far the development of the conceptual framework has been restricted to a single process or element (e.g., the foundations of an office block). The issue of joining sub-networks needs to be addressed.

Joining of sub-networks

Following on from the reasoning outlined above, it would seem feasible that identical logic is applicable to all the other elements/components of a building. However, given that a choice of networks exists from within a family of relevant networks, it is evident that the problem of coupling chosen networks together needs attention. The concept of linking appropriate (chosen) networks is illustrated in Fig. 4.4.5.

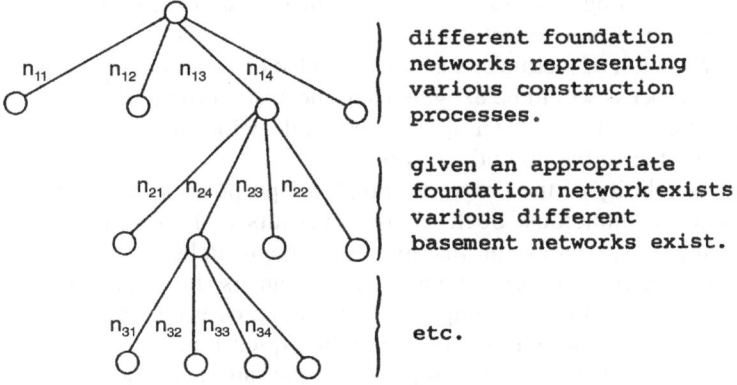

Figure 4.4.5 Linking of appropriate component networks.

Indeed, it is not inconceivable that various sub-networks can be attached to preceding activities at different points in time and nodal positions, depending on the dependencies created by **that** project. It is here that the writers consider the field of expert systems and artificial intelligence to be particularly important, creating the ability to link nodules of knowledge. The groundwork for this already exists within the field of mathematical **nebulae**.

Expanding the system

The conceptual framework outlined above is not restricted to the networks contained in the data-base. It would seem quite feasible, given the input data $(x_1, x_2, \ldots, x_n)$, to extract the **most** appropriate network and amend it (via an editing procedure) to represent more accurately the process being modelled. The logical extension of this is the fact that the network data-base can grow as more networks are added.

Use of the system

It is often argued, for example by Raftery (1984), that the type of cost model utilised by the cost advisor must match the stage of the design process and the level of information available. The issue of the model-data interface has been resolved, at least in part, by Newton (1983) who developed an expert system capable of use throughout all the stages associated with the design process.

The model conceptualised by the writers would appear also to be suitable throughout the design process, with norms, heuristics and historical information used in the early stages when exact data are lacking, giving way to the incorporation of detailed information in the latter stages of design. Thus, the cost modelling process is seen as cyclical, becoming more tangible with the provision of more exact data.

Optimisation of factors and cost would also be possible insofar as the optimisation process would be exogenous to the system (being, in fact, a discrete choice) done at the managerial level by the decision-maker who is presented with a number of Pareto optimal solutions.

The suitability of the proposed model for purposes other than cost determination are numerous. Such a model permits cash flow projections, and labour and other factor of production requirements to be scheduled. Its usefulness is not restricted to project planning as it can also be employed as a control device. Arguably the most important advantage over conventional methods, however, is the model's ability to permit the exploration of comparative design economics, producing cost comparisons that are generated in a process-related manner.

CONCLUSION

Conceptually, the development of modelling systems for the purposes of design economics must spring from the desire for a closer representation of reality. This implies on the one hand the acceptance and appropriate treatment of variability and uncertainty, and on the other hand a more thorough representation of construction processes, each taking place within a suitable simulation environment. Expert systems, representing an important set of applications of artificial intelligence, and used in conjunction with operational/resource-based networks and techniques, offer the potential for a process-based algorithmic modelling system which could make a significant contribution to decision-making for construction projects.

REFERENCES

Baxendale, T. (1984) Construction resource models by Monte Carlo Simulation. *Const. Mgmt. and Econ.*, 2, 201–217.

Bennett, J. and Fine, B. (1980) Measurement of complexity in construction projects. *Occasional Paper No. 8*, Department of Construction Management, University of Reading.

Bennett, J. and Ormerod, R. N. (1984) Simulation applied to construction projects. *Occasional paper No. 12*, Department of Construction Management, University of Reading.

Bowen, P. A. and Edwards, P. J. (1985) A conceptual understanding of the paradigm shift in modelling techniques used in the economics of building. *QS Occasional*

Paper Series, No. 6, Department of Quantity Surveying and Building Economics, University of Natal.

Brandon, P. S. (1982) Building cost research – need for a paradigm shift, in *Building Cost Techniques: New Directions* (ed. P. S. Brandon), E. & F. N. Spon Ltd., London, pp. 5–13.

Brandon, P. S. and Newton, S. (1986) Improving the forecast. *Chartered Quantity Surveyor*, May, 24–26.

Department of the Environment (1971) Cost consequences of design decisions: cost of contractors operations (COCO). *Directorate of Quantity Surveying Development*, London.

Diekmann, J. E. 91983) Probabilistic estimating: mathematics and applications. *J. Const. Eng. and Mgmt.*, Am. Soc. Civ. Eng., 109, 297–308.

Hardcastle, C. (1984) The relationship between cost communications and price prediction in the evaluation of building design, in *CIB Third International Symposium on Building Economics* proceedings, Ottawa, July, 3, 112–123.

Legard, D. A. (1983) Probabilitistic analysis of an idealised model for construction projects. *Const. Mgmt. and Econ.*, 1, 31–45.

Mathur, K. (1982) A probabilistic planning model, in *Building Cost Techniques : New Directions* (ed. P. S. Brandon), E. & F. N. Spon Ltd., London, pp. 181–191.

Newton, S. (1983) *Analysis of construction economics: a cost simulation model.* Unpublished Doctoral Thesis, Department of Architecture and Building Science, University of Strathclyde.

Raftery, J. J. (1984) *An investigation of the suitability of cost models for use in building design.* Unpublished Doctoral Thesis, Department of Surveying, Liverpool Polytechnic.

Science and Engineering Research Council (U.K.) (1984) *Specially Promoted Programme in Construction Management, Newsletter*, 6, Summer.

Spooner, J. E. (1974) Probabilitistic estimating. *J. Const. Div.*, Am. Soc. Civ. Eng., 100, 65–77.

Taylor, R. G. (1984) A critical examination of quantity surveying techniques in cost appraisal and tendering within the building industry. *QS Occasional Paper Series*, No. 3, Department of Quantity Surveying and Building Economics, University of Natal.

White, A. (1982) The critical path method and construction contracts: a polemic. *Const. Mgmt. and Econ.*, 3, 15–24.

Wilson, A. J. (1982) Experiments in probabilistic cost modelling, in *Building Cost Techniques: New Directions* (ed. P. S. Brandon), E. & F. N. Spon Ltd., London, pp. 169–180.

4.5 Automatic resource based cost-time forecasts*

V. K. Marston and R. M. Skitmore

ABSTRACT

There is a need for improved methods of time and cost forecasting for housing refurbishment projects. What is needed is a resource based non-deterministic approach. Recent advances in two fields appear to offer a way forward: firstly, much progress has been made in the automation of construction planning through expert system technology; secondly a number of construction plan based simulation models have been developed to provide probabilistic cost-time forecasts for construction work.

This paper reviews these advances and, by outlining a system currently under development at Salford University, shows how they can be applied to housing refurbishment cost-time forecasts.

Keywords

Cost, Time, Expert Systems, Simulation, Housing

INTRODUCTION

There is a huge stock of housing in the United Kingdom awaiting refurbishment. The 1988 Housing Act seeks to achieve a re-distribution of low cost housing ownership, with a reduction in local authority ownership and an expansion of the role of housing associations, and a move towards market related rents (1). In this new climate of ownership, effective forecasts of construction cost and time are essential, particularly to the assessment of development feasibility, the evaluation of alternatives, budget setting, and financial planning and control. Furthermore, to be truly effective, such predictions need to be accompanied by some evaluation of risk.

*1990, *Transactions of the 34th Annual Meeting* of the American Association of Cost Engineers, Symposium M – Project Control, M.6.1 – M.6.6. Reprinted with the permission of AACE International, Morgantown, USA.

Housing refurbishment projects are characterised by

- the sensitivity of cost and time to operational factors such as access, temporary works requirements, and continuity of working, and
- high levels of uncertainty centred on such matters as the condition of the existing structure and the availability of work areas.

The most suitable cost-time forecasting techniques to handle these special implications must therefore be

- resource based, to reflect the actual production process, and
- non-deterministic, to allow the inclusion of uncertainties and risk assessment.

The first part of this paper reviews some recent advances in cost-time forecasting systems of this type, both generally and in the construction field. The second part describes a new system, currently under development at the University of Salford, which incorporates and extends this previous work specifically for housing refurbishment projects.

RESOURCE BASED COST-TIME FORECASTING: STATE OF THE ART

The central feature of all resource based cost-time forecasting systems is the construction plan. PERT/CPM has long been advocated for use in planning, with mixed results. A major difficulty has been the dynamic nature of the object process which can result in carefully worked out networks quickly becoming out of date – a common feature in construction. The emphasis of recent developments reflects this situation.

Two approaches seem to be available. Firstly, to automate the planning process as much as possible to speed up initial work and replanning responses to situational changes. Secondly, to model minor perturbations by means of computer based simulation studies.

Some recent advances in these two approaches – automation and simulation – are described below.

Automation

Current construction planning relies upon the manual formulation of plans and is usually peformed in an intuitive and unstructured fashion with considerable reliance on engineering judgement (2). The generation of PERT/CPM networks is painstakingly performed by engineers who work directly from the project drawings with few computer based aids other than general project templates or past project networks that can be adapted to the particulars of a new project.

Automated planning, however, has been a feature of artificial intelligence (AI) research since Newell and Simon in the early 1960's (3). This defines the planning task as predetermining a transformation process by choosing a set of actions that, when arranged chronologically, will eventually convert some initial situation into a 'goal' situation. Plans can be either linear, where each activity is treated strictly in sequence, or nonlinear (typical of construction operations), where activities may occur in parallel (4,5). NOAH (5), an early goal seeking nonlinear planning system, used a set of procedures termed 'critics' to identify unfavourable interactions between different parts of the plan, but without any facility to backtrack from 'bad' decisions. NONLIN (6), an extension of NOAH, was developed to detect interactions by analysis of the underlying goal structure and, with the inclusion of backtracking, applied to a construction project planning problem. More recently, MOLGEN (7), has been developed to use constraint generation and posting to reduce back-tracking in genetic applications. other systems that plan with temporal con-straints include DEVISER (8), for planning the operation of space probes, and NUDGE (9), a meetings schedule planner.

Several construction specific automated planning systems have been reported using knowledge based systems (KBS) technology.

PLATFORM (10), for off-shore platform construction, updates either network attributes (eg. durations) or topology as information concerning the actual duration of activities is received. Topology alteration is achieved by choosing among several alternative predefined sub-networks for major activities.

CONSAS (11), for medium-high-rise reinforced concrete buildings, analyses construction schedules by means of a knowledge base consisting primarily of scheduling decision rules, construction 'common sense' knowledge, and construction knowledge developed or used by the experts when planning a project. An important characteristic of the system is its utilisation of sophisticated existing software for the expert system (Personal Consultant Plus), project control (Primavera Project Planner) and data base (dBASE III Plus). A more substantial version is under development which replaces PCP with the more powerful ART, uses object oriented programming and allows interfaces to cost, schedule, and quality control modules of a broad project control system (11).

CONSTRUCTION-PLANEX (2) suggests technologies, generates activi-ties, determines precedences, estimates durations, and produces a complete plan, provisional schedule and cost estimate for the ground works and frame of a modular high rise building system. The program's knowledge base is derived from a large number of sources, each input as decision tables and converted into a network of frame schema. The system is also said to provide a good user interface and impressive graphics.

LIFT-2 (12) generates plans for heavy-lift operations using a hierarchy of goals, constraint satisfaction, and a trial and error facility in seeking the achievement of goals and sub-goals.

GHOST (13) is a prototypical network generator, forming part of MIT's CONPLAN system,* which relies on an input set of activities for schedule production. This is done incrementally by first assuming a naive nonlinear solution for correction by a set of knowledge based 'critics' which know about basic physics, construction norms, network redundancy, etc. GHOST does not extract activities from drawings or estimate durations.

MIRCI (14) is another prototypical system for construction planning. MIRCI comprises three main integrated software components (1) a KBS activity generator, using the Leonardo Level 3 shell, (2) data storage, using Dbase III, and (3) project planning software, using Pertmaster Advance. The activity generator operates at three levels (1) the executive level which identifies functional (design) elements, (2) an intermediate level which relates functional elements to their corresponding constructional activities, and (3) a final level (still under development) which takes into account all possible site conditions, available resources and weather conditions. The system is said to have a 'friendly user interface'.

Simulation

All the automatic planners described above are deterministic in nature, ie. the estimates contained in the system are treated as exactitude. Several approaches are available to take into account the likely differences between estimates and actual outcomes. One such approach is to adopt a probabilistic model under the assumption that all unaccounted variability is purely random (in the statistical sense). In simple probabilistic problems, solutions can often be obtained analytically by relatively straightforward pencil and paper work. For most practical applications however, including construction planning, solutions can only be obtained numerically by a long series of laborious calculations usually performed by computer. A particularly useful numerical technique of this kind is stochastic or Monte Carlo simulation, by which the random aspects of the problem are modelled by pseudo random number generation. By repeated sampling and problem solving, it is possible to generate a set of potential solutions, any of which may be correct. Inspection of these potential solutions provides an indication of the likely range of answers to the problem, the actual outcomes, the probability of a certain event occurring, and thus a measure of risk associated with potential events.

Several attempts have been made to develop probabilistic resource based models (15,16,17,18,19,20), one of the most advanced general purpose systems of this type being CPS (21), Reading University University's Construction Project Simulator. This system runs on a micro computer using stochastic

*Currently under development. Intended input to CONPLAN is (1) construction drawings, (2) materials specifications, (3) resources available, and (4) past approaches. Output should include (1) optimised networks, (2) durations, and (3) network analysis.

simulation to generate cost-time frequency distributions from probabilistic activity times and interruptions between activities. A comprehensive range of probability functions is available in the system, some of which are pre selected (eg. weather conditions) but mostly for user selection. Output is in the form of histograms or cumulative frequency curves. All activities, precedences, costs, time parameters, and many of the probability distribution parameters have to be input by the user.

Although it is theoretically possible to combine automatic planning and dynamic rescheduling with stochastic simulation (22), no such combined system is yet available.

Discussion

As yet, no one system has been developed which combines current advances in both automatic and non deterministic planning.

Of the automatic planning systems, MIRCI appears to be the most comprehensive general purpose system as it is capable of transforming design information into production information with a minimum of user involvement and without restriction to any specific construction system. It identifies activities and durations, determines activity relationships, establishes precedences, and devises a suitable network based on all known information concerning site conditions, available resources and weather conditions. It does however lack PLATFORM's ability to update attributes and topology in response to actual changes in activity durations – an important 'real time' feature necessary in simulation systems.

CPS on the other hand, though providing a good representation of several important variability aspects associated with construction work, is completely lacking in automatic facilities. This clearly a major drawback for its intended users who are not expected to be greatly skilled in construction planning. Also it has not yet been possible to acquire empirical parameter estimates for all the probability distributions included in the system, many estimates being currently obtained by subjective user assessment. Again the limited experience and skill of the intended users of the system is a clear constraining factor. A further point concerning CPS is that, to be completely non deterministic, cost as well as duration needs to be treated probabilistically.

Finally, a recent survey by Bell (23) has identified the importance of visually interactive modelling (VIM), and particularly visually interactive simulation (VIS) in problem solving aids. Of special relevance to simulations involving 'real time' dynamic changes, such as in replanning, is the work of Alemparte et al (24) and Palme (25) in using simulation driven moving displays. The reported success in the use of such techniques as a means of communicating the state of the simulated process to the user suggests that this may be an approach of great potential for project managers, who are known to rely on sets of key performance indicators and exception reporting mechanisms for status assessment (11).

THE HOUSING REFURBISHMENT SYSTEM

Our examination of previous work now indicates that the new system should satisfy the following revised criteria

1   resource based, to reflect the actual production process
2   automatic translation of design to construction process, to minimise demands on user technical expertise, by means of

   a   element generation
   b   activity type, duration and cost generation
   c   precedence generation
   d   network generation
   e   schedule generation

3   probabilistic, to allow the inclusion of uncertainties, with automatic generation of activity duration and cost probability distribution parameters
4   capable of stochastic simulation, for risk assessment, with

   a   'real time' dynamic automatic replanning in response to changes in activity durations and costs
   b   automatic generation of histograms and cumulative frequency curves
   c   backtracking facilities, for post stimulation analysis.

The experiential base of many of the decisions involved in planning/ replanning and probability assessment indicates that the expert system approach is likely to dominate the automatic functioning aspects of the system, at least until more fundamental research is completed.

   In our view, these criteria translate into a proposed new system for housing refurbishment time-cost forecasting containing the following operational characteristics

1   an expert system to support the generation of the project operational plan
2   a stochastic simulation facility to evaluate time and cost from the operational plan, incorporating the effects of both interferences to the production process and variable performance
3   an automatic KBS based control function to emulate production management control of the simulated operations
4   a user friendly VIS report and monitoring facility (cockpit) to permit real-time monitoring of the simulation and full status reporting

   Figure 4.5.1 illustrates a possible configuration of the new system. Design information is input to the system which automatically generates probabilistic cost-time data, via simulation iterations. User modifications to design can then be assessed in terms of cost-time until a satisfactory design solution is found.

Alternatively, the production control function (replanner) may be modified as a means of assessing potential production management strategies. Thus, by unifying the design and production functions, the user is able to investigate both aspects simultaneously to improve for instance, buildability.

As we know, most of the technology demanded by this system is already in existence. The probabilistic element is introduced into the activity network in a similar manner to PERT's use of beta models. increasing the usual single parameter durations and costs to a multi-parameter equivalent (eg. by adding upper and lower limits) provides the extra data needed. Estimates of the activity unit cost parameters are generated by a permanent system data base of

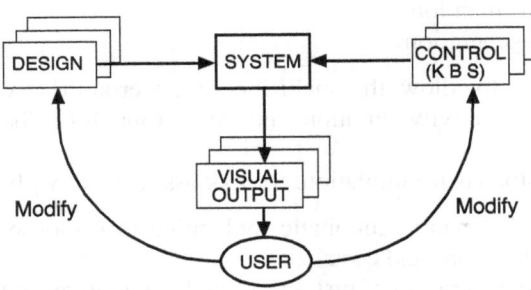

Figure 4.5.1 System configuration.

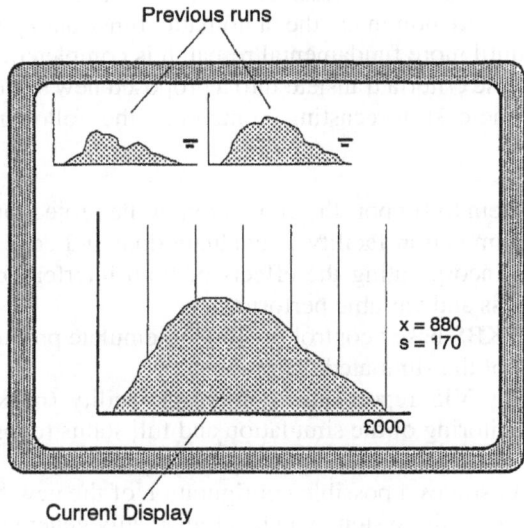

Figure 4.5.2 The cockpit.

costs, while the activity duration parameter estimates are generated by the activity generating knowledge base. *The most significant new feature is the introduction of dynamics to the topology of the network through the automatic control module.*

The remaining aspect of interest is the system's output format which is a dynamic visually interactive 'cockpit' (Figure 4.5.2). With each simulation iteration, the cost-time status is updated on the screen. When the user feels the system is sufficiently stable, the iteration sequence can be terminated. The result may be saved in a screen window at the user's discretion. Following a change in design or control information, the saved status indicators may be compared for an assessment of the cost-time consequences as an aid to selecting the best combined design/control strategy for the project.

Full backtracking facilities allow the user to trace the major factors contributing to each cost-time outcome, and also any 'fine tuning' of the knowledge base for special project characteristics.

CONCLUSION

The system outlined above is currently undergoing development by a team of researchers led by the authors in collaboration with the Salford University's Information Technology Institute, a commercial design organisation, construction company, and housing association in the Manchester area. The work is anticipated to reach completion within the next three years.

ACKNOWLEDGEMENTS

The authors wish to express their thanks to Mustafa Alshawi of Salford University, Chris Hogben of Hogben Associates, Building Surveyors and Housing Development Consultants, Manchester, and Colin Mawson and Chris Ryan of Wimpey Construction UK.

REFERENCES

1 Stafford, B. and Brandon, P. S., *The Potential for Expert Systems within Housing Associations: an overview*, University of Salford, 1988, pp. 5–7.
2 Hendrickson, C., Zozaya-Gorostiza, C., Rehak, D. and Baracco-Miller, E., Expert System for Construction Planning. *Journal of Computing in Civil Engineering*, American Society of Civil Engineers, vol 1, No 4, 1987, pp. 253–269.
3 Newall, A. and Simon, H., *Human Problem Solving*, Prentice Hall, 1972.
4 Nillson, N. J., *Principles of Artificial Intelligence*, Tioga Publishing Co., Palo Alto, California, 1980.
5 Sacerdoti, E., A Structure for Plans and Behaviour. *Tech Report* **109**, SRI Int, 1975.
6 Tate, A., Generating Project Networks. *Proceedings* 5th IJCAI, 1977, pp. 888–893.

7   Stefik, M., *Planning with Constraints*, PhD thesis, Stanford University, 1980.
8   Vere, S. A., Planning in Time: Window Cutoff for Activities and Goals. *IEEE Transactions* Pattern Analysis and Machine Intelligence, PAMI, 5(3), 1983, pp. 246–267.
9   Goldstein, I. and Roberts, B., NUDGE: a Knowledge-Based Scheduling System. *Proceedings 5th IJCAI*, 1977, pp. 257–263.
10 Levitt, R. E. and Kunz, J. C., Using Knowledge of Construction and Project Management for Automated Schedule Updating. *Project Management*, XVI(5), 1985, pp. 57–76.
11 Ibbs, C. W. and De La Garza, J. M., Knowledge engineering for a construction scheduling analysis system. In *Expert Systems in Construction and Structural Engineering*, H. Adeli (ed.), chapter 8, 1988, Chapman and Hall, pp. 137–159.
12 Bremdal, B. A., Control Issues in Knowledge-Based Planning Systems for Ocean Engineering Tasks. *Proceedings 3rd Int. Expert Systems Conference*, 1987, pp. 21–36.
13 Navinchandra, D., Sriram, D. and Logcher, R. D., GHOST: Project Network Generator. *Journal of Computing in Civil Engineering*, American Society of Civil Engineers, Vol 2, No 3, 1988, pp. 239–254.
14 Alshawi, M. and Jagger, D., An Expert System to Assist in the Generation and Scheduling of Construction Activities. *Proceedings, AI Civil-Comp.*, London, Sept, 1989.
15 Baxendale, T., Construction resource models by Monte Carlo Simulation. *Construction Management and Economics*, Vol 2, 1984, pp. 201–217.
16 Legard, D. A., Probabilistic analysis of an idealised model for construction projects. *Construction Management and Economics*, Vol 1, 1983, pp. 31–45.
17 White, A., The critical path method and construction contracts: a polemic. *Construction Management and Economics*, Vol 3, pp. 15–24.
18 Newton, S., *Analysis of construction economics: a cost simulation model*, Phd Thesis, 1983, Department of Architecture and Building Science, University of Strathclyde.
19 Thompson, P. A. and Willmer, G., CASPAR – A Program for Engineering Project Appraisal and Management. *Proceedings 2nd International Conference on Civil and Structural Engineering Computing*, Vol 1, London, December, 1985, pp. 75–81.
20 Pohl, J. and Chapman, A., Probabilistic Project Management. *Building and Environment*, Vol 22, No 3, 1987, pp. 209–214.
21 Bennett, J. and Ormerod, R. N., Simulation Applied to Construction Projects. *Construction Management and Economics*, Vol 2, No 3, 1984, pp. 225–263.
22 Bowen, P. A., Wolvaardt, J. S. and Taylor, R. G., Cost Modelling: a Process-Modelling Approach. In *Building Cost Modelling and Computers*, P. S. Brandon (ed.), 1987, E. and F. N. Spon, pp. 387–395.
23 Bell, P. C., Visual Interactive Modelling in Operational Research: Successes and Opportunities. *Journal of the Operational Research Society*, Vol 36, No 11, 1985, pp. 975–982.
24 Alemparte, D., Chheda, D., Seeley, D., Walker, W., Interacting with Discrete

Simulation Using On-Line Graphic Animation. *Computer Graphics*, Vol 1, 1975, pp. 309–318.

25 Palme, J., Moving Pictures Show Simulation to User. *Simulation* Vol 29, 1977, pp. 204–209.

Part 5 Dealing with uncertainty

This part contains a collection of papers relating the general them of uncertainty either in terms of validation or non-deterministic cost modelling. The papers dealing with validation use an essentially black box approach to quantify the performance of models through their accuracy, the more recent work concentrating on the human aspects involved. The modelling papers range from the straightforward use of probabilistic and stochastic simulation to the application of fuzzy set theory and the use of dynamic modelling to incorporate feedback.

Beeston's **One statistician's view of estimating** (ch 5.1) is truly seminal in a number of respects. His decomposition of variance, termed {distribution of variability' (which attempts to separate the accuracy of the model from other influences on accuracy, is the perhaps the most quoted and reappears continually even today. Beeston urges modellers to relax the principle of homogeneity much more than is usual in practice and analyse a large sample of previous buildings for estimating the element and function values. Although, when first published, the paper was too difficult for most readers, now we understand it better there are many questions to be asked. The main difficulty is that most of the details of the empirical analysis have been omitted and it would be useful to have these now for checking and replication purposes. The major interest is in the decomposition and its derivation. The first problem with this is the assumption of independence between the various factors assessed. The second is that it applies only to PSA contracts, which may be different to others. Thirdly, times change and there may be special effects in changing economic climates (cf De Neufville *et al.*, 1977). Finally, the distinction of *within* or *between* variability is unclear – does 4% cv for QS intuition mean 4% for each person or 4% between persons? Probably the main reason for the paper's popularity is that it addresses the subject at just about the right level for many serious readers – offering an easy introduction to the statistical content and intuitively feasible results, explained in mostly familiar lay terms. This is a paper almost as important as Freidman's (1956) is to bidding, and which, together with Barnes (1971), forms the foundation of the statistical approach to the analysis of forecasting accuracy.

Ashworth and Skitmore's **Accuracy in estimating** (ch 5.2) is just a literature review on accuracy, mainly in the construction industry, of pretender cost forecasts. Its novelty is that it is more comprehensive than most accounts on the subject and also draws in a lot of 'by the way' comments on the influence of human factors. The major conclusion from the work is that accuracy is around 15–20% cv in early stages of design and improves to around 13–18% cv prior to tender, and that big differences can be expected with different implementors. This is really a prelude to Skitmore *et al.*'s (1990) experimental work on human factors. The review has been updated, expanded and largely reproduced by Ogunlana and Thorpe (1988).

Morrison's **The accuracy of quantity surveyors' cost estimating** (ch 5.3) main claim to fame is the number of projects retrospectively analysed (915) and for which pre-tender cost forecasts (priced BQ) and lowest bid values were available – giving an average coefficient of variation of accuracy of 12.77% (9.80% mean deviation) – over seven separate and predominantly public sector quantity surveying organisations. The differences between offices' accuracy was ascribed to the different types of projects undertaken, with supermarkets, schools and houses being done more accurately. This is attributed to the cost information available and experience of the implementators involved rather than some inherent characteristic of the project itself. Morrison also claims that accuracy is better, in percentage terms, for larger projects (a claim found in several other papers). He also comes to the surprising conclusion that cost forecasts using the BQ model are only slightly more accurate than when using the 'elemental cost analysis' model.

Wilson's **Experiments in probabilistic cost modelling** (ch 5.4) describes early attempts at modelling the uncertainty in sub-system cost forecasts and determine the effects on the evaluation of design alternatives by means of **controlled experiments** using the Delphi method. Assuming triangular distributions, he then attempts a Monte Carlo simulation to forecast total uncertainty with a presumed improvement on the PERT approach. The problem of interdependence is mentioned but not addressed. It is interesting to note that only the rate (element function value) is treated in a probabilistic manner – the quantity (element value) is treated as deterministic (see also Bennett and Ormerod ch 4.3 for a more sophisticated later work). The paper contains many astute remarks, especially in the introduction, with states the author's position with great clarity on several pivotal matters including the role of **purpose** in modelling, a definition of **macro** and **micro** models, the separation of monetary costs from other types of cost, a definition of architecture, the volume of design decisions, the novel view of product vs process approaches as deductive and inductive and the need to move from one approach to the other over time, a coherent statement of the role of uncertainty the Central Limit Theorem applied to swings and roundabouts (cf Barnes 1971), and the important aspect of relativity in micro modeling (cf Bathurst and Butler, 1973).

Flanagan and Norman's **The accuracy and monitoring of quantity surveyors' price forecasting for building work** (ch 5.5) is in two parts (1) an analysis of two LA pretender cost forecasts and (2) a proposed monitoring system for bias. The second is the more important. The analysis in (1) is a little different in treating the contract value (low bid) as an independent variable (with the cost forecast as the dependent variable) to test for systematic changes in bias with contract value. In fact the same results can be achieved more easily by using the more conventional approach of taking the forecast/low bid ratio as the dependent variable. Also the use of number of bidders as a proxy for market conditions is rather confusing as the number of bidders *per se* are known to influence the level of lowest bid and therefore accuracy of forecast. The proposals in (2) are the first to attempt a positive use of feedback for debiasing. These adopt a well known method (CUSUM) without modification but nevertheless represent a considerable advance in the treatment of feedback – an area palpably lacking in this field. Unfortunately, no further development of this, or any other, approach has been reported (except Skitmore and Patchell, ch 1.5).

REFERENCES

Barnes, N. M. L., 1971, *The Design and Use of Experimental Bills of Quantities*, PhD Thesis, University of Manchester Institute of Science and Technology.

Bathurst, P. E. and Butler, D. A., 1973, *Building Cost Control of Techniques and Economics*, Heinemann.

Neufville, De R., Hani, E. N. and Lesage, Y., 1977, Bidding model: effects of bidders' risk aversion, *Journal of the Construction Division*, 103, no CO1, March, 57–70.

Ogunlana, O. and Thorpe, T., 1988, Design phase cost estimating: the state of the art, *International Journal of Construction Management and Technology*, 34–47.

5.1 One statistician's view of estimating*

D. T. Beeston

SUMMARY

This paper assesses the level of estimating performance which should be attainable by present methods and another which might be achievable by improved methods. These two levels would produce respectively about 85% *and 90% of estimates within 10% of the lowest tender.*

To improve estimating performance with traditional methods the basic historic data used in cost planning and estimating should whenever possible be drawn from several buildings rather than one, even if this means sacrificing some degree of comparability.

In practice no useful gain in accuracy can be achieved by aiming the estimate at the second or third lowest or the average tender.

The most promising way to achieve a great improvement in estimating accuracy in the long urn is to devise methods for directly estimating con-tractors' costs. Such methods would also provide the best means of giving cost advice to designers.

INTRODUCTION

This paper was originally published by the Property Services Agency (DOE).

Mr Beeston's predicted results may differ from subscribers' views of the accuracy of their own estimating. It would be interesting and worthwhile for subscribers to check these statistical results by a survey of their own esti-mating performance.

Subscribers should be sure to include all estimates in their survey including those where the inaccuracy can be explained. The accuracy of the estimates should be viewed from the client's objective position not the quantity sur-veyor's subjective one.

*1974, *Cost Study F3*, BCIS 1974/5-123-9 to 16, © Building Cost Information Service, Royal Institution of Chartered Surveyors.

This paper is intended primarily to establish a reasonable aim for estimating performance and to give a little guidance on the use of historic data for estimating and cost planning. At the same time it may act as a useful source document to those working in the field of research into building prices.

Evaluation of specific methods of estimating and cost planning is beyond the scope of this paper but the reader may find it useful to appraise his own methods. The paper is intended more to encourage an objective way of thinking about current methods than to compare them.

In the interest of brevity and clarity I have not described the analyses and sensitivity tests upon which the various coefficients of variation are based. Such detail would have made a long paper very much longer. Again in the interests of clarity most of the figures are rounded off in the test even though the calculations leading to them were made to a sufficient degree of accuracy.

No knowledge of statistical theory is required to understand the reasoning. Only one statistical measure is used. this is the coefficient of variation and its meaning is described below. Other statistical techniques and measures, notably formal analysis of variance and sampling theory, were used in the analyses from which these results are drawn but the argument has been arranged so as not to need their use.

VARIABILITY OF PRICE DATA

Cost planning and estimating rely heavily on historic price data. The reliability of these data can be measured by the consistency which they exhibit when many prices are obtained relating to the same item description. The item may be as small as an entry in a Bill of Quantities or as large as the value of the whole building.

Even if repeat prices are not obtainable this concept of variability is still useful. Any price should be looked upon as a member of a huge imaginary family of prices for, as nearly as we can define it, the same item. The prices in the family have different values any of which could by chance have occurred instead. The one that did occur is no more 'correct' than any of the others. If there is a 'right' price it is some sort of average of all the possible prices in the family but if we have only one price we do not know where it stands in relation to the average.

If we know more than one price in the family the average of those that we know is an estimate of the average of the whole family and the more repeat prices we have the better the estimate of the average.

The variability between prices may be due to known or unknown causes. Known cases can sometimes be isolated and allowed for thus creating new families within the original one.

Statistical theory helps in dealing with the situation of variable data and in my work in the Directorate of Quantity Surveying Development (now part of the Property Services Agency) in the Department of the Environment I have studied several aspects of price variability which, when brought together, give a picture of the way prices behave.

Some of the figures needed in the following argument are not measurable directly but I have obtained them by approaching them from more than one direction and making a few reasonable guesses. Sensitivity tests, some of which are quoted, give reason for confidence in the conclusions.

THE COEFFICIENT OF VARIATION

The measure of variability of most value statistically is the coefficient of variation. This is *the standard deviation* (root-mean-square of deviations from the arithmetic mean) *expressed as a percentage of the mean.*

Prices tend to be distributed about their mean in a lopsided (skew) way because *high prices tend to be further above the mean than low prices are below* it. When statistical analyses are being carried out this must be allowed for. Although it appears to have been ignored in the figures quoted here, more rigorous analysis has been made to ensure that the figures do not mislead. In no case would it be wrong to use the figures given here as they stand.

An idea of the way the prices are distributed is given by the general rule that *two-thirds* of the prices are likely to fall within the coefficient of variation above and below the mean.

When variability is due to a *combination of* several distinct *causes* each of which has its own coefficient of variation the overall coefficient can be obtained in the work described here by adding the squares of the individual coefficients and taking the square root of the total. There are sometimes dangers in this procedure but they do not invalidate anything here.

VARIABILITY OF WHOLE BUILDING PRICES

The ideal data for an analysis aimed at measuring the inherent variability of whole building prices would be a large number (more than 10) of repeat prices independently obtained for the same design to be built in the same part of the country at about the same time. Such data can hardly ever be found but the nearest approach to come to my notice has produced a coefficient variation of just under 10% for lowest tenders for identical superstructures in the same region. *In every case the estimated cost was the same apart* from small adjustments for different dates of tender, so there is evidence here of the size of the inherent (as far as the quantity surveyor is concerned) variability of whole building prices. Another batch of buildings only slightly less standard produced the same coefficient of variation. Both types of buildings were smaller than the average PSA project and other work shows that a reduction of 10% in the coefficient of variation would be expected for buildings of typical size so an 'identical building' figure of 9% will be assumed. Further adjustment for the effect of slightly differing locations leads to a final coefficient of variation of just under $8^1/_2\%$.

A similar effect to identical designs can be achieved by standardising whole building prices to allow for different building composition. This can be done by comparing the prices in the Bills of Quantities item by item with those in a standard schedule of prices for work of the same description. A method based on this principle can be looked upon as standardising for building composition and is in use in PSA (reference 1). When this is done for a large number of projects the coefficient of variation of the standardised prices is about *11%*.

Before this figure can be compared with that for identical buildings the effects of different locations and types of buildings must be removed, and this is dealt with under the heading 'Causes of variability' where the 11% coefficient is reduced to just over $8^1/_2\%$, thus confirming well enough the figure for identical buildings.

This amount of variability is less than that which is recorded when the costs per square metre of whole buildings or elements are analysed for a number of buildings. This is reasonable because cost per unit area is a cruder price standardising method than the use of rates in Bills of Quantities as described above.

VARIABILITY OF BQ RATES

The coefficients of variation of BQ rates for similarly described items of measured work in Bills of Quantities are much greater than for whole, identical buildings. They are also different from each other. Typical figures for selected items in the various trades are as follows. They are coefficients of variation from bill to bill for the same item. They are not whole trade figures but are averages for several items in the trade.

Excavator	45%	Joiner	28%
Drainlayer	29%	Roofer	24%
Concreter	15%	Plumber	23%
Steelworker	19%	Painter	22%
Bricklayer	26%	Glazier	13%
Carpenter	31%	All trades	22%

It would be wrong to read much into differences between these figures. They are fit only to classify the BQ rates for trades into broad categories of reliability.

It is interesting to see the high figure for 'excavator' because it is a major trade common to both building and civil engineering. It fits well with the 40% coefficient of variation for a typical civil engineering bill item, related to its true cost, quoted in paragraph 2.4 of reference 4. It is clearly much greater than a typical building item.

METHODS OF ESTIMATING

It is time to take a first look at the effect of these figures on the quantity surveyor's work.

When he uses a price relating to one building and applies it to another, with
or without adjustment, he should remember that the original price should be
looked upon as having been chosen by chance from a hypothetical family of
which other members would have had different values and which could have
been chosen by chance instead. The variability of the prices in the family is
such that two-thirds of them lie within a range covered by twice the coefficient
of variation. Thus the price under consideration is only twice as likely to lie
within this range as outside it.

The use of data from previous cost analyses of Bills of Quantities is
the commonest identifiable technique used by quantity surveyors when cost
planning and estimating. If the cost analysis selected relates to a very similar
design to the one for which cost data are required and the price is updated
using a price index its reliability will be that of the original data combined with
that of the index, which has a coefficient of variation of 1% in the case of the
PSA quarterly tender price index.

Although simple adjustments to the price for time, location, building types
and value are well worth making, it is shown later how hard-won is
every reduction in coefficient of variation by adjustment of factors causing
variation. On the other hand taking the average of several appropriately
adjusted price analogues will reduce the variability by a factor equal to the
square root of the number of price analogues. This is potentially a much
more powerful way of reducing the coefficient of variation. For example the
average of four prices has half the coefficient of variation of one, if all four are
equally good analogues. If they are not the improvement will be less, but just
as it is difficult to make much reduction in variability by adjustment of prices,
so the increase in variability cased by mismatching analogues will also not be
great.

For this reason, if choosing buildings or parts of buildings for price analogues, I would much rather use the average of simply adjusted prices from
several reasonably similar buildings or items than just one, however exact
the similarity of the chosen single example. Although it would lead to an
improvement in estimating performance overall, such a method is seldom
of use in the most difficult cases because of the dearth of remotely similar
buildings for use as analogues. As it is in difficult cases that improved methods
would be most useful I shall not assume that this method will be available for
the purpose of this analysis of potential estimating performance.

The methods of adjustment referred to so far are those which allow for the
particular characteristics of the building chosen as the analogue. There is one
method of appraisal of price, however, which may reduce variability
more than the adjustments so far considered. This is the method used in the

Directorate of Quantity Surveying Development (reference 1) where the prices are standardised by comparing a sample of the rates in the Bill of Quantities with a standard schedule of rates. An index is attached to the cost analysis to indicate how the prices in the Bill of Quantities compared with those in the Schedule.

The reliability of this method of adjustment cannot be directly measured but the method is accepted by quantity surveyors as the best available so it can be assumed that the variability is less than that resulting from adjustments for characteristics of the buildings.

All that has been said about using rates from Bills of Quantities is true for cost data related to areas or volumes of the building or elements of the building except that in such cases the variability is greater so the advantage of taking several analogues is greater. As it is easier to obtain such data it should be possible to obtain larger numbers of analogues.

The use of price analogues is only one way of estimating. Methods using standard price rates for work measurement items or groups are subject to the level of variability experienced in the indexing process refereed to in reference 1. As explained later under 'Sensitivity', the number of items being priced or the number of groups into which they are formed is unlikely to have a great effect on the observable estimating variability. This should be a little better than that obtainable from the careful use of a single price analogue.

CAUSES OF VARIABILITY

A few of the causes of variability in building prices have been studied and *location, building function* and *contract value* have proved to be important causes. Unfortunately, although they are important it is an uncomfortable statistical fact that when causes of variability are removed (for example by adjustment for them) the *effect on the amount of variability is very little*. The reason for this is that in the case of the factors considered in this paper the combination and separation of coefficients of variation is by squares. The following figures illustrate this.

The location of a building is responsible for a variability in price amounting to a coefficient of variation of 5.2% but when the effect of location is allowed for using adjustments derived from over 1,000 buildings the overall coefficient is only reduced from 10.9% to 9.6%. Removing the effect of both building function and contract value brings the result down only to 8.6%, even though they jointly have a coefficient of variation of 4.2%. In tabular form the explanation is as follows (Table 5.1.1). Adding and subtracting can only be done when the coefficients of variation are squared.

The question naturally follows of how far this process of removing causes of variability can be taken. There is no indication of this yet for the whole range of buildings but for the special field of local authority housing a study carried out at University College, London (reference 2) showed that location

Table 5.1.1 Explaining the causes of variability

Factor	Coefficient of variation % (CV)	(CV)²
Total	10.9	118
– Location	–5.2	–27
Remainder	9.6	92
– Function & Value	–4.2	–18
Remainder	8.6	74

and building type accounted for two-thirds of the very large variance (compared with a quarter in our case) with only a very small amount being explained by the other variables which they studied. They were left with a coefficient of 12% due to unidentified causes.

The search for causes of variability goes on but the prospects of reducing the coefficient of variation due to unexplained causes to much below 8% are poor. To achieve even this level will require the identification of factor which jointly account for a coefficient of variation of 3%.

Causes of variability for which the quantity surveyor cannot make adjustments combine to produce what I shall call the 'QS observable estimating variability'. This is the familiar variability of the difference between the estimate and the lowest tender.

So that the coefficient of variation shall be unaffected by the values of the contracts the differences between the estimate and the lowest tender must be measured by the ratio between tender and estimate expressed as a percentage from which 100 may be subtracted.

To allow for the probability that there is an acquired ability among quantity surveyors to make *'intuitive' adjustments* to prices I shall assign a figure of 7% to the expected observable estimating variability instead of the 8% left unaccountable for above. The reduction from 8% to 7% would leave a remainder representing an ability to allow for unidentified causes which alone would have a coefficient of variation of 4%: the same as building function and contract value combined.

I now summarise the distribution of variability so far deduced:

a.   Location factors   5%
b.   Building function and value   4%
c.   Other identifiable causes   3%
d.   'QS intuition'   4%
e.   QS observable estimating variability   7%

All causes ($\sqrt{a^2 + b^2 + c^2 + d^2 + e^2}$)   11%

A coefficient of variation of 7% for the QS observable estimating variability is the performance which can be expected using present methods in the best way under average circumstances.

VARIABILITY OF TENDERS

An important contributor to the QS observable estimating variability is the variability of the lowest tender. To deduce this we can start with the variability of tenders between contractors bidding for the same contract. This is available only for whole building prices and the average coefficient of variation of these is usually about 5.2% *but with the recent uncertainty of the market it has risen to 6%.*

Taking 5.2% as the long-run average we can deduce the variability which would be exhibited by the lowest tenderers in a series of identical contracts in which the only differences in circumstances are those which also occur between tenderers for the same contract.

The long-run average number of contractors bidding for PSA projects worth more than £50 000 is just under 8 and, under present rules, does not depend much on contract value. Because bids are normally distributed about their average the coefficient of variation of the lowest tenders in the circumstances described in the previous paragraph should be 61% (reference 3) of that for the whole list. 61% of 5.2% is 3.2%.

The calculation of the theoretical coefficient of variation of the lowest tender assumes that the order in which the bids fall can be regarded as random. An indication that the order of the contractors' bids cannot be predicted better than by chance is given by the fact that when contractors from a reserve list are called upon to tender in substitution for others who drop out I have shown that they take their place in order randomly. They were not expected to produce low bids but in fact stood the same chance of being lowest as those on the original list. Thus it is reasonable to treat the ordering of the bids as a random process from the quantity surveyor's point of view.

The theoretical coefficient of variation of the lowest tender of 3.2% cannot be directly checked from experience because it could only be measured from a series of repeated, independent sets of 8 bids for identical contracts in identical circumstances. There is, however, an indirect way.

The coefficient of variation of the mean tender is 1.8% which is 5.2% divided by the square root of the number of tenders (eight). This can now be combined with the observed coefficient of variation of the lowest tender expressed as a percentage of the mean tender f or the contract. In an analysis of 74 building contracts this coefficient of variation was 2.4%. The combined coefficient is 3.0% which agrees well enough with the theoretical figure of 3.2% derived above to justify using it for further deductions.

Going through the same procedure for the second and third lowest tenders completes Table 5.1.2.

Table 5.1.2 Calculating coefficients of variation

	Coefficient of variation measured about the mean	Total coefficient of variation
Lowest tender	2.4%	3.0%
Second lowest	2.2%	2.9%
Third lowest	1.6%	2.4%
Mean tender	0	1.8%

ESTIMATING VARIABILITY

QS estimating variability is the variability in the price estimates which would be recorded by a number of quantity surveyors performing the same typical task of adjusting for differences between one building and another. The quantity surveyors are assumed to be fully informed and using currently acceptable methods.

Although this is potentially measurable it has not, as far as I know, been measured except for a contractor's estimators. The experimental results cannot be applied directly to QS work but provided a valuable guide to potential performance as well as being useful as they stand. The contractor's variability in forecasting his own true directly incurred costs is understood to have been found by one contractor in a careful analysis of the extent of agreement among his estimators to have a coefficient of variation of 4%. This was in civil engineering work where the variability of cost data is greater than in building (see 'variability of BQ rates' above).

A figure of 6% is given in paragraph 2.4 of reference 4 for the coefficient of variation of civil engineering contractors' observable variability in estimating costs. The figure for civil engineering would be expected to be higher than that for building but as well as this it is inflated by the variability in the final measurement of actual costs. This inflation was not present in the figure of 4% quoted in the previous paragraph because the variability measured was that of estimators among themselves when estimating for the same project. This 4% coefficient of variation is numerically equal to that for the variability of estimates of true (but unmeasured) cost.

However, the evaluation of actual costs is unlikely to be as variable as estimating so a reasonable split of the coefficient of variation of 6% quoted in the previous paragraph could be as follows (Table 5.1.3).

Taking the two figures for the coefficient of variation of the variability of contractors' estimating for civil engineering work, 4% and 5% from the paragraphs above and allowing for the smaller variability in building work, a reasonable figure for the likely variability between contractors' estimators is a coefficient of just under 4%.

Comparing the figures just given with those in Table 5.1.2, it may seem

Table 5.1.3 Calculating observable variability

	Coefficient of variation
a. Due to variability in estimating	5%
b. Due to variability in measurement of actual costs	3%
Total observable variability in estimating costs $(\sqrt{a^2 + b^2})$	6%

strange that the variability of the lowest tender is less than the contractors' estimating variability. The explanation is that in a series of repeated sets of bids the lowest (or second or third lowest) would not always be from the same contractor (because of the contractors' estimating variability). choosing the lowest produces less variability than choosing the same contractor each time.

ESTIMATING THE LOWEST TENDER

To apply the figures so far obtained to the problem of estimating the tender the concept of aiming at a moving target is needed. The variability of the lowest tender can be represented by a target and the quantity surveyor's estimate by a shot fired at it. If the target is first imagined to be stationary the distribution of repeated estimates around it (the 'fall of shot') will have a coefficient of variation equal to the QS estimating variability. This assumes that the quantity surveyors estimating are using the best methods.

Incidentally the shots are assumed to fall symmetrically in a group centred on the target. If they do not the centre of the group could easily be moved by adding to or subtracting from estimates in future a constant proportion: the equivalent of adjusting the sight of a weapon.

If the target (the lowest tender) is now imagined to move randomly with a coefficient of variation equal to that of the variability of the lowest tender (3%) the distribution of 'miss distances' will be the result of combining this with the QS estimating variability. This will produce the expected variability of the difference between the estimate and the lowest tender. This combined coefficient of variation is the QS observable estimating variability and ought to be equal to 7% (item e in the list under 'causes of variability', above). This leads to a figure for the QS estimating variability of $6\frac{1}{2}\%$ ($\sqrt{7^2 \square 3^2}$) using present methods in the best way under average circumstances.

Perhaps by adopting improved methods the QS estimating variability can be reduced to close to that of the contractors. It can hardly go lower because at estimating stage the information available to the quantity surveyor is incomplete and because a successful contractor should always know more about his own costs than anyone else. I shall therefore assume that the QS estimating variability can be reduced to 5%. This would lead to a coefficient of variation of differences between estimates and lowest tenders of 6% (ie $(\sqrt{3^2 \square 5^2})$. If this could be achieved the quantity surveyor would find that 60% of his estimates fell within 5% of the lowest tender, 90% within 10% and all within 20%.

The percentages for this and other coefficients of variation would be as follows (Table 5.1.4).

Table 5.1.4 Percentages for coefficients of variation

Coefficient of variation	Within 5%	Within 10%	Within 30%
4%	79%	99%	100.0%
6%	60%	90%	100.0%
7%	52%	84%	100.0%
10%	38%	68%	99.7%
15%	26%	50%	95%

When considering estimating performance in practice it is very difficult to have much idea of what proportions are falling within the various limits of the lowest tender unless objective records are kept. When this is done it usually indicates a very much worse performance than was subjectively believed.

For example I have found that until it was demonstrated with actual results quantity surveyors who would readily accept that, say, only 40% of their estimates were within 5% of the lowest tender found it hard to believe that as many as 30% were more than 10% away. In fact, because of the shape of the normal distribution of differences, the first pair of figures almost inevitably implies the others.

Claims for consistently good performance usually prove to be unfounded when judged by 'client standards'. When, after it is received, valid reasons are identified explaining why a tender is very different from the estimate it is often excluded from subjective consideration of estimating performance. This may sometimes be justified form a professional point of view but not from that of the client for whom the estimate was just as much wrong as if an error had been committed.

If the QS estimating variability could be reduced to almost the same level as that of the contractors *it would be difficult to achieve further improvement without helping the contractor to reduce his own estimating variability by giving him fuller information upon which to base* his estimate and by helping him to improve his estimating methods. I believe that it would reduce the estimating variability of both the QS and the contractor if they used as far as possible the same improved methods although there might be objections to this on other grounds.

If the drastic reduction in QS estimating variability mentioned above is to be made it may be necessary to use a new approach to prices. This could be through direct *evaluation of contractors' costs*. If this assesses costs by a method related to the way in which they arise it will be especially suitable for giving cost advice to designers.

ESTIMATING TENDERS OTHER THAN THE LOWEST

Although the second and third and more especially the average tenders offer a more stable target to the estimator than the lowest tender (Table 5.1.2) the usefulness of estimating them is doubtful and the following paragraphs show that the effect on estimating accuracy is small with currently achievable levels of performance.

The QS observable estimating variability whose coefficient of variation of 7% was broken down in 'Estimating the lowest tender' would be only slightly affected by reducing the coefficient of variability of the tender from 3% to one of the other figures given in Table 5.1.2. The greatest reduction would result from aiming the estimate at the mean tender when the combined coefficient would fall from 7.0% to 6.6%. Aiming at the third lowest tender would only reduce it to 6.8%.

To test the theory I compared the extent to which the estimate was correlated with the various tenders and the mean tender. The comparison was not affected by the average difference between estimate and tender but was concerned only with how consistent the performance was: the size of the group and not its position on the target, as explained in 'Estimating the lowest tender'. No measurable differences were apparent between results for the various tenders and the mean tenders even in an analysis covering 140 contracts.

The practical use of aiming the estimate at any but the lowest tender is obscure but it is often suggested and sometimes done; so perhaps the above analysis of its effect will be useful to some readers.

SENSITIVITY

I have tried the effect of changing to an appropriate extent the figures upon which the reasoning depends and am satisfied with the robustness of the argument and the conclusions. The number of calculations involved is too great to give here but one aspect of sensitivity is worth exploring a little because it may be concerning some readers with a knowledge of statistical theory. Some may have been shocked by my apparently cavalier method of combining and breaking down coefficients of variation by use of the property of additiveness of variance with no apparent regard for the effect of interactions produced by correlations.

If the operation of one cause of variability tends to affect that of another whether positively or negatively the two causes are said to be associated and their effects correlated. For example, the function of a building is associated with its value because, for instance, office buildings tend to cost more than workshops. For this reason I have not separated building function from value in the analysis (Table 5.1.1).

The only causes of variability used in the analysis which are at all likely to be associated to anything but a negligible extent are the QS estimating variability and that of the lowest tender. However, correlation between them can only exist if the quantity surveyor has a greater ability to predict the lowest tender than I have allowed for in 'causes of variability'. In other words, if QS estimating variability is less than I have deduced. I believe that I have allowed more than enough in the list for identifiable causes and 'QS institution' but it is here that those who claim greater accuracy in estimating must disagree. If their performance is better than that given, their claim will be justified.

Incidentally, *correlations between rates in the same Bill of Quantities are extremely strong* but have some useful effects. The sampling of items for index purposes and the grouping of items in estimating methods using measured quantities have only small effects on the total coefficients of variation. The high correlations are caused by rates in a bill which has a high or low total price level being high or low almost throughout.

CONCLUSIONS

The study of the variability of building price data leads to the following conclusions with implications for conventional methods of cost planning and estimating.

When using historic data, if the only available means of adjustment of price are for character of building and 'intuition', great improvement in reliability can be obtained if the price data are drawn from several buildings instead of one, even if this means casting the net wider and sacrificing some comparability.

Using currently available methods of estimating in the best possible way the coefficient of variation of the ratio of the lowest tender to the estimate should be about 7%. Such performance would produce about 85% of estimates within 10% of the tender (Table 5.1.4).

Direct methods of improving the quantity surveyor's own estimating performance can be hoped to reduce its coefficient of variation to 6%. Further improvement could be achieved by helping the contractor to estimate his own costs more accurately. In my view the quantity surveyor would probably benefit by using basically the same improved methods of estimating as the more sophisticated contractors.

The only advantage claimed for aiming the estimate at any but the lowest tender – a gain in accuracy – is negligible.

ACKNOWLEDGEMENTS

Although I remain responsible for the opinions expressed in this paper I am indebted to the members of the Directorate of Quantity Surveying Development who have been kind enough to offer helpful comments. I am grateful to the Construction Industry Research and Information Association for permission to quite from their Report 34 (reference 4).

REFERENCES

1   R. S. Mitchell 'A Tender-based Building Price Index' *Chartered Surveyor* Volume 104 No 1, July 1971, pages 34–36.

2   M. E. A. Bowley & W. J. Corlett '*Report on a Study of Trends in Building Prices*' Ministry of Public Building and Works Research and Development Paper.

3   M. G. Kendall & A. Stuart '*The Advanced Theory of Statistics*' Volume 1, page 336.

4   N. M. L. Barnes & P. A. Thompson '*Civil Engineering Bills of Quantities*' CIRIA *Report* 34.

5.2 Accuracy in estimating*

A. Ashworth and R. M. Skitmore

INTRODUCTION

Several methods of estimating the cost or price of construction projects are in current use in the construction industry. These are well established and due to a measure of conservatism in both estimators and quantity surveyors changes are slow to come about. There is, however, a movement towards greater use of the computer.

An important consideration with any method of estimating is the accuracy by which anticipated costs can be predicted. Any improvement in prediction will be welcomed because existing estimating methods are poor when judged by this criterion. Builders' estimating has, in the past, had its own mystique, with the result that certain processes discussed both in the classroom and in the estimator's office are found to have little practical relevance.

What makes an estimator of quantity surveyor good at forecasting the right price? – always assuming there is a right price! To what extent does human behaviour influence or have a part to play? These and some of the other aspects of affective estimating will be examined in this paper.

ESTIMATING METHODS

Traditional estimating

Estimating methods can be classified as approximate, analytical or operational.

Approximate methods do not give results that are necessarily inferior to any of the other methods. However, inaccuracy will be enhanced where there is inadequate information provided by the designer.

Civil engineering works are typically estimated using the operational method. Although approximate methods are largely supposed to be employed

*1983, *Occasional Paper* No 27, Chartered Institute of Building.

by quantity surveyors and the other methods by contractors' estimators, some overlap inevitably occurs in practice. For example, a quantity surveyor in estimating the costs of a new process or system, may have to resort to price analysis where no other data are available. On the other hand the costs of speculative housing are often determined by contractors using an approximate method. Each method has its own particular rules for application and all rely heavily upon previously recorded or published cost information. In practice, the selection of the method adopted is influenced by tradition, familiarity and the information and time available for estimating. it can be show, for example, that accuracy in estimating is likely to be improved by the familiarity of the method chosen.

(a) Approximate estimating

Approximate estimating methods can allow the calculation of cost on the basis of a single variable describing the proposed building project. For example, the unit method describes cost in terms of an occupancy factor, whereas the superficial floor area method or the cube rules quantify the building by either area or volume. Agreed rules of measurement have been prepared in each case to allow for some comparability between different schemes. However, it should be understood that the easier the method of measuring the more difficult it will be to be precise in pricing.

Other methods of approximate estimating include the storey enclosure method and approximate quantities. The former method[1] was considered to be revolutionary for its time, but needless to say, it did not succeed and has found very little application in practice. Approximate quantities is a method favoured by both the quantity surveyor for pre-contract estimating and also by the contractor's estimator when bills of quantities are not supplied.

The advantages of all methods include ease of application and speed of estimating. They are well tried and do provide some element of accuracy.

Substantial research has been undertaken into price forecasting by the professional quantity surveyor culminating in cost planning[2]. Cost planning involves the use of elemental and comparative estimating methods and is a more rigorous approach.

It gives a much more detailed analysis and it also claims the advantage of being able to effect cost control.

Research into alternative and improved methods still continues. For example, Southwell[3] recognising that floor area was the major factor correlated with price, devised a formula allowing an increase in the accuracy of price prediction. He also showed that perimeter was the second most important factor and this has been incorporated into the formula.

(b) Analytical estimating

Analytical estimating is traditionally adopted by contractors' estimators to determine the individual rates for measured work items in a bill of quantities. Each individual measured item is separately analysed into its constituent parts of labour, materials and plant. Each part is then costed by means of a detailed analysis using outputs, gang sizes, material quantities and the plant hours required. The importance of the size, type and shape of the particular project are of particular significance. This method, in theory, relies heavily upon the feedback of information. However, the complexity of modern construction and the variety of processes used limit the availability of reliable feedback information. In practice, the estimator will use his own standard outputs and couple these with an expectation of future performance. The latter is particularly subjective in approach and is an important part of the estimator's work.

(c) Operational estimating

Operational bills of quantities were developed at the Building Research Establishment[4] during the early 1960's. It was claimed that the format assisted in the planning of the project and also enabled a more realistic method of estimating to be adopted. In addition it also provided the contractor with more useful contract management information.

The true operational bill allows the estimator to price the work in stages of construction rather than on the basis of individual items, trades or elements. It represents a considerable deviation from established processes and, possibly for these reasons alone, it failed to obtain either a fair assessment or sufficient practical experience to make it a useful alternative.

Although an alternative operational format bill was devised as an intermediary document to a full operational bill, this too has failed to find support from professional quantity surveyors. The present layout of bills of quantities makes it almost impossible to apply operational estimating in practice.

It is argued, however, that this approach to estimating, is based on how costs are incurred on site. The segregation of the work is both essential for proper estimating and also for the efficient cost control of the work during construction. Conventional bills of quantities do not allow the estimator to price the work in an organised manner and the calculation of individual unit rates is largely irrelevant, although involving the contractor in considerable time and expense.

Computer-aided estimating

The use of computers for estimating can be divided into two categories:

– computerised traditional methods;
– computerised statistical methods

(a) Computerised traditional methods

The computer used in conjunction with traditional methods retrieves data either from a central storage system or from the firm's own previously recorded data[5]. It may be used as a form of source book that can be regularly updated and can also supply very comprehensive information. Some programs allow the user to undertake price analysis from a very wide base, or prepare target cost planning on the analysis of previous schemes.

Some systems have proved cumbersome in use and of little advantage. In computerising manual processes the difficulties encountered often appear to outweigh the advantages. Often the computer process employed required the user to change his method of working to suit the machine. The modern approach, however, seeks to fit the computer program within the task of the user.

(b) Computerised statistical techniques

Cost modelling is one of the newest techniques to be used for forecasting the estimated cost of a proposed construction project[6]. Although there is only scant evidence of its use in practice there has been considerable interest as well as research into this method.

A mathematical model or formula is constructed that best describes the data collected in terms of cost or price. A technique suitable to this problem is multiple linear regression analysis. Developing a cost model in practice includes the collection of suitable data, a decision on the type of model to be constructed, preparation for the computer and analysing the final model. Before it can be used in practice, several statistical tests need to be applied in order to measure its effectiveness.

Cost models have been developed both for some of the pre-contract elements of a building[39] and also for use by the contractor's estimator[8]. Models have incorporated both existing data currently used in practice and also new data derived directly from construction sites. Advantages include speed of application, easy updating, greater availability of information and earlier forecasts. Although improved accuracy is also claimed as an advantage, there is little evidence to show its superiority over traditional methods. A disincentive towards the adoption of cost models is their radical approach towards building estimating.

ACCURACY OF CONTRACTORS' CONSTRUCTION COST ESTIMATES

Any assessment of the accuracy of cost prediction methods must ultimately rely upon quantitive techniques, such as a simple comparison with recorded expenditure. The construction industry's predilection for design flexibility, even during the construction process, and difficulties in accurately accounting for expenditure, together with contractors' understandable reluctance to reveal their profit levels, has resulted in a noticeable lack of reliable data.

This section presents a review of the efforts that have been made to establish a value for estimating accuracy.

General views

It has been said that contractors should estimate with an error of considerably less than 10% of their total final cost[9] and generally of the order of □ 5%, given a set of quantities and sub- contractors' quotation[10]. Experience in process engineering contracts suggest that □ 5% is a reasonable figure[11], although a figure of 10 to 25% accuracy has been quoted in fluid flow processing[12]. An opinion survey taken at a seminar in Loughborough also confirmed the view that a □ 5% accuracy is generally appropriate, notwithstanding the lack of supportive data[13].

Mathematical models (Monte Carlo simulations) have indicated higher figures in comparison with cost and profit levels of a leading contractor. □ 8% to □ 11%[14], usually rounded to □ 10% has been quoted[15], together with figures between □ 5% and □ 15%[16]. The models are rather crude in that all bids are assumed to be drawn at random from a uniform aggregate distribution, although it has been claimed that a close resemblance to reality exists.

Beeston[17] reported a coefficient of variation (cv) of 4% found by one civil engineering contractor after careful analysis of the extent of agreement among his estimators when estimating the same project. The cv statistic is more appropriate for such error type distributions but the limited sample size and lack of experimental control makes the resulting figure of questionable reliability.

Willenbrock's[18] research showed a mean 3% increase in actual over-estimated costs but no details of dispersion around the mean. Other researchers suggest a cv of 5.5% for engineering services 'from experience' and 'depending on the type of job being figured'[19], or a cv of 2% from an analysis of a limited sample of contractor's cost and estimating data[6].

Gates[20] has analysed the bidding performance of a large highway construction company on 110 projects to the Connecticut State Highway Department between 1963 and 1965. The data, consisting of actual/ estimated construction cost ratios, appear to be normally distributed with a cv of approximately 7.5%. No allowance appears to have been made for any extra payments caused by design changes etc.

Barnes[21] uses the ratio of the actual total cost to the estimated total cost multiplied by the ratio of the tender sum to the final account as his measure of estimating accuracy. Analysis of data collected for 160 completed British construction contracts indicate a cv of 5.8%. The method has the advantage that no adjustment for design changes etc. is necessary but suffers from the dependence on the assumption that the actual profit margin achieved on originally tendered work is similar to that on extra work, although Barnes claims that the calculation is not very sensitive to the assumption. Barnes, in a more recent joint publication with Bennett, subsequently used a cv of 7% for the inherent uncertainty in building, a figure which is presumably exclusive of any errors the estimator may make. the origin of the 7% is said to be from some previous research by Barnes, but this has not been traced. Barnes[22] has also made reference to examples varying from 2 to 15% cv but with no indication of source.

The data used in the above analyses do, however, have questionable features, apart from the reconciliation difficulties mentioned. One feature is that the use of data is limited to won and completed contracts. The old adage that the biggest mistakes win most jobs suggests that the data are likely to be biased in favour of low estimates. To gain an overall impression of estimating variability one would, ideally, need to know what the actual cost would have been for all jobs tendered.

One approach to this problem has been to assume that all tenderers for the same contract should incur the same actual cost, an assumption based on the similarity of construction methods and work forces between contractors. In this case knowledge of the actual cost of all jobs won and lost is a prerequisite. In the highly competitive environment of contracting it is virtually impossible to obtain this sort of data, but Ashworth[23] has devised an experiment whereby the information can be generated in a laboratory type situation. His experiment consisted briefly of obtaining estimates of the number of man-hours of brick-laying needed for nine different projects from nine different estimator. As the projects had already been completed the actual results were known to the investigator, so that the difference between estimated and actual man-hours on each project could easily be calculated. analysis of the results of the experiment show the mean error for each estimator to be between −3% and +46%, with standard deviations ranging from 17 to 30 with a distinct correlation between the mean error and standard deviation. Coefficients of variations for each project ranged from 13% to 20%, considerably different to Gates' and Barnes' figures.

However, some further comments are appropriate before direct comparison can be made. Ashworth's experiment measures only estimates of labour output, probably the most uncertain part of a project. On an average project with say 40% nominated work, 20% domestic sub-contracted and the remaining 40% divided between materials, preliminaries and labour, the contractor's direct labour content is not likely to exceed 20% of the contract value. An error of 25% in estimating labour is, therefore, only likely to produce an error in the total value of work of the order of 5%. In addition, bricklaying is a more unpredictable activity than, for instance, the finishing trades, and estimating errors in this area may not be typical. The experimental method is also open to question in its obvious deficiencies in recreating the competitive environment in which estimators normally work.

Further criticisms can be levelled at these approaches in the reliability of the actual cost data. Booking errors have been shown by Fine[15] and others to be a major source of distortion, and interdependence between the estimates and actual costs may interfere with meaningful interpretation of results.

Other approaches

Where no data are available for profit levels or costs, then estimates of the likely accuracy of the cost estimate can only be made by comparison with cost estimates from other sources. such sources will provide a direct comparison, (for example, the designer's estimate), or an inferred comparison (in the case of other contractors' bids).

Watson's[24] analysis of 44 tenders based on drawings and specification only and 37 tenders based on bills of quantities purports to show an improvement in estimating accuracy in the latter group due to a small cv of the tender price around the designers' estimate, despite the larger difference in means. Similarly, McCaffer's[25] comparison between bids and designers' estimates for 132 buildings and 168 road contracts has been used to test relationships with estimating accuracy and contract size and number of firms bidding. Other research[26] has indicated a correlation with market conditions. It is not certain though whether the contractors' bids are being judged by the designer's estimate or whether the reverse is taking place. The difference between the two figures is a result of the combined variations in both estimates, an inescapable conclusion, even where the designer's estimates is claimed to be very accurate as in the system sometimes employed by the Belgium Government in detecting errors in bids[27]. The dispersion of bids for a contract has been theorised to emanate from three variables – the cost estimate distributions, the mark-up distributions and the degree of 'proximity' of bidders[28]. Only if the mark-up cv's are zero and all bidders cost estimates are drawn from the same distribution, will the dispersion of bids coincide with the variability of cost estimates. In reality, the dispersion of bids must always exceed the cost estimate variability, and research has shown that the excess is likely to be too great for an assessment of estimating variability to be made[28]. Several researchers have calculated the average cv for bids for individual contracts which are given in Table 5.2.1. It can be seen that general variability of cost estimating in the industry is highly unlikely to exceed cv of 6.5% for building contracts. The variability of cost estimates can be equated with accuracy of cost estimates provided the central tendency of the bids bears a close

Table 5.2.1 Mean coefficients of variation of construction bids

Source	Data source	Mean CV
Fine and Hackemar[14]	'Adequate' sample of construction contracts	5%
Beeston[17]	Large sample PSA building contracts	5.2%
Grinyer and Whittaker[76]	153 Government construction contracts	8.04%
Skitmore[28]	269 building contracts	8.4%
Barnes[21]	159 construction contracts	6.5%
McCaffer[28]	185 Belgium Building contracts	6.5%
AICBOR[77]	213 motorway contracts	6.8%
McCaffer[25]	16 Belgium bridge contracts	7.5%
McCaffer[25]	384 Belgium roads contracts	8.4%

resemblance to the actual cost. The little evidence available suggests that this may not be the case, the lowest estimate being the best predictor of the actual cost[28]. Morrison and Stevens[29], in recognition of this, have computed a mean accuracy figure of 5% to 7.5% through simulations, though variations caused by mark-ups and proximity have been ignored.

ACCURACY OF DESIGNERS' CONSTRUCTION PRICE FORECASTS

Early stage estimates

The accuracy of the designer's price forecast is, unlike the contractor's cost estimate, thought to be largely affected by the amount of information available. Figure 5.2.1 illustrates the concept.

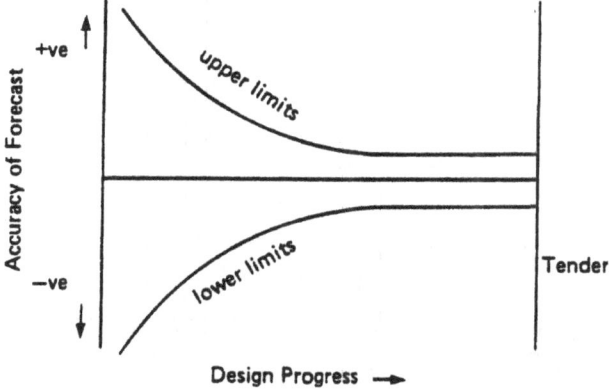

Figure 5.2.1 Chronology of price forecast accuracy over project pre-tender time cycle (adapted from Barnes' Figure 6.4[30]).

The values of forecasting accuracy thought to exist for a given state of design progress varies between authors. Barnes[30] suggests +20% to –40% at the commencement of feasibility studies improving to +10% to –20% at the commencement of detail design. Keating's[12] process plant equivalent of the first stage forecast, called the 'preliminary cost study or order of magnitude', is accorded a range of □ 25%, with □ 15% being considered appropriate at the following stage, called 'appropriation grade'. Park[31] refers to 'order of magnitude' forecasts as being □ 30%, whilst 'semi-detailed' forecasts fall in the range □ 10%. Marr's[32] 'standards' for construction cost forecasting split the pre-tender stage into planning, budget, schematics and preliminaries in which corresponding 'adequate' degrees of accuracy are stated as 20–40%, 15–30%, 10–20% and 8–15% reducing to 5–10% prior to a construction. Finally, the chemical industry suggest four stages – order of magnitude estimate, study estimate, budget authorisation estimate, project control estimate, attaching the corresponding ranges of accuracy of over 30%, □ 30%, □ 20% and □ 10% respectively[33]. McCaffrey's[34] survey of quantity surveyors forecasting accuracy for 15 schools gives figures for four stages – forecast, brief, sketch plan, detail design – of cv's of 17%, 10%, 9% and 6% respectively; quite an improvement on the textbook suggestions. Greig's[35] interviews with 32 client organisations produced accuracy figures for forecasts made in the early and late stages of design as shown in Figure 5.2.2.

These results are extraordinary in producing cv's of roughly 6–7% in the

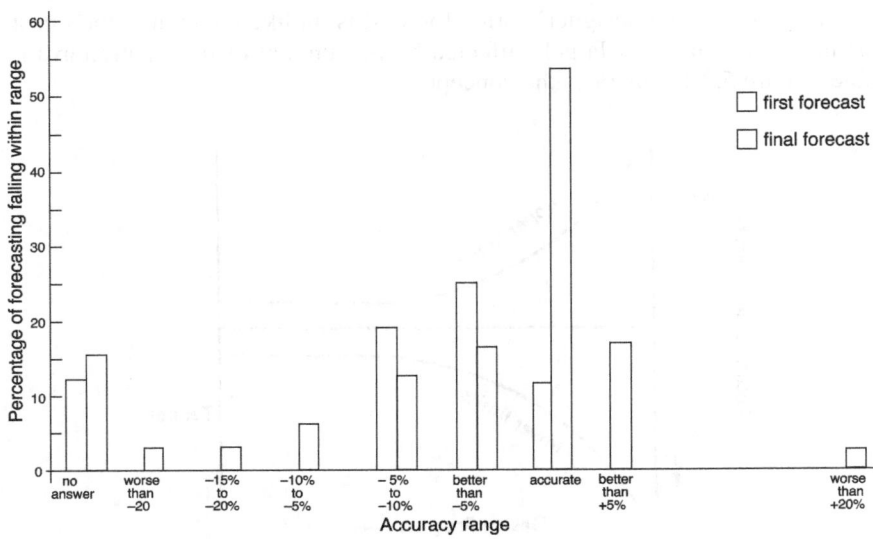

Figure 5.2.2 Accuracy of forecasts made through design stages (from Greig[35] pp102, 106).

early stages and less than 5% prior to tender. They are confirmed, however, by Jupp's opinion survey of 49 quantity surveying practices, suggesting an accuracy of □ 5% in the majority of cases[36]. Flanagan's[37] data obtained from the records of two county council's cost planning departments of forecasts made at inception and outline stage suggested a cv of about 15%, 25-30% of predictions falling with □ 5%, a distinct contrast to Greig's 53.1 out of 87.5.

The remainder of the literature deals with forecasts made immediately prior to receipt of bids, an exercise of questionable value but, for the purposes of this review completing the survey of forecasting accuracy.

Forecasts based on detailed design

Beeston's[17] mixture of deduction, induction and sometimes pure assumption based on an analysis of a large amount of data in the form of bids for Property Services Agency contracts suggests a cv of 7% to be the best possible figure for quantity surveyors observable forecasting variability 'using present methods in the best way under average circumstances'; □ 10% has been considered by others to be normal[38]. When drawings and quantities are to hand, many experienced (forecasters) can get closer than this'[78]. 'Studies at Loughborough' indicate a cv in excess of 15% for forecasting the price of office buildings, 20% for road works, 26% for H & V services and 34% for electrical services[39], whilst an attitude survey reveals a figure of □ 9% for frames structures[40].

Park's[31] analysis of engineers' cost estimates on nearly 100 projects published in *Engineering News-Record* in 1971 suggests a cv somewhere between 10 and 15%. NASA's space shuttle programme, involving 273 construction projects let between 1974 and 1981 valued at over $220 000 000 produced engineers' forecasts on average 7.8% higher than the low bids, unfortunately no cv being available[41]. Analysis of 22 recent projects, however, show the engineers forecast to be an average of 7.49% higher than the low bid, with a cv of 16.77%[42].

Handscomb Roy Associates[43], who describe themselves as an integrated organisation of quantity surveyors, engineers and construction experts, claim to have a cv of 7.71% from an analysis of forecasts for 62 North American construction projects between 1973 and 1975. Considering the lack of detailed price data and the uncertain conditions at the time, this result represents a truly remarkable degree of accuracy. No details are provided, however, of the methods used in compiling the forecasts or if the projects are truly representative of all project forecasts for the firm.

An analysis of forecasts for 915 UK construction projects between 1973 and 1979 from seven quantity surveying offices, predominantly in the public sector, reveals a mean deviation ranging from 7.57% to 11.76% (mean 9.80%), representing a cv of approximately 13%[44].

McCarfer's[25] analysis of 132 building contracts and 168 roads contracts in

Belgium showed that designers overestimated on average by 5.17% and 1.45% respectively (cv 13.13% and 18.37%).

Jupp et al[36] reported on the only attempt to measure forecasting accuracy under experimental conditions. Three surveyors were required to price bills of quantities from a controlled data source of five prices bills of quantities for similar previous projects. Cv's of 18% and 24% were recorded, depending on the number of previous bills and the updating index used, rather higher than expected from the analysis of existing records detailed above.

Factors influencing forecasting accuracy

The effects of differing availability of design information have been discussed above but other factors influencing forecasting accuracy have been investigated by researchers.

Jupp's experiment with differing amounts of price data suggest a minimum of two previous bills of quantities are needed as a reasonable data base, although no statistical significance has been attached to the results. Similarly, the type of index used for updating historical price data is said to influence accuracy. The BCIS tender price index is the most useful, but the sample sizes used discourage reliable conclusions.

McCaffer's[25] analysis showed a better accuracy with buildings than roads and Morrison and Stevens concluded that housing and school projects are associated with higher degrees of accuracy.

Building size, in terms of value, produced the tentative conclusion from Moris and Stevens[44] that 'accuracy improves by a small factor as projects increase in size'. McCaffer[25] on the other hand, concluded that building size had no effect on accuracy. In the absence of further data, it would seem unwise to attribute any significant effect of building size on forecasting accuracy. (It is possible that such effects that have been noted are confounded with forecaster's expertise as noted below).

Market conditions have long been thought to have an effect on forecasting accuracy, being particularly stressed in Ferry and Brandon's[45] 'Cost planning of building'. Data obtained from the Commonwealth of Massachusetts' Department of Public Works on 691 highway projects between 1966 and 1974 show a distinct difference between accuracy during 'good' and 'bad' years ('good' and 'bad' years being defined as years with the greatest and least activity for contractors) – Figure 5.2.3.

As can be seen, the number of bidders is also a significant factor, the designer's estimate appearing to increase relatively to the number of bidders. McCaffer[25] has found a similar phenomenon with his Belgium data. Flanagan[37] also noted a decrease in estimating error commensurate with an increase in the number of bidders. The influence of market conditions and number of bidders is attributed to 'competitiveness' and the action of random variables, particularly the contractor's cost estimate.

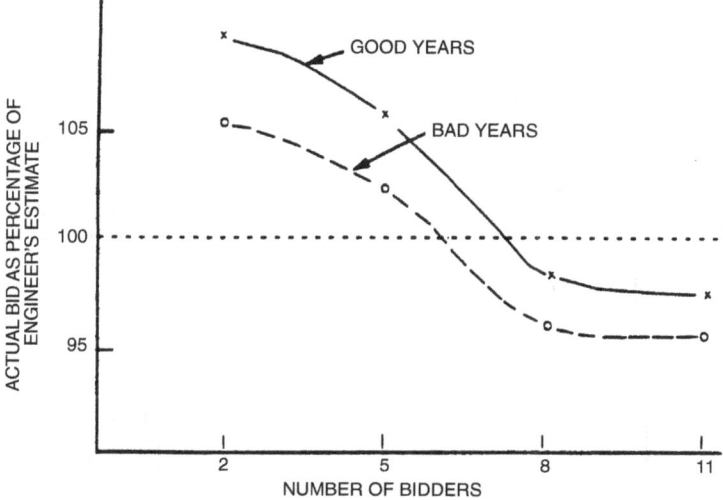

Figure 5.2.3 Effect of market conditions on forecasting accuracy (source: Neufville *et al.*[26] p65 Figure 6).

Another source of forecasting variability has been attached to what might be called 'personal' factors. Familiarity with a particular type of building or one client has been associated with up to 40% improvement in forecasting accuracy[29]. The results of Jupp's[36] experiments suggest a distinct difference in accuracy levels between forecasters, although no actual figures have been made available. In general, quantity surveyors 'intuitive' forecasting ability is postulated to account for a maximum of 4% of the total accuracy percentage[17].

Accuracy

The accuracy figures obtained from the literature are summarised in Figure 5.2.4. Apart from McCaffer, Flanagan and Greig, no supporting evidence is available for the preliminary design stage. Of the three, McCaffer's figure is based on a rather small sample of data and all three have been obtained from existing records and could, therefore, be of questionable validity. Greig's results are particularly surprising in view of the accuracy quoted at detailed design stage. Of the results shown at detailed design stage only Jupp's figure of 18% to 24% cv is obtained under reasonable experimental conditions. Morrison and Stevens' approximate 13% cv is probably the next most reliable, being based on nearly 1000 contracts, although mainly from the public sector. McCaffer's similar result (13.13% cv) for a reasonable sized sample of Belgium building data offers some confirmatory evidence. Greig's data obtained from clients seems, as previously noted, abnormally low.

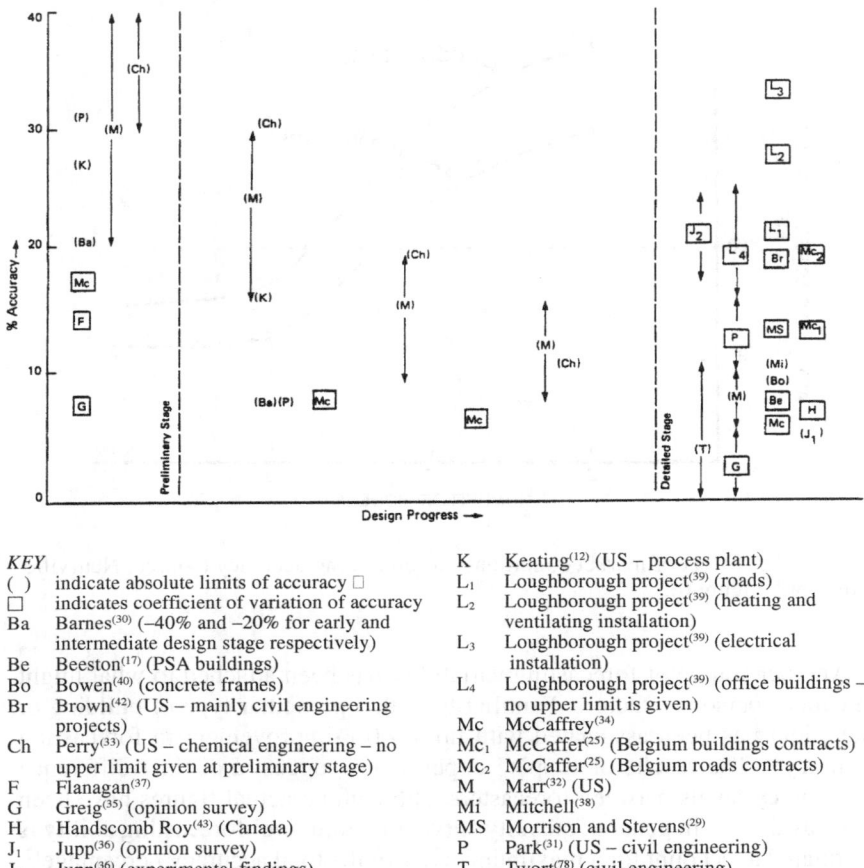

KEY

( )   indicate absolute limits of accuracy □
□   indicates coefficient of variation of accuracy
Ba   Barnes[30] (−40% and −20% for early and
     intermediate design stage respectively)
Be   Beeston[17] (PSA buildings)
Bo   Bowen[40] (concrete frames)
Br   Brown[42] (US – mainly civil engineering
     projects)
Ch   Perry[33] (US – chemical engineering – no
     upper limit given at preliminary stage)
F   Flanagan[37]
G   Greig[35] (opinion survey)
H   Handscomb Roy[43] (Canada)
J₁   Jupp[36] (opinion survey)
J₂   Jupp[36] (experimental findings)

K   Keating[12] (US – process plant)
L₁   Loughborough project[39] (roads)
L₂   Loughborough project[39] (heating and
     ventilating installation)
L₃   Loughborough project[39] (electrical
      installation)
L₄   Loughborough project[39] (office buildings –
     no upper limit is given)
Mc   McCaffrey[34]
Mc₁ McCaffer[25] (Belgium buildings contracts)
Mc₂ McCaffer[25] (Belgium roads contracts)
M   Marr[32] (US)
Mi   Mitchell[38]
MS   Morrison and Stevens[29]
P   Park[31] (US – civil engineering)
T   Twort[78] (civil engineering)

Figure 5.2.4 Accuracy of designers price forecasts.

McCaffrey's figure based on a small sample also seems low, as does Handscomb Roy's 7.71% cv arguably biased towards the more satisfactory forecasts. Beeston's 7% is an 'ideal' figure based largely on theoretical considerations. The remaining figures either lack supporting data or are not strictly applicable to UK building contracts.

From the above considerations it would seem that a suitable accuracy of forecasting in the early design stages would be of the order of 15% to 20% cv, improving perhaps to around 13% to 18% cv at detailed design stage immediately prior to receiving tenders. The absence of reliable data collection methods tends to preclude any more confident prediction but it would appear at this stage that the absence of detailed design information has little effect on forecasting accuracy.

Other factors affecting forecasting accuracy have been shown to include the extent of the data base, updating index used, type of building, market conditions, the number of bidders and personal factors such as familiarity with the work and the level of individual expertise.

The research undertaken to date has produced little reliable evidence for the confident assessment of forecasting accuracy in general and, more importantly, completely failed to provide any indication of the influence of the main factors involved for the benefit of forecasters. It is evident that a great deal more research will be necessary, preferably using suitable experimental techniques to exclude observer bias, before sufficient understanding of existing forecasting performance is gained in order to embark on any work of a prescriptive nature.

PREDICTING CONSTRUCTION COSTS – THE CONTRACTOR'S VIEWPOINT

Experience in estimating construction costs suggests that there is a disparity between what is considered to be 'good practice' and what estimators are actually able to do when faced with the realities of the construction process. Major discrepancies include the following;

(a) Accuracy

It is a common belief of both quantity surveyors and estimators that they are able to predict the costs of a proposed construction project more accurately than can be proved by empirical tests. Most estimators, for example are convinced that they can consistently estimate to at least within □ 5%. Research indicates that this may be overly optimistic. By definition alone, it is impossible for an estimator to predict the costs of a building accurately in advance.

Contractors are unable to determine with any degree of accuracy the likely cost of a one-off project designed by others[45A]. If this is to imply that the estimator's performance is poor – then the quantity surveyor working in professional practice or in a government department – is even worse. It has been suggested that just as accounts are guesses of past costs, estimates are guesses of future ones. In order for the quantity surveyor to improve his estimating ability the quality of the contractor's estimator must first be transformed.

(b) Feedback I

It is generally presumed that all contractor's estimators calculate the costs of proposed projects based upon information described as site feedback. Investigation, however, in the practice of estimating has shown that if feedback ever does reach the estimator it is more often than not ignored. It is assumed that the data relates to a one-off occurrence, and is, therefore, not relevant to the estimator's future work. Estimators' standard outputs are contained and secretly guarded in their 'black book'. They are only rarely ever amended or revised. It has been suggested by estimators that their outputs are unlikely to change even over very long periods of time. There may be trends in the industry, but changes in outputs are more likely to be influenced by revisions to working rules or improved methods in the construction process. A major reason given why estimators disassociate themselves from site feedback is due to the poor recording systems employed by contractors and hence a lack of confidence in the data provided.

(c) Feedback 2

The recording of the performance of construction site operations is particularly difficult owing to the wide variety of work expected of each contractor[46]. The usual method is to adopt a classification system and attempt to record costs in accordance with it. However, a complex classification system is required to cover the majority of possible cases. Even a four digit code system may be insufficient to capture the performance data in such a way to enable it to be synthesised to represent future projects. Tests have shown that the poor reliability of recording and the subsequent misallocation increases substantially when the number of cost codes exceed fifty. For example, it should be straightforward to record the costs of bricklayers precisely and, therefore, accurately. When it becomes necessary to allocate costs to individual bricklaying operations, misallocation occurs resulting in inaccurate and unreliable feedback data.

(d) Purpose of the estimator

The basic reason for estimating is to obtain work. Indeed it is the prime reason for estimating and tendering. One might then be forgiven for assuming that one of the functions of the estimator is to obtain work by placing in bids at the appropriate level. The real function, indeed the major role of the estimator, is to determine the contractor's costs for a project. He has to interpret the design in financial terms. It should be the function of management to determine the price for the job by adding the appropriate mark up necessary to maximise future profits whilst at the same time secure work.

(e) Right price

There is no such thing as the right price for a particular item of measured work. Price will vary for many reasons, not least of which is the type and size of the proposed project. Estimating by both quantity surveyors and contractors' estimators relies heavily on some form of data. The reliability of these data can be measured by the consistency which they exhibit when many such prices are obtained for a similar described item[17]. Any price should be looked upon as a member of a population of prices for the same item. Variability between apparently identical items of work can be considerable. Some of this variation can be easily explained, whereas some appears beyond any understanding. Identical bill items from different contracts can vary by as much ass 200%[13], whereas, for individual trades the percentage may be around 30%. The right price or rate will also mean different things to different people. In one sense the right place might be assumed to be the average of all possible prices from a population of identical items from similar projects.

(f) 80/20 syndrome

Many of the items included in a bill of quantities for a building project are of little cost significance. The suggestion has been made that somewhere in the region of 80% of the cost of a project is represented by only 20% of the bill items. The inverse of this argument must also be true; that 20% of the cost is represented by 80% of the items. There has in more recent years been some attempt to remove some of these low cost significant items from measurable work, by revising the say quantity surveyors measure building work using the Standard Method of Measurement. These low cost significant items are rarely estimated using normal estimating procedures, but are priced in a very sub-jective way on an ad-hoc basis. The removal of these minor items from bills of quantities might go some way towards improving the accuracy of the remaining cost important items and so make the whole process more realistic.

(g) American system

Some comparisons have been made in recent years between the construction processes in the UK with those used in the USA[47]. Although the American approach towards construction estimating is fundamentally different, there is no evidence to suggest in practice that it is better. American opinion indicates that they do not consider themselves to be better estimators than those in the UK. Indeed, rather than for UK practice to be amended, there is some indi-cation that the Americans see inherent weaknesses in their own approach and advantages in adopting the UK system.

The influence of the market

Accurate prediction of future costs is a prerequisite in the successful operation of the 'cost-plus-markup' pricing policies, but as Oxenfeldt observes, there is a tendency in any industry for sellers to misperceive (or miscalculate) their costs[48].

Construction of buildings in a market-orientated economy is a result of a decision making process which is influenced by assessment of factors which are difficult to express in measurable terms, or are variable because of the inherent inability to make accurate forecasts. In these respects assessments of the value of building contrasts dramatically with the value of manufactured components, the cost factors of which are much more readily determined[49]. The long-range and short-term traditional methods of financial forecasting as used in the steady state industries are of questionable value, or even inapplicable, in the construction environment[50].

The construction process is characterised by complexity [51] and uncertainty[52]. Complexity is often such that even the most common types of strategic decisions which face management, such as pricing a bid, involve an immense number of variable factors[10]. The volume and interdependence of these factors is such that, even if they could all be identified (which is unlikely) in a quantitative and reliable form, the task of evaluating the interaction alone is 'an exercise beyond the realm of mental capacity'[53]. The human mind is hopelessly inadequate to cope with even a small part of the available information[31]. Further, Jepson[79] makes the point that, in a state of complete uncertainty, forecasting fails by definition.

In the event it is clear that many of the factors affecting costs 'can only be determined after they have occurred'[50]. The estimator cannot price at estimated costs because he has no reliable means of knowing what his costs are going to be[54].

There is some evidence to show that estimators escape from the real problems of their trade by presenting socially acceptable forecasts[15], that the cost estimate is merely a target figure for site management[55]. Thus 'tenders are not attempts at accurate estimating' but rather 'to obtain the contract at the highest possible price'[3] in consideration of market conditions[45]. Estimators are skilled in arriving at a price for a project but they do not know what it will cost them to build[54].

Pricing methods so closely linked to the market-price have been shown to predominate in construction contract bidding, the goals and price policies of firms being explained in terms of general learning theory[56]. Here the reward of winning a contract or punishment of not winning a contract is seen to have a significant influence on a contractor's bidding (and estimating) behaviour.

Prices, costs and value judgements

As a consequence of the theoretical considerations and observed behaviour of contractors in a competitive environment it is not surprising to find references to prices being completely unrelated to performance[57] or costs of construction[58]. Detailed analytical estimating methods will be inappropriate[59], a rough budget figure for feasibility being all that is needed, achieved by building up a system of pricing which will obtain a reasonable number of contracts and generate an adequate profit. This involves the synthesis of global rates, irrespective of location, extent and complexity of the work, these being adjusted for market conditions and whether the contractor wants the job[58]. These adjustments are necessarily a matter of judgement and, in view of the degree of uncertainty involved, largely value judgements[55].

It is not surprising, therefore, that the industry has received little formal instruction and only scant advice on tendering[60] for as Fine states 'estimating is like witchcraft: it involves people in foretelling situations about which they can have little real knowledge'[15].

Realists will acknowledge that a large measure of subjectivity is involved in estimating in practice, although researchers have generally preferred to disregard or degenerate its action. Recent contributions on the use of decision theory and utility theory in construction estimating recognise the need for subjective assessment[61] and the applicability of the extensive literature on bidding strategy is in question because of the many subjective factors involved[62].

Pricing decisions, as Mulvihill and Paranka observe, are the ones most often involved with problems that concern delicate and unpredictable human relations. It should be recognised that, as a topic for study, pricing should be handled by behavioural studies[63]. There are, however, certain characteristics regularly attributed to estimating expertise and upon which a general consensus appears to exist.

Estimating expertise

It has been shown that a consistent disparity of performance exists amongst estimators[8]. Proficiency in estimating is said to be a result of skill[16], experience[64], judgement[53], knowledge[61], intuition[61], feel[65], academic background, personality, enthusiasm[66], hunch[64] and a 'feeling in the back of the head'[53].

Conclusion

It is apparent that the dominating presence of uncertainty in the construction process militates against the production of accurate forecasts of costs by numerical analysis and synthesis alone, in the conventional objective sense. The estimator is necessarily forced into a position of making value judgements despite exhortations to do otherwise.

Subjectivity is a topic most appropriate studied by behavioural scientists, the following quotation from Blakeslee's 'The Right Brain – A new understanding of the unconscious mind and its creative powers' may provide, perhaps, an indication of the processes involved;

> 'If we take the plans for a house we want built to an experienced contractor, he may glance over them for about ten minutes and tell us what it will cost and how long it will take. This is **intuitive** judgement. Another approach he could

use would be to add up every item on the bill of materials, calculate the price one item at a time, then schedule each stage of the construction and estimate building time. With experience, the intuitive judgement can be as accurate as the methodical one. If you ask him to explain the intuitive estimate, he might say something like 'gut feel' or 'experience'. In actual fact, the intuitive approach is a result of right-brain thinking. Just as it can recognise a face in a crowd at one glance, the right brain can analyse large masses of data and make a judgement in one step'.

Blakeslee[68]

FORECASTING CONSTRUCTION PRICES – THE DESIGNER'S VIEWPOINT

A regular feature of the construction industry is the need for prospective building owners to be informed of their likely financial commitment in purchasing new construction prior to commissioning extensive design work. A further commensurate need is that the subsequent completed design attracts a price which approximates closely to this anticipated commitment. It is necessary, therefore, that a reliable forecast is made, as soon as possible, of the region in which bids are likely to fall and institute some system for monitoring the progress of the design to ensure the forecast remains realistic.

The responsibility for this activity lies with the designers and/or cost consultant, normally the quantity surveyor in design cost (price) relationships.

The quantity surveyor has a remarkably difficult task. He has little design information with which to produce the initial forecast; he does not know who the contractor will be; and he has no usable cost information at all. The main source of the monetary data will be through analysis of prices of past projects. The problem is considered, therefore, to centre on the quality and quantity of two data sources;

(a) the design data available for the current project;
(b) the price information on past projects.

It is known that the impact of design decisions is not linear[45]. The initial decisions on construction type, size, shape and general quality level have a far greater significance in determining price level than subsequent detailed design decisions. It has been said by many that 90% of the price of the project is set during the preliminary design stage[69], so it may be fair to assume that lack of design data is not a significant problem, providing basic decisions have been made.

The value of price data needs to be considered in respect of the contractor's approach to pricing. It has been postulated that prices are closely related to market forces and, therefore, to an extent divorced from actual costs, thereby justifying the price forecaster's use of price base data. The subjectivity applied by contractors is equally pertinent then to the forecasters.

Franks[70] and Southwell[3] have both questioned the use of detailed price synthesis by forecasters as being excessively refined in comparison with contractors' estimating methods, value in the last resort being purely subjective[71]. There are undoubtedly many factors affecting tender prices which defy objective measurement or calculation and it is to these areas that professional skill and expertise should be directed[72].

Forecasting expertise

Accurate forecasting of construction prices involves rather more than numerical synthesis[36]. Qualities demanded of forecasters include experience[45] intuition[17] and 'feel' for costs[73].

Systems development

In view of the large subjective element displayed in forecasting it would seem counter-productive to deliberately exclude the human factor in the development of any forecasting system. Morrison and Stevens[29], otherwise commendable proposals for a construction price computer data base, imply professional intuition to be a source of error, although the project supervisor has commented encouragingly that the data base ought to provide an aid to the forecaster's intuitive approach[73].

Some existing systems have wisely provided for intuitive adjustment[74, 75].

Conclusion

The efforts of the construction price forecasters can be seen to be similar in many ways to the contractor. The forecaster, in predicting the contractor's prices, needs to consider the market forces, and his historical price data is eminently suited to aid this task. It is for the forecaster, in the same way as the contractor, to be familiar with general price levels in the market, the fluidity of which is likely to elude the rather crude, static numerical analysis methods employed in systems currently in development.

It will be wise, until such time as the forecaster's expertise has been studied and evaluated, to avoid taking any measure which may reduce its influence.

SUMMARY AND CONCLUSIONS

This paper has reviewed the general state of the science and art of estimating costs and prices for construction work.

Traditional methods of estimating have been described and categorised as either approximate, analytical or operational. Although not mutually exclusive, the approximate and analytical methods are associated with the designer's price forecast and the constructor's cost predictions respectively for 'work in pace' items, whilst operational methods are associated with the constructor's predictions of resource costs.

Relatively recent developments in computer based systems, for the benefit of both designers and constructors. have resulted in the computerisation of existing manual methods and in attempts to develop mathematical predictive models through statistical analyses.

In order to assess the effectiveness of any estimating techniques some measure of performance must be obtained. The difficulties of obtaining a suitable measure has been discussed in some detail and reference to published research findings is made to determine accuracy levels considered appropriate generally in the industry. From this it would appear that contractors estimate with an accuracy in the order of 6% cv whilst designers estimate with an accuracy in the order of 15% t 20% cv n the early stages of design, improving to 13% to 18% cv immediately prior to receiving tenders. The surprisingly slight increase in accuracy of designers' estimates over the design period, together with the lack of any reliable empirical evidence of the means whereby effective estimates can be generated, suggests the need for serious fundamental research in this area.

Various discrepancies between recommended and actual estimating practice have been outlined, leading to the consideration of subjective influences involved in estimating expertise.

It is becoming increasingly apparent that accurately predicting costs is a problem common to all industries but in construction there are extra difficulties due to complexity and uncertainty. The complexities, it is argued, are beyond human analytical capabilities and the uncertainties render forecasting virtually impossible. The result is that contractors' estimators produce socially acceptable figures in the light of the market, leading to a process reminiscent of learning theory. Consequently, prices and costs are only tenuously related, rendering analytical methods excessively refined. A budget figure is, therefore, produced by synthesising global rates and adjusted for market conditions by value judgements.

It is also apparent that contractors' estimators performance is considerably influenced by personal characteristics including experience, intuition, feel, personality and enthusiasm, none of which is normally associated with simply analytical ability.

The designer's forecasting difficulties, on the other hand, centre obstensibly on the lack of reliable design and price information. These two problems, however, are shown not be serious because of the major price determining effects of early design decisions and the tendency of competing contractors to charge market driven acceptable prices. The small difference in relative forecasting accuracy at early and late design stages supports this view.

The designer can, therefore, be seen as being in a similar situation to the contractor in which knowledge of general price levels is the main factor. Such knowledge is to a large extent acquired by experience and other subjective attributes similar to those required of the contractor.

REFERENCES

1 James, W. A new approach to single price rate approximate estimating. *Chartered Surveyor* 1954

2 RICS. *An Introduction to Cost Planning* 1976

3 Southwell, J. *Building Cost Forecasting*. RICS. 1971

4 Skoyles, E. R. Operation bills – (various publications, see *CIOB Bibliography* No. 82)

5 Ryan, T. M. Computers for estimating. *Building Technology and Management* 1981 **19** January, pp15–16

6 Ashworth, A. Cost modelling in the construction industry. *Quantity Surveyor* 1981 **37** July, pp132–134

7 Case, E. Consideration of variability in cost estimating. *IEEE Transactions on Engineering Management* 1972 **EM19** November

8 Ashworth, A., Neale, R. H. and Trimble, E. G. An analysis of the accuracy of some builders' estimating. *Quantity Surveyor* 1980 **36** April, pp65–70

9 Rubey, H. and Milner, W. W. *Construction and Professional Management*. Macmillan. 1966

10 Park, W. R. The strategy of contracting for profit. Prentice-Hall. 1965

11 Liddle, C. Process engineering – the QS role. *Chartered Quantity Surveyor* 1979 **2** October, pp58–60

12 Keating, C. J. Looking back at cost estimating and control. *Transactions of the American Association of Cost Engineers* 1977, pp13–20

13 Moyles, B. F. An analysis of the contractor's estimating process. Submitted in partial fulfilment of the requirements for Msc. Loughborough University of Technology. 1973

14 Fine, B. and Hackemar, G. Estimating and bidding strategy. *Building Technology and Management*. 1970 **8** September, pp.8–9

15 Fine, B. Tendering Strategy. *Building* 1974 **227** October 25, pp115–21

16 Hackemar, G. C. Profit and competition: estimating and bidding strategy. *Building Technology and Management* 1970 **8** December, pp6–7

17 Beeston, D. T. *One statistician's view of estimating*. Cost study 3. Building Cost Information Service 1974. RICS. p13

18 Willenbrock, J. H. A comparative study of expected monetary value and expected utility value bidding strategy models. Thesis submitted in partial fulfilment of the requirements for the degree of PhD. Pennsylvania State University 1972

19 Morin, T. L. and Clough, R. H. Opbid: competitive bidding strategy model. *ASCE Journal of the Construction Division*. 1969 **95** July, pp85–106

20 Gates., M. Bidding strategies and probabilities. *ASCE Journal of the Construction Division* 1967 **93** March, pp75–107

21 Barnes, N. M. L. The design and use of experimental bills of quantities for civil

engineering contracts. PhD Thesis No.P1263. 1971 University of Manchester Institute of Science and Technology.

22 Barnes, N. M. L. Measurement theory. *Chartered Quantity Surveyor* 1979 December pp117–118. (Correspondence)

23 Ashworth, A. Regression analysis for building contractors. An assessment of its potential. Thesis submitted in partial fulfilment for the award of Master of Science. 1977. Loughborough University of Technology

24 Watson, B. The influence of tendering methods on engineering services costs. *Chartered Quantity Surveyor* 1979 **1** May, pp149–151

25 McCaffer, R. Contractor's bidding behaviour and tender price prediction. A Doctoral Thesis submitted in partial fulfilment of the requirements for the award of Doctor of Philosophy. September 1976. Loughborough University of Technology

26 Neufville, R. et al. Bidding models: effects of bidders' risk aversion. *ASCE Journal of the Construction Division* 1977 **103** March, pp57–70

27 Cauwelaert, F. V. and Heynig, E. Correction of bidding errors: The Belgian solution. *ASCE Journal of the Construction Division* 1978 **105** March, pp19–23

28 Skitmore, R. M. Bidding dispersion – an investigation into a method of measuring the accuracy of building cost estimates. thesis submitted for the degree of Master of Science. University of Salford. 1981

29 Morrison, N. and Stevens, S. *Cost Planning and Comuters*. A research study. Department of Construction Management. University of Reading. 1981

30 Barnes, N. M. L. Financial control of construction. *In:* Wearne, S. H. ed *Control of Civil Engineering projects*, Chapter 5. Edward Arnold. 1974

31 Park, W. R. *Cost Engineering Analysis*. John Wiley & Sons, 1972

32 Marr, K. F. Standards for construction cost estimates. *Transactions of the American Association of Cost Engineers* 1977 (Paper B–6). pp77–80

33 Perry, J. H. ed. *Chemical Engineers Handbook* McGraw-Hill. 1963

34 McCaffer, R. and McCaffrey, M. J. *The Reliability of Cost Data* Quantity Surveyors Research and Development Forum, contribution to confrence held at the Polytechnic of the South Bank, 16 June 1981

35 Greig, M. D. Construction cost advice. 'Is the customer satisfied? A study of construction cost forecasting and levels of client satisfaction. Submitted in partial fulfilment of the requirements for the degree of Master of Science. Herriot-Watt University. 1981

36 Jupp, B. C. and McMillan, V. Quantity Surveyors Research and Development Forum. *The Reliability of Cost Data* Contribution to conference held at the Polytechnic of the South Bank, 16 June 1981

37 Flanagan, R. Tender price and time prediction for construction work. Thesis submitted to the University of Aston in Birmingham for the degree of Doctor of Philosophy. 1980

38 Mitchell, R. F. Assessing the economics of building, *CIB Symposium*, Dublin. 1974

39 McCaffer, R. Some examples of the use of regression analysis as an estimating

tool. *Quantity Surveyor* 1975 **32** December, pp81–86

40 Bowen, P. A. An investigation into the feasibility of producing an economic cost model for framed structures. Msc Thesis, 1980. Herriot-Watt University

41 Brown, J. A. Bid cost of shuttle facilities – construction bidding cost of KSC's space shuttle facilities. Paper presented at *23rd Annual AACE Meeting* Cincinnati, Ohio, July 15–18 1979

42 Brown, J. A. Personal communication dated 30/11/81

43 Handscomb Roy Associates. Measurement of estimating performance. *Newsletter* 1976 **3** June

44 Morrison, N. and Stevens, S. *Construction Cost Data Base* 2nd annual report of research project by University of Reading, Dept. of Construction Management. 1980

45 Ferry, D. J. and Brandon, P. S. *Cost Planning of Buildings*, 4th ed. 1980. Crosby Lockwood

45A Fine, B. Paper presented to Builders Conference. 1968

46 Trimble, E. G. Regression analysis – new uses for an established technique. Internal paper. Loughborough University of Technology. 1970

47 RICS. UK and US construction industries: a comparison of design and contract procedures. Study by Department of Construction Management, University of Reading. 1979

48 Oxenfeldt, A. *et al*, Insights into pricing. From *Operations Research and Behavioural Science* Wadsworth Publishing Company. 1961. p71

49 Tucker, S. N. *Symposium: Quality and Cost in Building*. Vol. 1, Investment appraisals, construction and use costs, the life-cycle concept for buildings. CIB W.55, 15–17 Sept. Institut de Recherche sur L'envirionement Cosntruct, Lousanne, Switzerland, pp93–105, Economic design making in the building development project.

50 Woollett, J. M. A methodology of financial forecasting in the building industry. PhD dissertation, University of Texas at Austin. 1978

51 Bennett, J. and Fine, B. Measurement of complexity in construction projects. SRVC Research Report GR/A/1342.4. 1980. Department of Construction Management, University of Reading

52 Tavistock Institute of Human Relations. *Interdependence and Uncertainty*. A study of the building industry. Tavistock Publications. 1966

53 Edelman, F. Art and science of competitive bidding. *Harvard Business Review* 1965 **43** July/August, pp54–66

54 Shaw, W. T. Do builders estimate? Or do they price? *Surveying Technician* 1973 **2** August, pp16–18

55 Relf, C. T. Study of the building timetable: the significance of duration parts 1 and 2 – Final report of a research project sponsored by the Department of the Environment, on the basis of work carried out by a team under the direction of Professor D. A. Turin. UCERG Building Economics Research Unit, University College London. 1974

56 Niss, J. F. Custom production, theory and practice: with special emphasis on the

goals and pricing procedures of the contract construction industry. Thesis submitted in partial fulfilment of the requirements for a degree of doctor of Philosophy in Economics. University of Illinois. 1965

57 Mayer, Colloquium held in Paris on Optimum Economics and Devolution of Public Tendering for works. *Bouwirtschaft* 1969 **23** December 4, pp1230–2

58 Skoyles, E. R. The commonability of national methodologies for controlling financial resources in the construction industry of western Europe. *Quantity Surveyor* 1979 **35** August, pp497–503

59 Benedict, J. Time-wasting analytical build-up sheet. *Building Trades Journal* 1971 **163** October 22 pp26–32

60 Tassie, C. R. Aspects of tendering: converting a net estimate into a tender in the practice of estimating, pp6–8 from *The Practice of Estimating*. Chartered Institute of Building.

61 Fellows, R. F. and Longford, D. A. Decision theory and tendering. *Building Technology and Management* 1980 **18** October, pp36–39

62 Curtis, F. J. and Maines, P. W. Competitive bidding. *Operational Research Quarterly* 1974 **25** pp179–181

63 Mulvihill, D. F. and Paranka, S. *Price Policies and Practices* John Wiley, 1967

64 Grant, G. The estimator's tasks and skills. *Building Trades Journal* 1974 **168** December 13, pp52, 54

65 McCaffer, R. Computer aided estimating – its evolution, difficulties and future pp9–13. In *The Practice of Estimating* Chartered Institute of Building

66 Lorenzoni, A. B. Recipe for good estimating – it works! *Transactions American Association of Cost Engineers* 1974. pp13–16

67 Portsmouth Polytechnic. Department of Surveying. *Acceptable Levels of Risk*, SMM Development Unit, 1974

68 Blakeslee, T. R. *The Right Brain* – a new understanding of the unconscious mind and its creative powers. Macmillan. 1980, p25

69 Osborn, C. L. Production and scheduling garden apartments. *Transactions of the American Association of Cost Engineers* 1978 (paper C–7). pp135–142

70 Franks, J. An exercise in cost (price?) estimating *Building* 1970 **217** June 12, pp133–134

71 Stone, P. A. *Building Design Evaluation* Costs In use. 3rd ed. E. & F Spon Ltd. 1980

72 Azzaro, D. W. Measuring the level of tender prices *BCIS Cost Study* No 5, 1976/77. RICS

73 Responsible for organisation. *Chartered Quantity Surveyor* 1982 **4** January, pp158–159

74 Barrett, A. C. A Swedish cost plan method *Building Economist* 1970 9 August, pp50–59

75 Powell, J. and Chisnall, J. Getting the early estimates right. *Chartered Quantity Surveyor* 1981 **37** March, pp279–281

76 Grinyer, R. H. and Whittaker, J. D. Managerial judgement in a competitive bidding model. *Operational Research Quarterly* 1973 **24 (2)**, pp181–191

77 Associated Industrial Consultants Limited and Business Operation Research Limited. Report of the Joint Consulting Team on Serial Contracting for Road Construction. 1967. Ministry of Transport

78 Twort, A. C., Fallacies about selective tendering *Civil Engineering and Public Works Review* 1969 February, p137

79 Jepson, W. B. Financial control of construction and reducing the element of risk. *Contract Journal* 1969 April, pp862–864

5.3 The accuracy of quantity surveyors' cost estimating

N. Morrison

ABSTRACT

This paper examines the accuracy of cost estimates prepared by quantity surveyors during the design stages of construction projects. Since these estimates influence the feasibility and form of projects, the ability of quantity surveyors to fulfil the objectives of cost planning is considered. It concludes that the present methods of estimating used by quantity surveyors produce results which are not sufficiently accurate to meet all the objectives of cost planning. The most likely area for improvement lies in developing methods of using large cost data bases.

Keywords

Cost estimating, cost planning, accuracy, variability, quantity surveyor

INTRODUCTION

Bennet *et al.* (1981) state that cost estimating is the key activity in the quantity surveying profession's cost planning service. They further conclude that if cost estimating is effective, then cost planning can achieve its objectives. Without good estimating, cost planning is a frail and ineffective on-cost.

The quantity surveying profession has developed a number of estimating techniques designed to cope with the many and varied instances in which predictions of cost are required during the development of a building design. These range in detail from simple lump sum evaluations and single unit methods to the measuring and pricing of very detailed approximate quantities or even pricing full bills of quantities. Quite naturally we would expect the accuracy of price predictions to increase as the level of detail in which estimates are prepared and the knowledge of proposed designs are increased.

*1984, *Construction Management and Economics*, 2, E & F N Spon, 57–75.

However, few attempts have been made in the past at quantifying the range of performance which is achieved, achievable or acceptable.

Since the notion of acceptability must ultimately be related to clients' specific objectives and is probably, therefore, project dependent, this paper confines itself to three major questions:

1   Is it possible to measure the cost estimating performance of quantity surveyors from actual practice, and if not, can it be reasonably quantified?
2   Is this performance likely to enable the objectives of the cost planning process to be achieved? and if not,
3   Can performance be improved and if so how?

THE DEFINITION OF 'ESTIMATING ACCURACY'

In preparing this paper it has been assumed that the accuracy of an estimate is measured by the deviation from the lowest acceptable tender received in competition for the project. Unfortunately, few estimates are prepared in the early design stages for schemes which subsequently remain unaltered prior to the invitation of tenders, and in consequence, the lowest tender cannot realistically be compared with the estimate as each relates to a different scheme. It has, therefore, been assumed that cost estimates produced during the various design stages of a construction project have the objective of predicting the tender price level (i.e. lowest tender) which might be expected to be achieved if that same scheme were assumed to be fully detailed and competitive tenders could be invited on the relevant contract particulars. The importance of this assumption will be discussed at some length later in the paper, but for now it is worth noting that the decision to measure quantity surveyors' estimating performance against lowest tenders means that the assessment of accuracy must in part be dependent upon the variability of these lowest tenders.

THE ACCURACY OF QUANTITY SURVEYORS' ESTIMATES IN PRACTICE

Irrespective of the definition of accuracy assumed it has proved impossible to measure performance at any stage of the design process other than the tender stage. It has been found to be common for quantity surveyors to produce 'pre-tender' estimates by pricing bills of quantities either before or during the tender period. Whilst such estimates are prepared too late in the design process to form part of an effective cost planning system, many quantity surveyors believe them to be a better basis against which to judge the acceptability of lowest tenders than estimates produced earlier in the design process.

It is not the intention of this paper to consider the merits of such estimates; however, they do form an important body of data when considering the accuracy of estimating. In consequence, samples of data originating from seven separate quantity surveying organizations have been collected. These comprise the accepted tender (prior to any post-tender reductions) and the quantity surveyors' pre-tender estimates for each project. It should be noted that:

1   Each of the offices from which data was collected stated that the intention of a pre-tender estimate was to predict he lowest tender which would be received.
2   In the vast majority of casts, the estimates were produced by pricing the bills of quantities, usually during the tender period.
3   The data collected predominantly emanates from the public sector as it was found to be much easier to gain access to data in public authorities than private practices. However in each public sector office some of the estimates were produced by the authorities' own staff and some by consultants employed by the authorities. Consequently the results relate to both public and private practice.
4   The projects which formed the sample taken from each office represented all of the work for which estimates were produced over a given period of time. The only exception to this rule is office F where the projects collected are a random sample of the work undertaken in 1978.
5   For comparison purposes, all the data collected has been rebased to an average 1978 tender price level by means of the BCIS tender price index.

The overall estimating accuracies achieved by the seven offices forming the sample are presented in Table 5.3.1. The mean error achieved in estimating was found to be significantly different from zero in all except one case. The values for this variable range from −5.76% to + 5.86% (with Prime Cost (P.C.) and provisional sums included). This finding suggests that the offices were failing in their objective of predicting lowest tenders since in six out of seven instances there was a distinct tendency to either underestimate or overestimate the average level of these tenders. This performance indicates that the six offices concerned were not modifying their pricing policy in reaction to the estimating results achieved on previous projects. It was found that none of the offices in the sample monitored their estimating performance on any consistent basis, and, consequently, it was not surprising to find that none of them, when questioned, actually recognized that a consistent error trend had developed in their estimating performance.

Table 5.3.1 shows that the mean deviation of the total estimates from the total tenders within the entire sample was 9.80%. However, on the 557 projects on which it was possible to establish the value of P.C. and provisional sums, the mean deviation was found to rise to 11.97% when these fixed sums were subtracted from both the estimates and tenders.

When the effect of P.C. and provisional sums is ignored, the estimating performances of offices C, D and E are remarkably consistent. Offices A and B

Table 5.3.1 A comparison of seven quantity surveying offices' estimating accuracy

Office	Number	Including PC and provisional sums			Number	Excluding PC and provisional sums		
		Mean error (%)	Mean deviation (%)	Coefficient of variation (%)		Mean error (%)	Mean deviation (%)	Coefficient of variation (%)
A	62	−5.76	10.30	11.68				
B[a]	115	4.38	10.43	12.22				
C	62	2.89	8.17	10.88				
D	222	2.61	8.27	11.50	221	3.48	11.13	15.11
E	89	−0.33	8.17	11.29	81	0.81	12.86	16.49
F	310	5.86	11.76	15.52	204	5.43	13.29	16.62
G	55	3.72	7.57	9.37	51	4.31	8.88	10.64
Total	915	3.17	9.80	12.77	557	3.88	11.97	15.45

[a]Excludes projects with a value of less than £30 000.

show similar coefficients of variation to offices C to E, but the mean deviations achieved are some 2% poorer. Closer analysis reveals that this is almost entirely due to the pronounced tendency of offices A and B to err in one particular direction. Office A consistently underestimates and office B consistently overestimates. Offices F and G, however, display performances which are significantly different from the other offices.

Whilst the results of office F are substantially poorer than those achieved by the other six offices, there was no apparent difference in the manner in which estimates were produced in the office. However, a study of this office's cost planning procedures revealed that cost planning played a very limited role during the development of building designs and that the pre-tender estimate was often only the second estimate to be produced for a project. Consequently, there is evidence to suggest that the office's practices and procedures which do not place as much emphasis on the production of estimates as other offices have led to the quantity surveyors being less experienced in this field.

Office G on the other hand produces estimates which are more accurate than any other office and the coefficient of variation at 9.37% (including P.C. and provisional sums) is approximately 1.5% better than that achieved by the next most accurate office. All of the data which forms this sample is taken from supermarket projects for one client. Consequently, there is evidence to suggest that the improved performance by this office might be attributable to the experience which the office has built up over the years on this type of project and the large amount of relevant cost data which the practice has at its disposal for estimating purposes. Further evidence to support this conjecture has been provided by the results achieved by other offices on building types which form a large proportion of their work. Office A, for example, has been shown to perform some 20–25% better on school projects and office D was found to estimate the cost of housing projects with a mean deviation which was half that achieved on the other projects in the sample drawn from that office.

However, in a number of other instances where we might have expected an improved performance due to similar imbalances in the type of work undertaken by the offices, no such improvement was detectable. We can therefore conclude that estimating accuracy does not automatically improve simply because an office handles many similar projects. It is apparent that in such instances, the achievement of an increase in accuracy is dependent upon the means by which knowledge and experience gained on previous projects is related to future work. In those offices where an improved performance was detected, it was noticeable that either a central library of information or an index system by which the quantity surveyors could familiarize themselves with the data at their disposal had been constructed. In two of these offices this familiarization of individual quantity surveyors with the type of work had been formalized to the extent that specialist groups of surveyors had been formed to deal solely with one particular project type. In three of the samples collected, it has been shown that better estimating performances are achieved on higher value projects. In one of these samples this improvement was almost certainly linked with the type of project which formed the sample, but, nevertheless, there is some evidence to suggest that the estimating performance of quantity surveyors improves as project value increases. In no instance was the reverse trend observed.

It is apparent from the figures presented in Table 5.3.1 that the estimating performance achieved over the entire sample is very poor. Furthermore, the estimates which form the sample have been prepared in the greatest level of detail possible. The mean deviation of approximately 12% and a coefficient of variation of 15.45% are, therefore, likely to represent the best performance currently achieved by quantity surveyors.

In answer to the first question posed at the beginning of this paper we may, therefore, state that it has been possible to establish, with some degree of certainty, the performance of quantity surveyors when using the most detailed estimating technique available to them. However, it has not proved possible to measure directly the actual performance achieved when using less detailed techniques. In consequence, it must be assumed that the only practical manner in which the likely accuracy of such techniques can be established is by analysis and interpolation of the available data.

The first stage of such an analysis is to attempt to establish the factors which combine to produce the estimating performance described above which has been measured in actual quantity surveyors practice and then to quantify the effect which each of these factors exerts. This will serve two purposes:

1   It will highlight those factors whose influence upon estimating accuracy is likely to be affected by differing estimating techniques thereby enabling closer analysis at a more detailed level, and

2 It will highlight the relative importance of the various factors thereby providing some indication of those areas where greatest improvement in overall estimating accuracy may be achieved.

THE FACTORS WHICH INFLUENCE ESTIMATING ACCURACY

It has already been suggested earlier in this paper that estimating accuracy is to some extent dependent upon the variability of the object which is to be estimated, i.e. the lowest tender. Given that there must exist some mean tender price level (which will undoubtedly vary with location, market conditions etc.), then lowest tenders will vary around this mean to some as yet undefined extent. It will be shown that lowest tenders must fall randomly around this mean and are, therefore, not capable of being predicted exactly. To minimize the effect of this variability quantity surveyors must attempt to predict the mean price level of lowest tenders, but their ability to do this is influenced by a number of further factors. Since quantity surveyors prefer to use data taken from previous lowest tenders when preparing estimates it is clear that the price level of the data used by the quantity surveyor will influence the price level of the resultant estimate. Also, different estimating techniques have differing inherent errors. The inherent error in a technique is linked to the detail in which the estimate is produced, and the variability of the cost data it employs. The suitability of the cost data used when estimating will have an important bearing on accuracy. It might be expected that the cost information derived from an office development would be more suitable than that taken from a warehouse project for the purposes of predicting the cost of another office development. However, even within the same building type, such factors as size, complexity, height and specification combine to produce a wide range of building costs. In addition, quantity surveyors make a number of adjustments to the cost data selected for pricing purposes. These adjustments are designed to convert the selected cost data from one time, location and prevailing market situation to the anticipated time, location and market situation surrounding the new project.

In considering the influence that each of the above factors exerts on overall estimating performance, the standard treatment for adding normal distributions has been used. This states that:

$$S = (S_1^2 + S_2^2 + \ldots S_n^2)^{1/2} \tag{1}$$

where S is the resultant standard deviation and S_1, S_2 etc. are the standard deviations of the distributions being added. (This expression may also be used in terms of coefficients of variation where the mean of individual samples are equal.)

THE VARIABILITY OF LOWEST TENDERS

In attempting to establish the variability of lowest tenders it may be considered that the tenders invited for any particular project represent a sample taken from the 'family' of prices which could be obtained for that project. The 'family' represents the hypothetical distribution which would be formed if every suitable contractor able to undertake the project were invited to submit a price.

If six contractors are chosen at random from the total pool, it would be unlikely that the value of the lowest tender would be exactly the same as that derived if six different contractors were chosen. Consequently, the variability of the lowest tender can be measured from the distribution of lowest tenders which would result from repeated random sampling of six (or whatever number) tenders from the 'family'. [Beeston (1973) concludes that random sampling is justified since he was able to demonstrate that contractors called from a reserve list in substitution for others who drop out take their place randomly in the tender list despite not being expected to produce the lowest tender.]

It is evident that the variability in lowest tenders, as described above, cannot be established by measurements from the field sine it can only be measured from a series of repeated, independent sets of bids for identical contracts in identical circumstances. To establish the variability it is therefore necessary to adopt a theoretical approach. This requires two steps:

1   Establishing the nature of the hypothetical 'family' of tenders which could be submitted for a particular project, and
2   Taking sample tender lists from this [family' and establishing the variability of the resultant series of lowest bids.

Table 5.3.2 shows the pattern of tender ranges found in 423 competitively let projects. However, the figures presented in this table somewhat dampen the true variability of tenders since they are based on total tender sums. Further work suggests that when the value of items which are common to each tender (i.e. P.C. and provisional sums) is removed, the real tender ranges increase as follows:

Up to £100 000   Range = 33.75%
£1 000 000 and over   Range = 18.61%
All projects   Range = 26.83%

If it is assumed that the average value of lowest tenders is represented by a notional 100 units, then from above, it can be seen that the average tender range is 26.83 units. If it is further assumed that the 'family' of tenders is normally distributed as Beeston (1973) asserts, then by use of a simple

Table 5.3.2 Range of competitive tender lists

Tender value of projects (£)	Number of projects	Average number of tenderers	Range of all tenders (%)	Range of first three tenders (%)
Up to 100 000	133	6.0	24.03	9.88
100 000–200 000	88	6.4	19.47	6.96
200 000–300 000	66	6.8	18.08	7.05
300 000–500 000	58	7.1	16.07	5.99
500 000–1 000 000	44	7.1	13.41	4.82
1 000 000 and over	34	6.2	13.25	5.14
100 000 and over	290	6.6	16.82	6.25
All projects	423	6.4	19.10	7.32

Note: All tenders have been rebased to an average 1980 tender level.

Table 5.3.3 Characteristics of tender lists

Project size (£)	Mean value of tenders (units)	Coefficient of variation of tenders (%)	Mean deviation of lowest tender from mean of lowest tenders (%)	Coefficient of variation of lowest tenders (%)
0–100 000	120	11.3	6.37	7.85
1 000 000 and over	110	6.8	3.70	5.07
All projects	115.5	9.3	5.01	6.59

computer-based model it is possible to find the characteristics of the distribution which recreates this overall range of tenders. Table 5.3.3 shows the mean and coefficient of variation of the distribution of tenders which have to be assumed to achieve an average lowest tender of 100 units and the overall ranges of tenders shown above. In addition, the characteristics of the distributions of lowest tenders are presented.

The coefficients of variation of lowest tenders presented in Table 5.3.3 represent the maximum possible values of this variable since they are based upon the assumption that contractors' tenders differ from one another for totally random rather than predictable reasons. Lowest tenders cannot be more variable than the figures in Table 5.3.3 suggest since this would imply larger tender ranges than those observed in practice. It is, therefore, necessary to establish the extent to which the spread of tender is due to random rather than predictable effects.

It is a commonly held view among quantity surveyors and contractors that the majority of the range of tenders found on live projects is attributable to the prevailing market conditions. Barnes (1974) challenges this assertion when he states that 'many of the misunderstood aspects of the process team from the disregard of its probabilistic nature. For example, we tend to think that contractors when submitting tenders vary their figures considerably, owing to market pressures, buying jobs and other mysterious influences. In fact, the majority of the spread of tender prices is due to the probability distribution which is a measure of the accuracy with which the contractor is able to predict the cost of the work'.

The accuracy of contractors' estimating has not been widely researched. Barnes and Thompson (1971) state that 'there is a wide variation in the prediction of net cost by contractors when estimating: the accuracy achieved in a sample of 159 contracts was such that one in three of the estimates diverged from the actual total net cost by more than 6%'. This observation was derived from analysis of civil engineering projects, and Beeston (1973) comments that 'the figure for civil engineering would be expected to be higher than that for building but as well as this it is inflated by the variability in the final measurement of actual cost'. Beeston continues by stating that one building contractor found that the extent of agreement among his estimators had a coefficient of variation of 4%. He concludes that a reasonable figure for the likely variability between contractors' estimators is a coefficient of variation of just under 4%.

Perhaps the most thorough research into building contractors' estimating accuracy has been undertaken by Barnes on behalf of the Builders' Conference (Barnes, 1972). In a study of 228 projects taken from 10 different contracting organisations, Barnes found that the average divergence between estimated costs and achieved costs on these projects was −0.25%. He continued by explaining that this 'means that if all contractors contributing to the study had added 0.25% to their estimates of net cost before they submitted tenders for work they would have made the net profit which they expected to make'. However, when Barnes came to consider the divergence between estimate and cost on individual projects rather than the sample as a whole, he found that 'in one contract out of three the estimate of net cost was more than 10% wrong. That figure was affected by a group of contracts where one contractor was virtually out of control in his estimating, and I think a figure of 7% is probably a more realistic, normal performance'.

Barnes therefore concludes that the coefficient of variation in contractors' estimating is 7% if one particularly poor contractor is excluded from his analysis. This finding conflicts with Beeston's prediction of 4%. Since such a wide range of values have been proposed for the accuracy of contractors' estimating accuracy, further research has been undertaken in an attempt to clarify the position.

Flanagan (1980) recognized this need and was able to gain access to a national contractors' cost book to examine the relationship between the true cost at the end of a project and the estimated cost at tender stage. In a sample of 64 projects which constituted all of the work completed by this particular contractor in the period 1970 to 1974, the percentage of overheads and profit added to the original tender and that actually achieved were deducted from the final account to leave a notional tender value of measured work and the actual value in the final account. It is not possible to compare the actual estimate with the final account since the latter contains the value of fluctuations and variations which would not have been contained in the original tender.

In Table 5.3.4 the results derived from the analysis of the data have been summarized. These show that over the whole sample of 64 projects, the mean deviation is 5.49% with a coefficient of variation of 8.22%. However, it is evident from this table that the contractor's performance is significantly better on larger projects. For those projects with a final account value exceeding £100 000 the mean deviation was found to be 3.75% with a coefficient of variation of 6.59%.

Using Flanagan's data it has therefore been possible to establish the estimating accuracy in 64 successful tenders submitted by one contractor. This measurement of the ability of this contractor to estimate accurately his costs as expressed by the coefficient of variation is approximately 1% higher than Barnes' assumption quoted earlier and significantly greater than Beeston's prediction. However, since this and Barnes' figures have been derived from actual measurement, more confidence should be placed in these figures than in Beeston's prediction. However, an analysis of 1338 further projects has led to the conclusion that the sample of 64 projects quoted above contains a relatively high proportion of smaller projects and, in consequence, the average coefficient of variation derived is likely to be higher than if a more representative sample could have been selected. Similarly, Barnes' measurement of the coefficient of variation is somewhat clouded by the performance of one particular contractor. Bearing these points in mind, it has therefore been concluded that the best measure of the average coefficient of variation in contractors' ability to estimate their own costs (on projects for which they are the lowest tenderer) is that given in Table 5.3.4 for those projects with a value of £100 000 or more. This figure is 6.6%.

Table 5.3.4 Summary of national contractor's estimating accuracy

Final account value (£)	Number of projects	Mean error (%)	Mean deviation (%)	Coefficient of variation (%)
Up to 50 000	14	−1.54	7.72	9.62
50000–100 000	10	−3.72	9.31	11.25
100 000–200 000	10	0.23	1.89	2.85
200 000–300 000	6	1.85	4.21	5.25
300 000–500 000	9	−0.11	4.60	6.75
500 000–1 000 000	5	−6.37	6.37	11.55
1 000 000 and over	10	−1.35	3.28	5.04
100 000 and over	40	−0.82	3.75	6.59
All projects	64	−1.43	5.49	8.22

If the simple assumption is made that this estimating performance by contractors on contracts which they win is also achieved by the unsuccessful tenderers, then by means of a model similar to that described earlier in this section, it can be found that this factor alone would create an average overall range of tenders of approximately 22%. If it is considered that the measured range of tenders is just under 27% (with P.C. and provisional sums removed), then this conclusion provides substantial evidence in support of Barnes' contention that the majority of the spread of tender prices is due to the probability distribution which is a measure of the accuracy with which the contractor is able to predict the cost of the work.

If this conclusion is linked to Barnes' (1972) finding that the lowest tender is on average 0.25% below actual cost, then it must be further concluded that, on average, contractors significantly overestimate cost. This would suggest that contractors have pitched their pricing level at a point where they are able both to win competitive tenders and make profits. Since contractors can only judge their ability to predict costs on those projects on which they are successful, it has been found, in a separate study (Department of the Environment, 1980), that they are unwilling to admit to overestimating on projects which they fail to secure, preferring to question the price level at which the successful contactor has won the project. However, the conclusion is supported by Ashworth et al. (1980) and Roderick (1977) who have both provided evidence that contractors significantly overestimate the labour content of projects.

From the research outlined so far, it may be concluded that whilst such factors as contractors' overheads and profit level and the construction method chosen must exert some degree of influence upon the variability of tenders received in competition, the vast majority of the spread of tenders results from contractors' inability to predict their costs. In consequence, the model described earlier is validated since it has been possible to prove that the majority of the spread of tenders is due to random rather than predictable influences.

Whilst it has been possible to conclude that the coefficient of variation of lowest tenders received in competition is on average approximately 6.6%, it has also been found that the coefficient varies significantly with the project value. The coefficient for projects with a contract value in excess of £1m has been found to be approximately 5.10% whereas a figure of 7.85% has been calculated for projects with a value of less than £100 000.

The importance of the variability in the lowest tenders received in competition lies in the fact that quantity surveyors' estimates of these tenders cannot, except by chance, be more accurate than this variability. If quantity surveyors were able to estimate consistently at a price level equivalent to the mean of the lowest tenders, then the accuracy in estimating would be identical to the variability in lowest tenders. To improve on this performance quantity surveyors would have to follow consistently the nature of the lowest tender in relation to the mean lowest tender, i.e. increase their estimate when the lowest tender is above the mean price level of lowest tenders and decrease their estimate when the lowest tender is below the mean price level of the lowest tenders. Since any lowest tender's relationship to the mean of lowest tenders is random, it must be concluded that quantity surveyors can only follow this relationship by coincidence. Therefore, the best performance which can be achieved by quantity surveyors attempting to predict the value of the lowest tender is equal to the level of variability found in the lowest tenders as quoted above. Because the price level of the lowest tender varies from one project to another, quantity surveyors should attempt to estimate at the mean price level of lowest tenders in order to minimize their estimating error. The remaining factors which influence estimating accuracy affect quantity surveyors' ability to do this.

THE SOURCE OF COST DATA USED IN ESTIMATING

In practice it is most usual for quantity surveyors to use cost data generated from previous lowest tenders (Department of the Environment, 1979). The effect of this on the overall variability of the estimate can be significant. As already show, if the rates are taken from an exactly similar previous project tendered in exactly similar circumstances, the likelihood that the source of data will be at a price level equal to the mean of lowest tenders will be given by the variability in lowest tenders, i.e. a coefficient of variation of 6.6%.

The effect that the source of cost data has exerted on the variability in the quantity surveyors' estimating accuracy measured earlier in this paper is difficult to assess since the estimates have, in some cases, been based on differing sources of data. At one office, all of the estimates were produced by pricing the bills of quantities from a standard schedule of prices, whereas all of the other offices priced the bills with rates taken primarily from a single previous bill. However, since the primary source of data would not always provide a suitable rate for certain measured items, prices were also taken from sources such as price books, specialist subcontractors and suppliers, and other priced bills. In these latter instances there is nothing to suggest that the variability in the price data is not close to the 6.6% coefficient of variation. However, the office which had produced a standard schedule of prices by analysis of many previous lowest tenders may well have reduced the coefficient of variation of their cost data since their averaged rates are more likely to represent the mean price level of lowest tenders. It has therefore been estimated that the coefficient of variation of the price data about its mean value is 5% based on a weighted average of the data sources used in the samples.

THE INHERENT ERROR ATTACHED TO THE ESTIMATING TECHNIQUE

The majority of the quantity surveyors' estimates which formed the sample outlined previously were prepared by pricing draft bills of quantities. Skinner (1982) provides figures which indicate that the bill of quantities for a project valued at £200 000–300 000 would on average contain 1100 measured items. Whilst it is recognized that for each individual project the number of measured items is heavily dependent upon such factors as the type of scheme, the manner in which bills of quantities are presented and the architectural licence granted to the designers, the average number quoted above does indicate the general level of detail in which bills of quantities are prepared. Quantity surveyors' pre-tender estimates are therefore typically formed by 1100 subestimates (hereinafter termed 'item estimates'). To discern the variability of the total estimate formed it has been necessary to take three steps:

1   Establish the variability displayed by the individual item estimates. It is important to recognize that a seemingly perfect overall estimate could be composed of may wildly inaccurate item estimates, which, when combined, cancel one another out.
2   Establish the size distribution of items which typically form the overall estimate, and
3   Build a model incorporating the results of 1 and 2 above in order to establish the likely variability in the total estimates produced.

The second of these three points presents few problems. Skinner (1982) has provided substantial evidence (based on 80 case studies) regarding the relationship between the number and value of items contained in bills of quantities. His results show that on average 86.3% of value lies in the largest 20% of measured items.

Consideration of the first of the above points is however less straightforward. A search of published literature found no evidence of research into the relationship between the item estimates produced by quantity surveyors and those inserted in bills of quantities by contractors. In consequence, two detailed case studies were conducted. These case studies showed that in the instance of an accurate overall estimate, the spread of individual item estimates were extremely wide (Fig 5.3.1). The most obvious characteristic of the distribution of item errors is the marked skew to the right. The reason for this is that quantity surveyors cannot underestimate individual items by less than 100% without applying negative item rates. It is reasonable to conclude that this situation does not occur in practice. On the other hand, there is no boundary to the extent to which an item can be overestimated.

A detailed examination of the distribution of item errors reveals that the majority of the errors which form the 'tail' to the right of Fig. 5.3.1 occur on the very low value bill items. Fig. 5.3.2 displays the distribution of item errors for those items with a tendered value in excess of £50. In this distribution, only 1.78% of items fall outside the range −100% to +100% compared with 8.20% over the whole sample.

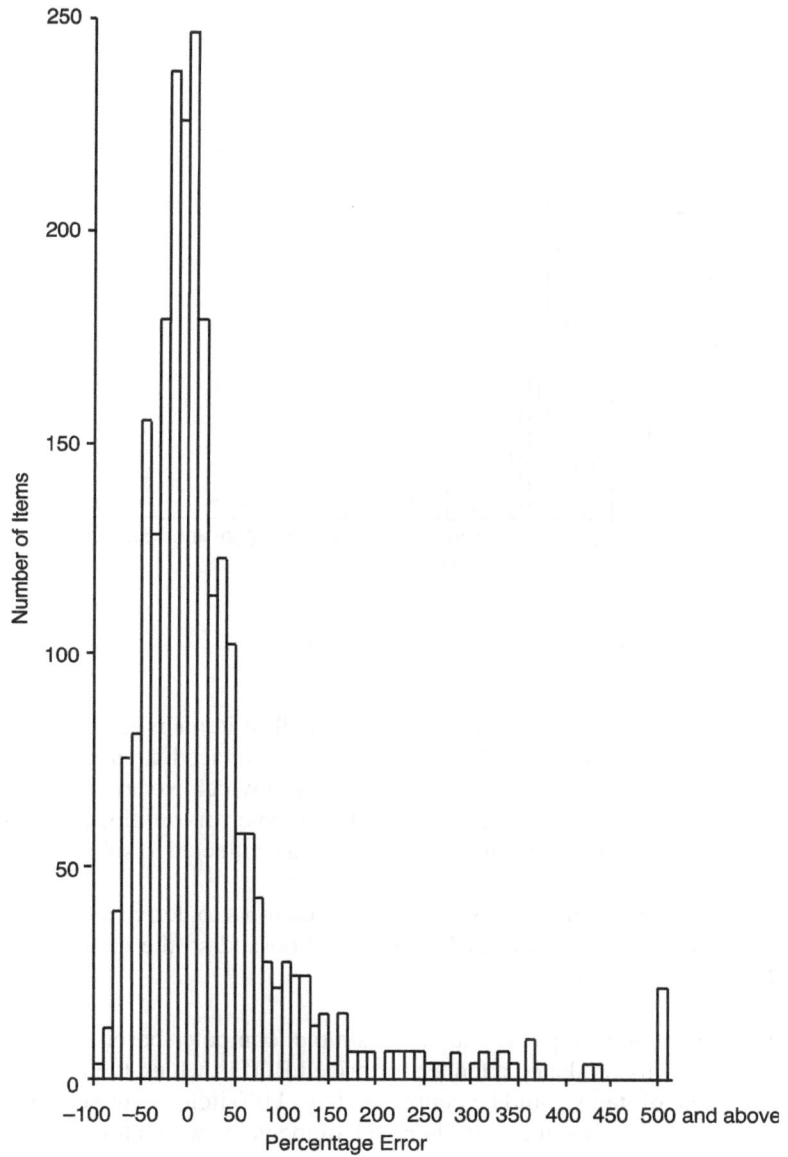

Figure 5.3.1 Estimating error in all items.

In building the simple model described in step 3 above, it has been assumed that quantity surveyors' item estimating errors follow the distribution shown in Fig. 5.1.2. This decision has been taken for three reasons:

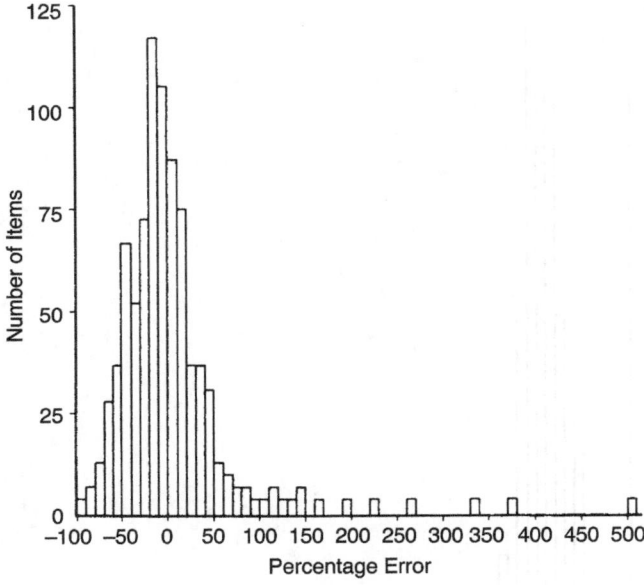

Figure 5.3.2 Estimating error in items greater than £50.

(a) The omission of the effect of the very smallest items is justified as such items do not form part of coarser estimates produced at earlier stages of the design process and it is these estimates in which we are interested.
(b) Whilst the case studies seem to show that item errors continue to improve as item value increases, insufficient data was collected to substantiate this contention.
(c) The data shown in Fig. 5.3.2 has defied attempts to derive a statistical fit to a normal, log-normal or beta distribution and so the distribution has been used in its 'raw' form.

Having completed steps 1 and 2, a simple computer-based model was constructed. This produced a tender (or priced BQ) in which the number and relative value of items could be controlled (e.g. 1100 items with 86% of value contained in 20% of the number). For each of the items which formed the total tender, an estimate was generated by randomly sampling from the distribution of item errors, e.g. if the tender value of an item was £200 and a value of −17.9% was randomly chosen from the distribution, then the value of the estimate for this item would be

$$£200 \times \frac{(100 - 17.9)}{100} = £164.20 \tag{2}$$

This process was repeated for all items and a total estimate was generated and could be compared with the total tender.

In 200 repeated experiments, based on the parameters bracketed in the previous paragraph, the mean deviation of the total estimates was found to be 1.52% with a coefficient of variation of 1.85%. This figure therefore represents the ability of the quantity surveyor to predict the mean price level of lowest tenders when pricing the typical bill of quantities, given that he is able to use a source of data whose total is equal to the mean price level of lowest tenders.

THE SUITABILITY OF COST DATA

If the three components of quantity surveyors' estimating error so far evaluated, namely, the variability of lowest tenders, the inherent variability of the estimating technique and the variability of data taken from previous lowest tenders, are added together, then the total variability (S) is given by the equation

$$S^2 = 6.6^2 + 1.85^2 + 5.0^2 \tag{3}$$

therefore, $S = 8.5\%$

This combined variability is therefore still some distance short of the coefficient of variation of 15.5% measured in practice. However, to this point we have only considered the influence attributable to the use of what may be termed 'perfect cost data', i.e. generated on a previous project exactly similar to the new project which wad let to competitive tender in exactly similar circumstances. The remaining error in quantity surveyors' estimating is attributable to the fact that such perfect data does not exist in practice. This has two principal effects:

1   Because two exactly similar buildings are rarely built, the accuracy of an estimate will become dependent upon the suitability and quantity of the price data available, and
2   Because the circumstances in which cost data is generated cannot be controlled, it is necessary to make adjustments to take account of such factors as time, location, market conditions, site conditions etc.

Data collected from quantity surveyors' offices relating to the estimating performance achieved on particular project types which form a substantial proportion of the work undertaken provides conflicting evidence. In two instances (office G and housing projects at office D) (Table 5.3.1), estimating performances with a coefficient of variation of approximately 10.5% were observed. At office D, this performance was found to be nearly twice as accurate as the other work undertaken. However, at other offices where an improved performance on certain project types might have been expected, no such improvement was found.

If it is considered that the two examples where improved quantity surveyors' performance have been observed represent the maximum improvement possible, then by deducting the variation calculated to be attributable to the factors already considered from the value of 10.5% it is possible to establish the likely influence attributable to the adjustments for time, location, market conditions etc. When account is taken of the value of the projects contained within the subsamples mentioned above, it has been concluded that the variability due to this factor is given by a coefficient of variation of 6.9%. If the remaining variability in quantity surveyors' estimating performance is attributed to the all-embracing category 'imperfections in the cost data employed', then it can be calculated that this factor has a coefficient of variation of 11.0%.

We can now summarize the influences which combine to produce the quantity surveying profession's average coefficient of variation in estimating lowest tenders of 15.5% when pricing bills of quantity as:

		Coefficient of variation (%)
1	The variability in lowest tenders received in competition	6.60
2	The variability due to using cost data from previous lowest tenders	5.00
3	The inherent variability of the estimating technique (pricing bills of quantities)	1.85
4	The variability due to making adjustments to the chosen cost data	6.90
5	The variability due to imperfections in the cost data employed	11.00

It must be stressed that these figures are not presented as being the result of rigorous research, they merely represent a 'best guess' based on what little data is available coupled with one or two assumptions considered reasonable by the researcher. The fact that we cannot rely on the accuracy of the individual coefficients of variation does not prevent us from drawing some broad conclusions concerning the causes of the poor estimating performance observed in practice. However, before we examine these causes and their possible remedy, it is interesting to consider the effect that the above factors are likely to exert upon the accuracy of estimates prepared at earlier stages within the design process.

THE ACCURACY OF ESTIMATES PRODUCED DURING THE DEVELOPMENT OF CONSTRUCTION DESIGN

Quantity surveyors can do nothing to influence the variability of lowest tenders given that the contractors chosen to tender have been carefully vetted on their ability to complete the project to the stated programme and are willing to submit a competitive tender. The effect that variability of lowest tenders exerts on estimating accuracy is, therefore, totally independent of the methods and cost data employed. Similarly, if it is assumed that cost data is taken from lowest tenders in the same manner as previously outlined, irrespective of the cost estimating technique used, then the variability induced into estimates because the price level of the data varies about the mean of lowest tenders will also remain unchanged. The variability due to making adjustments to the cost data chosen should remain constant irrespective of the cost estimating technique being used. Such adjustments are designed to alter the price level of historic data to the price level expected for the new project. Quantity surveyors' perception of this price difference should not be influenced by the estimating technique they intend to use since this does not affect such factors as the time, location, market conditions etc. which surround the historic data or the new project.

Since we would not expect the suitability of cost data taken from the same source to be any better or worse when differing estimating techniques are used, we reach an exceptionally important conclusion. Given that quantity surveyors are consistent in the manner in which cost data is selected for use in cost estimating, only the inherent variabilities of the different estimating techniques will cause the performance of these techniques to differ from one another. Consequently, if we deduct the inherent variability of pricing bills of quantities from the coefficient of variation of 15.5% observed in practice, we must conclude that the best performance that quantity surveyors will achieve, assuming that they do not alter the manner in which cost data is selected and manipulated, is described by a coefficient of variation of 15.4%.

Earlier in this paper it was concluded that the inherent variability of an estimating technique is dependent upon three factors:

1   The number of subestimates which form the total estimate
2   The size distribution of the 'items' forming the estimate, and
3   The variability of subestimates

The influence exerted by these three factors when we consider less detailed methods of estimating than pricing bills of quantities are significantly different. We would expect a reduction in the number of subestimates forming the total estimate to cause the variability of this total to increase since the incidence of 'cancelling errors' reduces. However, as the number of items decreases, so the size distribution of the items forming the total shifts. An analysis of 22 cost analyses relating to office buildings taken from the BCIS reveals that on average, only 45% of value lies in 20% of the number of elements (total of 19 builder's work elements considered), whereas it has been shown earlier in this paper that 86% of value typically lies in 20% of the items which form a bill of quantities. The effect of this more even distribution of value among the items is to dampen the increased variability due to the number of items being decreased. Similarly, the accuracy with which packages of work or elements as opposed to bill items are estimated by quantity surveyors is likely to increase as the number of packages reduces.

In an attempt to quantify the effect that the change in the above factors exerts on estimating accuracy, we have considered an estimate prepared by elemental cost analysis. It has been found that such estimates are typically formed form 13 individual element estimates. Unfortunately, it has not been possible to collect data concerning the accuracy of quantity surveyors' element cost predictions as it has been impossible to find an estimate prepared by elemental cost analysis which could be compared with an actual tender. However, analysis of the two pre-tender estimates described earlier shows that the coefficient of variation of the estimates from those items with a value in excess of £1000 was 25.25%. (These items accounted for approximately 58% of the total value.) From the same data it is possible to calculated that the coefficient of variation of the total element estimates is 15.07%. However, we would not expect estimates of these elements produced by means of a single price prediction to be as accurate as the above figure indicates since this has been calculated from element estimates produced by pricing many sub-elements or items.

By repeating the use of the model described earlier in this paper for the revised criteria outlined above, we find that the inherent variability of elemental estimates based on 13 subestimates would be expressed by a coefficient of variability of 5.3% if we assumed that the accuracy of sub-estimates has a coefficient of variation of 15 (if we assume 25%). This range of 5.3% to 8.9% about which we can be n o more precise, compares with the figure of 1.85% calculated for pricing bills of quantities. As we might expect, therefore, the inherent variability of this much less detailed approach to estimating is far poorer than that of the detailed pre-tender estimate. However, when we substitute the value of 1.85% in our total equation with 5.3 and 8.9% then the overall accuracy of elemental estimates is shown to have a coefficient of variation which lies in the range 16.3 to 17.8%. If we compare this range with the coefficient of variation of 15.5% measured for pre-tender estimates then we must draw the rather startling conclusion that estimates prepared by elemental cost analysis are not much poorer than those prepared by pricing full bills of quantities.

In reaching the above conclusion, we have, as far as possible, completed the first objective of this paper which was to define the level of estimating accuracy achieved in quantity surveying practice. Whilst we have not considered the performance of the most simple estimating techniques such as single unit methods including the superficial floor area method, it is, however, reasonable to conclude that such techniques will provide somewhat poorer performances than those measured and described above.

The second of our stated objectives is to consider the effectiveness of cost planning in the light of the estimating performance measured in practice. Seeley (1972) considers the aims of cost planning to be threefold:

1   To give the building client good value for money
2   To achieve a balanced and logical distribution of the available funds between the various parts of the building, and
3   To keep the total expenditure on a building within the sum of money that the client has agreed to spend

The definition of value for money is a subject in its own right. However, in terms of cost planning, many of the constituents of value for money such as 'operational efficiency', 'aesthetics' and 'amenity' are unquantifiable. For the purpose of this work, we have consequently considered value for money as being economy of construction and layout within the client's brief.

It is, therefore, evident that to fulfil the objectives of cost planning requires two types of estimate, the comparative and the absolute. The measurements of estimating performance which we have taken consider absolute estimates, i.e. those designed to predict total costs. In this respect we must conclude that the present estimating performance of quantity surveyors is unlikely to achieve the objective of containing the total expenditure on a building within the sum of money that the client has agreed to spend. However, this conclusion cannot be extended to cover the other objectives of cost planning since these are dependent upon good comparative estimating. Even though absolute estimating performance has been shown to be poor, comparative estimates based on similar techniques and consistent data are not necessarily invalidated. The right conclusion may be drawn from two or more estimates whose absolute accuracy may be questionable.

IMPROVING COST ESTIMATING

The analysis of the factors contributing to the estimating accuracy achieved by quantity surveyors when pricing bills of quantities shows clearly that the largest constituents of the measured variability are due to imperfections in the cost data used and adjustments made to this data to adjust for time, location, market conditions etc. It is, therefore, clear that the greatest scope for improving upon present estimating performance lies in the selection, manipulation and application of cost data. In particular two aspects of current quantity surveying practice have been identified as possible contributors to this performance.

These are:

1   Quantity surveyors prefer to use cost data from projects with which they have had personal experience. This practice largely prevents the experience gained by an office or practice on previous projects from being used to advantage on future projects since individual surveyors become involved with relatively few projects in their careers.
2   Quantity surveyors prefer to use a single source of cost data when estimating. Since this single source is often the priced bill of quantities from a previous lowest tender, it is likely to be a poor indication of the mean price level of lowest tenders.

Speaking more generally, the advent of micro-chip technology will provide the quantity surveying profession with the means to alter radically the manner in which cost data is captured, sorted, manipulated and disseminated. Research into the impact of this technology is therefore required in order that the profession does not fall into the trap of simply mechanizing existing manual systems but takes the opportunity to rethink current practice in the light of the information revolution.

CONCLUSIONS

When pricing bills of quantities, quantity surveyors' ability to predict the value of lowest tenders is expressed by a coefficient of variation of 15%. Our research shows that when using very simple estimating techniques this performance is only likely to decrease to 16.3–17.8%. Contrary to expectations, the accuracy of estimating does not improve appreciably during the development of a design as more detailed information becomes available. This level of performance is unlikely to fulfil clients' requirements for controlling the cost of construction within predetermined limits, although it does not prevent other aspects of the cost planning process from being effective. The most likely area of improvement on the current performance of quantity surveyors lies in improved methods of selection, manipulation and application of cost data.

REFERENCES

Ashworth, A., Neale, R. H. and Trinble, E. G. (1980) An analysis of some builders estimating. *The Quantity Surveyor* 65–70.
Barnes, N. M. L. (1972) On getting paid the right amount. Minutes of proceedings at Buildings' Conference luncheon.
Barnes, N. M. L. (1974) Financial control of construction. In *Control of Engineering Projects* (edited by S. H. Wearne). Edward Arnold, London.
Barnes, N. M. L. and Thompson, P. A. (1971) Civil engineering bills of quantities. CIRIA Report no. 34.

Beeston, D. T. (1973) *One Statistician's View of Estimating* Property Services Agency, London

Bennett, J. Morrison, N. A. D. and Stevens, S. D. (1981) Cost planning and computers. Property Services Agency Library, London.

Department of the Environment (1979) *Construction Cost Data Base* Property Services Agency Library, London.

Department of the Environment (1980) *Construction Cost Data Base* Property Services Agency Library, London.

Flanagan, R. (1980) Tender price and time prediction for construction work. PhD thesis, University of Aston, Birmingham.

Roderick. I. F. (1977) Examination of the use of critical path methods in building *Building Technology and Management* 16–9.

Seeley, I. H. (1972) *Building Economics*, MacMillan, London

Skinner, D. W. H. (1982) Major and minor cost factors in building measurements. Unpublished paper.

5.4 Experiments in probabilistic cost modelling

A. J. Wilson

INTRODUCTION

This paper discusses some preliminary results from a major research project into the design, development and testing of models for predicting the economic consequences of building design decisions. Such economic response models generally have one of the following purposes.

(i) To predict the total price which the client will have to pay for the building. (Tender Price models)
(ii) To allow the selection, from a range of possible design alternatives, at any stage in the design evolution, of the optimum design according to some predefined criterion of economic performance.
(iii) To predict the economic effects upon society of changes in design codes and regulations.

It has been found that the purpose of the model has a considerable influence upon the most appropriate approach to model formulation. It will prove useful to characterise models according to purpose and the classification used here is to identify those under (i) above as macro models, and those under (ii) and (iii) as micro models. This paper is essentially concerned with micro models.

It is possible to propose a great many economic criteria of performance which could be adopted tin the economic modelling of building design decisions. That adopted here is monetary cost to the client. We are thus dealing with micro cost models.

APPROACHES TO MODELLING

Architecture is concerned with the creation of space. A good building design will capture and articulate space to satisfy both quantitatively and

*1982, in Brandon P. S. (ed.) *Building Cost Techniques – New Directions*, E & F N Spon, 169–80.

qualitatively the demands of the processes to be accommodated. The total number of design decisions which must be taken is enormous. They vary, for example, from the choice of structural frame type to the position of light switches, from the number of storeys to the type of window fastening. It is convenient to regard the decisions as design variables which simply take different values in different buildings. Since it is these decisions which alone determine the nature of the building, it is they which give rise to the cost of the building.

Thus,   Cost, $C = f_1 (v_1, v_2, v_3 \ldots v_N)$ (1)

where v_1, v_2, etc. are the design variables.

However, the cost of building work is actually incurred and price is usually expressed, not in terms of design variable, but in terms of the resources of all kinds which the design decisions commit.

Thus, $C = f_2 (\Box Rj)$ (2)

where Rj are the resources committed.

The central task of cost modelling is the reconciliation of equations (1) and (2).

It is possible to recognise two different, although not mutually exclusive, approaches to the construction of cost models.

deduction
induction

Deductive methods involve the analysis of cost data over whichever design variables are being considered, with the objective of deriving formal mathematic expressions which succinctly relate a wide range of design variable values to cost. This approach draws heavily upon the techniques of statistics; correlation and least squares regression, in particular. Disadvantages of this approach arise from the not inconsiderable limitations of these statistical techniques, and on the total dependence upon the suitability of the cost data used.

Inductive methods, on the other hand, involve, not analysis of a set of given cost data, but rather the synthesis of cost of individual discrete design solutions from the constituent components of the design. Whilst deductive methods are, perhaps, more important in the early stages of design, inductive methods are more important in the later stages. Deductive methods arise largely from equation (1), inductive methods arise largely from equation (2). Inductive methods require the summation of cost over some suitably defined set of subsystems appropriate to the design. The most detailed level of subsystem definition would be the individual resources themselves, but several other levels of aggregation are in common use, e.g. operational activities and constructional elements.

This paper is primarily concerned with inductive, micro, cost models. Such concentration upon taxonomy may be thought excessive, and even pedantic, but it is suggested that it may be conducive to the continued development of economic modelling in building design.

UNCERTAINTY

Whilst the uncertainties implicit in any industrial cost estimating have long been recognised at least qualitatively, it is only in the fairly recent past, with the stimulus of computer based cost modelling that there has been real movement towards attempting to quantity them. In his seminal paper on probabilistic estimating[1] Spooner commented upon the lack of data on uncertainty, in 1974. Unfortunately, little has been added in the intervening years.

A major problem is the unsuitable form of existing cost data. In particular, there appears to be widespread confusion in the building industry between cost and price. It is a recognised phemomenon in economics that the selling price of any product is determined by the market for the product, and not by the manufacturer's input costs. Costs are certainly important, since they determine profitability, but their relationship to selling price is often extremely tenuous. thus, an analysis of tender prices does not reflect the variability of costs, but rather reflects the exigencies of the market place. Further, bills of quantity rates, insofar as they represent anything, reflect this market abstraction.

In the absence of objective data some writers in the area of probabilistic estimating have made assumptions which are unproven but convenient. Prime amongst these, for example, is the assumption that cost uncertainties are always symmetrical, i.e. that they can be accounted for by an expression of the form £x□a%. It was the intuitive feeling that this was an oversimplification which, amongst other things, promoted the research described in this paper.

Much of the author's experience of modelling has been devoted to micro models, i.e. cost models for fairly tightly defined design optimisation problems[2,3], and these are the types of cost model primarily under consideration in the present research. Such models have two important characteristics.

Firstly, inductive micro cost models do not include the large number of subsystems which is the case in macro (tender price) models. One excuse widely used by analysts for ignoring cost uncertainty in macro models is the 'swings and roundabouts' argument. More formally, this is expressed by the Central Limit Theorem of statistics which suggests that the summation of the mean costs of each subsystem will tend towards the man cost of the total system and no matter how asymmetrical the uncertainty in each subsystem, the final composite uncertainty in total cost will tend to the symmetry of the

Normal Distribution. Whilst such assumptions may be valid in macro models with large numbers of constituent subsystems, to assume that it is the case for micro models with only a few subsystems seems rather foolhardy.

The second characteristic of micro models is that since, in the search for optimal or at least improved design solutions, we are comparing alternative solutions to the same design problem, we are comparing alternative solutions to the same design problem, we are interested only in those components which are   likely to differ. this allows us to avoid incorporating some of the more troublesome items of any estimate such as profit, overheads and tactical marketing considerations which have significance only in the case of macro cost models. They are likely to be constant for each design alternative, provided we take care in our model formulation.

EXPERIMENTAL METHOD

The objective then is to obtain data indicating the distribution of uncertainty around subsystem cost estimates, and to determine the effects of such uncertainties upon the evaluation of alternative design solutions. It is a vast task. It will take a great deal of time and manpower to achieve fully. This paper describes the strategy adopted and presents early results.

Controlled experiments of the kind possible in the scientific world are generally impractical in economics and impossible in building economics. So, unfortunately, there appeared no way of obtaining unbiased objective prime source data on cost uncertainty. Further, no secondary sources of data, i.e. data acquired for other purposes, seemed suitable, largely for the reasons discussed above.

The experimental approach adopted to the acquisition of data was the Delphi method which is often used in the field of technological forecasting.[4] Essentially, it comprises a structured approach to exploiting the specialised knowledge and judgement of experts in a particular field. In a commercial environment, such as the one we are dealing with, it is difficult to define 'an expert'. Eventually, though, a panel of eight competent contractor's estimators was constituted, where competence was judged on the basis of performance, position, experience and standing. It is difficult to be dogmatic about the optimum size of such a panel. Certainly the statistical view of 'the more the better' seems appropriate, to be tempered only by administrative convenience and compatibility of experience. To call the assembled estimators a panel is, perhaps, misleading, since at no time did they meet each other, or indeed, know each other's identity. The anonymity is an advantage of this experimental approach. The weakness, of course, so the Delphi method is its reliance upon subjective judgement, but if the experiment is carefully controlled, this can become a virtue.

The constructional element chosen as the vehicle for the first experiment was a 150mm thick, suspended, cast *in-situ*, reinforced concrete, floor slab.

This may be unexciting, but it is an important, common item in modern construction and it does require a number of resource types.

Round one in the experiment consisted of identifying how the estimators themselves went about pricing such a floor slab. The differences in approach are considerable. Whilst all the estimators worked at a detailed level, they often identified different constituent items as significant. After some feedback between experimenter and estimators, a pro forma of 17 items necessary to the pricing of the slab was adopted. (There still remained discrepancies in interpretation, especially with regard to the need for skilled and unskilled labour in certain of the constituent tasks.)

The second round was the linchpin of the experiment in that its purpose was to obtain the uncertainty data. The problems facing the experimenter are considerable since he must design a questionnaire which will elicit the required information without leading the participant.

It is important to realise that the way the questions are framed very much influences the response of the participants. It soon became apparent that it was necessary to make assumptions about the shape of the uncertainty probability distribution in order that the appropriate questions could be asked, and the correct parameters sought. The distribution function will, in practice, be continuous, but it seemed unlikely that the estimators would be able to associate quantitative levels of probability with their estimates. (This is a common criticism of the PERT procedure.) Thus, it was necessary to choose a distribution function which could be characterised by a few significant parameters. The four distributions most often proposed for this type of problem are the uniform, normal, triangular and beta distributions. That selected was the triangular distribution. It requires only three point estimates to define it; the probable value, and the limiting minimum and maximum values. In addition to having the support of other workers in the field[1], it has the considerable merit of simplicity.

The estimators were then asked to fill in a questionnaire which, for each of the seventeen identified constituent items in the cost synthesis of the concrete floor slab, sought their estimates of minimum, probable and maximum values. The questionnaire reminded the participants that the values sought were to be net of overheads, profit and tactical considerations.

The third round of the Delphi experimental method consists, in this case, of an analysis of the results of round two, the presentation of those results in a suitable format, and their return to the participating estimators for comment and amendment. At this stage the estimators see the estimates of their peers for the first time. (Although it is not possible for them to associate names and values) The purpose here is a movement towards consensus, to give them an opportunity to change their minds in the face of the collected body of results, and , if necessary, to make apparent their reasons for dissent. This feedback is essential to guard against the experimenter influencing the analysis of results.

RESULTS

A study of this kind necessarily generates a large number of results, and it is inappropriate in this paper to attempt to publish them in full. Thus a few typical results have been selected for discussion.

Figure 5.4.1 illustrates the results of Round Two for seven of the sub-systems. In Figure 5.4.2 these same results have been processed using a very simple statistical analysis. There are, of course, always dangers in reading too much into what is necessarily a very small sample, but it is suggested that the results do allow a few general conclusions.

The variability in pricing all of the items is considerable, but it does appear that estimators are constrained by the traditional practice of single figure estimating, and when the opportunity is presented to them expressing the uncertainties in their estimates they rise to the occasion. A more complex statistical analysis upon the total set of results revealed no consistent bias amongst individual estimators, and so the results for each item can be averaged across the estimators with some confidence. Two phenomena are displayed in Figure 5.4.2 and it is important that they are distinguished. Firstly, there is the variability in each of the point estimates measured by the coefficient of variation, and secondly, there is the perceived range of uncertainty around each item indicated by the difference between the minima and maxima.

Looking firstly at the variability of the three point estimates in each category, some general conclusions are possible. As might be suspected, the variability in estimating materials prices is consistently less than that for labour prices. Perhaps more surprisingly, the variability in estimating the maximum and minimum values is not always greater than that for the probable value. It does not seem that the opportunity to express the range of uncertainty reduces the variability of the single point (probable) estimate from the levels which are generally accepted by the industry.

Perhaps the most striking feature of the results emerges when we look at the distribution of the maxima about the probable estimates. There is a pronounced and consistent skewness to the right in all cases, for both labour and material items. Whilst the range of the uncertainty about the probable value (a_1 to a_2) is considerable ($> 100\%$ in one case), the asymmetry (a_2/a_1) averages at 2.6. The skewness appears greater in the case of the labour items as indeed does the uncertainty range, an unsurprising result perhaps, but the extent of the skewness in material prices is, perhaps, surprising.

MONTE CARLO SIMULATION

Having examined the uncertainty around the individual subsystem costs we must now investigate the way in which these individual uncertainties combine to influence the uncertainty surrounding the total cost of our concrete slab.

Figure 5.4.1 Results of round 2 for 7 subsystems.

ESTIMATOR SUBSYSTEM		A	B	C	D	E	F	G	H
READY MIXED CONCRETE (£/m³)	Min.	24.34	24.00	32.00	28.00	25.36	26.50	26.75	27.00
	Prob.	25.86	26.00	34.00	29.50	28.58	28.60	27.35	28.00
	Max.	27.96	30.00	34.00	35.00	35.60	30.60	28.50	32.00
LABOUR; PLACING & FINISHING CONCRETE (man.hrs/m³)	Min.	0.46	0.50	0.38	0.36	0.45	0.67	0.50	0.40
	Prob.	0.51	0.75	0.45	0.40	0.56	0.74	0.50	0.45
	Max.	0.66	1.125	0.75	0.45	0.90	0.80	0.60	0.60
BAR REINFORCEMENT (£/Tonne)	Min.	245.40	240.00	290.00	243.25	243.25	235.00	247.66	238.00
	Prob.	247.66	300.00	320.00	275.68	275.68	240.00	247.66	262.00
	Max.	286.18	440.00	380.00	303.54	300.00	400.00	247.66	304.00
LABOUR – STEEL FIXING (man.hrs/Tonne)	Min.	17.00	37.00	22.00	26.00	16.00	25.00	32.00	35.00
	Prob.	25.00	43.00	26.00	37.00	30.00	32.00	32.00	53.00
	Max.	37.00	54.00	34.00	73.00	66.00	52.00	38.00	143.00
PLYWOOD FORMWORK (£/m²)	Min.	3.11	5.30	4.00	3.40	2.90	3.10	3.50	3.90
	Prob.	3.49	5.55	6.00	3.50	3.49	3.42	3.50	3.90
	Max.	4.52	6.50	6.00	4.50	3.72	3.90	3.85	9.68
LABOUR – MAKING, FIXING & STRIKING FORMWORK (man.hrs/m²)	Min.	1.00	2.00	3.60	2.48	1.50	2.90	2.00	1.39
	Prob.	1.65	2.50	4.00	2.48	1.60	3.10	2.25	1.57
	Max.	2.27	3.50	5.00	2.48	2.10	3.20	3.00	2.34

Figure 5.4.2 Summary statistics.

				a_1	a_2	a_1/a_2
	PROBABILITY — triangular distribution diagram with a_2, a_2, MIN., PROB., MAX.					
SUBSYSTEM		Mean	Coeff. of Var.	a_1 Mean	a_2 Mean	a_1/a_2 Mean
1. READY MIXED CONCRETE (£/m³)	Min. Prob. Max.	26.70 28.50 31.70	9% 9% 9%	1.74 (6%)	3.22 (11%)	1.85
2. LABOUR FOR CONCRETE WORK (man.hrs/m²)	Min. Prob. Max.	0.47 0.55 0.74	21% 24% 28%	0.08 (15%)	0.19 (35%)	2.38
3. STEEL BAR REINFORCEMENT (£/Tonne)	Min. Prob. Max.	247.80 271.10 332.70	7% 10% 20%	23.27 (9%)	61.59 (23%)	2.65
4. LABOUR FOR STEEL FIXING (man.hrs/Tonne)	Min. Prob. Max.	26.30 34.80 62.10	30% 27% 57%	8.50 (24%)	27.38 (79%)	3.22
5. PLYWOOD FORMWORK (£/m²)	Min. Prob. Max.	3.65 4.11 5.33	21% 26% 38%	0.46 (11%)	1.22 (30%)	2.65
6. LABOUR FOR FORMWORK (man.hrs/m²)	Min. Prob. Max.	2.11 2.39 3.21	41% 35% 32%	0.29 (12%)	0.82 (34%)	2.83

The pronounced asymmetry of the uncertainties seemed to suggest that the often used, simple analysis of variance was inappropriate. The technique adopted was Monte Carlo Simulation.

A micro computer problem was written which would randomly select costs for each subsystem in accordance with the triangular probability distribution obtained experimentally for each subsystem. These values were then

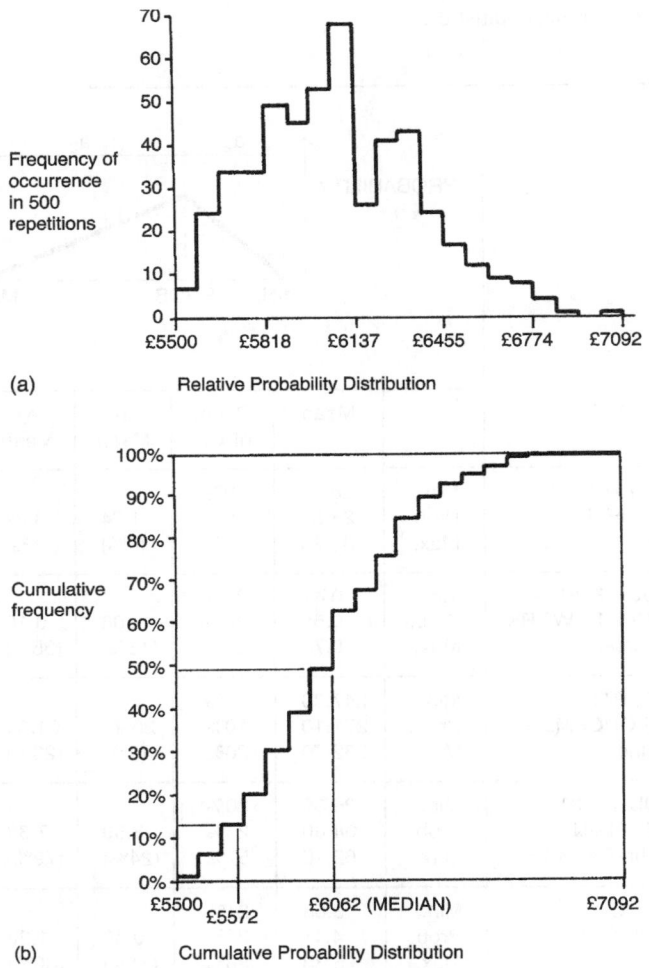

Figure 5.4.3 Cost distribution for 100 m² concrete slab.

combined in a simple inductive cost model aggregating the subsystems of Figure 5.4.2, with suitable coefficients for hourly wage rates, etc., incorporated, to give a single deterministic cost for 100m² of suspended concrete slab. This whole process is then repeated a large number of times to enable a frequency diagram to be plotted (Figure 5.4.3a). When using Monte Carlo simulation there is often some doubt as to the 'randomness' of the computer generated random numbers. In our trials we have found those produced by a typical micro computer standard function to be more than adequate. Another issue facing the investigator is the number of iterations necessary for a reliable result. The error of convergence reduces in proportion to the square root of the number of iterations, so the more iterations, the better, but there is a diminishing return on accuracy. The number of iterations carried out in this case and illustrated in Figure 5.4.3 was 500.

Because of the use of random numbers, each 500 simulations will result in a different relative probability distribution. This non-repeatability tends to disturb traditional cost estimators, and care is necessary in interpreting such probabilistic cost distributions. Of most value perhaps to the estimator is the cumulative probability diagram of Figure 5.4.3b, but the relative diagram of Figure 5.4.3a is useful as an indication of the resulting skewness of the total cost profile. Both Figures 5.4.3a and 5.4.3b are, in reality, continuous curves, but it is computationally more convenient to divide the output into equal band widths and this accounts for the stepwise appearance.

Diagrams such as that of Figure 5.4.3b bring a new dimension to estimating. For each proposed cost value of our concrete slab, we can now associate a probability of occurrence. The likelihood that the cost will be less than £5 500 is zero, the likelihood that the cost will be between £5 500 and £7 092 is 100%. (The monetary amounts shown in the figures are not significant since any monetary value has a transient validity, but they may be regarded as indices.) The median cost of our slab is £6 062 – this is the value which has a 50% chance of being exceeded.

It is illuminating to interpret the estimators' most probable values in terms of Figure 5.4.3b. By substituting the average probable estimates for each subsystem in the inductive cost model used, the value of £5 500 was obtained. Inspection of the graph shows that this has an 87% chance of being exceeded. A strange interpretation of most probable!

There is one major omission in the analysis described so far and in the results illustrated in Figure 5.4.3. Each subsystem has been regarded as an independent entity with respect to the choice of random values. No account has been taken of the obvious interdependence of some of the items. For example, bad weather would adversely affect all three labour items, to continue to regard them as randomly unrelated would be inaccurate. The omission of this phenomenon, (known statistically as covariance), is partly because its treatment is more appropriate to a further paper, and partly because experiments have not yet revealed the correct way of dealing with it.

CONCLUSIONS

It is hoped that this modest paper has emphasised the need for the collection of more data revealing the nature and distribution of uncertainties in the cost estimation of building work. The paper has suggested on experimental approach to the collection of such data, and has shown how the data can be incorporated in the context of the expanding discipline of design cost modelling. It is hoped that the inappropriateness of some analyses of cost uncertainty have been highlighted, particularly those which fail to make allowance for the extent of the skewness of uncertainty distribution.

The work is continuing, primarily in the general area of data acquisition for probabilistic design cost models, and in the specific area of the treatment of covariance.

ACKNOWLEDGEMENTS

The author is indebted to the contractors' estimators who freely gave of their time and expertise to take part in this study, and to John Raftery for his computational assistance.

The financial assistance of the Nuffield foundation in the ongoing research is gratefully acknowledged.

REFERENCES

Spooner, J. E. 'Probabilistic Estimating' *J. Con. Divn. A.S.C.E.,* March, p. 65. (1974)

Wilson, A. J., Templeman, A. B. and Britch, A. L. 'The Optimal Design of Drainage Systems', *Engineering Optimisation,* 1, p.111. (1974)

Wilson, A. J. and Templeman, A. B. 'An approach to the Optimal Thermal Design of Office Buildings', *Building Science* Spring. (1976)

Jones, H. and Twiss, B. C. *Forecasting Technology for Planning Decisions,* Macmillan Press. (1978)

5.5 The accuracy and monitoring of quantity surveyors' price forecasting for building work

R. Flanagan and G. Norman

ABSTRACT

This paper examines the performance of two public sector quantity surveying departments when forecasting the lowest tender price for proposed projects at the design stage. The reliability of any price forecast is dependent upon professional skill and judgement and the availability of historical price data derived from competed projects, the quantity surveyor also requires an effective feedback mechanism that provides information on the accuracy of previous forecasts. A simple feedback mechanism is developed in this paper which can be used to assess forecasting performance and give an early warning of bias and identify any patterns that may emerge.

Keywords

Price forecasting, accuracy, monitoring, quantity surveyor

INTRODUCTION

One of the most important tasks performed by the consultant quantity surveyor is the forecasting of the tender price at the design stage for a proposed building. In common with any forecasting process, however, this will inevitably involve error for a number of reasons:

1   Forecasting is not an exact science, there is no infallible way to predict the future. Forecasting construction prices is heavily dependent upon the availability of historical price data, professional expertise and judgement.
2   The historical price data on which the forecasts are based may be defective. For example, the data may be inaccurate because they are based upon tender prices rather than final account prices. Also, data limitations may

*1983, *Construction Management and Economics*, 1, E & F N Spon, 157–80.

force reliance upon price data derived from a sample of buildings that is not perfectly matched to the proposed building.

3   The limited information available about the proposed building at the early design stage may mean that the quantity surveyor must rely upon assumed or default values for many of the project details. In essence, ambiguity in the design and ambiguity in the price foreast go hand in hand.

4   The supposition that forecasts will be accurate if only the quantities and unit price rates can be determined precisely ignores the variability of unit price rates contained in bills of quantities.

It must also be recognized that good forecasting requires effective feedback mechanisms that provide information on performance of previous forecasts. Designing such feedback mechanisms is difficult enough in any area in which forecasts are made, but confronts particularly severe problems in the building industry. Specifically, there will often be significant delays between the preparation of a price forecast at the inception stage and receipt of tender information against which to check that forecast.

Against this background the objectives of this paper are twofold. Firstly, we present an analysis of forecasting performance by consultant quantity surveyors. Price predictions made at the design stage are compared with tender prices submitted by contractors for a range of projects in order to identify whether there exist any relationships between prediction and received bids. Secondly, we outline a simple feedback mechanism that can be used to assess forecasting performance and, in particular, give early warning of constant bias that may characterize forecasts.

THE MEANING OF ACCURACY

The starting point for any analysis of the accuracy of the quantity surveyor's price prediction measured against the contractor's lowest bid for building projects must be a consideration of what is meant by the term accuracy. A naive definition of accuracy would be simply the absence of error, that is, the smaller the error the higher the accuracy and vice-versa. This is, largely, true but merely restates the problem by requiring a definition of error. To define error mathematically we suppose that there exists a number X which we know to be 'exact'. In describing X we use another number x which is an approximation of the 'exact' value. Thus the 'exact' number might be p while the approximation x is 22/7; or X might be the actual volume of concrete in a ready mix concrete truck, whereas x is the volume of concrete shown as being delivered on the delivery ticket; or X might be the mean price per square metre of the gross floor area of a series of ten factories constructed by a contractor and x the mean price per square metre of a sample of three of these factories.

Whence do these errors arise? There are several simple sources of error, for example, measurement errors and rounding errors. Most importantly for our purposes, however, errors may arise because crucial information is not available. We are concerned in this paper with price prediction; in other words with forecasts of lowest bids as an input to the client's decision-making process. As indicated in the Introduction, no matter how good the data on which those forecasts are based, if the estimators do not have adequate feedback mechanisms, errors are likely to arise and to be compounded. what we mean by this is that the forecaster must have an adequate means for assessing previous performance, analysing errors and detecting trends in errors if these exist.

We can develop this theme more adequately by analogy. Consider a marksperson shooting at a target, but being given no information about performance on previous shots, in other words shooting without any feedback mechanism. The better the marksperson, the more closely grouped we would expect the shots to be. In this sense, a good marksperson will exhibit high precision, or reproducibility. the shooting will be accurate, however, only if the shots fall on the target (i.e. the bullseye).

Thus to develop the analogy further, if the marksperson is using a rifle with a consistent, but unknown bias, the shots will be very closely grouped on the wrong target. The shooting is precise, but not accurate. Now introduce a feedback mechanism which allows the marksperson knowledge of previous performance. With such feedback we would expect adjustment of the bias and convergence on the 'true' target.

These considerations are of particular relevance wen we apply them to price prediction. Firstly, it must be recognized that neither of the two quantities with which we are immediately concerned in prediction, the quantity surveyor's price prediction and the contractor's low bid on a project, are 'true' values. Both are, in fact, estimates and both, therefore, have associated error bounds. Secondly, we must consider what types of feedback mechanisms are necessary in price prediction and how error bounds might be identified. We concentrate below on the estimates that are produced and their accuracy. Later we return to the design of some feedback mechanisms that facilitate identification of any consistent bias in prediction errors.

THE ACCURACY OF THE QUANTITY SURVEYOR'S PRICE PREDICTION

The quantity surveyor is concerned with forecasting tender prices for building work. We shall, therefore, look at the relationship between the quantity surveyor's price prediction made at the design stage and the low bids received from contractors. In doing so we are assuming that the direction of causation is from the bid to the prediction, i.e. that the estimate is in some sense the dependent variable. This is reasonable because we can generally conclude that tenders submitted by contractors are based upon cost considerations plus desired overheads and profit, whereas price predictions are based upon previous tenders.

In view of the confidentiality of the data there is no significant available published documentation analysing a quantity surveyor's performance in relation to price prediction. As a result, a case study was undertaken to analysis the consultant quantity surveyor's forecasting performance over a period of time. Since it provided impossible to obtain sufficient data on projects from any one private practice, two County Councils were approached, County Council A located in the north west, and county Council B in the south east of England. The Councils were selected because:

(a) They both had fully documented records;
(b) The price predictions for projects were being undertaken by in-house cost planning departments using cost planning techniques;
(c) There was similarly in the functional types of projects being constructed;
(d) Selective tendering was used to obtain tenders.

In order to test performance of the two Councils, each councils' last price predictions prior to tenders being sought were related for each project to the lowest tender prices received from the contractors. Both Councils were using the superficial floor area method of prediction at the inception and outline proposal sages of a project, and cost planning with approximate quantities at the scheme and detail design stages.

Council A was considered on a sample of 103 projects and Council B on a sample of 63 projects. The sample for council A consists of all documented projects over the period 1975 to early 1978, and for council B all documented projects during the period 1971 to early 1977. The functional types of building in our samples are shown in Table 5.5.1.

It is appreciated that these data are now somewhat out of date, an inevitable consequence of their confidential nature. This would be important if our intention was to derive precise, quantitative relationships for future price prediction. Our concern, however, is purely with assessing accuracy in price prediction and with indicating how proposed feedback mechanisms work. In

Table 5.5.1 Analysis of project type

Types of project	Council A (103 projects)	Council B (63 projects)
Schools (all types)	66	31
Police station	3	3
Fire station	2	5
Hostels	–	5
Children's nurseries	–	7
Old people's homes	6	–
Day care centres	13	–
Other	13	12

Table 5.5.2 Accuracy of estimate to low bid

| | Relationship of price prediction to low bid (%) | | | | | |
	□5%	□10%	□15%	□20%	□25%	Over □25%
Council A (103 projects)	25.27	48.57	63.13	78.66	86.42	100.00
Council B (63 projects)	28.57	57.14	76.19	85.71	93.65	100.00
Council A Schools only (66 projects)	27.14	52.60	68.46	92.10	100.00	—
Council B Schools only (31 projects)	32.33	39.16	74.48	92.11	100.00	—

these circumstances, our analysis and conclusions are not affected to any significant degree by the age of the data.

Similarly, if we had been able to obtain data from other local authority sources a very different functional breakdown from that in Table 1 would have been generated. Again, this does not invalidate our analysis. We are concerned with the application of techniques that allow us to assess accuracy, and the techniques we employ are robust enough to be unaffected by the samples to which they are applied.

Table 5.5.2 highlights the relationship of the quantity surveyor's prediction to the low bid measured in percentage bands of 5%. The percentages relate to the total number of projects undertaken by the Councils. In view of the large proportion of schools both Councils were building, we also looked independently at the quantity survey's performance on schools projects. However, the performance by both councils is not significantly better for schools than for other types of projects.

While Table 5.5.2 implies that both Councils have been making significant estimating errors, the interesting question to be resolved is whether there is any pattern to these errors. Regression analysis was used to test the performance of both Councils. If the Councils were estimating perfectly we would expect to find:

$$\text{ESTIMATE} = \text{LOW BID} \tag{1}$$

The regression analysis used a least squares technique to estimate an equation of the form:

$$\text{ESTIMATE} = A + B \ \square \ \text{LOW BID} \tag{2}$$

From Equation 1 it can be seen that perfect accuracy would lead to the estimated parameters of Equation 2 being such that:

$A = 0.0$ (constant)
$B = 1.0$ (regression coefficient)

The results are summarized in Tables 3 and 4. The estimated equations are[41]:

$$\text{Council A ESTIMATE} = -3.7076 + 1.1146 \ \Box \ \text{LOW BID} \qquad (3)$$

$$\text{Council B ESTIMATE} = 14.7982 + 0.8328 \ \Box \ \text{LOW BID} \qquad (4)$$

R^2 in both cases is significant at the 95% confidence level.

The interpretation of Equation 3 is that Council A was overestimating by approximately 11.5% on all the projects analysed. For Council B interpretation is somewhat more awkward, since the constant term in Equation 4 cannot be ignored. A graphical analysis may be of some help. In Fig 5.5.1 line E represents Equation 2 and line B Equation 4.

Table 5.5.3 and 5.5.4 indicate first, that the constant term for Council A is not significantly different from zero. Secondly, the slope coefficient (1.1146) is significantly different from unity at the 95% level. For Council B the slope coefficient (0.8328) is also significantly different from unity and the constant term is significantly different.

Equation 4 then indicates that Council B was overestimating on small jobs, up to £88 000 value [$14.7982 \ \Box \ (1.0 - 0.8328)$] and underestimating on jobs above this value. In addition, the percentage underestimate increased in the value. Put another way, for every £1000 000 increase in project size, Council B's estimate increased by only £83 280.

Examination of scatter diagrams (Figs. 5.5.2 and 5.5.3) of the observations for both Councils suggests that a simple linear regression as in Equations 3 and 4 might not be the correct specification of estimating performance. It would appear that both councils exhibited a reasonable measure of accuracy when forecasting prices for projects up to approximately £300 000 value, but above this the estimates were less successful. Council B, for example, had a tendency to consistently underestimate on projects over £300 000 in value.

A curvilinear regression was undertaken on Council B's data, but this did not capture sufficiently the underestimation on very large projects. A more appropriate hypothesis is that estimating performance on small jobs differs in some distinct way from that on large jobs. The projects for both Councils were, therefore, analysed by project size on the assumption that estimating performance on projects up to £300 000 differed from performance on projects over £300 000. The results are summarized in Tables 5.5.5–8.

---

[41]All figures in this and subsequent equations are in multiples of thousands of pounds.

Table 5.5.3 Council A – all projects[†]

Number of projects	Mean of all estimates	Mean of all low bids	Multiple R (correlations)	R^2	Variable	Coefficient	t-statistic	95% confidence interval	
								R	R^2
103	193 045	176 518	0.9938	0.9877	Low bid	1.1146[b]	90.0145	1.0900	1.1392
					Constant	−3.70750[c]	−0.9987	−11.0721	3.6567

Table 5.5.4 Council B – all projects

Number of projects	Mean of all estimates	Mean of all low bids	Multiple R (correlation)	R^2	Variable	Coefficient	t-statistic	95% confidence interval	
								R	R^2
63	167 294	183 102	0.9839	0.9681	Low bid	0.8328[b]	42.6647	0.7938	0.8719
					Constant	14.7982[a]	2.9091	4.6227	24.9729

[†]In Tables 5.5.3–5.5.12 inclusive the following legend applies:
[a]Significantly different from zero at 95% confidence level.
[b]Significantly different from unit at 99% confidence level.
[c]Not significantly different from zero at 95% confidence level.
[d]Not significantly different from unity at 99% confidence level.

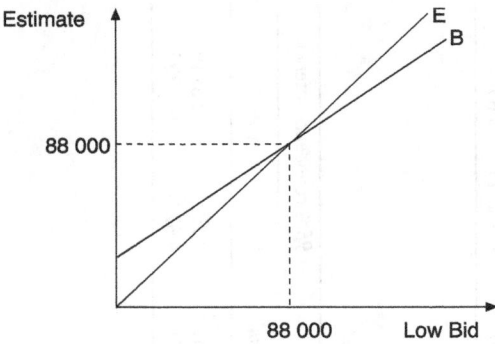

Figure 5.5.1 The relationship between the estimate and the lowest bid using Equation 2 (line E) and Equation 4 (line B).

The break point was chosen from visual inspection of the data rather than form a statistical analysis. Subsequent testing indicated that it was an appropriate choice.

For projects costing less than £300 000, the estimated equations for the two Councils are:

$$\text{Council A ESTIMATE} = 0.4753 + 1.0668 \,\square\, \text{LOW BID} \tag{5}$$

$$\text{Council B ESTIMATE} = -3.3577 + 0.9825 \,\square\, \text{LOW BID} \tag{6}$$

Equation 5 indicates that Council A was overestimating on small projects by about 7%. For Council B, however, Equation 6 is not statistically different from the 'perfect estimating' Equation 2. to all intents and purposes, council B was estimating perfectly on small projects: Equation 6, if taken literally, implies underestimation of something less than 2%.

For projects greater than £300 000, a somewhat different pattern emerges. The estimated equations are now:

$$\text{Council A ESTIMATE} = 24.2168 + 1.1462 \,\square\, \text{LOW BID} \tag{7}$$

$$\text{Council B ESTIMATE} = 118.4474 + 0.6449 \,\square\, \text{LOW BID} \tag{8}$$

For Council A, overestimation is now rather more severe. Every £100 000 increase in project size led to an increase of £114 620 in the estimate. In percentage terms, overestimation exceeded 10% for projects above £600 000 and exceeded 12% for projects above £1 000 000. Council B, on the other hand, exhibited consistent and serious underestimation. Every £100 000 increase in project size led to an increase of only £64 490 in the estimate. Underestimation

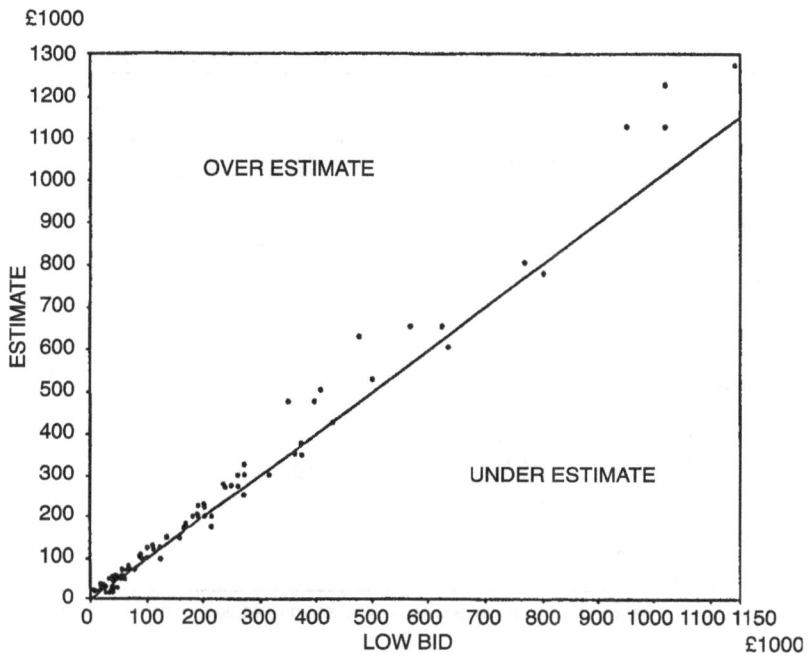

Figure 5.5.2 Council A scatter diagram of estimate against low bid.

exceeded 10% for projects above approximately £480 000, exceeded 20% for projects in excess of £750 000 and exceeded 24% for projects in excess of £1 000 000. Graphically, Council A was performing as in line A in Fig 5.5.4 and Council B as in line B.

The results would appear to confirm our hypothesis regarding the effect of project size on estimating performance. It also highlights the very different performance of the two Councils. Neither Council would appear to have been particularly good at predicting the price of the large projects, Council B having a particularly poor performance record.

An immediate question arises, of course. Can specific reasons be advanced to explain these differences in estimating performance? Quite clearly, any attempt to answer this question requires detailed analysis of estimating methods used, staff expertise, and so on. We were unable to undertake such an analysis, for obvious reasons, but the results of our statistical investigation were communicated to the two Councils.

One important point should be made at this stage. For Council B severe underestimation was apparent on only 11 of the 63 projects. But as the Lorenz curve in Fig 5.5.6 indicates, these 11 projects accounted for 46.72% of total proposed expenditure. Similarly for Council A, significant

£1000

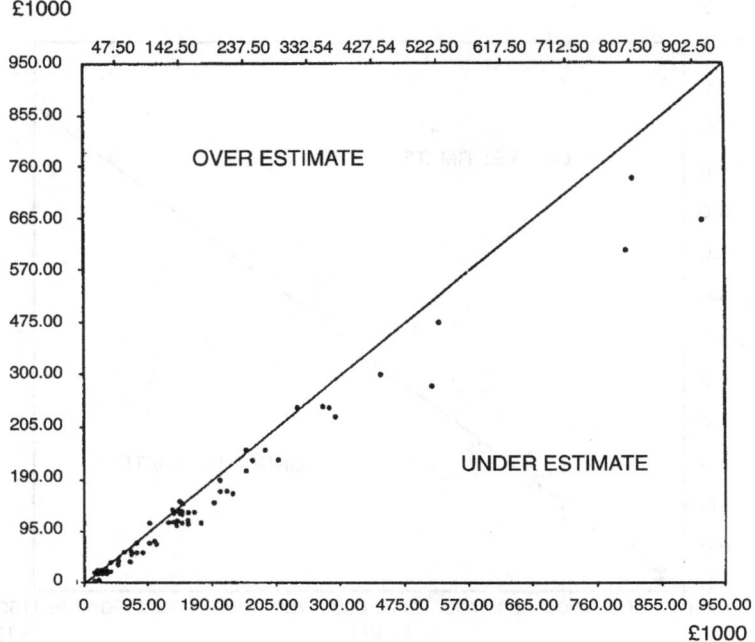

Figure 5.5.3 Council B scatter diagram of estimate against low bid.

overestimation occurred on only 19 out of 103 projects, but as Fig 5.5.5 shows, these accounted for 63.17% of total planned expenditure. In other words, significant estimating errors occurred on relatively few projects, but these projects tended to be concentrated on the large size ranges, and to account for a large proportion of the total construction budget.

Because of Council A's tendency towards overestimation, we investigated whether there might be a relationship between the second lowest bid and Council A's estimate. Table 5.5.9 summarizes the results for all the projects, Table 10 for projects under £300 000 in value and Table 5.5.11 for projects over £300 000.

The estimated equations are:

All projects: ESTIMATE = −6.4342 + 1.0930 □ SECOND BID   (9)

All projects up to
£300 000:   ESTIMATE = −0.5529 + 1.0235 □ SECOND BID   (10)

Projects over
£300 000   ESTIMATE = −24.8413 + 1.1299 □ SECOND BID   (11)

Table 5.5.5 Council A – projects less than £300 000 value

Number of projects	Mean of all estimates	Mean of all low bids	Multiple R (correlation)	R^2	Variable	Coefficient	t-statistic	95% confidence interval R	R^2
84	86 155	80 316	0.9915	0.9830	Low bid	1.0068[b]	68.8714	1.0360	1.0976
					Constant	0.4753[a]	0.2702	−3.0246	3.9752

Table 5.5.6 Council A – projects greater than £300 000 value

Number of projects	Mean of all estimates	Mean of all low bids	Multiple R (correlations)	R^2	Variable	Coefficient	t-statistic	95% confidence interval R	R^2
19	665 612	601 832	0.9771	0.9548	Low bid	1.1462[b]	18.9433	1.0186	1.2739
					Constant	−24.2168[a]	−0.6102	−107.9489	59.5172

Table 5.5.7 Council B – projects less than £300 000 value

Number of projects	Mean of all estimates	Mean of all low bids	Multiple R (correlations)	R^2	Variable	Coefficient	t-statistic	95% confidence interval R	R^2
52	106 179	111 484	0.9836	0.9675	Low bid	0.9825[d]	38.1981	0.9308	1.0342
					Constant	-3.3577[c]	-0.9772	-10.2629	3.5476

Table 5.5.8 Council B – projects greater than £300 000 value

Number of projects	Mean of all estimates	Mean of all low bids	Multiple R (correlation)	R^2	Variable	Coefficient	t-statistic	95% confidence interval R	R^2
11	450 545	515 148	0.9436	0.8903	Low bid	0.6449[b]	8.5465	0.4741	0.8156
					Constant	-118.4474[a]	2.8661	24.9592	211.9355

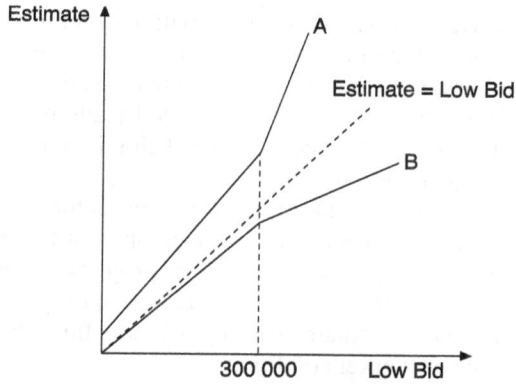

Figure 5.5.4 Estimate against low bid for projects of £300 000.

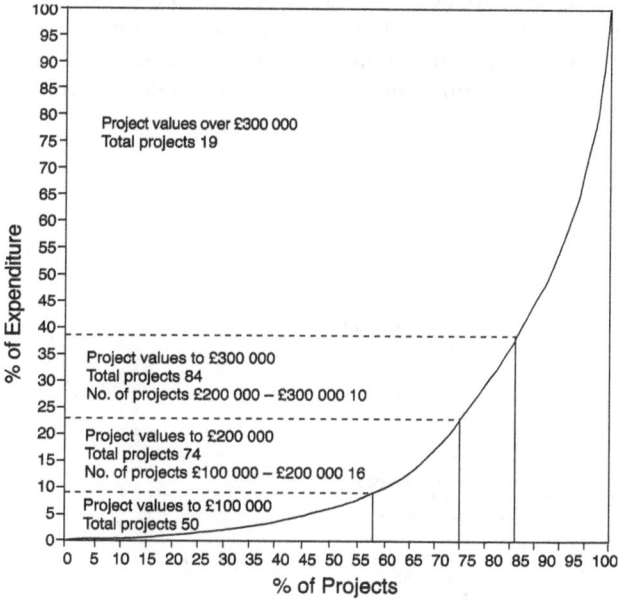

Figure 5.5.5 Council A Lorenz curve of total expenditure. Total expenditure £18 181 354; total projects analysed 103; time span March 1975– March 1978.

On all projects, the Council was overestimating on average by 5.78% with respect to the mean second lowest bid. On projects up to £300 000 the average overestimation was 8.26%. In fact, Equation 10 is not statistically different from the perfect estimating equation. Thus, Council A would appear to have been estimating perfectly on small projects with respect to the second lowest bid, but to be overestimating on large schemes even when the second bid is taken as the 'true' price. At this stage in our analysis, therefore, we can conclude that there is a significant gain to be made by attempts to improve estimating performance on large projects, even if this is at the expense of time spent estimating project costs for small projects.

The analysis has, however, ignored a number of factors that may well be of importance. In particular, no account has been taken of the impact of market condition on estimating performance. The projects being analysed are spread over the period 1971 to 1977, a period during which activity in the construction industry fluctuated sharply. We now turn, therefore, to a consideration of the role of market conditions.

A major problem to be faced in any statistical attempt to assess the role of market conditions in price prediction is that no definitive measure is available of the thing we have called 'market conditions'. The standard approach in these circumstances is to identify a measurable quantity that can be taken as an indicator, or proxy for the variable we are actually trying to measure. In other words, we attempt to identify a proxy variable whose values, and changes in values, constitute an indirect measure of the variable we would like to have measured.

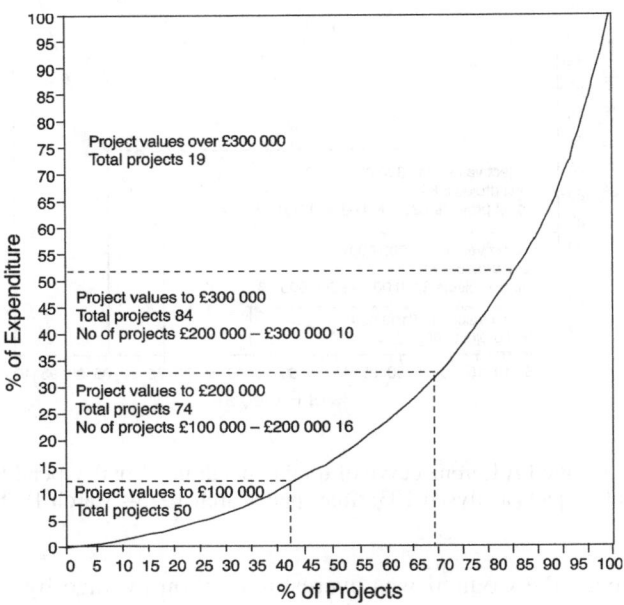

Figure 5.5.6 Council B Lorenz curve of total expenditure. Total expenditure of £11 420 412; total projects analysed 63; time span March 1971–March 1977.

Many possible proxy variables for market conditions can be envisaged, but data limitations, as usual, limit our choice. Notwithstanding the Code of Procedure for Single Stage Selective Tendering, we took the number of tenders received for particular projects as a manifestation of market conditions, our argument being that the more depressed the market the greater the number of firms who would wish to compete for a particular project.[42] Further analysis was then undertaken to identify the effect on estimating performance of the number of tenders received, the hypothesis being that the greater the number of bids received for a project – the slacker the market – the lower the bid price would be. Lack of data meant that analysis had to be confined to Council B. The results are shown in Table 5.5.12.

The regression equation derived from Table 5.5.12 is:

$$\text{ESTIMATE} = -25.5299 + 7.9190 \;\square\; \text{NO. OF BIDS} + 0.8124 \;\square\; \text{LOW BID} \qquad (12)$$

This can be written in an alternative fashion:

$$\text{LOW BID} - \text{ESTIMATE} = 25.5299 - 7.9190 \;\square\; \text{NO. OF BIDS} + 0.1876 \;\square\; \text{LOW BID}$$
$$(13)$$

to show that an increase by one in the number of bidders reduced the estimating error by nearly £8000. In other words, the estimating error became smaller as the number of bidders for a scheme increased. It should be noted that this conflicts with the policy of the Council, which intuitively felt that the greater the number of bidders the higher the bid price they received.

Equation 13 supports our hypothesis, given that the numbers of tenders received on a particular project is a reflection, admittedly imperfect, of prevailing market conditions; Council B's estimating performance has been better the more depressed the market.

To develop this idea further the accuracy of the estimate to the low bid was related to the time scale 1971 to 1978. Fig. 7 shows the error between the Council's estimate and the low bid expressed as a percentage over time and Table 5.5.13 summarizes the overall results.

During the period 1971 to 1974 there was a tendency to consistently underestimate by a significant amount. The consistency leads to the conclusion that the Council was not responding quickly to the prevailing market conditions. During 1975 and 1976 two factors led to a different pattern emerging. The downturn in the market activity accounted for a slower rate of price inflation and the Council changed their system of cost planning. These moves gave rise to a dramatic improvement in the Council's estimating performance. The

[42]As stated, other proxy variables could be used as an indirect measure of market conditions. For example, it might be possible to use the rate of change of a price index such as the Building Cost Information Service (BCIS) tender price index.

Table 5.5.9 Council A – All projects (second lowest bid)

Total number of projects	Mean of all estimates	Mean of all second lowest bids	Multiple R (correlation)	R^2	Variable	Coefficient	t-statistic	95% confidence interval R	R^2
103	193 045	182 505	0.9945	0.9891	Second lower bid	1.0930[b]	95.8625	1.0707	1.1156
					Constant	-6.4342[c]	-1.8361	-13.3860	0.5175

Table 5.5.10 Council A – projects less than £300 000 value (second lowest bid)

Total number of projects	Mean of all estimates	Mean of all second lowest bids	Multiple R (correlation)	R^2	Variable	Coefficient	t-statistic	95% confidence interval R	R^2
84	86 155	84 133	0.9905	0.9812	Second lower bid	1.0235[d]	65.3242	0.9924	1.0547
					Constant	-0.5529[e]	-0.2964	-4.2633	3.1576

Table 5.5.11 Council A – projects greater than £300 000 value (second lowest bid)

Total number of projects	Mean of all estimates	Mean of all second lowest bids	Multiple R (correlation)	R²	Variable	Coefficient	t-statistic	95% confidence interval	
								R	R²
19	665 612	614 846	0.9813	0.9624	Second lower bid	1.1299[b]	20.8643	1.0094	1.2365
					Constant	−24.8413[c]	−0.6884	−100.9711	51.2885

Table 5.5.12 Council B – all projects and number of bidders

Total number of projects	Mean of all estimates	Mean of all low bids	Mean number of bids for each project	Multiple R (correlation)	R²	Variable	Coefficient	t-statistic	95% confidence interval	
									R	R²
63	167 294	183 102	5.5645	0.9901	0.9802	Low bid	0.8124[a]	51.1881	0.7807	0.8442
						Number of bids				
						Constant	−25.5299[a]	−3.2595	−41.2029	−9.8570

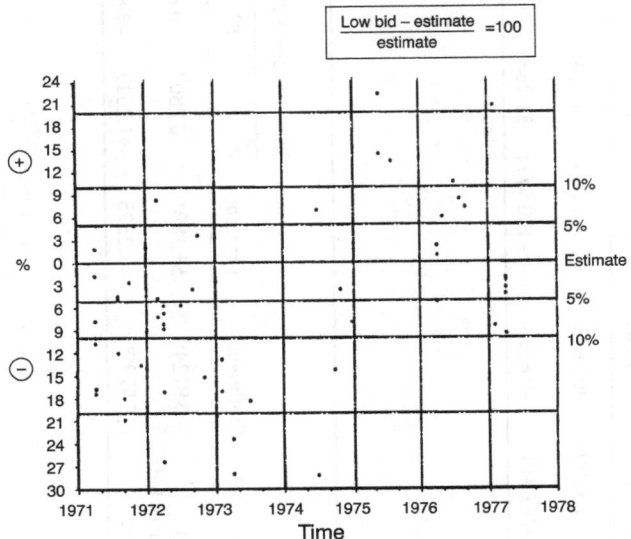

Figure 5.5.7 Council B estimate accuracy against time.

results for 1977 were not so encouraging, but the Council subsequently recognized that their pricing strategy had not responded quickly enough to the changed market conditions.

More generally, this section has indicated that the quantity surveyor's price prediction can exhibit quite high degrees of inaccuracy. Secondly, while there is no consistency in inaccuracy across organizations, there does appear to be consistency in inaccuracy within a particular organization. This points to the need for the quantity surveying profession to pay greater attention to the analysis of its performance when undertaking price predictions for proposed buildings. For both county councils the detrimental trend could have been detected and remedied by closely monitoring performance. Going back to the marksperson analogy at the beginning of the section, there is a need for a feedback mechanism that will allow a particular organization to monitor its performance, and, in terms of the analogy, adjust its sights to take account of identifiable bias. We now consider such a mechanism in the next section.

Table 5.5.13 Council B – estimating performance 1971–77

	1971	1972	1973	1974	1975	1976	1977
Percentage of total projects where low bid exceeded quantity surveyor's estimate	85	87	100	80	–	14	86

A PROPOSED FEEDBACK MECHANISM

A feedback mechanism intended to assess forecasting performance should have a number of characteristics. First, it must be simple to use and capable of giving an early indication of error. Secondly, it must be capable of identifying any consistent bias in forecasts. Thirdly, it should be capable of highlighting situations in which, although there is no **consistent** bias, forecasts are becoming increasingly inaccurate.

The feedback mechanism explained below has been designed with these characteristics very much in mind.

Trend control analysis

The objective of trend control analysis is to identify whether estimating errors exhibit a consistent trend over time. A trend control chart is built up as follows:

1   List existing projects in chronological order
2   Enter the projects along the horizontal axis of the trend control chart, spacing the projects equally along this axis as in Fig 5.5.8.
3   For each project calculate the percentage deviation between estimate and low bid (positive if estimate exceeds low bid, negative if estimate is less than low bid).
4   Cumulate the percentage deviations and plot the cumulated deviation against the last project used.

An example may be useful to illustrate these various stages. Table 5.5.14 lists three (hypothetical) projects with estimates and low bids. Percentage deviation for each project is given in column 4 and cumulative deviation in column 5. The points plotted on the trend control chart (Fig. 5.5.8) are given in column 6.

Since the projects are evenly spaced on the horizontal axis, the slope of a line between a point on the curve and the initial point is the average deviation for all projects up to that point. The slope is the total deviation divided by the number of projects. For example, the slope of the line from the origin to the

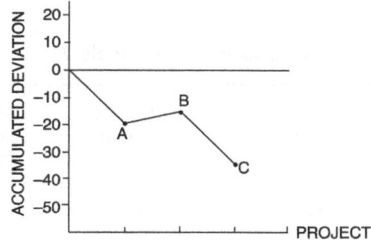

Figure 5.5.8 Accumulated deviation against project for example 1.

Table 5.5.14 Trend control chart: hypothetical example 1

Project	Estimate (£'000)	Low bid (£'000)	Percentage deviation	Cumulative deviation	Plotted point
(1)	(2)	(3)	(4)	(5)	(6)
1	20	25	−20	−20	A
2	105	100	+5	−15	B
3	80	100	−20	−35	C

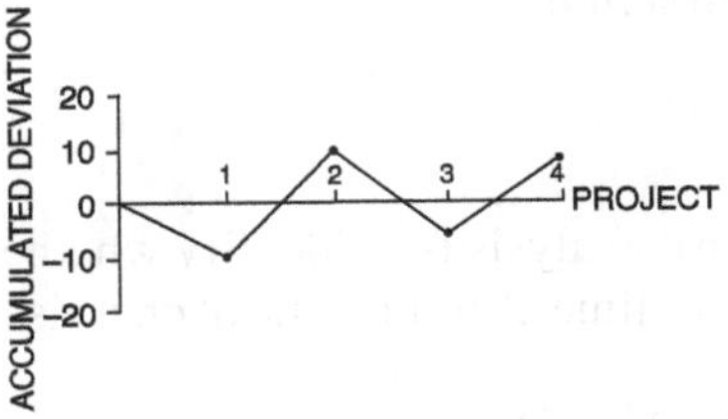

Figure 5.5.9 Accumulated deviation against project for example 2.

point C (Fig 5.5.8) is −11.66%, which is the average percentage estimating error for the three projects in Table 5.5.14. Similarly, the slope of the line from point A to point C is −7$\frac{1}{2}$% which is the average percentage deviation for all projets completed after project 1.

The line joining the various points identifies the trend of estimating performance. The nearer the trend line to horizontal the closer the estimate is to the low bid. An upward trend line shows a tendency towards over-estimating, whilst a downward trend line indicates a tendency towards under-estimating. This form of plot enables the surveyor to identify a consistent trend within a time frame.

Consider the example in Table 5.5.15, for which the trend control chart is Fig 5.5.9. In this case no consistent trend can be identified, and the average percentage deviation over the four projects is 2%. Clearly, however, esti-mating performance as illustrated for Table 5.5.15 cannot be described as 'good' for no low bid was within 10% of the estimate. This indicates that the

Table 5.5.15 Trend control chart: hypothetical example 2

Project	Estimate (£'000)	Low bid (£'000)	Percentage deviation	Cumulative deviation	Plotted point
(1)	(2)	(3)	(4)	(5)	(6)
1	90	100	+10%	−10	A
2	120	100	+20%	+10	B
3	170	200	−15%	−5	C
4	113	100	+13%	+8	D

trend control chart needs to be supplemented by another chart that indicates the deviation of estimates from the low bid, and we now describe that chart.

Deviation control analysis

A deviation control chart is constructed as follows:

(a) Perform elements 1 and 2 for the construction of the trend control chart.
(b) Take the percentage deviations obtained from step 3 and plot the cumulative squared deviation against the last project used.

This exercise is performed in Tables 5.5.16 and 5.5.17.

It is now possible to derive the standard deviation of the estimate measured against the low bid between any two points in time. The construction of the deviation control chart is such that the slope of a line from the origin to any point on the curve gives the variance of estimate about low bid for all projects up to that point. The variance is the square of the standard deviation, thus a slope scale can be constructed to allow the standard deviation to be read off.

It should be noted that the slope in the slope chart will not carry □ signs, since we cannot derive from the deviation control chart the sign of the standard deviation; squaring the deviation removes the sign. This must, therefore, be obtained from the trend control chart.

Bringing the trend control and deviation control charts together, we can indicate from the former whether any consistent trend is emerging in estimating performance, and from the latter the deviation of the estimate from low bid. Clearly, the two control charts must be used together if estimating

Table 5.1.16 Deviations controls chart: hypothetical example 1

Project	Percentage deviation	Squared deviation	Cumulative squared deviation
1	−20	400	400
2	+5	25	425
3	−20	400	825

Table 5.1.17 Deviations controls chart: hypothetical example 2

Project	Percentage deviation	Squared deviation	Cumulative squared deviation
1	−10	100	100
2	+20	400	500
3	−15	225	725
4	+13	169	894

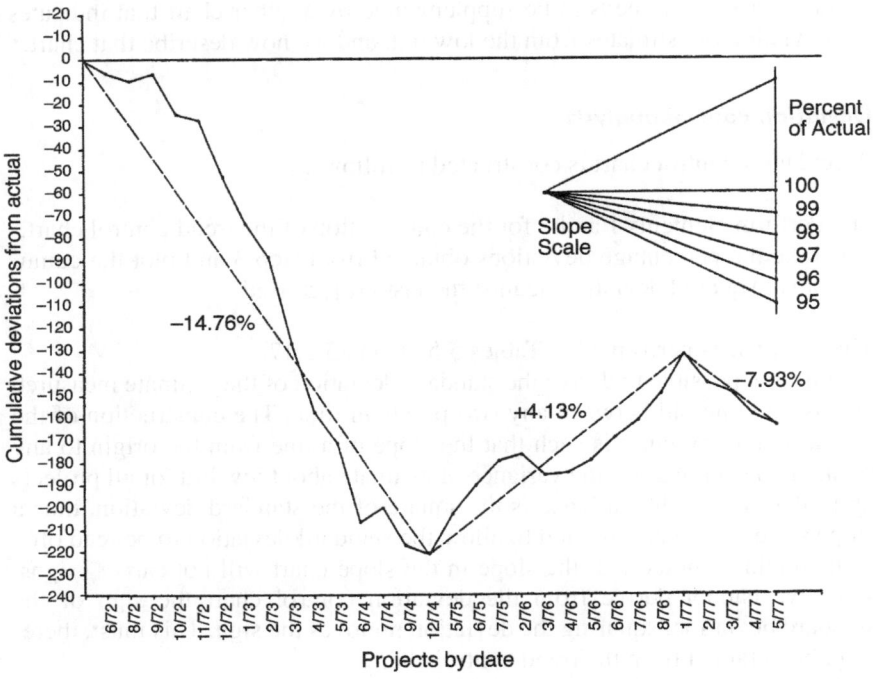

Figure 5.5.10 Trend control chart for Council B estimates 1973–7.

performance is to be properly assessed. The deviation control chart assesses the accuracy of estimating and the trend control chart indicates whether any inaccuracy is consistent in its direction.

Trend control and deviation control charts have been constructed for council B's projects and are presented in Figs 5.5.10 and 5.5.11.

The trend control chart shows up the consistent underestimating up to December 1974, followed by a period of overestimating and finally a period of further underestimating. The deviation control chart shows the sharp jump in deviations in 1973 and 1974, and then a deviation of approximately □ 11% over the period June 1974 to May 1977.

CONCLUSIONS

A number of major conclusions can be drawn from our analysis of the estimating performance of Council A and Council B. First, there need be no consistency in estimating errors between different organizations. Our analysis has shown significant differences in performances between two organizations in the same sector, commissioning similar work. The only common factor

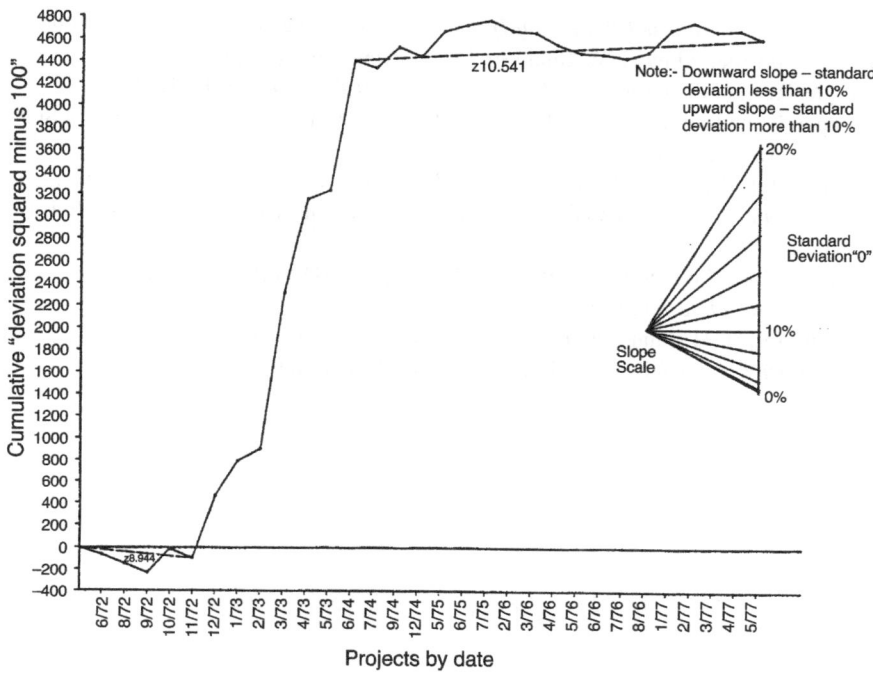

Figure 5.5.11 Deviation control chart for Council B estimates 1972–7.

is that both perform badly on large projects.

Secondly, there is a need for estimating performance to be monitored consistency. The custom in the building industry appears often to have been to take the forecast, as being 'correct' and the tenders, when they differ from the forecast, as being 'wrong'. We would argue that this is very far from being the case, particularly in situations where forecasts differ from low bids in some consistent fashion. The quantity surveyor must constantly monitor the performance of estimates, in order to identify quickly any patterns that may emerge. This paper presents one simple method whereby such monitoring can be undertaken.

A list of further reading material is included in the references.

REFERENCES

Bauman, H. C. (1961) *Industrial and Engineering Chemistry* 53, 52A–54A.

Bragg, G. M. (1974) *Principles of Experimentation and Measurement* Prentice-Hall, Englewood Cliffs, NJ.

Chapanis, A. (1961) Theory and methods for analysing errors in man–machine systems. In *Selected Papers on Human Factors in the Design and Use of control Systems* (edited by H. W. Sinaiko). Dover. New York, pp. 1179–1202.

Doyle, R. C. (1977) *American Association of Cost Engineers Bulletin* 19, 93–7.

Draper, N. and Smith, H. (1981) *Applied Regression Analysis.* John Wiley and Sons, New York, Chichester.

Hemphill, R. B. (1968) *A Method for Predicting the Accuracy of a Construction Cost Estimate*, paper presented to American Association of Cost Engineers, Houston, June.

National Joint Council for the Building Industry (1981) *Code of Procedure for Single Stage Selective Tendering.*

Wonnacott, T. H., and Wonnacott, R. J. (1981) *Regression: A Second Course in Statistics*, John Wiley and Sons, New York, Chichester.

Selected Bibliography

This section contains a list of further publications relevant to the field at the time of writing (1992). The list is by no means exhaustive but does give a general coverage. The quality of these papers is uneven however and some care is urged, especially where papers have not undergone any review process.

Ahuja, H. N., 1982, A conceptual model for probabilistic forecast of final cost, *Proceedings of the PMI/INTERNET Symposium,* Toronto, Project Management Institute, 23–31.

Allman, I., 1988, Significant items estimating, *Chartered Quantity Surveyor*, Sept, 24–5.

Ashworth A., 1981, Cost modelling in the construction industry, *Quantity Surveyor*, 37, July, 132–4.

Ashworth, A., 1983, *Building Economics and Cost Control: Worked Solutions*, Butterworths.

Ashworth, A., Skitmore, R. M., 1983, 'Accuracy in estimating', *Occasional Paper* No 27, Chartered Institute of Building.

Ashworth, A., Skitmore, R. M., 1986, 'Accuracy in cost engineering', *Transactions 9th International Cost Engineering Congress*, Oslo, Norway, Paper A3.

Atkin, B., 1987, 'A time/cost planning technique for early design evaluation', In *Building Cost Modelling and Computers*, ed. P. S. Brandon, E & F N Spon, 145–54.

Avery, D. W., 1982, 'Problems of locality in construction cost forecasting and control', In *Building Cost Techniques: New Directions*, ed. P. S. Brandon, 159–68, E & F N Spon.

Avery, D. W., 1987, 'Steps towards cost modelling for the control of small factory building cost', In *Building Cost Modelling and Computers*, ed. P. S. Brandon, E & F N Spon, 155–64.

Apcar, H. V., 1960, 'Structural frameworks for single-storey factory buildings', *Factory Building Studies*, No. 7, Department of Scientific and Industrial Research Building Research Station, HMSO.

Azzaro, D. W., 1976, 'Measuring the level of tender prices', *Chartered Surveyor – Building and Quantity Surveying Quarterly*, Winter 1976/77, 17–9.

Banks, D. G., 1974, 'Uses of a length/breadth index and a plan shape ratio, *Building and Quantity Surveying Quarterly*, 2 (1).

Barnes, M., 1989, 'Introducing MERNA', *Chartered Quantity Surveyor*, Jan, 19.

Barrett, A. C., 1970, 'Preparing a cost plan on the basis of outline proposals', *Chartered Surveyor*, May, 507–20.

Basden, A., 1987, 'An expert system for estimating construction costs', *Technical Report*, Dept of Civil Engineering, University of Salford.

Baxendale, T., 1984, 'Construction resource models by Monte Carlo simulation', *Construction Management and Economics*, 2, 201–17.

B.C.I.S., 1969, *Standard Form of Cost Analysis*, The Building Cost Information Service, The Royal Institution of Chartered Surveyors, Surrey.

Beeston, D. T., 1973, 'Cost consequences of design decisions: cost of contractors' operations (COCO)', *Report*, Directorate of Quantity Surveying Development, Property Services Agency, Department of the Environment.

Beeston, D. T., 1974, 'One statistician's view of estimating', *Cost Study F3*, BCIS 1974/5–123–9 to 16, Building Cost Information Service, Royal Institution of Chartered Surveyors.

Beeston, D. T., 1978, 'Cost models', *Chartered Surveyor – Building and Quantity Surveying Quarterly*, Summer 1978, 56–9.

Beeston, D. T., 1982, 'Modelling and Judgement', *Cost Study F29*, Building Cost Information Service, Royal Institution of Chartered Surveyors.

Beeston, D. T., 1983, *Statistical Methods for Building Price Data*, E & F N Spon

Beeston, D. T., 1986, 'Risk in estimating', *Construction Management and Economics*, 4 (1).

Beeston, D. T., 1987, 'A future for cost modelling', in *Building Cost Modelling and Computers*, ed. P. S. Brandon, E & F N Spon, 15–24.

Bennett, J., 1981, *Cost Planning and Computers*, Property Services Agency.

Bennett, J., 1987, *Report on a Test of Griffiths, Laxtons, Spons and Wessex Price Books*, Thomas Skinner Directories, West Sussex.

Bennett, J. and Ferry, D., 1987, 'Towards a simulated model of the total construction process', In *Building Cost Modelling and Computers*, ed. P. S. Brandon, E & F N Spon, 377–86

Bennett, J. and Fine, B., 1980, 'Measurement of complexity in construction projects', *Occasional Paper*, 8, Department of Construction Management, University of Reading.

Bennett, J. and Ormerod, R. N., 1984, 'Simulation applied to construction management', *Construction Management and Economics*, 2, 225–63.

Bennett, Sir T., 1946, 'Cost investigation', *Journal of the RICS*, XXV (XI), May, 492–509.

Bindslev, B., 1985, 'On conceptual isomorphism in the resource table of the SfB system', *Paper for the CIB W-74 Meeting*, Stuttgart 14–16, Oct (extracts from '*Projektstyring med Mikrodatamat*' Kunstakademiets Arkitektsole, Kobenhavn 1985).

Birrell, G. S., 1980, 'Construction cost estimating in the design process', *ASCE J Const Div*, 106 (C04), 551–6.

Bowen, P. A., 1982, 'An alternative estimating approach', *Chartered Quantity Surveyor*, Feb, 191–4.

Bowen, P. A., 1982, 'Problems in econometric cost modelling', *The Quantity Surveyor*, May, 83–5.

Bowen, P. A., 1984, 'Applied econometric cost modelling', *Proceedings, 3rd Int Symp on Build Econ*, CIB W–55, Ottawa, 3, 144–57.

Bowen, P. A. and Edwards, P. J., 1985, 'Cost modelling and price forecasting: from a deterministic to a probabilistic approach', *Occasional Paper* No 5, Department of Quantity Surveying and Building Economics, The University of Natal, Durban, South Africa.

Bowen, P. A. and Edwards, P. J., 1985, 'Cost modelling and price forecasting: practice and theory in perspective', *Construction Management and Economics*, 3, 199–215.

Bowen, P. A., and Edwards, P. J., 1985, 'A conceptual understanding of the paradigm shift in modelling techniques used in the economics of building', *Occasional Paper* No 6, Department of Quantity Surveying and Building Economics, The University of Natal, Durban, South Africa.

Bowen, P. A., Wolvaardt, J. S. and Taylor, R. G., 1987, 'Cost modelling: a process-modelling approach', In *Building Cost Modelling and Computers*, ed. P. S. Brandon, E & F N Spon, 387–95

Bowley, M. E. A. and Corlett, W. J., *Report on a Study of Trends in Building Prices*, Ministry of Public Building and Works Research and Development.

Braby, R. H., 1975, 'Costs of high-rise buildings', *Building Economist*, 14 (2), 84–6.

Bradley, J., Brown, L. H. and Feeney, M., 1974, 'Cost optimisation in relation to factory structures', *Engineering Optimisation*, 1, 125–38.

Brandon, 1977, *A Framework for Cost Exploration and Strategic Cost Planning in Design*, Msc Thesis, University of Bristol, Oct.

Brandon, 1978, 'A framework for cost exploration and strategic cost planning in design', *Chartered Surveyor Building and Quantity Surveying Quarterly*, 5(4), Summer.

Brandon, P. S., 1982, 'Building cost research – need for a paradigm shift? In *Building Cost Techniques: New Directions*, ed. P. S. Brandon, 5–13, E & F N Spon.

Brandon, P. S., 1987, 'Models, mortality and the ghost in the machine', In *Building Cost Modelling and Computers*, ed. P. S. Brandon, E & F N Spon, 1–14.

Brandon, P. S., 1990, 'The development of an expert system for the strategic planning of construction projects', *Construction Management and Economics*, 8 (3), 285–300.

Brandon, P. S. and Newton, S., 1986, 'Improving the forecast', *Chartered Quantity Surveyor*, May, 14–26.

Broadbent, G., 1982, Design, economics and quality, In *Building Cost Techniques: New Directions*, ed. P. S. Brandon, E & F N Spon 41–60.

Brown, H. W., 1987, 'Predicting the elemental allocation of building costs by simulation with special reference to the costs of building services', In *Building Cost Modelling and Computers,* ed. P. S. Brandon, E & F N Spon, 397–406.

Browning, 1960, *Estimating and Cost Planning*, B T Batsford,

Bryan, S, 1991, 'Assembly pricing in construction cost estimating', *Cost Engineering*, Vol 33 no 8, 17–21.

Buchanan, J. S., 1973, *Cost Models for Estimating: Outline of the Development of a Cost Model for the Reinforced Concrete Frame of a Building*, Royal Institution of Chartered Surveyors.

Building Research Establishment, 1962, 'The economics of factory buildings', *Factory Building Studies*, 12, HMSO.

Case, K. E., 'On the consideration of variability in cost estimating', *IEEE Transactions on Engineering Management*, EM–19 (4), Nov, 114–8.

CEEC, 1986 *Unit Rates for Material and Work Items.*

Clapp, M. A. and Mayer, J. F., 1983, 'Factors affecting tender price indices', *Building Research Establishment Note*, N66/83, April, BRE, Watford.

Co-ordinated Building Communication (CBC), 1964, 'Application of SfB to contract documentation', *Architects' Journal*, March 64 – October 65 (22 articles).

Cost Research Panel, 1961, 'Report of the work of the cost research panel', *Chartered Surveyor*, 93 (8), Feb, 438–41.

Cusack, M. M., 1987, 'Optimisation of time and cost', *Project Management*, 3 (1), 50–4.

Davis, Belfield and Everest, 1997, 'Initial cost estimating: the cost of factories', *Architects' Journal*, 166 (14 Sept 77), 509–12.

Department of Education and Science, 1972, *Building Bulletin* No 4: Cost Study, 3rd Ed., HMSO.

Diekmann, J. E., 1983, 'Probabilistic estimating: mathematics and applications', *J Const Eng and Mgmt*, ASCE, 109, 297–308.

Dodge, 1974, *Manual for Building Construction, Pricing and Scheduling*, Dodge Building Cost Services, McGraw-Hill Information Systems.

Drake, B., 1983, 'Cost data', *Paper presented to a Seminar on Cost Data*, Royal Institution of Chartered Surveyors, Jan.

Drake, B., 1984, 'Using cost data', *Chartered Quantity Surveyor*, Jan, 214–5.

Ferry, D. J. and Brandon, P. S., 1980, 'Introduction to cost modelling', *Cost Planning of Buildings*, 4th ed, ch. 7, 100–14, Granada.

Fine, B., 1970, 'Simulation technique challenges management', *Construction Progress*, Jul, 33-4.

Fine, B., 1974, 'Tendering Strategy', *Building*, 25 Oct, 115–21.

Fine, B., 1982, 'The use of simulation as a research tool', In *Building Cost Techniques: New Directions*, ed. P. S. Brandon, 61–70, E & F N Spon.

Fine, B., 1987, 'Kitchen sink economics and the construction industry', In *Building Cost Modelling and Computers*, ed. P. S. Brandon, E & F N Spon, 25–30.

Flanagan, R., 1980, *Tender Price and Time Prediction for Construction Work*, PhD thesis, University of Aston, Birmingham.

Flanagan, R. and Norman, G., 1978, 'The relationship between construction price and height', *Chartered Surveyor B and QS Quarterly*, Summer, 69–71.

Flanagan, R. and Norman, G., 1983, 'The accuracy and monitoring of quantity surveyors' price forecasting for building work', *Construction Management and Economics*, 1, 157–80.

Forbes, W. S., 1974, 'Design cost information', *Building*, 27/9/74, 89–90.

Fortune, C. and Lees, M., 1996, 'The relative performance of new and traditional cost models in strategic advice for clients', *RICS Research Paper Series* Volume Two, Number Two, March, Royal Institution of Chartered Surveyors.

Franks, J., 1970, 'An exercise in cost (price?) estimating', *Building*, 12 June, 133–4.

Gates, M., 1970, *Review of Existing and Advanced Construction Estimating Techniques*, edited from video tape, Construction Estimating, West Hartford, Connecticut, USA.

Gilmore, J., Skitmore, R. M. 1989, 'A new approach to early stage estimating', *Chartered Quantity Surveyor*, 11 (9), 36–8.

Gould, P. R., 1970, 'The development of a cost model for the heating ventilating and air-conditioning installation of a building'. MSc Project Report, Loughborough University of Technology.

Gray, C., 1981, 'Analysis of the preliminary element of building production costs', MPhil Thesis, University of Reading.

Gray, C., 1982, 'Analysis of the preliminary element of building production costs', in *Building Cost Techniques – New Directions*, P. S. Brandon, ed. E & F N Spon, 290–306.

Gray, C., 1986, '"Intelligent" cost and time analysis', *Occasional Paper*, 18, Department of Construction Management, University of Reading.

Green, D. H., 1982, 'Data bases used by Property Services Agency Quantity Surveyors', In *Building Cost Techniques – New Directions*, ed. P. S. Brandon, E & F N Spon, 85–89.

Greven, E. D. W., 1978, 'The use of a building cost-element model in the early design stage', *Proceedings of the C.I.B. W–65 2nd Symposium on Organisation and Management of Construction*, Haifa, Israel, Session 11, III, 157–176.

Hanscomb Associates, 1984, *Area Cost Factors*, Report of the US Army Corps of Engineers, Hanscomb Associates, Atlanta, Georgia, USA.

Hardcastle, C, 1982, 'Capital cost estimating and the method of presenting information for pricing of construction work', In *Building Cost Techniques – New Directions*, ed. P. S. Brandon, E & F N Spon, 373–86.

Hardcastle, C., 1984, 'The relationship between cost communications and price prediction in the evaluation of building design', *Proceedings 3rd Int Symp on Build Econ*, CIB W–55, Ottawa, 3, 112–23.

Hardcastle, C., Brown, H. W. and Davies, A. J., 1987, 'Statistical modelling of Civil Engineering Coasts in the Petro-Chemical Industry', In *Building Cost Modelling and Computers*, ed. P. S. Brandon, E & F N Spon, 183–92.

Hemmett, J., 1987, 'Building cost modelling and computers', In *Building Cost Modelling and Computers*, ed. P. S. Brandon, E & F N Spon, 103–8.

Hemphill, R. B., 1968, 'A method for predicting the accuracy of a construction cost estimate', *Transactions*, American Association of Cost Engineers, Houston, Texas.

Holes, L. G., 1987, 'Holistic resource and cost modelling', In *Building Cost Modelling and Computers*, ed. P. S. Brandon, E & F N Spon, 193–202.

Holes, L. G. and Thomas, R., 1982, 'General purpose cost modelling', In *Building*

Cost Techniques – New Directions, ed. P. S. Brandon, E & F N Spon, 220–7.

Institute of Quantity Surveyors, 1970, 'A research and development report: design cost planning', *The Quantity Surveyor*, 27, 3.

Jackson, M. J., 1986, *Computers in Construction Planning and Control*, Allen and Unwin.

Jagger, D. M., 1982, 'Cost data bases and their generation', In *Building Cost Techniques – New Directors*, ed P. S. Brandon, E & F N Spon, 90–106.

Jagger, D. M., 1982, 'The quantification of building work at resource level', MPhil Thesis, Liverpool Polytechnic.

James, W., 1954, 'A new approach to single price-rate approximate estimating', *RICS Journal*, XXXII (XI), May, 810–24.

James, W., 1955, 'The art of approximate estimating', *Chartered Surveyor*, LXXXVIII (4), Oct., 214–18.

Johnson, V. B. and Partners (eds), 1987, *Laxton's National Building Price Book*, 159th ed, Thomas Skinner Directories, Sussex.

Jones, C. F., 1987, 'Computers and modelling for cost effective building procurement', In *Building Cost Modelling and Computers,* ed. P. S. Brandon, E & F N Spon, 203–12.

Jupp, B. C., 1980, 'Cost data estimating processes in strategic design of building', *Report* describing research undertaken for the Building Research Establishment.

Jupp, B. C., 1984, 'Reliability of detailed price data for estimating during detailed design', *Report*, Bernard C Jupp and Partners, London.

Jupp, B. C. and McMillan, V., 1981, 'The reliability of cost data', Contribution to Conference held at the Polytechnic of the South Bank, London, June.

Kelly, J. R., 1980, 'A methodology for determining the variable costs of building elements at sketch design stage', *Papers Quality and Cost in Buildings Symposium, 1, Investment Appraisals, Construction and Use Coats, the Life Cycle Costing of Buildings, CIB W–55*, 15–17 September, Institute de Recherche sur l'environement Construct, Lausanne, Switzerland, 35–71.

Kouskoulas, V. and Koehn, E., 1974, 'Predesign cost estimation function for buildings', *ASCE J of Const Div*, Dec, 589–604.

Lavelle D. J., 1982, 'Development of an interactive tender price prediction system', MSc Project Report, Department of Civil Engineering, Loughborough University of Technology.

Lee, B. S. and Knapton, K. M., 1974, 'Optimum cost design of a steel-framed building', *Engineering Optimisation*, 1, 139–53.

Lichtenberg, S., 1981, 'Real world uncertainties in project budgets and schedules', *Proceedings of the INTERNET Symposium*, Boston, USA, 179–93.

Liddle, C. J. and Gerard, A. M., 1975, *The Application of Computers to Capital Cost Estimation*, Institution of Chemical Engineers.

Lipson, S. L., Russell, A. D., 1971, 'Cost optimisation of structural roof systems', *Journal of the Structural Division*, ASCE, 103 (ST8), 2057–71.

Lu Qian, 1988, 'Cost estimation based on theory of fuzzy sets and prediction techniques – an expert system approach', In *Construction Contracting in*

China, Department of Civil and Structural Engineering, Hong Kong Polytechnic, 113–25.

McCaffer, R., 1975, 'Some examples of the use of regression analysis as an estimating tool', *Quantity Surveyor*, Dec, 81–6.

McCaffer, R., 1976, *Contractor's Bidding Behaviour and Tender Price Prediction*, PhD Thesis, Loughborough University of Technology, September.

McCaffer, R. and McCaffrey, M. J., 1979, 'Computer aided tender price prediction for buildings', *Report on Current Research*, Loughborough University.

McCaffer, R., McCaffrey, M. J. and Thorpe, A., 1984, 'Predicting the tender price of buildings during early stage design: method and validation', *J. Opl Res. Soc.*, 35 (5), 415–24.

McCaffrey, 1978, 'Tender–price prediction for UK buildings – a feasibility study', Msc Project Report, Department of Civil Engineering, Loughborough University of Technology.

Marston, V. K., 1988, 'Design uncertainty and estimating', *Proceedings of the 3rd Euro Cost Eng Forum*, Association of Cost Engineers, March.

Marston, V. K. and Skitmore, R. M., 1990, 'Housing refurbishment: cost–time forecasts by intelligent simulation', *Proceedings of the International Symposium on Property Maintenance Management and Modernisation*, 1, Longman Singapore Publishers, 463–72.

Marston, V. K. and Skitmore, R. M., 1990, 'Automatic resource based cost–time forecasts', *Transactions of the 34th Annual Meeting of the American Association of Cost Engineers*, Symposium M – Project Control, M.6.1 – M.6.6. AACE, Morgantown, USA.

Marston, V. K., Skitmore, R. M., Alshawi, M., Retik, A. and Koudounas, C., 1992, *The Use of Intelligent Simulation in Cost and Time Forecasts for Housing Rehabilitation Works*, Department of Surveying, University of Salford.

Mathur, K., 1982, 'A probabilistic planning model', In *Building Cost Techniques – New Directions*, ed P. S. Brandon, E & F N Spon, 181–91.

Means, R. S., 1972, *Building Construction Cost Data*, Robert Snow Means, Dunbury, Massachusetts.

Meijer, W. J. M., 1987, 'Cost modelling of archtypes', In *Building Cost Modelling and Computers*, ed. P. S. Brandon, E & F N Spon, 223–32.

Meyrat, R. F., 1969, 'Algebraic calculation of cost price', *BUILD International*, Nov, 27–36.

Maver, T., 1979, 'Cost performance modelling', *Chartered Quantity Surveyor*, 2 (5), Dec, 111–15.

Mayer, J. F., 1984, 'Factors affecting tender price indices – stage II', *Building Research Establishment Note*, N36/84, Feb, BRE, Garston, Watford.

Moore, G. and Brandon, P. S., 1979, 'A cost model for reinforced concrete frame design', *Chartered Quantity Surveyor*, Oct, 40–4.

Morris, A. S., 1976, 'Regression analysis in the early forecasting of construction prices', Msc Thesis, Loughborough University of Technology.

Morris, A., 1982, 'Algorithms in price modelling the effect of incorporating provision for physically handicapped people into existing houseplans', In *Building Cost Techniques – New Directions*, ed. P. S. Brandon, E & F N Spoon, 126–34.

Morrison, N., 1984, 'The accuracy of quantity surveyors; cost estimating', *Construction Management and Economics*, 2, 57–75.

Morrison, N. and Stevens, S., 1980, *Construction Cost Data Base*, 2nd annual report of research project by Department of Construction Management, University of Reading, for Property Services Agency, Directorate of Quantity Surveying Services, DOE.

Morrison, N. and Stevens, S., 1980, 'A construction cost data base', *Chartered Quantity Surveyor*, Jun, 313–5.

Nadel, E., 1967, 'Parameter cost estimates', *Engineering News Record*, 16 Mar, 112–23.

National Building Elements Committee, 1973, *National Standard Building Elements and Design Cost Control Procedures*, An Foras Forbartha, Dublin.

Newton, S., 1982, 'ACE: analysis of construction economics', Internal Report, University of Strathclyde, April.

Newton, S., 1982, 'Cost modelling: a tentative specification', In *Building Cost Techniques – New Directions*, ed. P. S. Brandon, E & F N Spon, 192–209.

Newton, S., 1983, *Analysis of Construction Economics: a Cost Simulation Model*, PhD Thesis, Department of Architecture and Building Science, University of Strathcyde.

Newton, S., 1984, 'An interactive computer-based cost simulation model,' *Proceedings 3rd Int Sump on Build Econ*, CIB W–55, Ottawa, 3, 58–67.

Newton, S., 1987, 'Computers and cost modelling: what is the problem?', In *Building Cost Modelling and Computers*, ed. P. S. Brandon, E & F N Spon, 41–8.

Newton, S., 1988, 'Cost modelling techniques in perspective', *Transactions 32nd Ann Meeting of the AACE and the 10th Int Cost Engineering Congress*, New York, July, ed B. Humphrys and S. Pritchard, B 7. 1–B. 7.7, AACE.

Newton, S, 1990, 'An agenda for cost modelling research', *Construction Management and Economics*.

Nisbet, J., 1961, *Estimating and Cost Control*, Batsford.

Nisbet, J., 1965, 'Cost planning and cost control', *Architects Journal*, 3, 10 & 24 November.

Nott, C. M., 1955, 'A method of cost analysis and cost planning: the effect on estimating and bills of quantities', *Journal of the RICS*, XXXIV (XI), May, 910–27.

Park, W. R., 1963, 'Pre-design estimates in Civil Engineering projects', *Journal of the Construction Division*, ASCE, 89 (CO2), 11–23.

Patchell, B. R. T., 1987, 'The implementation of cost modelling theory', In *Building Cost Modelling and Computers*, ed. P. S. Brandon, E & F B Spon, 233–42.

Pegg, I., 1984, 'The effect of location and other measurable parameters on tender levels', *Cost Study F33*, BCIS 1983–4–219-5 to 17, Building Cost Information Service, Royal Institution of Chartered Surveyors.

Pegg, I., 1987, 'Computerised approximate estimating from the BCIS on-line data base', In *Building Cost Modelling and Computers* ed. P. S. Brandon, 243–9, E & F N Spon.

Perry, J. H., ed., 1973 *Chemical Engineers Handbook*, McGraw Hill, New York.

Portsmouth Polytechnic, 1974, 'Acceptable levels of risk, Standard Method of Measurement Development Unit', *Report to the SMM Working Party*, Royal Institution of Chartered Surveyors.

Powell, J., Chisnall, J., 1981, 'Getting early estimates right', *Chartered Quantity Surveyor*, Mar, 279–81.

Property Services Agency, 1977, *Early Cost Advice (B & CE Elements) – Offices, Sleeping Quarters,* March, HMSO.

Property Services Agency, 1980, *Schedule of Rates for Building Works*, HMSO.

Price Davies, B., 1922, 'Analysis of building costs', *Transactions* The Surveyors Institute, London, 168–210.

Raftery, J., 1984, 'Models in building economics: a conceptual framework for the assessment of performance', *Proceedings 3rd Int Symp on Build Econ,* CIB W–55, Ottawa, 3, 103–11.

Raftery, J., 1984, *An Investigation into the Suitability of Cost Models for use in Building Design*, PhD Thesis, Department of Surveying, Liverpool Polytechnic.

Raftery, J., 1987, 'The state of cost/price modelling in the construction industry: a multicriteria approach', In *Building Cost Modelling and Computers,* ed. P. S. Brandon, 49–71, E & F N Spon.

Raftery, J., 1991, 'Models for construction cost and price forecasting', *Proceedings of the 1st National RICS Research Conference*, 10–11 January, E & F N Spon.

Raftery, J. and Wilson, A. J., 1982, 'Some problems of data collection and model validation', *Paper for Research Seminar*, Liverpool Polytechnic, March.

Relf, C. T., 1974, 'Study of the building timetable: the significance of duration', (ARCHIT.BA.1.UCE), parts 1 and 2 – *Final Report of a Research Project sponsored by the Department of the Environment,* on the basis of work carried out by a team under the direction of Prof. D. A. Turin, UCERG Building Economics Research Unit, University College, London.

Reynolds, G., 1978, 'On the cost model', *Building*, 20 October, 91.

Reynolds, G., 1978, 'More on the cost model', *Building*, 27 October, 67.

Reynolds, G., 1980, 'Predicting building costs', *Chartered Quantity Surveyor*, 3 (4), 137–9.

R.I.B.A., 1968, *Construction Indexing Manual*, RIBA, London.

Robertson, D., 'Research based information: developments in information technology', In *Building Cost Techniques – New Directors,* ed. P. S. Brandon, E & F N Spon, 107–14.

Ross, E., 1983, 'A database and computer system for tender price prediction by approximate quantities', MSc Project Report, Loughbourough University of Technology.

Royal Institute of British Architects, 1954, *RIBA Rules for Cubing Buildings for Approximate Estimates*, D/1156/54. RIBA.

Royal Institution of Chartered Surveyors, 1957, 'The use of elemental bills of quantities', *The Chartered Surveyor*, April.

Royal Institution of Chartered Surveyors Junior Organisation, 1964, 'The effect of shape and height on building cost', *The Chartered Surveyor*, May.

Royal Institution of Chartered Surveyors, 1975, 'Extent to which cost is affected by detail', *Report by the Research Group* appointed by the London Branch Quantity Surveyors Committee, Royal Institution of Chartered Surveyors.

Royal Institution of Chartered Surveyors, 1976, *An Introduction to Cost Planning* Royal Institution of Chartered Surveyors, Junior Organisation Quantity Surveyors Standing Committee.

Royal Institution of Chartered Surveyors, 1982, *Pre-contract Cost Control and Cost Planning*, Quantity Surveyors Practice Pamphlet, No 2, Surveyors Publications, London.

Runeson, G., 1988, 'An analysis of the accuracy of estimating and the distribution of tenders', *Construction Management and Economics*, 6 (4), 357–70.

Runeson, G. and Bennett, J., 1983, 'Tendering and the price level in the New Zealand building industry', *Construction Papers*, 2, pt. 2, 29–35.

Russell, A. D. and Choudhary, K. T., 1980, 'Cost Optimisation of Buildings'. *American Society of Civil Engineers, Journal of the Structural Division*, January, 283–300.

Schofield, D., Raftery, J. and Wilson, A., 1982, 'An economic model of means of escape provision in commercial buildings', In *Building Cost Techniques – New Directions*: ed. P. S. Brandon, E & F N Spon, 210–20.

Seeley, I. H., 1972, 'Approximate estimating', *Building Economics*, ch 6, 100–19, MacMillan.

Selinger, S., 1988, 'Computerised parametric estimating', *Proceedings, British–Israeli Seminar on Building Economics*, Technion, Haifa, 160–7.

Shealy, H. F., 1986, 'Right of way clearing for a modern rapid transit railway', *Engineering*, 28 (4), 12–8.

Sidwell, A. C. and Wootton, A. H., 1984, 'Operational estimating', *Proceedings, Organising and Managing Construction, 3: Developing Countries Research*, ed. V. K. Handa, Ontario: University of Waterloo, 1015–20.

Skitmore, R. M., 1981, 'Why do tenders vary?', *Chartered Quantity Surveyors*, December, 128–9.

Skitmore, R. M., 1983, 'Heuristics and building price forecasting', *Proceedings 2nd South East Asian Survey Congress*, Technical Paper Q4, 1–8.

Skitmore, R. M., 1986, *Towards an Expert Building Price Forecasting System*, Surveyors Publications.

Skitmore, R. M., 1987, *Construction Prices: The Market Effect*, University of Salford.

Skitmore, R. M., 1988, 'Fundamental research in bidding and estimating', *Transactions CIB British–Israeli Seminiar on Building Economics*, Haifa, May, eds P. S. Brandon and A. Warszawski, 130–155, Israel Building Research Station.

Skitmore, R. M., 1991, 'Which estimating technique?', in *Investment, Procurement and Performance in Construction*, eds P. Venmore-Rowland, P. Brandon and T. Mole, E & F N Spon, 276–289.

Skitmore, R. M., 1991, 'Design economics: Subject C, Rapporteur's Paper', *Building Economics and Construction Management*, vol 7, *Post Symposium Papers*, (ed. V. Ireland), The International Council for Building Research Studies and Documentation, CIB W-55/65, The University of Technology, Sydney, Australia, 69-77.

Skitmore, R. M. and Patchell, B., 1990, 'Developments in contract price forecasting and bidding techniques', *Quantity Surveying Techniques: New Directions*, ed. P. S. Brandon, BSP Professional Books, 75–120.

Skitmore, R. M. Stradling, S. G., Tuohy, A. P., and Mkzwezalamba, H., 1990, *The Accuracy of Construction Price Forecasts*, University of Salford, Department of Surveying.

Skoyles, E. R., 1964, 'Introduction to operational bills', *The Quantity Surveyor*, 21 (2), 27–32.

Skoyles, E. R., 1977, 'Prices or costs', *The Quantity Surveyor*, Apr.

Slough Estates, 1976, 'Industrial investment: a case study in factory building', *Report*, Slough Estates.

Smith, M. and Skitmore, R. M., 1991, 'Automatic BQ Pricing', *The Professional Builder*, 6 (2), 14–21.

Smith, R. M., 1982, 'Communicative Bills of Quantities – do they influence costs?', In *Building Cost Techniques – New Directions*, ed. P. S. Brandon, E & F N Spon, 362–72.

Southgate, A., 1988, 'Cost planning; a new approach', *Chartered Quantity Surveyor*, Nov, 35–6: Dec., 25.

Southwell, J., 1970, 'Advice on the cost of buildings at the feasibility stages of design', *Research Report* to RICS Research & Information Group of QS Committee, Bath University.

Southwell, J., 1971, 'Building cost forecasting', *Selected Papers* on a systematic approach to forecasting building costs, presented to the Quantity Surveyors (Research and Information) Committee, Royal Institution of Chartered Surveyors.

Spooner, J. E., 1974, 'Probabilistic estimating', *J Con Divn ASCE*, March, 65.

Stacey, N., 1979, 'Estimates of uncertainty', *Building*, 19 Oct, 63–4.

Stichting Bouwresearch, 1979, *Rapport A20–1* – een rekenmodel van de samehang tussen gebouwonderdelen, Rotterdam, Bouwcentrum boekhandel.

Sweett, C., 1957, 'Cost research', *Chartered Surveyor*, 90 (2), Aug, 90–3.

Taylor, R. G., 1984, 'A critical examination of quantity surveying techniques in cost appraisal and tendering within the building industry', *QS Occasional Paper Series*, 3, Department of Quantity Surveying and Building Economics, University of Natal.

Taylor, R. G. and Bowen, P. A., 1986, 'Generating reliable quantities-price data', *Building Economics*, 373–6.

Thorpe, A., 1982, 'Stability tests on a tender price prediction model', Msc Thesis, Department of Civil Engineering, Loughborough University of Technology.

Townsend, P. R. F., 1978, 'The effect of design decisions on the cost of office development', *Chartered Surveyor – Building and Quantity Surveying Quarterly*, Summer, 53–6.

Tregenza, P., 1972, 'Association between building height and cost', *Architects' Journal*, 156 (44), 1031–2.

Trimble, E. G., 1970, 'Regression analysis – new uses of an established technique', Internal Paper, Loughborough University of Technology.

Venning, H. J., Every, C. T., Thackray, M. H., James, W., Coffin, B. G., Honeybell,G., Rae, W. D. and Waghorn, G. F. H., 1942, 'Cost investigation: discussion', *Journal of the RICS*, XXVV ((XII), June, 560–4.

Weight, D. H., 1987, 'Patterns cost modelling', In *Building Cost Modelling and Computers*, ed. P. S. Brandon, 257–66, E & F N Spon.

Wilderness Group, 1964, 'An investigation into building cost relationships of the following design variables: storey height, floor loading, column spacings, number of storeys', *Report to the Royal Institution of Chartered Surveyors.*

Wiles, R. M., 1976, 'A cost model for lift installations', *The Quantity Surveyor*, May.

Wilson, A. J., 1978, 'Cost modelling in building design', *Chartered Surveyor – Building and Quantity Surveying Quarterly*, Summer.

Wilson, A. J., 1982, 'Experiments in probabilistic cost modelling', In *Building Cost Techniques – New Directions*, ed. P. S. Brandon, E & F N Spon, 169–80.

Wilson, A. J., 1987, 'Building design optimisation', *Proceedings, 4th Int Symp on Building Economics*, Copenhagen, CIB W–55, keynotes volume, ed. D. O. Pederson and J. R. Kelly, 152–162.

Wilson, A. J. and Templeman, A. B., 1976, 'An approach to the optimal thermal design of office buildings', *Building Science*, Spring.

Wilson, A. J., Templeman, A. J., Templeman, A. B. and Britch, A. L., 1974, 'The optimal design of drainage systems', *Engineering Optimisation*, 1, 111

Wolochowicz, Z., 1975, 'Directions of price modelling in building,' *Inwestycie i Budownictwo*, 3.

Wootton, A. H., 1982, 'The measurement of construction site performance and input values for use in operational estimating', In *Building Cost Techniques – New Directions*, ed. P. S. Brandon, E & F N Spon, 307–23.

Zahry, M., 1982, 'Capital cost prediction by multi-variate analysis,' MSc Thesis, School of Architecture, University of Strathclyde.

Index

ABACUS 283
Accuracy 12, 46–7, 59–60, 64, 102,
 116, 141–2, 163–5, 183, 193, 202,
 204, 235–6, 239, 245–8, 252,
 308–9, 331, 396–8, 412–33,
 438–58, 471–93
 decomposition of variance 13, 395,
  454
 ex ante measures 8, 164
 ex post measures 7, 164, 206
 influencing factors 422–3, 427–9
 see also reliability
activity diagrams 49–50
Analysis of Variance 164
analytical estimating 414
approximate
 estimating models 26, 46, 55, 59–60,
  103–4, 163, 166–82, 254, 413
 quantities method 260–2, 438, 474
artificial intelligence 374, 381, 386
automation 60, 164, 246, 252, 309, 335,
 384–91

bar charts 349–51
base Schedule of Rates 294
BCIS 'on line' 60, 104
benefit 17
bibliography 495–506
BICEP system 60
bidding 1, 7, 25, 45, 50, 53, 73–8, 87,
 183, 274, 310–11, 328, 395, 416,
 473
Bill of quantities
 cost model 374, 396, 438
 production 329–30
bivariate analysis 239
black box validation 6, 7, 18, 32, 396

 see also white box validation
BLCC model 91
Box's method 222
building surfaces 326

CAD 252
Calculix system 60
cancelling errors 455
capability of performing models
 121–2
Cartesian dimensions 329
cash-flow forecasts 331, 382
causation
 direction 473
cause of variability 403–5
Central Limit Theorem 396, 462
chain model 38
Chiltern method 59
CIRCE 325, 328
CI/SfB 327
COCO 14, 23, 50, 307, 310–18, 377
coefficient of variation 46, 164–5, 206,
 235, 246–7, 340–1, 395–6
 399–410, 416, 441, 443, 445–9,
 453–6, 458, 565
comparability (of projects) 399
comparison of methods 54–61
computer-aided estimating 414–15
conceptual framework 37
content-free systems 325
contingencies 369
control charts 73
conventional cost predictions 325
 see also traditional models
core rates 258, 261
cost
 allowances 272

analysis 236, 238, 311
benefit trade-off 18
book 446
checking 311
components 274
consultants 18
control 310
data bases 438, 443
entry 352–5
estimating 13
evaluation 18, 252
fixed and variable 220
forecasting 18
function 184
implications of design alternatives
    260, 310
information 18, 396
limit 235, 291, 310–11
modelling 307
net 446
operational 275
parameters 283–4
performance modelling 252, 281–92
planning 438–9, 441, 457–8, 474
relationships 259
value trade-off 252
/worth ratios 328
costing procedure 218
covariance 469–70
CPS 50, 60, 104, 308–9, 331–73, 387
CPU 60
crystal and black velvet stage 169
cuboids 328
CUSUM method 397

data
    accuracy 339, 399
    adjustments 331
    availability 184, 190, 220, 311, 332,
        337, 341, 462, 471, 473
    bills of quantities 236, 294, 440
    compatibility 40
    confidentiality 474
    demands 324
    imperfections 457
    interference data 345–7
    libraries 244
    limitations 63–4, 74–5, 164, 235,
        471
    model interface 41
    models 328
    preparation and entry 65

    segmentation 165, 244
    selection criteria 294–5
    sources 339, 449, 457–8
    specific/non-specific 88
    storage 325
    suitability 462
    transformations 39, 66, 165, 206,
        241–4, 296
    variability 40
    weather data 359–60
    Wilderness data 258, 263
debiasing 9, 25, 65, 73–8, 397, 471–2,
    489–92
decision
    environment 39
    simulation 310
    support systems 5
    tables 386
    tree structure 312
    variables 217
Delphi method 396, 463
descriptive primitives 25, 87–91
design
    algorithm 217
    configurations 164
    cost relationships 255
    decision 37, 461
    hypothesis 252
    information 13–14, 18
    optimisation 42, 215, 462
    process 86
    space 233
    stage 13–14, 45, 382, 396, 419–22,
        438–9, 455–8, 471–2, 474
    support 334
    variables 18, 251, 253, 260, 461
determinism 308, 374–5, 384, 396
    single value methods 334
deviation control analysis 491–2
DHSS 272
distribution of variability 395
DOE 272, 398
dynamic modelling 395

early cost advice 109, 253
economic
    climate 395
    response model 37, 460
economics
    of asymmetric information 25
    of building design 37
educational issues 123–6

elemental estimating 59–60, 103, 311, 375, 396, 456
elements 3, 5, 9, 10, 19, 236, 238, 309, 327, 374
   BCIS 165, 252
   of construction operations 219
   constructional 462
   distribution of cost 311
   as independent variables 164
   interdependent 252
   product and process-based 9–10
   unit rates (see functional values)
   useful 21, 252
   values 7, 10, 12–3, 19, 165, 251–2, 308, 395–6
ELSIE 60, 104
empirical
   analysis 252
   identification 252
   research 16, 439
errors
   measurement 472
   rounding 472
estimating
   methods 412–13
   variability 406–7
experimental work 396
expert systems 60, 86, 91, 381, 384
external inferences 331

factor methods 59
family of prices 443
feedback 471–4, 489–92
flow chart 238, 311
forecast 19
formal modelling 13, 309, 377
'front-end' approaches 60
functional
   components 255, 258, 261
   relationships 19, 96
   values 9–14, 19, 40, 55, 163, 165, 238, 251–2, 308, 324, 395–6
fuzzy sets 12–13, 60, 395

gambler's fallacy 43
geometric mean 294
geometrical considerations 311
graphic interfaces 308
graphical analysis 476
guesstimating 169

Hand method 59

heuristics 7, 9, 19, 382
hierarchical
   linked bar charts 331
   tree structure 308
historical determinism 86, 103, 376–7
Holes' system 60
holistic considerations 376
homogeneity of database contracts 60, 254, 279, 332, 337, 449, 453, 472
housing
   projects 442
   refurbishment system 389–91
human factors 396
hunches of experienced practitioners 263
hypothetical steel framed buildings 255

ignorance
   and variability 27
illusions of certainty 333
implementation 7, 16, 19
   environment 24
implementor bias 13
improving cost estimating 457–8
incidence of use 25, 102, 113
   by computing facility 137–141
   by size of organisation 132–7
   by type of organisation 127–32
independence assumption 239, 308, 395–6
inexplicability 308, 374
inflation 46
inferential–relational methods 86
information systems
   criteria 376
inherent
   error 449–53
   uncertainty 13
   variability 453–4, 456
in-place quantities 45–6
intelligent front end 309
interdependence 376
interference 336, 342–7
interpolation method 103
intuition 45, 91, 190, 404, 429–30

judgement 9, 25, 42, 46, 103, 105, 117–18, 144–7, 181, 184, 190, 251, 253–4, 261, 296, 396, 428–9, 431, 463, 471

knowledge-based systems 108, 113,
      309, 386
   see also expert systems
knowledge gap 105

Lang method 59
learning curve 274–5, 338
   effect of interruptions 338–9
light industrial buildings 216
linch-pin 376
locality
   cost indexes 186
   variable 186
Lorenz curve 483

macro and micro-
   approaches 89
   climates 50
market
   conditions 445, 484
   price 20
mark-up 50
mathematical and typological features
      54–61
mean deviation 440–1
means-ends 375
method of least squares 190
modal value 241
model
   age 43
   blindness 31
   causal 47
   cognitive 3
   computer-based 281
   conceptual 6
   context 37
   data interface 381
   deductive 37, 41, 66, 396, 461
   definition 2
   descriptive 46–8, 281
   deterministic 42
   dynamic 251, 274, 281
   economic response 460
   empirical 6
   estimation 11
   fitting 5, 7, 20
   functional values 9–11
   identification 5
   inductive 37, 42, 66, 396, 461
   isomorphic 2
   limitations 31, 36
   macro and micro 396, 460

new generation 282
   performance 39–43
   predictive 281
   probabilistic 460–70
   purpose 2, 20, 36, 46, 308–9, 396
   quality 20
   realistic 36, 48–51
   selection 65–6
   static 251, 274, 281
   stochastic 42
   structuring 5, 7, 20, 309
   symbolic 5
   tasks 9
   technical 3, 9, 21
   theoretical 274
   validation 6
modellers 20
modelling
   criteria 245
   environment 24, 38
   methods 244

NAG routines 243–4
NDIQR system 78–80
nebula theory 13, 309, 374, 381
negative scalars 328
network
   (activity) cost modelling system
      375
   planning models 308
   and related models 377
non-linear
   mixed integer mathematical
      programming 221
   modelling 16
   plans 386

objective functions 215
object oriented programming 386
operational
   activities 462
   estimating 414
optimisation 25, 90, 164, 215–34, 382
   algorithm 225
   models 375
organisational
   issues 127
   trends 110
Orthogonal building assumption 308,
      327
output interpretation 43

paradigm shift 23, 30, 86, 102, 122, 375–6
parameters 252, 293–306
parametric methods 59
pay-off 16, 20
pay-off analysis 17
perceptions
    and reality 309
perfect
    cost data 453
    estimating 478, 484
performance criterion 217
PERT
    -COST 60
    -CPM networks 385
    -like networks 374
    procedure 464
point estimates 465
predesign multilinear cost estimating function 184–98
predictive equation 289
previous projects 441
price
    and cost 259, 462
    and height 272–80
    index variable 187
    skew distribution 400
    variability 399–401
pricing policy 440
primary work package 341
principal item method 103
principle of homogeneity 13, 395
probabilistic
    approaches 376–7, 384, 395
    estimating 462
    networked process models 309
probability
    distribution shape 42, 165, 241–2, 337, 340, 357–9, 380, 396, 416, 443–4, 451–2, 462–9
problem
    complexity 221
    decomposition 222
    hierarchy 222
process models 9–14, 20, 54, 104, 115, 164, 307, 377–82
    conceptual framework 380
    logical necessity 308
product models 9–14, 20, 23, 54, 104
project indices 294
prototype 2, 7, 9, 20
proxy variables 485

Quality index 188
QS observable estimating variability 404

randomised items and rates 59
rate table 220
realistic models 14, 32, 375
regression
    absolute error regression 242
    analysis 12, 45, 47, 61–73, 86, 91, 104, 164–5, 191, 201–14, 252, 276, 476–87
    curvilinear 476
    example 68–73
    forward, backward, step-wise and best subset 63
    L1.5 regression 242
    maximum number of variables 205
    model 242, 272, 375
    parameter estimation 66, 242–4
    reliability analysis 67
    weighted regression 242
regularities of buildings 308, 324, 326
relativity 396
reliability 56–7, 64, 184, 235–6, 239
    confidence interval 296–7
resampling method 239
research
    funding 28
    programme 17, 25, 27–30, 36, 85, 251, 253, 256, 263–4, 375, 460
    scientific vs. technological 29
resource
    allocation 312
    collectives 308, 325
    commited by design decisions
    consumption rates 325
    input 355–7
retention in use 119, 143–4
rigour 251
risk
    analysis 42, 105, 108, 115
    evaluation 384
    models 13, 60
robustness 242, 475
rule 21

scalars 327–8
school projects 441
sensitivity analysis 32
significant items method 59, 103
simplified Bills of Quantities theory 59

simulation 23–4, 32–4, 49, 60, 86, 103,
    331, 334–6, 375, 387–8
  of construction planning methods 50,
    312, 385
  independence assumption 42, 239,
    380, 469
  and management control 309
  Monte-Carlo 42, 90, 104, 239, 334,
    361–3, 380, 387, 390, 416, 465–9
  results 363–7
  stochastic 42, 86, 308–9, 332, 334,
    375–6, 395, 452–3
single price-rate estimating 169–70,
    251, 375, 438
smooth models 14, 21, 46, 308, 381
spread of tenders 445, 448
stakeholders 3, 23, 33, 53
standard error of estimate 193
Standard Form of Cost Analysis 236,
    327
statistical
  approach 396
  inferences 294
  methods 113, 237, 251, *see also*
    regression
  modelling 239
  theory 399
stochastic variability 376–8
Storey Enclosure model 14, 55, 60, 103,
    163, 171–82, 254
strategic cost advice 109, 111–12
strategic models 26, 103, 105
structural
  optimisation 215
  validation 7, 12, 307
supermarket projects 441
swings and roundabouts 462
systems viewpoint 86

target analogy 473, 488
taxonomy of classification 88, 103
technique-bound problem domain 87
technological forecasting 463

technology index variable 188
tender price prediction system 235–49
terminology 1, 2, 17, 37
theoretical
  development 16
  quantitative estimates 254
time models 219
traditional models 26, 36, 47, 54, 103,
    113, 375, 398
  weaknesses 311–12, 386
Trend control analysis 489–91
Type index 187

uncertainty
  quantification 46
understanding of models 120–1
units
  abstract, finished work, as-built 88
  see also elements
univariate analysis 239
unrelatedness 308, 374, 376
usage
  cost/price 89
  see also models purpose

validity 6, 12, 14, 96, 312
validation 239, 332, 395
value
  for money 457
  related models 116
  as a tool 118–19, 142–3
variability
  in tenders 405–6, 439, 443–5, 448,
    453–5
  of unit price rates 472
verdict value 147–50
visually interactive modelling 388

white box validation 6, 7, 12, 21
Wilderness
  curves 261–2, 266–71
  study 23, 164, 166, 251, 253–71